Selected Titles in This Series

274 **Ken-ichi Maruyama and John W. Rutter, Editors,** Groups of homotopy self-equivalences and related topics, 2001

273 **A. V. Kelarev, R. Göbel, K. M. Rangaswamy, P. Schultz, and C. Vinsonhaler, Editors,** Abelian groups, rings and modules, 2001

272 **Eva Bayer-Fluckiger, David Lewis, and Andrew Ranicki, Editors,** Quadratic forms and their applications, 2000

271 **J. P. C. Greenlees, Robert R. Bruner, and Nicholas Kuhn, Editors,** Homotopy methods in algebraic topology, 2001

270 **Jan Denef, Leonard Lipschitz, Thanases Pheidas, and Jan Van Geel, Editors,** Hilbert's tenth problem: Relations with arithmetic and algebraic geometry, 2000

269 **Mikhail Lyubich, John W. Milnor, and Yair N. Minsky, Editors,** Laminations and foliations in dynamics, geometry and topology, 2001

268 **Robert Gulliver, Walter Littman, and Roberto Triggiani, Editors,** Differential geometric methods in the control of partial differential equations, 2000

267 **Nicolás Andruskiewitsch, Walter Ricardo Ferrer Santos, and Hans-Jürgen Schneider, Editors,** New trends in Hopf algebra theory, 2000

266 **Caroline Grant Melles and Ruth I. Michler, Editors,** Singularities in algebraic and analytic geometry, 2000

265 **Dominique Arlettaz and Kathryn Hess, Editors,** Une dégustation topologique: Homotopy theory in the Swiss Alps, 2000

264 **Kai Yuen Chan, Alexander A. Mikhalev, Man-Keung Siu, Jie-Tai Yu, and Efim I. Zelmanov, Editors,** Combinatorial and computational algebra, 2000

263 **Yan Guo, Editor,** Nonlinear wave equations, 2000

262 **Paul Igodt, Herbert Abels, Yves Félix, and Fritz Grunewald, Editors,** Crystallographic groups and their generalizations, 2000

261 **Gregory Budzban, Philip Feinsilver, and Arun Mukherjea, Editors,** Probability on algebraic structures, 2000

260 **Salvador Pérez-Esteva and Carlos Villegas-Blas, Editors,** First summer school in analysis and mathematical physics: Quantization, the Segal-Bargmann transform and semiclassical analysis, 2000

259 **D. V. Huynh, S. K. Jain, and S. R. López-Permouth, Editors,** Algebra and its applications, 2000

258 **Karsten Grove, Ib Henning Madsen, and Erik Kjær Pedersen, Editors,** Geometry and topology: Aarhus, 2000

257 **Peter A. Cholak, Steffen Lempp, Manuel Lerman, and Richard A. Shore, Editors,** Computability theory and its applications: Current trends and open problems, 2000

256 **Irwin Kra and Bernard Maskit, Editors,** In the tradition of Ahlfors and Bers: Proceedings of the first Ahlfors-Bers colloquium, 2000

255 **Jerry Bona, Katarzyna Saxton, and Ralph Saxton, Editors,** Nonlinear PDE's, dynamics and continuum physics, 2000

254 **Mourad E. H. Ismail and Dennis W. Stanton, Editors,** q-series from a contemporary perspective, 2000

253 **Charles N. Delzell and James J. Madden, Editors,** Real algebraic geometry and ordered structures, 2000

252 **Nathaniel Dean, Cassandra M. McZeal, and Pamela J. Williams, Editors,** African Americans in Mathematics II, 1999

251 **Eric L. Grinberg, Shiferaw Berhanu, Marvin I. Knopp, Gerardo A. Mendoza, and Eric Todd Quinto, Editors,** Analysis, geometry, number theory: The Mathematics of Leon Ehrenpreis, 2000

(Continued in the back of this publication)

Groups of Homotopy Self-Equivalences and Related Topics

CONTEMPORARY
MATHEMATICS

274

Groups of Homotopy Self-Equivalences and Related Topics

Proceedings of the Workshop on
Groups of Homotopy Self-Equivalences
and Related Topics
September 5–11, 1999
University of Milan, Gargnano, Italy

Ken-ichi Maruyama
John W. Rutter
Editors

American Mathematical Society
Providence, Rhode Island

This volume contains the proceedings of the Workshop on "Groups of Homotopy Self-Equivalences and Related Topics," which was held at the University of Milan, Gargnano, Italy, September 5–11, 1999. The conference was made possible by a grant from the National Institute for Advanced Mathematics "Francesco Severi" (INdAM).

2000 *Mathematics Subject Classification.* Primary 55P10, 55P15, 55P20, 55P35, 55P45, 55P60, 55P62, 55R05, 55R35.

Library of Congress Cataloging-in-Publication Data

Workshop on Groups of Homotopy Self-Equivalences and Related Topics (1999 : Università di Milano)

Groups of homotopy self-equivalences and related topics : proceedings of the Workshop on Groups of Homotopy Self-Equivalences and Related Topics, September 5–11, 1999, University of Milan, Gargnano, Italy / Ken-ichi Maruyama, John W. Rutter, editors.

p. cm.

ISBN 0-8218-2683-2 (alk. paper)

1. Homotopy equivalences—Congresses. I. Maruyama, Ken-ichi, 1954– II. Rutter, John W., 1935– III. Title.

QA612.72.W67 1999

514′.24—dc21 00-069977

♾ The paper used in this book is acid-free and falls within the guidelines
established to ensure permanence and durability.
Visit the AMS home page at URL: `http://www.ams.org/`

10 9 8 7 6 5 4 3 2 1 06 05 04 03 02 01

Contents

Foreword

From September 5 to 11, 1999, a workshop on groups of homotopy self-equivalences and related topics was held at the Gargnano conference centre of the University of Milan, located in the town of Gargnano (Lake Garda). The workshop was attended by some 30 mathematicians from Canada, England, Germany, Italy, Japan, Korea, the Republic of South Africa, Singapore, Slovenia, Spain and the United States. This volume contains papers presented to the workshop. It contains two specially written articles—a review of work done in the area of homotopy self-equivalences since the Montreal conference of 1988, including a bibliography, and a list of problems.

The workshop was made possible by a grant from the National Institute for Advanced Mathematics "Francesco Severi" (INdAM). Thanks are due to Martin Arkowitz and Renzo Piccinini who were main organisers before and during the workshop.

Each research paper in this volume was carefully refereed and thanks are due to all who helped in this.

Ken-ichi Maruyama and John W. Rutter

Contemporary Mathematics
Volume **274**, 2001

Homotopy Self-equivalences 1988 -1999

John W. Rutter

ABSTRACT. We give here a summary of the results published in the area of 'Spaces and Groups of Homotopy Self-equivalences' between 1988, the year of the Montreal conference, and the Gargnano workshop in 1999.

Introduction

We give here a summary of the results published in the area of 'Spaces and Groups of Homotopy Self-equivalences' between 1988, the year of the Montreal conference (see Piccinini [1990]), and the Gargnano workshop in 1999. In general, because of space considerations, we omit the detail where results are technical. The bibliography which follows this summary is the list of references for it; our thanks to those who brought to our attention papers not included in the list circulated at the workshop, and especially to Kouzou Tsukiyama.

For more detail of earlier results in this subject area see the survey by Rutter [1997]. We follow here the uniform notation used in Rutter [1997, pp. 163-164]. An earlier recent survey was given by Arkowitz [1990], and recent bibliographies were given in [1] and by Rutter [1997].

We denote by $\mathcal{E}^*(X)$ and $\mathcal{E}(X)$ the groups of base point preserving and free homotopy classes of homotopy self-equivalences of X. The corresponding groups of diffeotopy classes of homeomorphisms are denoted by $\mathcal{H}^*(X)$ and $\mathcal{H}(X)$. Also we denote the kernels of the representations $\mathcal{E}^*(X) \to \operatorname{aut} \pi_*(X)$, $\mathcal{E}^*(X) \to \operatorname{aut} H_*(X)$, and $\mathcal{E}^*(X) \to \operatorname{aut} H^*(X)$, by $\mathcal{K}^*_\pi(X)$, $\mathcal{K}^*_H(X)$, and $\mathcal{K}^*_C(X)$ respectively, and the corresponding groups in the free case by $\mathcal{K}_\pi(X)$, $\mathcal{K}_H(X)$, and $\mathcal{K}_C(X)$ respectively. In the categories of spaces under A and/or over B, we have the groups $\mathcal{E}^A(X)$, $\mathcal{E}_B(X)$ and $\mathcal{E}^A_B(X)$ of homotopy classes under A and/or over B, respectively, of homotopy self equivalences. The corresponding spaces of homotopy self-equivalences are denoted $E^*(X)$, $E(X)$, etc. Eilenberg-MacLane spaces and Moore spaces in 'dimension' n are denoted $K_n(\pi)$ and $M^n(\pi)$ respectively.

2000 *Mathematics Subject Classification.* Primary 55-02, 55P10; Secondary 55P15, 55P60, 55P62, 55P91, 55R10, 57M20, 57N37, 57Q10.

Key words and phrases. self-equivalence, self homotopy, fibre equivalence, homotopy equivalence.

CONTENTS

1. Surfaces, manifolds

1.1. Surfaces. Let $\mathbf{T}_g$ be a compact orientable surface of genus g. Let G be a subgroup of $\mathcal{E}(T_g)$, then either G contains an abelian subgroup of finite index or G contains a non-abelian free group (see Ivanov [1992, Theorem 4]). $\mathcal{E}(T_g)$ does not contain an arithmetic subgroup of finite index for $g \geq 2$ (Ivanov [1988B]). The Ballman-Eberlien rank of $\mathcal{E}(T_g)$ is 1 for for $g \geq 1$ (Ivanov [1988A]). A presentation for the symplectic modular group $\mathrm{Sp}^+(2g, \mathbf{Z})$ ($g \geq 3$) on 3 generators and $3g + 5$ relators was given by Lu [1992, Theorem 3].

1.2. Higher manifolds. For orientable geometric 3–manifolds, the group of simple homotopy self-equivalence classes $\mathcal{S}(M)$ lies in the image of $\mathcal{H}(M) \to \mathcal{E}(M)$ (Turaev [1988], see also Kwasik and Schultz [1992, Theorem 1.1]).

In the case of a compact connected orientable 3–manifold

$$M = M_1\#...\#M_n\#(\#_k(S^1 \times S^2)),$$

McCullough [1990] and Hatcher and McCullough [1990] obtained results on finite generation and finite presentation of the homeotopy group $\mathcal{H}(M)$ and of $\mathcal{K}(M)$, the (orientation-preserving subgroup of the) kernel of $\mathcal{H}(M) \to \operatorname{out}\pi_1(M)$. McCullough [1991, Theorem page 2] and Kalliongis and McCullough [1992, Theorem 4.2.3] showed that $\mathcal{H}(M)$ is finitely presented in case M is a compact connected irreducible sufficiently large (in the sense of Swarup in the non-orientable case) 3–manifold.

Let X be a compact connected 4–manifold with finite-dimensional fundamental group and $f : X \to B$ where B is the second Postnikov stage of X. Hayat and Legrand [1996] obtained several exact sequences including

$$H^1(\pi_1, \pi_2) \to \mathcal{E}_B(X) \to \mathcal{E}(X) \to \mathcal{E}(B)^f \to 1$$

$$H^2(B, \pi_3) \to H_0(\pi_1, \pi_4) \to \mathcal{E}_B(X) \to H_1(\pi_1, \pi_3).$$

In case π_1 has periodic cohomology of period 4, they calculated $\mathcal{E}(X)$ algebraically, up to extension.

For a closed, compact, oriented manifold M, Baues [1996] studied the sequence

$$0 \to \overline{\mathcal{E}}^A(M) \to \mathcal{E} \to \hat{\mathcal{E}}(A, S^1) \to 1$$

where $A = M \backslash e^M$, S^1 is the boundary of A, $\overline{\mathcal{E}}^A(M)$ is the image of $\overline{\mathcal{E}}^A(M) \to \overline{\mathcal{E}}(M)$ and $\hat{\mathcal{E}}(A, S^1)$ is the image of $\mathcal{E}(A, S^1) \to \mathcal{E}(A)$. The group $\overline{\mathcal{E}}^A(M)$ is finite and abelian; special results hold if M is $2n$–dimensional and $(n-1)$–connected, or is $(n-1)$–connected where $m < 3n-1$. Where $n < 11$ and M is a connected sum of copies of $S^m \times S^n$, $\mathcal{E}(M)$ was calculated in many cases. Baues and Buth [1996] made substantial calculations of $\mathcal{E}^*(M)$ for simply-connected 5–manifolds.

Let G be a compact connected Lie group, then $\mathcal{E}(BG) \cong \operatorname{out} G$, the group of Lie group outer automorphisms (Jackowski, McClure and Oliver [1995] and Osse [1997]); also the finite homology type of $E(BG, 1)$ was determined as BZ where Z is the centre of G (Jackowski, McClure and Oliver [1992A] and [1992B]).

The $\frac{1}{2}$–local homotopy type of $E(M)/H(M)$, for a compact topological manifold M, is the loop space of the Hermitian K–theory Whitehead space of M (Fiedorowicz, Schwändl and Vogt [1992]).

In the PL–category, for $n \geq 2$, the mapping class group $\mathcal{M}(\#_p(S^1 \times S^n)) \cong \mathcal{E}(\#_p(S^1 \times S^n))^{or}$, the group of orientation-preserving homotopy self-equivalence classes, and there is the exact sequence

$$0 \to \bigoplus\nolimits_p \mathbf{Z}_2 \to \mathcal{E}(\#_p(S^1 \times S^n))^{or} \to \operatorname{out} F_p \to 1$$

(Cavicchioli, Hegenbarth and Spaggiari [1988]).

2. Localization

$\mathcal{K}_\pi^m(X) \to \mathcal{K}_\pi^m(X_P)$ P–localizes in case X is a connected simple finite dimensional CW–complex of finite type, and $\mathcal{K}_H^m(X) \to \mathcal{K}_H^m(X_P)$ P–localizes in case X is a simply-connected finite CW–complex (Maruyama ([1989, Theorem 0.1] and [1990A, page 291])). Here P is a set of primes, $\mathcal{K}_\pi^m(X)$ is the kernel of the representation $\mathcal{E}(X) \to \prod_1^m \operatorname{aut} \pi_r(X)$, and $\mathcal{K}_H^m(X)$ is the kernel of the representation $\mathcal{E}(X) \to \prod_1^m \operatorname{aut} H_r(X)$. Calculations of $\mathcal{K}_H(X)_P$ and $\mathcal{K}_\pi(X)_P$ for a number of H-spaces were given by Maruyama ([1990A, Example 3.1] and [1990B, 3.1]). Results on localization were also given by Møller ([1992, Corollary 3.2] and [1989, Theorem 4.3]). See also §8 of Rutter [1997].

Arkowitz [2] showed that for a minimal DGA $\mathcal{A}$, $\mathcal{E}(\mathcal{A}) \cong \operatorname{aut} H^*(\mathcal{A}) \rtimes \mathcal{K}_C(\mathcal{A})$ and give an algorithm for calculating $\mathcal{K}_\pi(\mathcal{A})$).

Let $\mathcal{E}_0(X) \subset \mathcal{E}(X)$ consist of the classes of rational equivalences, and let X be a space whose p–completion $(X)_p^\wedge$ has the homotopy type of $(BG)_p^\wedge$ for any prime p, where $G = S^3 \times \cdots \times S^3$. Ishiguro, Møller and Notbohm [1999] studied $\mathcal{E}_0(X)$; a submonoid of $\mathcal{E}_0(X)$ determines the decomposability of X.

3. $\mathcal{E}^*(X)$ finitely presented, nilpotent

Dror and Zabrodsky had shown that $\mathcal{K}_H^*$ is a nilpotent group in case X is a path-connected nilpotent finite dimensional CW-complex. Maruyama and Mimura [1990] extended this to suitable generalised homology theories; they showed also that it is not true for K–theory. Xu [1996] showed that if a subgroup G of $\mathcal{E}^*(X)$ acts on $E_i(X)$ nilpotently then G is nilpotent, where X is finite connected, nilpotent and has only p–torsion homology, and E is a connective spectrum.

Felix and Murillo [1997] showed that the nilpotency of the rationalization of $\mathcal{E}^*(X))$ is $\leq \operatorname{cat}(X) - 1$ and obtained a corresponding result for fibre homotopy equivalences; in [1998] they showed that, if cat (X) is finite, the kernel of $\mathcal{E}^*(X) \to \mathcal{E}^*(\Omega X)$ is nilpotent and they give a bound for the nilpotency class.

Arkowitz and Lupton [1996] studied the nilpotency class of $\mathcal{K}_\pi^*(X)$ and obtain bounds for it for certain finite complexes.

Roitberg [1991] proved that $\mathcal{E}^*(X) \cong \mathbf{R} \rtimes (\mathbf{Z}_2 \times \mathbf{Z}_2)$ is solvable and uncountable but not residually nilpotent for the simply-connected space $X = K(\mathbf{Z}, 2) \times S^3$.

D.W. Kahn [1990A], complementing his earlier result, also showed, essentially, for a finite connected stable complex, that the image of the stable homotopy (factored by torsion) representation $\mathcal{E}^*(X) \to \prod \operatorname{aut} \pi_r^S(X)_F$ has finite index.

Triantafillou [1992, Theorem 1.2 and Corollary 1.5], extending the Sullivan-Wilkerson result, showed that $\operatorname{stab}_f(\mathcal{E}(X))$ is finitely presented and of finite type where X is a finite nilpotent CW–space, Y is a nilpotent CW–space of finite type, and $f \in (X, Y)^*$. This was further extended by Maruyama [1996], who showed that the subgroups of $\mathcal{E}^*(X)$, $\mathcal{E}^*(Y)$ and $\mathcal{E}^*(X) \times \mathcal{E}^*(X)$ which leave S (pointwise) invariant are all finitely presented, where X and Y are finite nilpotent complexes and S is a finite subset of $(X, Y)^*$; also any nilpotent subgroup of $\mathcal{E}^*(X)$ is finitely presented. Triantafillou also showed [1995] that $\mathcal{E}(X)$ is finitely presented provided that X is a finite complex with finite fundamental group, the group of homotopy classes of tangential homotopy self-equivalences (also for tangential simple homotopy self-equivalences) of an oriented manifold with finite fundamental group is finitely presented, and the group of homotopy classes of diffeomorphisms of a smooth closed oriented manifold of dimension ≥ 5 with finite fundamental group is finitely presented. The results in general show that the appropriate group is commensurable with an arithmetic group.

By the result of Formanek and Procesi [1992], we have that the free group $\mathcal{E}^*(X)$ is not linear in the case where X is a bouquet of n circles ($n \geq 3$), though it is finitely presented.

Arkowitz and Lupton [1995, Appendix] showed that $\mathcal{K}_H^*(X)$ and all its subgroups are finitely generated, where X is a 1–connected finite complex.

4. Methods for calculation

4.1. Cell complexes. Rutter [1988A, Theorem 1 and Theorem 2] gave methods for calculating $\mathcal{E}^*(X)$, in case X is simply-connected, by moving progressively up the stages of a homology decomposition.

Let $\bar{\mathcal{E}}(X)$ be the subgroup of $\mathcal{E}(X)$ consisting of those classes which can be represented by a cellular map $f : X \to X$ which is a homotopy self-equivalence $f^r : X^r \to X^r$ on each skeleton X^r of X. Rutter [1990A, Theorem B] gave conditions (e.g. X is simply-connected finite complex with $X = \{\text{pt}\}$) which ensure that $\bar{\mathcal{E}}(X)$ has finite index in $\mathcal{E}(X)$.

Kahn [1992] discussed the twisted torus and characteristic classes of a homotopy self-equivalence.

4.2. Non-1-connected Postnikov pieces. Methods were given by Rutter [1992B, Theorem 2.4, Theorem 2.9] for calculating $\mathcal{E}^*(X)$ and $\mathcal{E}(X)$ for connected spaces by proceeding up the stages of a Postnikov decomposition. In the case of a non-simply connected space having precisely two non-trivial homotopy groups and k–invariant h, we have the group extension (obtained independently by Møller [1991, Theorem 5.6])

$$0 \to H^n_\phi(K_1(\pi_1); \pi_n) \to \mathcal{E}^*(X) \to \operatorname{stab}_h(\operatorname{aut} \pi_1 \times_\phi \operatorname{aut} \pi_n) \to 1\,,$$

where ϕ is the action of π_1 on π_n. In the special case where $X = L^\phi_n(\pi_n)$ is the generalized Eilenberg-McLane space, the sequence splits (Rutter [1992B, Lemma 2.1

and Theorem 2.2], see also [1992A]); the special case of this for the classifying space of $K_{n+1}(\pi_n)$ was considered by Møller [1988, Corollary 3.5].

Spectral sequences which can be used to calculate $\mathcal{K}^*_\pi(X)$ and $\mathcal{E}^*(X)$ were given by Didierjean [1992].

Didierjean and Legrand [1994] compared the first differential of the spectral sequences for the homotopy groups of the components of map(X, X) containing the trivial and the identity maps, respectively; the difference of these differentials involves a Whitehead product. They extend their results to fibre homotopy self-equivalences.

4.3. Algebraic complexes. Baues [1989, VI 6.13] showed that the kernel of $\mathcal{E}^*(X) \to \mathcal{E}^*(\{C, \partial\})$ is a solvable group where X is a finite dimensional CW complex and $\{C, \partial\}$ is the corresponding chain complex, and showed further [1991B, Proposition IV.8.12] that there is a short exact sequence

$$H^3_\varphi(X; \Gamma_3(X)/I(1) \rightarrowtail \mathcal{E}^*(X) \twoheadrightarrow \mathcal{E}^*(\{C, \partial\}),$$

where X is a 3–dimensional CW complex with one 0–cell. Here the cohomology has local coefficients and $\Gamma_3(X) = \mathrm{im}\,(\pi_3(X^2) \to (\pi_3(X^3)).$

Baues [1991B, §IV.3 and §IV.4], introduced 'quadratic chain complexes', and extended the results of Whitehead on faithful functorial algebraic representability from the homotopy category of 2–dimensional complexes to the homotopy category of 3–dimensional complexes. Let X be a (connected) CW complex with one 0–cell, then (Baues [1991B, §7 and §8]) $\mathcal{E}(X) \to \mathcal{E}(\sigma(X))$ is surjective if X is 4–dimensional and an isomorphism if X is 3–dimensional. Here $\sigma(X)$ is a quadratic chain complex associated with X.

4.4. Unions and products. In the case where X and Y are suspensions (or more generally are h–coloops), each element of the algebraic loop $(X \vee Y, X \times Y)$ can be uniquely represented by a matrix $\begin{pmatrix} f & g \\ h & k \end{pmatrix}$ where $f : X \to X$, $g : Y \to X$, $h : X \to Y$, and $k : Y \to Y$. Furthermore matrix multiplication gives this set $(X \vee Y, X \times Y)$ a monoidal structure. In case the multiplication is associative, we denote the group of units of this monoid by $U(X|Y)$. Rutter ([1988A, Example 2.6], [1990B]) calculates $\mathcal{E}^*(X \vee Y)$ up to extension in many general cases where X and Y are Moore spaces by using the representation onto $U(X|Y)$ or that onto $\mathcal{E}^*(X) \times \mathcal{E}^*(Y)$.

Dually Booth and Heath [1990], extending results of Yamanoshita, give conditions on X and Y for which $\mathcal{E}^*(X \times Y) \to \mathcal{E}^*(X) \times \mathcal{E}^*(Y)$ is a surjective homomorphism which induces a semi-direct product $\mathcal{E}^*(X \times Y) \cong (X, E(Y))^* \rtimes (\mathcal{E}^*(X) \times \mathcal{E}^*(Y))$. Here $E(Y)$ is the topological monoid of non-pointed homotopy self-equivalences of Y. Heath [1996] gives further results on $\mathcal{E}^*(X \times Y)$, especially where X and Y are suspensions, including Moore spaces, and determines $\mathcal{E}^*(S^m \times S^n)$ $(m \neq n)$ generally as a semi-direct product; see also §17.5.1 of Rutter [1997]. Where $\mathcal{E}^*(X \times Y) \to \mathcal{E}^*(X) \times \mathcal{E}^*(Y)$ is a well-defined (surjective) homomorphism, Pavešić [1999] showed that $\mathcal{E}^*(X \times Y) = \overline{\mathcal{E}}^X(X \times Y) \, . \, \overline{\mathcal{E}}^Y(X \times Y)$, where, for example, $\overline{\mathcal{E}}^X(X \times Y)$ is the image of $\mathcal{E}^X(X \times Y) \to \mathcal{E}^*(X \times Y)$ and $\overline{\mathcal{E}}^X(X \times Y) \cong (X, E(Y))^* \rtimes \mathcal{E}^*(Y)$. He also extended earlier results, including cases where Y, say, is an H–space, or a co–H–space, or especially an Eilenberg-MacLane space. Similar results hold for $\mathcal{K}^*_\pi(X \times Y)$.

5. Calculations made

5.1. Cell complexes. For the Moore space $M^n(\pi)$ $(n \geq 3)$ a form for the extension due to Hartl for the exact sequence of groups

$$0 \to G \to \mathcal{E}^*(M^n(\pi)) \to \operatorname{aut}\pi \to 1$$

was given by Baues [1989, page 269]. In general the sequence is not split. Results on $\mathcal{E}^*(M^2(\pi))$ were given by Baues and Hartl [1996]; the extension is trivial in case π has no 2–torsion and is given in case π is a product of cyclic groups.

Let X be an 2–connected 4–manifold. Cochran and Habegger [1990, Theorem 3.1] showed that a basic group extension used for calculating $\mathcal{E}^*(X)$ is split. See also Baues [1989, V Theorem 8.7].

For the suspension of $\mathbf{C}P^3 = S^2 \cup_\eta e^4 \cup e^6$ we have $\mathcal{E}(S\mathbf{C}P^3) \cong \mathbf{Z} \rtimes (\mathbf{Z}_2 \times \mathbf{Z}_2)$, $\mathcal{E}(S^2\mathbf{C}P^3) \cong \mathcal{E}(S^3\mathbf{C}P^3) \cong \mathbf{Z}_2 \times \mathbf{Z}_2 \times \mathbf{Z}_2$, and $\mathcal{E}(S^k\mathbf{C}P^3) \cong \mathbf{Z}_2 \times \mathbf{Z}_2$ $(k \geq 4)$ (Yamaguchi [1991, Theorems A, B and C]). Yamaguchi determined the rings $(S^k\mathbf{C}P^3, S^k\mathbf{C}P^3)$ and the representations $(S^k\mathbf{C}P^3, S^k\mathbf{C}P^3) \to \operatorname{end}H_*(S^k\mathbf{C}P^3)$ $(k > 2)$. Mukai [1993] determined the stable groups $\mathcal{E}^*_s(\mathbf{C}P^n)$ for $n \leq 7$.

$\mathcal{E}^*(\mathbf{H}P^3) \cong \mathbf{Z}_2 \oplus \mathbf{Z}_2$ and $\mathcal{E}^*(\mathbf{H}P^3) \to \mathcal{E}^*(\mathbf{H}P^2)$ is onto (Iwase, Maruyama and Oka [1991, Theorem 0.4]); the authors showed that $\mathcal{E}^*(\mathbf{H}P^4)$ is $\mathbf{Z}_2$ or 1. Takahashi [1996] calculated $\mathcal{E}^*(S\mathbf{H}P^n)$ $(n = 2, 3)$ and subgroups such as $\mathcal{K}^*_\pi$ and $\mathcal{K}^*_H$; he also calculated the set of self-maps.

$\mathcal{E}^*(\mathbf{R}P^3 \# \mathbf{R}P^3) = \mathbf{Z}_2$ (Baues [1991B, Theorem page 252]).

$\mathcal{E}^*(\mathbf{C}P^\infty \times S^3) \cong \mathbf{R} \rtimes (\mathbf{Z}_2 \times \mathbf{Z}_2)$ (Roitberg [1991, Theorem 2.1]).

Shitanda [1993A] and [1993B] calculated $\mathcal{E}^*(\mathbf{H}P^\infty \times S^n)$ and $\mathcal{E}^*(K_m(\mathbf{Z}) \times S^n)$, and the monoid of self-maps in each case.

Let $X = \vee S\mathbf{C}P^2$ be a bouquet of r copies of $S\mathbf{C}P^2 = S(S^2 \cup_\eta e^4)$. Then $\mathcal{E}^*(S^kX)$ is isomorphic with a certain subgroup of $\mathrm{GL}(\mathbf{Z}, r) \times \mathrm{GL}(\mathbf{Z}, r)$ $(k \geq 1)$. The kernel of $\mathcal{E}^*(X) \to \mathrm{GL}(\mathbf{Z}, r) \times \mathrm{GL}(\mathbf{Z}, r)$ can also be calculated (cf. Yamaguchi [1990] and Unsold [1989]). Yamaguchi subsequently used these results to provide information on the classification of 2–connected 8–dimensional Poincaré complexes.

Mimura and Ōshima [1999] gave further presentations for the groups $\mathcal{E}^*(\mathrm{SU}(3))$ and $\mathcal{E}^*(\mathrm{Sp}(2))$.

Llerena and Rutter [1998] determined $\mathcal{E}^*(X \vee Y)$ as a quotient of a semi-direct product, where X and Y are $(n-1)$–connected $2n$–manifolds. The group does not in general itself inherit the semi-direct product structure.

Arkowitz and Maruyama [1998] studied $\mathcal{K}^*_H(X)$, $\mathcal{K}^*_\pi(X)$, $\mathcal{K}^*_C(X)$ and intersections of these groups; they gave examples where $\mathcal{K}^*_H(X) \neq \mathcal{K}^*_C(X)$, though they showed these to be equal for a large class of spaces including Poincaré complexes. All three groups were described for Moore spaces M^n $(n \geq 3)$. They also found a bound for the nilpotency of $\mathcal{K}^*_\pi(X) \cap \mathcal{K}^*_C(X)$

5.2. Non-simply-connected spaces. Rutter [1988B] gave a geometrical interpretation of Olum's results on $\mathcal{E}^*(M_q)$ using obstruction theory, and Baues [1991A] recovered these results using algebraic models of the homotopy category of Moore spaces. Here $M_q = S^1 \cup_q e^2$ is the pseudo-projective plane.

Rutter (see [1997, §15.2]) calculated $\mathcal{E}^*(X)$ for the $(2n+1)$–dimensional lens spaces of combinatorially aspherical groups, including one-relator groups, and their even dimensional skeleta.

6. Structural properties

We consider here structural properties including realizability, rigidity, finiteness, residual finiteness, abelianness and rank for the group $\mathcal{E}(X)$ and $\mathcal{E}^*(X)$.

6.1. Realizability. Any finite group occurs as $\pi_0(\text{Diff}M)$ for some closed manifold M (Kojima [1988, Corollary page 297].

Let Y be a connected H–space of finite homotopical dimension for which each $\pi_i(Y)$ is a finite p–group, then the representation of $\mathcal{E}(Y)$ into the automorphism group of the primitives in mod p homology is onto a finite product of general linear groups over finite fields of characteristic p. Conversely any such product is such an image. Dually, let X be a finite p–torsion complex (a finite complex with p–torsion integral homology), then the representation of the stable group $\mathcal{E}^s(Y)$ into the automorphisms of a subgroup of $H_*(X) \otimes \mathbf{Z}_p$ is onto a finite product of general linear groups over finite fields of characteristic p. Conversely any such product is such an image. The kernel in each case is the unique maximal normal p–subgroup. In particular the group of units in any finite field can be realized. (Crabb, Hubbuck and Xu [1992, Theorem 1.13, Proposition 1.1 and Theorem 1.4].)

Maruyama [1994] gave spaces $X = S^4 \cup e^8 \cup e^8$ for which $\mathcal{E}^*(X) \cong \mathbf{Z}$.

6.2. Elements of order p and finiteness. Let X be a simply-connected finite complex which is a rational H–space of rank 3, then $\mathcal{K}_H(X)$ is a finite group (Maruyama [1990A, Example 3.2]). Let G be a compact connected Lie group and T a maximal torus, then $\mathcal{E}(G/T)$ and $\mathcal{K}_H(G/T)$ are finite (Papadima [1986] and Notbohm and Smith [1992, Proposition 3.1]). Let X be a finite nilpotent complex which is rationally equivalent to a (finite) bouquet $\bigvee S^{n_i}$ of spheres, then $\mathcal{K}^*_\pi(X)$ is a finite nilpotent group. Furthermore, in the case where X is also simply-connected, $\mathcal{E}^*(X)$ is finite if, and only if, $\text{rk}\,(\pi_{n_i}(X)) = 1$ for each i (Maruyama [1992, Theorem page 34]). Let $X = Y \cup e^n$ have the rational homotopy type of a formal 1–connected Poincare complex of dimension n which is not rationally a sphere, then $G = \ker(\mathcal{E}^*(X) \to \mathcal{E}^*(Y) \times \mathcal{E}^*(S^n))$ is finite (Aubry and Lemaire [1991 Theorem 2.7]). They also note that $G = \mathcal{K}_\pi \subset \mathcal{K}_H$ in case $X = S^3 \times S^3$. Arkowitz and Lupton [1995] considered finiteness properties for $\mathcal{K}^*_H$, $\mathcal{K}^*_\pi$ and $\mathcal{K}^*_H \cap \mathcal{K}^*_\pi$; the main results were for rational spaces having 2–stage ‘decompositions’, for example rationalizations of homogeneous spaces; specific criteria were given for flat manifolds and products of spheres.

Pavešić [1999] obtained structural properties for $\mathcal{E}^*(X \times Y)$.

6.3. Rank. Let X be a simply-connected finite nilpotent complex which is rationally equivalent to a (finite) bouquet $\bigvee S^{n_i}$ of spheres, then $\mathcal{K}^*_H(X)$ has finite rank. Furthermore $\mathcal{E}^*(X)$ has finite rank if, and only if, all the n_i are distinct. Formulae for these ranks are given (Maruyama [1992, Theorem page 34]).

6.4. Non-abelian nature. $\mathcal{K}_\pi(X)$ is in general non-abelian; see for example Maruyama [1989, Example 3.1] for SO(6) and SU(4).

6.5. Mislin genus. Roitberg [1996] showed that $\mathcal{E}^*(X)$ is not an invariant of the Mislin genus; his counter-example consists of two simply-connected 4–manifolds. A previous example consisting of infinite dimensional complexes was given by McGibbon and Møller [1992] (see §9).

6.6. Whitehead torsion, simple homotopy. Metzler [1990, page 126] gave further examples for which the Whitehead torsion $\tau : \mathcal{E}^*(X) \to \mathrm{Wh}\,\pi_1(X)$ is not surjective.

6.7. Residually finite. The group $\mathcal{E}_*(X)/WI(X)$ is residually finite provided that X is a path-connected CW complex satisfying certain finiteness conditions. Here $WI(X)$ consists of the classes represented by maps weakly homotopic to the identity. Conversely, given that X is an h–associative h–space with finite fundamental group and finitely-generated higher homotopy groups and $\mathcal{E}^*(X)/N$ is residually finite, then $N \supset WI(X)$ (Roitberg [1989]).

7. Homotopy type, homotopy groups

7.1. Homotopy type. The space $E(K_1(\pi))$ of (non-pointed) homotopy self-equivalences has the weak homotopy type of $\mathrm{out}\,\pi \times K_1(\mathfrak{Z}(\pi))$; the appropriate multiplicative structure on the simplicial monoid corresponding to $\mathrm{out}\,\pi \times K_1(\mathfrak{Z}(\pi))$ for which this is an h–weak-homotopy-equivalence was given by Hirashima [1991, Theorem 1].

Yamanoshita [1993] obtained a homeomorphism

$$E(\mathbf{R}P^2) \simeq \mathrm{SO}(3) \times (E^*(\mathbf{R}P^2)/\mathrm{O}(2));$$

the inclusions $H(\mathbf{R}P^2) \subset E(\mathbf{R}P^2)$ and $\mathrm{Diff}(\mathbf{R}P^2) \subset E(\mathbf{R}P^2)$ are not homotopy equivalences, since $H(\mathbf{R}P^2) \simeq \mathrm{SO}(3)$ and $\mathrm{Diff}(\mathbf{R}P^2) \simeq \mathrm{SO}(3)$.

For simply-connected spaces X having precisely two non-zero homotopy groups and these in consecutive dimensions, the homotopy type of the classifying space of $E(X)$ was obtained by Hayat-Legrand [1994, Théorème 2.5] provided that a certain obstruction vanishes.

Didierjean [1992] used her Gysin-type exact sequence sequence to determine an algorithm for obtaining information about the homotopy groups of $E(X)$ in a number of cases and in particular for a space having at most three non-vanishing homotopy groups; in the case where X has precisely two non-zero homotopy groups, the homotopy groups are determined with one exception [1992, page 156]. Didierjean [1992] and [1990] used this algorithm to obtain some partial results on the homotopy groups $\pi_i(E(X))$ in case $X = \mathrm{SU}(3)$ and $X = \mathrm{Sp}(2)$. A second quadrant spectral sequence having edge homomorphism $\pi_*(E^*(X)) \to E_\infty^{0,*} \to E_2^{0,*} \to \mathrm{end}_{\mathrm{id}}\,(\pi_*(E(X))$ was given by Dwyer, Kan, Smith and Stover [1994, Theorem 1.2]; here $\mathrm{end}_{\mathrm{id}}\,(\pi_*(E(X))$ is a function Π–algebra.

Møller [1990A] calculated the Samelson product in $E^A(X)$ where $A \to X$ is a cofibration with cofibre a Moore space, and in $E_B(X)$ where $X \to B$ is a fibration with fibre an Eilenberg-McLane space.

For a covering map $p : X \to Y$, let $E_p(X)$ and $E^p(Y)$ be the spaces of homotopy self-equivalences of X and Y, respectively, which compress, respectively lift, to homotopy self-equivalences of Y and X; and let $T_p(X)$ and $T^p(Y)$ be the corresponding spaces of self-homeomorphisms. Miller [1990] gave conditions, with counterexamples, that $E_p(X) \to E^p(Y)$ and $T_p(X) \to T^p(Y)$ are topological semigroup, respectively topological group, quotient maps.

Hansen [1990] showed that, for each $k \neq 0$, $T^*(S^2, 1)$ is, up to homotopy, a k–fold cover of the component of the space of self-maps of S^2 having degree k. Here $T^*(X)$ is the space of orientation-preserving homotopy self-equivalences.

7.2. Rational homotopy groups. In the case where M is the Grassmann manifold $G(n;p)$ of p–planes in $\mathbf{C}^n$, or the variety $U(n)/G$, where G is the product of n copies of $U(1)$, the groups $\pi_i(E(M)) \otimes \mathbf{Q}$ were calculated by Grivel [1994, Théorèmes 5.3 and 7.7]. More generally, in the case where X is 1–connected, $\dim \pi_*(X) \otimes \mathbf{Q} < \infty$ and $H^{\text{odd}}(X;\mathbf{Q}) = 0$, Hauschild [1993, Theorem B] gave $\pi_n(E(X,1))$ in terms of $\mathbf{Q}$–derivations of $H^*(X)$; furthermore he extended this to a result on the group of fibre homotopy self-equivalences.

7.3. Nilpotency/homotopy nilpotency. For a connected R–nilpotent space X with finite-dimensional homology, the classifying spaces of certain natural submonoids of $E(X)$ are also R–nilpotent (Dwyer [1989]). Here $R = \mathbf{Z}[1/p]$ where p is prime.

Salvatore [1997] computed the homotopy nilpotency of various spaces of homotopy self-equivalences, including $E(X,1)$, in terms of the nilpotency of the homology of subspaces of derivations; for example $E(X,1)$ is not nilpotent where X is a non-trivial union of copies of an odd sphere.

7.4. Minimal model. Felix and Thomas [1994] obtained a minimal model for $E(G/H,1)$, and formulae for the number of generators where G and H are suitable compact connected Lie groups.

8. Fibre and equivariant homotopy equivalences

In the case where $p : X \to Y$ is a regular cover with deck transformation group G, Sasao and Joon-Sim [1990] gave a number of exact sequences relating the various groups of homotopy self -equivalences, for example

$$1 \to A \to G \to \mathcal{E}(p) \to \mathcal{E}(Y),$$

where A is well defined.

In the case of a 2–stage G–space, that is a G–space induced by a G–map between Eilenberg-MacLane spaces, Møller [1990. Theorem 4.1] determined, up to extension, the group $\mathcal{E}^*(X)_G$ of G-homotopy self-equivalence classes.

Zhang [1992, Theorem 1.4] considered the case where a group G acts in a fibrewise way on X and gave conditions which ensure that the group $\mathcal{E}_B(X)_G$ of equivariant fibre homotopy equivalences is nilpotent.

An example where $\mathcal{E}_B(X)$ is uncountable was given by Tsukiyama [1992, Theorem A]. Let T be a maximal torus in a compact connected Lie group which is not a torus, then, for the fibration $G/T \to BT \to BG$, $\mathcal{E}_{BG}(BT)$ is uncountable although its image in $\mathcal{E}(BT)$ is finite.

Yamanoshita [1995] associated with any G–principal bundle $E \to B$ a Serre fibration with total space $\text{top}_G(E)$; he then determined the weak homotopy types of $\text{top}_{S^1}(S^3)$ and $\text{top}_{SO(2)}(SO(3))$.

Tsukiyama [1996] gave an example where $\mathcal{E}_B(X)_G \to \mathcal{E}(X)$ is not a monomorphism, where $X \to B$ is a principal G–bundle.

The exact sequence $\pi_1^X(X,1) \to \pi_1^X(B,f) \to \mathcal{E}_B(f) \to \mathcal{E}(X)$ for a Hurewicz fibration $f : X \to B$ and its dual was used by Baues, Hardie and Kamps [1997] to calculate $\mathcal{E}_{S^3}(S^5)$ and $\mathcal{E}^{S^5}(S^3)$ for $\eta^2 : S^5 \to S^3$.

Let $H \triangleleft G$ be closed subgroups of a compact Lie group E. Xia [1998] calculated $\mathcal{E}(E)_G$ and $\mathcal{E}^*(E)_G$ where $\mathcal{E}(E/H)_{G/H}$ is known, and related the relevant spaces.

Let $p : E \to B$ be a Hurewicz fibration and $\mathcal{E}(p)^G = R^{-1}(G)$, where $R : \mathcal{E}_B(E) \to \mathcal{E}(F)$ and G is a subgroup of $\mathcal{E}(F)$. Also let $E(p)^G$ and $E(F)^G$ be the corresponding submonoids of $E(p)$ and $E(F)$ respectively. In case $B = S^n$ there is (Felix and Thomas [1995]) an exact sequence

$$\pi_1(E(F)^G) \to \pi_n(E(F)^G) \to \mathcal{E}(p)^G \to G_\alpha \to 1;$$

conditions are given that $E(p)^G$ is nilpotent. A description of the 0–localization of this exact sequence is given in terms of the Sullivan model of F. More generally, with other restrictions, conditions are given that $\pi_1(E(p)^G)$ is finitely generated or that $E(p)^1 \to E(F)^1$ is a rational homotopy equivalence.

Piccinini and Spreafico [1998A] showed that two different G–bundles over B have 'conjugate' Gauge groups if, and only if, they differ by a certain action of a ZG–bundle; in case $G = U(n)$ or $O(n)$, equivalently they differ by a line bundle. They also considered which line bundles can be added without changing a given bundle. Analogously Crabb, Spreafico and Sutherland [1999] considered where adding a line bundle to a vector bundle results in an equivalent bundle; they obtained specific results in case the line bundle is the Hopf bundle over projective space. Piccinini and Spreafico [1998B] reviewed the classifying functor $G \to B_G$ for topological groups and obtained new results. With certain restrictions, the classification of $\mathcal{F}$–fibrations reduces to the classification of the associated principal fibrations (Pavešić [1993]). Pacati, Pavešić and Piccinini [1997] considered the Dold-Lashof-Fuchs construction in the category of k–spaces; in this category classifying spaces for numerable principal h–fibrations can be constructed in a simpler way. Pacati, Pavešić and Piccinini [1998] studied the construction of principal fibrations associated to $\mathcal{F}$–fibrations, and an inverse construction; they gave a classification theorem for numerable $\mathcal{F}$–fibrations. Ōshima and Tsukiyama [1998] considered bundle map theory using the category of weak Hausdorff k–spaces and extended some known results.

8.1. Equivalence of the theories. The group of G–equivalences is isomorphic to the group of top_{BG} equivalences in the following sense. Let G be a topological group having the homotopy type of a CW complex, and let $X \to B$ be a principal G–bundle, which is a Dold fibration, classified by $k : B \to BG$. Also let $T \to BG$ be the mapping track fibration of k. Then we have

$$\mathcal{E}(X)_G \cong \mathcal{E}_{BG}(T).$$

Using constructions similar to the categories of fibres, Booth [1990, Corollary 2.2] proves this result for topological spaces in the more situation where G is a grouplike topological monoid. He also obtains related results where, for example, BG is replaced by the classifying space of $\mathrm{map}(F, F)$ or $\mathrm{map}^*(F, F)$, where these spaces have the homotopy type of CW complexes.

9. Applications

For a given complex X, let $\mathcal{X}$ be the set of Postnikov conjugates of X, that is the set of homotopy types which have the same n–type as X for all n.

McGibbon and Møller (see [1992A, Theorem 1 and Theorem 2]) showed, for simply-connected spaces X satisfying certain conditions, that $\mathcal{X} = \{X\}$ if, and only if, $\mathcal{E}^*(X) \to \mathcal{E}^*(X_n)$ has finite cokernel for each n. In [1992B, Theorem 4] they showed that $\mathcal{X}$ is uncountably large, in the case where $X = \mathcal{B}G$, for almost every compact connected Lie group G; they also showed that $\mathcal{X}$ is either trivial or uncountable in the case where X is a nilpotent CW complex of finite type, and gave many examples.

Postnikov conjugates X and Y can have different groups of homotopy self-equivalences. An example was given by McGibbon and Møller [1992C, Example A] where $X = K(Z,3) \times S^3$, and $\mathcal{E}(X)$ is the (infinite) group of upper triangular matrices in $\mathrm{GL}(\mathbf{Z},3)$ but $\mathcal{E}(Y)$ is cyclic of order 2. On the other hand, suppose now that X and Y are 1–connected, of finite type and have the same n–type for all n and that the groups $\pi_1(E(X_n))$ and $\pi_1(E(Y_n))$ are all abelian (for example if the spaces X and Y are nilpotent of finite $\mathbf{Z}$–type), then the groups $\lim^1 \pi_1(E(X_n))$ and $\lim^1 \pi_1(E(Y_n))$ can differ only in their torsion summands; an example was given by McGibbon and Møller [1992C, Example B] where $X = \mathbf{C}P^\infty \times \Omega S^3 \times S^3$, $\lim^1 \pi_1(E(X_n)) = \mathbf{R}$ and $\lim^1 \pi_1(E(Y_n)) = \mathbf{R} \oplus \mathbf{Q}/\mathbf{Z}$.

Let $F \to E \to B$ be a CW fibration and let $\mathcal{E}(F)$ have no non-trivial perfect subgroup. Berrick [1992] showed that B is acyclic if, and only if, $H_*(F) \to H_*(E)$ is an isomorphism.

Let G be a CW topological group, M a space, and $\mathcal{Q}(M) = \mathrm{map}(M, M)$, then the set of G–homotopy equivalence classes of free G–spaces homotopy equivalent to M is classified by $[B_G, B_{\mathcal{Q}(M)}]$ (Booth [1992]).

Division of Pure Mathematics, University of Liverpool, Liverpool L69 3BX, England.

Contemporary Mathematics
Volume **274**, 2001

Bibliography on $\mathcal{E}(X)$ 1988-1999

John W. Rutter

ABSTRACT. This bibliography lists articles published in the area of 'Spaces and Groups of Homotopy Self-equivalences' between 1988, the year of the Montreal conference, and the Gargnano workshop in 1999.

[1] List of papers on or relevant to groups of self-homotopy equivalences. In *Groups of self-equivalences and related topics (Montreal, PQ, 1988), Lecture notes in Math. 1425*, pages 208–214. Springer, Berlin, 1990.
[2] Martin Arkowitz. Formal differential graded algebras and homomorphisms. *J. Pure Appl. Algebra*, 51(1-2):35–52, 1988.
[3] Martin Arkowitz. The group of self-homotopy equivalences—a survey. In *Groups of self-equivalences and related topics (Montreal, PQ, 1988), Lecture notes in Math. 1425*, pages 170–203. Springer, Berlin, 1990.
[4] Martin Arkowitz and Gregory Lupton. Equivalence classes of homotopy-associative comultiplications of finite complexes. *J. Pure Appl. Algebra*, 102:109–136, 1995.
[5] Martin Arkowitz and Gregory Lupton. On finiteness of subgroups of self-homotopy equivalences. In *The Čech centennial (Boston, MA, 1993), Contemporary Math. 181*, pages 1–25. Amer. Math. Soc., Providence, RI, 1995.
[6] Martin Arkowitz and Gregory Lupton. On the nilpotency of subgroups of self-homotopy equivalences. In *Algebraic topology: new trends in localization and periodicity (Sant Feliu de Guíxols, 1994), Progress in Math. 136*, pages 1–22. Birkhäuser, Basel, 1996.
[7] Martin Arkowitz and Ken-ichi Maruyama. Self-homotopy equivalences which induce the identity on homology, cohomology or homotopy groups. *Topology Appl.*, 87(2):133–154, 1998.
[8] M. Aubry and J.-M. Lemaire. Sur certaines équivalences d'homotopies. *Ann. Inst. Fourier (Grenoble)*, 41(1):173–187, 1991.
[9] H. J. Baues, K. A. Hardie, and K. H. Kamps. The self-equivalence groups in certain coherent homotopy categories. *Tsukuba J. Math.*, 21(1):213–228, 1997.

2000 *Mathematics Subject Classification.* 55P10.

[10] Hans Joachim Baues. *Algebraic homotopy.* Cambridge University Press, Cambridge, 1989.
[11] Hans Joachim Baues. On the homotopy category of Moore spaces and an old result of Barratt. In *Algebraic topology Poznań 1989, Lecture notes in Math. 1474*, pages 207–230. Springer, Berlin, 1991A.
[12] Hans Joachim Baues. *Combinatorial homotopy and 4-dimensional complexes.* Walter de Gruyter & Co., Berlin, 1991B. With a preface by Ronald Brown.
[13] Hans Joachim Baues. On the group of homotopy equivalences of a manifold. *Trans. Amer. Math. Soc.*, 348(12):4737–4773, 1996.
[14] Hans Joachim Baues and Joachim Buth. On the group of homotopy equivalences of simply connected five manifolds. *Math. Z.*, 222(4):573–614, 1996.
[15] Hans-Joachim Baues and Manfred Hartl. On the homotopy category of Moore spaces and the cohomology of the category of abelian groups. *Fund. Math.*, 150(3):265–289, 1996.
[16] A. J. Berrick. Homologous fibres and total spaces. *Osaka J. Math.*, 29(2):339–346, 1992.
[17] P. I. Booth. Free G-spaces, principal G-fibrations and maps between classifying spaces. *Rend. Mat. Appl. (7)*, 12(4):901–919 (1993), 1992.
[18] P. I. Booth and P. R. Heath. On the groups $\mathcal{E}(X \times Y)$ and $\mathcal{E}_B^B(X \times_B Y)$. In *Groups of self-equivalences and related topics (Montreal, PQ, 1988), Lecture notes in Math. 1425*, pages 17–31. Springer, Berlin, 1990.
[19] Peter Booth. Equivalent homotopy theories and groups of self-equivalences. In *Groups of self-equivalences and related topics (Montreal, PQ, 1988), Lecture notes in Math. 1425*, pages 1–16. Springer, Berlin, 1990.
[20] A. Cavicchioli, F. Hegenbarth, and F. Spaggiari. Topological properties of high-dimensional handles. *Cahiers Topologie Géom. Différentielle Catég.*, 39(1):45–62, 1998.
[21] Tim D. Cochran and Nathan Habegger. On the homotopy theory of simply connected four manifolds. *Topology*, 29(4):419–440, 1990.
[22] M. C. Crabb, J. R. Hubbuck, and Kai Xu. Fields of spaces. In *Adams Memorial Symposium on Algebraic Topology, 1 (Manchester, 1990)*, pages 241–254. Cambridge Univ. Press, Cambridge, 1992.
[23] M. C. Crabb, M. Spreafico, and W. A. Sutherland. Enumerating projectively equivalent bundles. *Math. Proc. Cambridge Philos. Soc.*, 125(2):223–242, 1999.
[24] G. Didierjean. Homotopie des espaces d'équivalences. In *Groups of self-equivalences and related topics (Montreal, PQ, 1988), Lecture notes in Math. 1425*, pages 32–39. Springer, Berlin, 1990.
[25] G. Didierjean and A. Legrand. Équivalences d'homotopie et crochet de Whitehead. *Canad. J. Math.*, 46(2):253–273, 1994.
[26] Geneviève Didierjean. Homotopie de l'espace des équivalences d'homotopie. *Trans. Amer. Math. Soc.*, 330(1):153–163, 1992.
[27] W. G. Dwyer. R-nilpotency in homotopy equivalences. *Israel J. Math.*, 66(1-3):154–159, 1989.
[28] W. G. Dwyer, D. M. Kan, and J. H. Smith. Towers of fibrations and homotopical wreath products. *J. Pure Appl. Algebra*, 56(1):9–28, 1989.
[29] W. G. Dwyer, D. M. Kan, J. H. Smith, and C. R. Stover. A π-algebra spectral sequence for function spaces. *Proc. Amer. Math. Soc.*, 120(2):615–621, 1994.

[30] Y. Félix and J. C. Thomas. The monoid of self-homotopy equivalences of some homogeneous spaces. *Exposition. Math.*, 12(4):305–322, 1994.

[31] Y. Félix and J. C. Thomas. Nilpotent subgroups of the group of fibre homotopy equivalences. *Publ. Mat.*, 39(1):95–106, 1995.

[32] Yves Félix and Aniceto Murillo. A note on the nilpotency of subgroups of self-homotopy equivalences. *Bull. London Math. Soc.*, 29(4):486–488, 1997.

[33] Yves Félix and Aniceto Murillo. A bound for the nilpotency of a group of self homotopy equivalences. *Proc. Amer. Math. Soc.*, 126(2):625–627, 1998.

[34] Z. Fiedorowicz, R. Schwänzl, and R. Vogt. Hermitian A_∞ rings and their K-theory. In *Adams Memorial Symposium on Algebraic Topology, 1 (Manchester, 1990), London Math. Soc. lecture notes 175*, pages 67–81. Cambridge Univ. Press, Cambridge, 1992.

[35] Edward Formanek and Claudio Procesi. The automorphism group of a free group is not linear. *J. Algebra*, 149(2):494–499, 1992.

[36] Pierre-Paul Grivel. Algèbres de Lie de dérivations de certaines algèbres pures. *J. Pure Appl. Algebra*, 91(1-3):121–135, 1994.

[37] Vagn Lundsgaard Hansen. The space of self-maps on the 2-sphere. In *Groups of self-equivalences and related topics (Montreal, PQ, 1988), Lecture notes in Math. 1425*, pages 40–47. Springer, Berlin, 1990.

[38] Allen Hatcher and Darryl McCullough. Finite presentation of 3-manifold mapping class groups. In *Groups of self-equivalences and related topics (Montreal, PQ, 1988), Lecture notes in Math. 1425*, pages 48–57. Springer, Berlin, 1990.

[39] Volker Hauschild. Deformations and the rational homotopy of the monoid of fiber homotopy equivalences. *Illinois J. Math.*, 37(4):537–560, 1993.

[40] C. Hayat and A. Legrand. Whitehead γ functor and homotopy. *Algebra Colloq.*, 3(2):147–156, 1996.

[41] Claude Hayat-Legrand. Représentation des groupes d'extension. Applications. *Compositio Math.*, 90(3):351–366, 1994.

[42] Philip R. Heath. On the group $\mathcal{E}(X \times Y)$ of self homotopy equivalences of a product. *Quaestiones Math.*, 19(3-4):433–451, 1996.

[43] Yasumasa Hirashima. Multiplicative structures on monoids of self homotopy equivalences of $K(G,1)$-spaces. *Osaka J. Math.*, 28(1):141–152, 1991.

[44] Kenshi Ishiguro, Jesper Møller, and Dietrich Notbohm. Rational self-equivalences of spaces in the genus of a product of quaternionic projective spaces. *J. Math. Soc. Japan*, 51(1):45–61, 1999.

[45] N. V. Ivanov. The rank of teichmüller modular groups. *Mat. Zametki*, 44(5):636–644, 701, 1988A. English translation: Math. Notes Acad. Sci. USSR 44 (1988) 829-832.

[46] N. V. Ivanov. Teichmüller modular groups and arithmetic groups. *Zap. Nauchn. Sem. Leningrad. Otdel. Mat. Inst. Steklov. (LOMI)*, 167(Issled. Topol. 6):95–110, 190–191, 1988B. English translation: J. Soviet Math. 52 (1990) 2809-2818.

[47] Nikolai V. Ivanov. *Subgroups of Teichmüller modular groups.* American Mathematical Society, Providence, RI, 1992. Translated from the Russian by E. J. F. Primrose and revised by the author.

[48] Norio Iwase, Ken-ichi Maruyama, and Shichirô Oka. A note on $\mathcal{E}(\mathbf{HP}^n)$ for $n \leq 4$. *Math. J. Okayama Univ.*, 33:163–176, 1991.

[49] Stefan Jackowski, James McClure, and Bob Oliver. Homotopy classification of self-maps of BG via G-actions. I. *Ann. of Math. (2)*, 135(1):183–226, 1992A.

[50] Stefan Jackowski, James McClure, and Bob Oliver. Homotopy classification of self-maps of BG via G-actions. II. *Ann. of Math. (2)*, 135(2):227–270, 1992B.

[51] Stefan Jackowski, James McClure, and Bob Oliver. Self-homotopy equivalences of classifying spaces of compact connected Lie groups. *Fund. Math.*, 147(2):99–126, 1995.

[52] Donald W. Kahn. Representations of the stable group of self-equivalences. In *Groups of self-equivalences and related topics (Montreal, PQ, 1988), Lecture notes in Math. 1425*, pages 58–70. Springer, Berlin, 1990A.

[53] Donald W. Kahn. Some research problems on homotopy-self-equivalences. In *Groups of self-equivalences and related topics (Montreal, PQ, 1988), Lecture notes in Math. 1425*, pages 204–207. Springer, Berlin, 1990B.

[54] Donald W. Kahn. Twisted tori and the characteristic classes of a self-equivalence. *Bol. Soc. Mat. Mexicana (2)*, 37(1-2):293–302, 1992. Papers in honor of José Adem (Spanish).

[55] John Kalliongis and Darryl McCullough. Homeotopy groups of irreducible 3-manifolds which may contain two-sided projective planes. *Pacific J. Math.*, 153(1):85–117, 1992.

[56] Sadayoshi Kojima. Isometry transformations of hyperbolic 3-manifolds. *Topology Appl.*, 29(3):297–307, 1988.

[57] Sławomir Kwasik and Reinhard Schultz. Vanishing of Whitehead torsion in dimension four. *Topology*, 31(4):735–756, 1992.

[58] Irene Llerena and John W. Rutter. The group of homotopy self-equivalences of a union of $(n-1)$-connected $2n$-manifolds. *Kodai Math. J.*, 21(3):330–349, 1998.

[59] Ning Lu. A simple presentation of the Siegel modular groups. *Linear Algebra Appl.*, 166:185–194, 1992.

[60] Ken-ichi Maruyama. Localization of a certain subgroup of self-homotopy equivalences. *Pacific J. Math.*, 136(2):293–301, 1989.

[61] Ken-ichi Maruyama. Localization of self-homotopy equivalences inducing the identity on homology. *Math. Proc. Cambridge Philos. Soc.*, 108(2):291–297, 1990A.

[62] Ken-ichi Maruyama. Localizing $\epsilon_\sharp(X)$. In *Groups of self-equivalences and related topics (Montreal, PQ, 1988), Lecture notes in Math. 1425*, pages 87–90. Springer, Berlin, 1990B.

[63] Ken-ichi Maruyama. Finiteness properties of self-equivalence groups of rational co-H-spaces. *Manuscripta Math.*, 76(1):33–43, 1992.

[64] Ken-ichi Maruyama. Finite complexes whose self-homotopy equivalence groups realize the infinite cyclic group. *Canad. Math. Bull.*, 37(4):534–536, 1994.

[65] Ken-ichi Maruyama. Finitely presented subgroups of the self-homotopy equivalences group. *Math. Z.*, 221(4):537–548, 1996.

[66] Ken-ichi Maruyama and Mamoru Mimura. Nilpotent subgroups of the group of self-homotopy equivalences. *Israel J. Math.*, 72(3):313–319 (1991), 1990.

[67] Darryl McCullough. Topological and algebraic automorphisms of 3-manifolds. In *Groups of self-equivalences and related topics (Montreal, PQ, 1988), Lecture notes in Math. 1425*, pages 102–113. Springer, Berlin, 1990.

[68] Darryl McCullough. Virtually geometrically finite mapping class groups of 3-manifolds. *J. Differential Geom.*, 33(1):1–65, 1991.

[69] C. A. McGibbon and J. M. Møller. On infinite-dimensional spaces that are rationally equivalent to a bouquet of spheres. In *Algebraic topology (San Feliu de Guíxols, 1990), Lecture notes in Math. 1509*, pages 285–293. Springer, Berlin, 1992A.

[70] C. A. McGibbon and J. M. Møller. On spaces with the same n-type for all n. *Topology*, 31(1):177–201, 1992B.

[71] C. A. McGibbon and J. M. Møller. How can you tell two spaces apart when they have the same n-type for all n? In *Adams Memorial Symposium on Algebraic Topology, 1 (Manchester, 1990), London Math. Soc. lecture notes 175*, pages 131–143. Cambridge Univ. Press, Cambridge, 1992C.

[72] Wolfgang Metzler. Die Unterscheidung von Homotopietyp und einfachem Homotopietyp bei zweidimensionalen Komplexen. *J. Reine Angew. Math.*, 403:201–219, 1990.

[73] Andy Miller. Projecting homeomorphisms from covering spaces. In *Groups of self-equivalences and related topics (Montreal, PQ, 1988), Lecture notes in Math. 1425*, pages 114–132. Springer, Berlin, 1990.

[74] Mamoru Mimura and Hideaki Ōshima. Self homotopy groups of Hopf spaces with at most three cells. *J. Math. Soc. Japan*, 51(1):71–92, 1999.

[75] Jesper Michael Møller. Self-maps on twisted Eilenberg-Mac Lane spaces. *Kodai Math. J.*, 11(3):372–378, 1988.

[76] Jesper Michael Møller. Self-homotopy equivalences of $H_*(-;\mathbf{Z}/p)$-local spaces. *Kodai Math. J.*, 12(2):270–281, 1989.

[77] Jesper Michael Møller. Equivariant self-homotopy equivalences of 2-stage G-spaces. In *Groups of self-equivalences and related topics (Montreal, PQ, 1988), Lecture notes in Math. 1425*, pages 133–146. Springer, Berlin, 1990.

[78] Jesper Michael Møller. Samelson products in spaces of self-homotopy equivalences. *Canad. J. Math.*, 42(1):95–108, 1990A.

[79] Jesper Michael Møller. Self-homotopy equivalences of group cohomology spaces. *J. Pure Appl. Algebra*, 73(1):23–37, 1991.

[80] Jesper Michael Møller. The normalizer of the Weyl group. *Math. Ann.*, 294(1):59–80, 1992.

[81] Juno Mukai. On stable homotopy of the complex projective space. *Japan. J. Math. (N.S.)*, 19(1):191–216, 1993.

[82] Dietrich Notbohm and Larry Smith. Rational homotopy of the space of homotopy equivalences of a flag manifold. In *Algebraic topology (San Feliu de Guíxols, 1990), Lecture notes in Math. 1509*, pages 301–312. Springer, Berlin, 1992.

[83] Hideaki Ōshima and Kouzou Tsukiyama. Bundle map theory in the category of weak Hausdorff k-spaces. *Mem. Fac. Sci. Eng. Shimane Univ. Ser. B Math. Sci.*, 31:27–55, 1998.

[84] Akimou Osse. λ-structures and representation rings of compact connected Lie groups. *J. Pure Appl. Algebra*, 121(1):69–93, 1997.

[85] Claudio Pacati, Petar Pavešić, and Renzo Piccinini. The Dold-Lashof-Fuchs construction revisited. *Rend. Sem. Mat. Fis. Milano*, 65:35–52 (1997), 1995.

[86] Claudio Pacati, Petar Pavešić, and Renzo Piccinini. On the classification of $\mathcal{F}$-fibrations. *Topology Appl.*, 87(3):213–227, 1998.

[87] Petar Pavešić. A note on the classification of $\mathcal{F}$-fibrations. In *Proceedings of the Eleventh International Conference of Topology (Trieste, 1993), Rend. Istit. Mat. Univ. Trieste*, volume 25, pages 413–418 (1994), 1993.

[88] Petar Pavešić. Groups of self-homotopy equivalence. *Boll. Unione Mat. Ital. Sez. A Mat. Soc. Cult. (8)*, 1(suppl.):55–58, 1998.

[89] Petar Pavešić. Self-homotopy equivalences of product spaces. *Proc. Roy. Soc. Edinburgh Sect. A*, 129(1):181–197, 1999.

[90] R. A. Piccinini, editor. *Groups of self-equivalences and related topics, Proceedings of the conference held at the Université de Montréal, Montreal, Quebec, August 8–12, 1988, Lecture notes in Math. 1425*, Berlin, 1990. Springer-Verlag.

[91] Renzo A. Piccinini and Mauro Spreafico. *Conjugacy classes in gauge groups, Queen's papers in Pure and Appl. Math. 111*. Queen's University, Kingston, ON, 1998A.

[92] Renzo A. Piccinini and Mauro Spreafico. The Milgram-Steenrod construction of classifying spaces for topological groups. *Exposition. Math.*, 16(2):97–130, 1998B.

[93] Joseph Roitberg. Weak identities, phantom maps and H-spaces. *Israel J. Math.*, 66(1-3):319–329, 1989.

[94] Joseph Roitberg. Note on phantom phenomena and groups of self-homotopy equivalences. *Comment. Math. Helv.*, 66(3):448–457, 1991.

[95] Joseph Roitberg. Genus and symmetry in homotopy theory. *Math. Ann.*, 305(2):381–386, 1996.

[96] John W. Rutter. The group of homotopy self-equivalence classes using an homology decomposition. *Math. Proc. Cambridge Philos. Soc.*, 103(2):305–315, 1988A.

[97] John W. Rutter. Homotopy classification of maps between pseudo-projective planes. *Quaestiones Math.*, 11(4):409–422, 1988B.

[98] John W. Rutter. On skeleton preserving homotopy self-equivalences of CW complexes. In *Groups of self-equivalences and related topics (Montreal, PQ, 1988), Lecture notes in Math. 1425*, pages 147–156. Springer, Berlin, 1990A.

[99] John W. Rutter. Homotopy self-equivalence groups of unions of spaces: including Moore-spaces. *Quaestiones Math.*, 13(3-4):321–334, 1990B.

[100] John W. Rutter. Whitney sums (fibre-joins) in over space theory and obstruction theory for cohomology with local coefficients. *Proc. Roy. Soc. Edinburgh Sect. A*, 115(3-4):359–365, 1990C.

[101] John W. Rutter. The group of homotopy self-equivalences of non-simply-connected spaces using Postnikov decompositions. *Proc. Roy. Soc. Edinburgh Sect. A*, 120(1-2):47–60, 1992A.

[102] John W. Rutter. The group of homotopy self-equivalences of non-simply-connected spaces using Postnikov decompositions. II. *Proc. Roy. Soc. Edinburgh Sect. A*, 122(1-2):127–135, 1992B.

[103] John W. Rutter. *Spaces of homotopy self-equivalences, Lecture notes in Math. 1662*. Springer-Verlag, Berlin, 1997. A survey.

[104] Paolo Salvatore. Rational homotopy nilpotency of self-equivalences. *Topology Appl.*, 77(1):37–50, 1997.

[105] Seiya Sasao and Joon-Sim Cha. On self-homotopy equivalences of covering spaces. *Kodai Math. J.*, 13(2):231–240, 1990.

[106] Yi Yun Shi and Zai Si Zuo. Some results on homotopy self-equivalences. *Kexue Tongbao (Chinese)*, 41(2):99–102, 1996.

[107] Yoshimi Shitanda. Monoid structure of endomorphisms of $\mathbf{HP}^\infty \times S^n$. *Yokohama Math. J.*, 41(1):7–15, 1993A.

[108] Yoshimi Shitanda. Phantom maps and monoids of endomorphisms of $K(Z,m) \times S^n$. *Publ. Res. Inst. Math. Sci. Kyoto University*, 29(3):397–409, 1993B.

[109] Hideo Takahashi. Note on homotopy classes of self maps; $[\Sigma\mathbf{H}\mathrm{P}^3, \Sigma\mathbf{H}\mathrm{P}^3]$. *Math. J. Okayama Univ.*, 38:159–196 (1998), 1996.

[110] G. Triantafillou. Isotropy groups of homotopy classes of maps. *Trans. Amer. Math. Soc.*, 334(1):37–48, 1992.

[111] Georgia Triantafillou. Automorphisms of spaces with finite fundamental group. *Trans. Amer. Math. Soc.*, 347(9):3391–3403, 1995.

[112] K. Tsukiyama. An uncountable group of fibre homotopy equivalences. *Bol. Soc. Mat. Mexicana (2)*, 37(1-2):545–548, 1992. Papers in honor of José Adem (Spanish).

[113] Kouzou Tsukiyama. An example of self-homotopy equivalences. *J. Math. Soc. Japan*, 48(2):317–319, 1996.

[114] V.G. Turaev. Homeomorphisms of geometric three-dimensional manifolds. *Math. Notes*, 43(4):307–312, 1988. Translated from *Mat. Zametki* 43(4):533-542, 575, 1988.

[115] Hans Michael Unsöld. A_n^4-polyhedra with free homology. *Manuscripta Math.*, 65(2):123–145, 1989.

[116] Jianguo Xia. Equivariant self-equivalences of principal fibre bundles. *Appl. Math. J. Chinese Univ. Ser. B*, 13(1):109–116, 1998. A Chinese summary appears in Gaoxiao Yingyong Shuxue Xuebao Ser. A 13 (1998), no. 1, 118.

[117] Kai Xu. Nilpotent subgroups of self equivalences of torsion spaces. *Israel J. Math.*, 93:189–194, 1996.

[118] Kohhei Yamaguchi. Self-homotopy equivalences and highly connected Poincaré complexes. In *Groups of self-equivalences and related topics (Montreal, PQ, 1988), Lecture notes in Math. 1425*, pages 157–169. Springer, Berlin, 1990.

[119] Kohhei Yamaguchi. Self-maps of $\Sigma^k\mathbf{C}\mathrm{P}^3$ for $k \geq 1$. *Kodai Math. J.*, 14(1):144–162, 1991.

[120] Tsuneyo Yamanoshita. On the space of self-homotopy equivalences of the projective plane. *J. Math. Soc. Japan*, 45(3):489–494, 1993.

[121] Tsuneyo Yamanoshita. On the group of S^1-equivariant homeomorphisms of the 3-sphere. *Publ. Res. Inst. Math. Sci.*, 31(5):953–958, 1995.

[122] Ping Zhang. On the weak nilpotency of H-spaces of sections. *Quart. J. Math. Oxford Ser. (2)*, 43(171):381–386, 1992.

Division of Pure Mathematics, University of Liverpool, Liverpool L69 3BX, England

Contemporary Mathematics
Volume **274**, 2001

Subgroups of the Group of Self-Homotopy Equivalences

Martin Arkowitz, Gregory Lupton, and Aniceto Murillo

Abstract. Denote by $\mathcal{E}(Y)$ the group of homotopy classes of self-homotopy equivalences of a finite-dimensional complex Y. We give a selection of results about certain subgroups of $\mathcal{E}(Y)$. We establish a connection between the Gottlieb groups of Y and the subgroup of $\mathcal{E}(Y)$ consisting of homotopy classes of self-homotopy equivalences that fix homotopy groups through the dimension of Y, denoted by $\mathcal{E}_{\#}(Y)$. We give an upper bound for the solvability class of $\mathcal{E}_{\#}(Y)$ in terms of a cone decomposition of Y. We dualize the latter result to obtain an upper bound for the solvability class of the subgroup of $\mathcal{E}(Y)$ consisting of homotopy classes of self-homotopy equivalences that fix cohomology groups with various coefficients. We also show that with integer coefficients, the latter group is nilpotent.

1. Introduction and Preliminaries

Let Y be a CW-complex of dimension N and $\mathcal{E}(Y)$ the group of homotopy classes of self-homotopy equivalences of Y. In this paper we present a sample of results about a number of subgroups of $\mathcal{E}(Y)$. We denote by $\mathcal{E}_{\#}(Y)$ the following proto-typical such subgroup:

$$\mathcal{E}_{\#}(Y) = \{f \in \mathcal{E}(Y) \mid f_{\#} = 1 \colon \pi_i(Y) \to \pi_i(Y), \text{ for all } i \leq N\}.$$

In Section 2, we give a way to construct elements in $\mathcal{E}_{\#}(Y)$. This is of interest since it provides a connection between the *Gottlieb group* of Y and certain subgroups of self-homotopy equivalences (Theorem 2.3). Next, in Section 3, we consider questions about the solvability and nilpotency of $\mathcal{E}_{\#}(Y)$. For example, we show that if Y is the cofibre of a map between two wedges of spheres, then $\mathcal{E}_{\#}(Y)$ is an abelian group (Corollary 3.5). This result generalizes into a simple upper bound on the solvability of $\mathcal{E}_{\#}(Y)$ in terms of a cone-length invariant of Y (Theorem 3.3). In Section 4 we dualize these results to obtain upper bounds for the solvability of the group of equivalences that fix cohomology with different coefficients (Theorem 4.2). We also show that the subgroup of self-homotopy equivalences which fix the integral cohomology of a finite complex is a nilpotent group (Proposition 4.9).

1991 *Mathematics Subject Classification.* Primary 55P10; Secondary 55P62, 55Q05.
Key words and phrases. Homotopy equivalences, Gottlieb group, cone-length, nilpotent group, solvable group.

We now review some standard material that we will use. A cofibration sequence

$$Z \xrightarrow{\gamma} Y \xrightarrow{j} X \xrightarrow{q} \Sigma Z,$$

where X is the mapping cone of γ, gives a homotopy coaction $c\colon X \to X \vee \Sigma Z$, obtained by pinching the 'equator' of the cone of Z to a point. The coaction induces an action of $[\Sigma Z, W]$ on $[X, W]$ for any space W (cf. [**Hil65**, Chap. 15] for details). This is defined as follows: If $\alpha \in [\Sigma Z, W]$ and $f \in [X, W]$, then f^α is the composition

$$X \xrightarrow{c} X \vee \Sigma Z \xrightarrow{f \vee \alpha} W \vee W \xrightarrow{\nabla} W.$$

The following properties of this action are well-known, and follow easily from the definitions:

(1) If $h\colon W \to W'$, then $h(f^\alpha) = (hf)^{h\alpha}$.
(2) If $\alpha, \beta \in [\Sigma Z, W]$, then $(f^\alpha)^\beta = f^{(\alpha+\beta)}$.

Next, consider the following portion of the Puppe sequence associated to the above cofibration sequence:

$$[\Sigma Z, W] \xrightarrow{q^*} [X, W] \xrightarrow{j^*} [Y, W].$$

As is also well-known, the orbits of the action are precisely the pre-images of j^*. That is, for $f, g \in [X, W]$, we have $f^\alpha = g$ for some $\alpha \in [\Sigma Z, W]$ if and only if $fj = gj$.

Next, we review some notation and terminology for groups. Suppose that G is a group and H and K are subgroups. Then $H \lhd G$ denotes that H is normal in G and $[H, K]$ denotes the subgroup generated by commutators of elements of H with elements of K. A *normal chain* for G is a sequence of subgroups

$$G = G_1 \supseteq G_2 \supseteq \cdots \supseteq G_{k+1} \supseteq \cdots$$

with $G_{i+1} \lhd G_i$ for $i \geq 1$. If $[G_i, G_i] \subseteq G_{i+1}$ for each i, then the sequence is called a *solvability series.* If, further, $G_{k+1} = \{1\}$, then we say that G is *solvable of class* $\leq k$ and write $\operatorname{solv} G \leq k$. Analogously, given a normal chain as above with each $G_i \lhd G$ and $[G, G_i] \subseteq G_{i+1}$, then it is called a *nilpotency series.* In this case, if $G_{k+1} = \{1\}$, then we say that G is *nilpotent of class* $\leq k$ and write $\operatorname{nil} G \leq k$. Clearly, we have $\operatorname{nil} G \geq \operatorname{solv} G$. In addition, we write the identity homomorphism of a group and the trivial homomorphism between two groups as $1\colon G \to G$ and $0\colon G \to H$, respectively. This notation is also used for sets with a distinguished element.

Finally, we fix our topological conventions and notation. By a *space*, we mean a connected CW-complex of finite type. Usually, we will be interested in finite-dimensional CW-complexes. When we discuss rational spaces, we will specialize to 1-connected CW-complexes. As is well known, such a space X admits a *rationalization*, which is denoted by $X_{\mathbb{Q}}$. Similarly, a map of 1-connected finite complexes $f\colon X \to Y$ admits a rationalization map $f_{\mathbb{Q}}\colon X_{\mathbb{Q}} \to Y_{\mathbb{Q}}$. A general reference for rationalization is [**HMR75**]. Furthermore, we do not distinguish notationally between a map and its homotopy class. We write $X \equiv Y$ to denote that the spaces X and Y have the same homotopy type. The identity map of a space X is denoted $\iota\colon X \to X$ and the trivial map between two spaces $*\colon X \to Y$.

2. A Connection with the Gottlieb Group

We consider the situation as in Section 1 of a mapping cone sequence

$$Z \xrightarrow{\gamma} Y \xrightarrow{j} X \xrightarrow{q} \Sigma Z,$$

and the induced action of $[\Sigma Z, W]$ on $[X, W]$ which yields $f^\alpha \in [X, W]$ for $\alpha \in [\Sigma Z, W]$ and $f \in [X, W]$. We are interested in the effect that f^α has on homology and homotopy groups. This is described in the following result.

PROPOSITION 2.1. *For the above cofibration sequence, suppose $f \in [X, W]$ and $\alpha \in [\Sigma Z, W]$.*

(1) *The induced homology homomorphism $(f^\alpha)_*\colon H_i(X) \to H_i(W)$ is given by $(f^\alpha)_*(x) = f_*(x) + \alpha_* q_*(x)$, for each $x \in H_i(X)$.*
(2) *Suppose that $(f, \alpha)\colon X \vee \Sigma Z \to W$ factors through the product $X \times \Sigma Z$. Then the induced homotopy homomorphism $(f^\alpha)_\#\colon \pi_i(X) \to \pi_i(W)$ is given by $(f^\alpha)_\#(x) = f_\#(x) + \alpha_\# q_\#(x)$, for each $x \in \pi_i(X)$.*

PROOF. (1) This follows directly from the commutative diagram

$$\begin{array}{ccccccc}
H_i(X) & \xrightarrow{c_*} & H_i(X \vee \Sigma Z) & \xrightarrow{(f\vee\alpha)_*} & H_i(W \vee W) & \xrightarrow{\nabla_*} & H_i(W) \\
& \searrow^{(1,q_*)} & \downarrow{\scriptstyle (p_{1*},p_{2*})} & & \downarrow{\scriptstyle (p_{1*},p_{2*})} & \nearrow_{+} & \\
& & H_i(X) \oplus H_i(\Sigma Z) & \xrightarrow[f_*\oplus\alpha_*]{} & H_i(W) \oplus H_i(W) & &
\end{array}$$

in which the vertical maps are isomorphisms induced by the two projections p_1 and p_2 and the top row is the homomorphism induced by f^α.

(2) Let $\sigma\colon S^i \to S^i \vee S^i$ denote the standard comultiplication. Write $f_\#(x) + \alpha_\# q_\#(x)$ as the composition

$$S^i \xrightarrow{(x \vee qx)\sigma} X \vee \Sigma Z \xrightarrow{(f,\alpha)} W$$

and $(f^\alpha)_\#(x)$ as the composition

$$S^i \xrightarrow{cx} X \vee \Sigma Z \xrightarrow{(f,\alpha)} W .$$

By hypothesis, we can factor (f, α) through the product as $(f, \alpha) = a \circ j\colon X \vee \Sigma Z \to X \times \Sigma Z \to W$, for some $a\colon X \times \Sigma Z \to W$. It is straightforward to prove that $j(x \vee qx)\sigma = jcx\colon S^i \to X \times \Sigma Z$, by checking that their projections onto each summand are homotopic. □

We now specialize to a mapping cone sequence of the form

$$S^{n-1} \xrightarrow{\gamma} Y \xrightarrow{j} X \xrightarrow{q} \Sigma S^{n-1} \equiv S^n,$$

i.e., $X = Y \cup_\gamma e^n$. Then we have an action of $\pi_n(X)$ on $[X, X]$. We consider elements of the form $\iota^\alpha \in [X, X]$, where ι is the identity map of X and $\alpha \in \pi_n(X)$. In general, these maps are not self-homotopy equivalences. However, by adding certain hypotheses, we obtain maps in $\mathcal{E}_\#(X)$, or some other subgroup of $\mathcal{E}(X)$. This approach is similar to that taken in [**AL91**], but instead of assuming $j_\#\colon \pi_\#(Y) \to \pi_\#(X)$ is onto, we shall consider restrictions on the homotopy element $\alpha \in \pi_n(X)$. Recall that the n'th *Gottlieb group* of X, denoted $G_n(X)$, consists of those $\alpha \in$

$\pi_n(X)$ for which there is an associated map $a\colon X\times S^n \to X$ such that the following diagram commutes:

$$\begin{array}{ccc} X\vee S^n & \xrightarrow{(\iota,\alpha)} & X\,. \\ {\scriptstyle j}\big\downarrow & \nearrow_{a} & \\ X\times S^n & & \end{array}$$

See **[Got69]** for various results on the groups $G_n(X)$.

Next, we introduce another subgroup of $\mathcal{E}(X)$. Define

$$\mathcal{E}_*(X)=\{f\in\mathcal{E}(X)\mid f_*=1\colon H_i(X)\to H_i(X),\ \text{for all } i\}.$$

We apply Proposition 2.1 and obtain the following consequence.

COROLLARY 2.2. *Let $X=Y\cup_\gamma e^n$ be a 1-connected CW-complex and $\alpha\in\pi_n(X)$.*

(1) *$\iota^\alpha\in\mathcal{E}_*(X)$ if and only if $\alpha_*q_*=0\colon H_n(X)\to H_n(X)$.*
(2) *Suppose $\alpha\in G_n(X)$, and X is of dimension N. Then $\iota^\alpha\in\mathcal{E}_\#(X)$ if and only if $\alpha_\# q_\#=0\colon \pi_i(X)\to\pi_i(X)$ for $i\le N$.*

PROOF. It is immediate from Proposition 2.1 that ι^α induces the identity on homology groups. Since X is a 1-connected CW-complex, ι^α is a homotopy equivalence. Hence $\iota^\alpha\in\mathcal{E}_*(X)$. This establishes (1), and (2) follows similarly. □

Hence, we are interested in finding situations in which $\alpha q\colon X\to X$ induces the trivial homomorphism, either in homology or homotopy. Our first result is an integral result. Following this, we shall focus on the rational setting, where more information can be obtained.

THEOREM 2.3. *Let $X=Y\cup_\gamma e^n$ be a 1-connected n-dimensional complex. Suppose that $q_\#=0\colon\pi_n(X)\to\pi_n(S^n)$. Then there is a homomorphism*

$$\Theta\colon G_n(X)\to\mathcal{E}_\#(X),$$

defined by $\Theta(\alpha)=\iota^\alpha$ for $\alpha\in G_n(X)$. This homomorphism restricts to

$$\Theta'\colon G_n(X)\cap\ker h_n\to\mathcal{E}_*(X)\cap\mathcal{E}_\#(X),$$

where $h_n\colon\pi_n(X)\to H_n(X)$ denotes the Hurewicz homomorphism.

PROOF. Let $\alpha\in G_n(X)$. Since $\pi_i(S^n)=0$ for $i<n$, the hypothesis gives that $q_\#=0\colon\pi_i(X)\to\pi_i(S^n)$ for $i\le n$. Hence, by Corollary 2.2, $\iota^\alpha\in\mathcal{E}_\#(X)$. Now suppose β is any element in $\pi_n(X)$. Since $\iota^\alpha\in\mathcal{E}_\#(X)$ and $\beta\in\pi_n(X)$, we have that $\iota^\alpha(\beta)=\beta$. Thus, by the properties of the action listed in the introduction, we have

$$\iota^\alpha\iota^\beta=(\iota^\alpha)^{\iota^\alpha(\beta)}=\iota^{\alpha+\iota^\alpha(\beta)}=\iota^{\alpha+\beta}.$$

Therefore Θ is a homomorphism.

Now suppose α is any element in $\ker h_n$. Then $\alpha_*\colon H_n(S^n)\to H_n(X)$ is zero. Since $H_i(S^n)=0$ for positive $i\ne n$, Corollary 2.2 implies that $\iota^\alpha\in\mathcal{E}_*(X)$. Thus Θ restricts to Θ' as claimed. □

REMARK 2.4. It is known that $G_n(X)\subseteq\ker h_n$ under certain hypotheses (cf. [**Got69**, Th.4.1]), so the homomorphism Θ and its restriction Θ' may agree.

We illustrate Theorem 2.3 with an example.

EXAMPLE 2.5. Take $X = S^2 \times S^3 = S^2 \vee S^3 \cup_{[i_1,i_2]} e^5$. This kind of example has been considered previously (cf. [**AM98, Saw75**]), but here we put it into the context discussed above.

As is well-known, S^3 is an H-space and therefore satisfies $G_i(S^3) = \pi_i(S^3)$ for all i. Further, the Gottlieb group preserves products so $G_5(X) = G_5(S^2) \oplus G_5(S^3)$. Since $\pi_5(S^3) = \mathbb{Z}_2$, there is at least a non-trivial element of order 2 in $G_5(X)$. Next, consider $q_\#\colon \pi_5(X) \to \pi_5(S^5)$. Since $\pi_5(X) = \pi_5(S^3) \oplus \pi_5(S^2)$ is a finite group and $\pi_5(S^5)$ is infinite cyclic, it follows that $q_\#$ is zero in this dimension. From Theorem 2.3, Θ defines a homomorphism from $G_5(X)$ to $\mathcal{E}_\#(X)$. Notice that $h_5\colon \pi_5(X) \to H_5(X)$ is zero, since $H_5(X)$ is infinite cyclic. Therefore, $G_5(X) \subseteq \ker h_5$ and $\Theta = \Theta'\colon G_5(X) \to \mathcal{E}_*(X) \cap \mathcal{E}_\#(X)$.

Note that the homomorphism Θ in Theorem 2.3 may have trivial image in $\mathcal{E}_\#(X)$. For Example 2.5, it follows from the computations in [**AM98**, §6] that Θ is actually injective. However, it seems to be difficult to give general conditions to guarantee that Θ is injective. Rather than doing this by placing strong hypotheses on our spaces, we turn now to the rational setting. For a 1-connected CW-complex X, the *rational Gottlieb group* of X is the Gottlieb group of the rationalization of X, that is, $G_n(X_\mathbb{Q})$. Notice that by [**Lan75**], we have $G_n(X_\mathbb{Q}) \cong G_n(X) \otimes \mathbb{Q}$ for each n. In contrast to the ordinary Gottlieb groups, much is known about the rational Gottlieb groups by results of Félix-Halperin [**FH82**]. For instance, a 1-connected, finite complex has no non-trivial rational Gottlieb groups of even degree, and has only finitely many non-trivial rational Gottlieb groups of odd degree (see [**Fél89**] for details). We only touch on these ideas here and avoid heavy use of rational techniques.

LEMMA 2.6. *Suppose we have a mapping cone sequence*

$$S^{n-1} \xrightarrow{\ \gamma\ } Y \xrightarrow{\ j\ } X$$

in which Y and X are 1-connected. If $\gamma_\mathbb{Q} \neq *\colon (S^{n-1})_\mathbb{Q} \to Y_\mathbb{Q}$, then $(j_\mathbb{Q})_\#\colon \pi_n(Y_\mathbb{Q}) \to \pi_n(X_\mathbb{Q})$ is surjective.

PROOF. This can be argued using the long exact homotopy and homology sequences, together with the relative Hurewicz theorem, for the pair (X, Y). Alternatively, Quillen minimal models can be used. We omit the details. □

REMARK 2.7. Notice we assert that $(j_\mathbb{Q})_\#$ is onto in degree n only, and not in all degrees. In the latter case, the cell attachment is called an *inert* cell attachment [**HL87**]. This is one of the hypotheses used in [**AL91**], but it is not satisfied by some of the examples we have in mind.

We will see that under our hypotheses, rational equivalences of the form ι^α are contained in a smaller subgroup of $\mathcal{E}(X)$ than $\mathcal{E}_\#(X)$. We introduce the following notation: For $r \leq \infty$, define

$$\mathcal{E}_{\#r}(X) = \{f \in \mathcal{E}(X) \mid f_\# = 1\colon \pi_i(X) \to \pi_i(X), \text{ for all } i \leq r\}.$$

Note that $f \in \mathcal{E}_{\#\infty}(X)$ if and only if f induces the identity homomorphism of *all* homotopy groups.

The following is our basic rational result.

THEOREM 2.8. *Let $X = Y \cup_\gamma e^n$ be a 1-connected CW complex with n odd and $\gamma_{\mathbb{Q}} \neq *\colon (S^{n-1})_{\mathbb{Q}} \to Y_{\mathbb{Q}}$. Then there is a homomorphism*

$$\Phi\colon G_n(X_{\mathbb{Q}}) \to \mathcal{E}_{\#\infty}(X_{\mathbb{Q}}),$$

defined by $\Phi(\alpha) = \iota^\alpha$ for $\alpha \in G_n(X_{\mathbb{Q}})$. This homomorphism restricts to

$$\Phi'\colon G_n(X_{\mathbb{Q}}) \cap \ker h_n \to \mathcal{E}_*(X_{\mathbb{Q}}) \cap \mathcal{E}_{\#\infty}(X_{\mathbb{Q}}),$$

where h_n denotes the rational Hurewicz homomorphism $h_n\colon \pi_n(X_{\mathbb{Q}}) \to H_n(X_{\mathbb{Q}})$.

PROOF. We proceed as in the proof of Theorem 2.3. First, we claim that $(q_{\mathbb{Q}})_\# = 0\colon \pi_i(X_{\mathbb{Q}}) \to \pi_i(S^n_{\mathbb{Q}})$, for all i. Since n is odd, we have $\pi_i(S^n_{\mathbb{Q}}) = 0$ for $i \neq n$. Hence we must only check that $(q_{\mathbb{Q}})_\# = 0$ in degree n. By Lemma 2.6, $(j_{\mathbb{Q}})_\#\colon \pi_n(Y_{\mathbb{Q}}) \to \pi_n(X_{\mathbb{Q}})$ is surjective. Given $x \in \pi_n(X_{\mathbb{Q}})$, write $x = (j_{\mathbb{Q}})_\#(y)$, for some $y \in \pi_n(Y_{\mathbb{Q}})$. Then $(q_{\mathbb{Q}})_\#(x) = (q_{\mathbb{Q}})_\#(j_{\mathbb{Q}})_\#(y) = \big((qj)_{\mathbb{Q}}\big)_\#(y) = 0$ since $qj = *$, and the claim follows.

Now a simple modification of the proof of Lemma 2.2 yields that $\iota^\alpha \in \mathcal{E}_{\#\infty}(X_{\mathbb{Q}})$, for each $\alpha \in G_n(X_{\mathbb{Q}})$. The remainder of the argument follows exactly as in the proof of Theorem 2.3. □

REMARK 2.9. Notice that, unlike Theorem 2.3, there is no restriction on the dimension of X in Theorem 2.8, and the attached cell need not be top-dimensional. If X is a 1-connected, finite complex, then there is no generality lost in assuming n odd since a 1-connected, finite complex has no non-trivial rational Gottlieb groups of even degree.

Although Theorem 2.8 is a rational result, we are able to 'de-rationalize' it to obtain the following integral consequence.

THEOREM 2.10. *Let $X = Y \cup_\gamma e^n$ be a 1-connected finite complex with n odd and $\gamma \in \pi_{n-1}(Y)$ not of finite order. If the homomorphism Φ from Theorem 2.8 is non-zero, then for each r with $\dim X \leq r < \infty$, there are elements of infinite order in $\mathcal{E}_{\#r}(X)$.*

PROOF. Suppose $\Phi(\alpha) = \iota^\alpha$ is not the identity element in $\mathcal{E}_{\#\infty}(X_{\mathbb{Q}})$. Since this latter is a $\mathbb{Q}$-local group, it contains no non-trivial elements of finite order, and hence $(\iota^\alpha)^k \neq \iota\colon X_{\mathbb{Q}} \to X_{\mathbb{Q}}$, for all k. By [**Mar89**], we have $\mathcal{E}_{\#r}(X_{\mathbb{Q}}) \cong \big(\mathcal{E}_{\#r}(X)\big)_{\mathbb{Q}}$, for each r with $\dim X \leq r < \infty$. From this it follows that for each r there is some positive integer p and some element $f \in \mathcal{E}_{\#r}(X)$ such that $f_{\mathbb{Q}} = (\iota^\alpha)^p$. Since $f_{\mathbb{Q}}$ is of infinite order in $\mathcal{E}_{\#r}(X_{\mathbb{Q}})$, the same must be true of f in $\mathcal{E}_{\#r}(X)$. □

3. Solvability of $\mathcal{E}_\#(Y)$

A result of Dror-Zabrodsky asserts that if Y is a finite complex, then $\mathcal{E}_\#(Y)$ is a nilpotent group [**DZ79**]. One can ask, therefore, whether there are reasonable estimates for the nilpotency, or perhaps the solvability, of $\mathcal{E}_\#(Y)$ in terms of the usual algebraic topological invariants of Y. Several results have been established that relate the nilpotency or solvability of $\mathcal{E}_\#(Y)$, or some similar group, to the *Lusternik-Schnirelmann category* of Y, or related invariants (cf. [**AL96, FM97, FM98, ST99**]). Some of these apply in a rational setting, and others in an integral setting. Typically, these results give an upper bound on the nilpotency or solvability of the group.

We begin by discussing a topological invariant which appears in our results.

DEFINITION 3.1. For any space X, a *spherical cone decomposition of X of length n*, is a sequence of cofibrations

$$L_i \xrightarrow{\gamma_i} X_i \xrightarrow{j_i} X_{i+1},$$

for $0 \leq i < n$, such that each L_i is a finite wedge of spheres, X_0 is contractible and $X_n \equiv X$. We define the *spherical cone-length* of X, denoted by $\mathrm{scl}(X)$, as follows: If X is contractible, then set $\mathrm{scl}(X) = 0$. Otherwise, $\mathrm{scl}(X)$ is the smallest positive integer n such that there exists a spherical cone decomposition of X of length n. If no such integer exists, set $\mathrm{scl}(X) = \infty$. If X is a finite-dimensional complex and we have a spherical cone decomposition of X of length n in which, in addition, $\dim L_i < \dim X$ for $i = 0, \ldots, n-1$, then this is called a *restricted spherical cone decomposition of X of length n*. We then define the *restricted spherical cone-length* of X, denoted $\mathrm{rscl}(X)$, using only restricted spherical cone decompositions in place of ordinary spherical cone decompositions.

REMARK 3.2. Spherical cone-length has been considered in [**Cor94, ST99**]. If we denote the Lusternik-Schnirelmann category of X by $\mathrm{cat}(X)$, then it is known that $\mathrm{cat}(X) \leq \mathrm{scl}(X)$. Note that a space X with $\mathrm{scl}(X) = 1$ is homotopy equivalent to a wedge of spheres and that a space X with $\mathrm{scl}(X) \leq 2$ is homotopy equivalent to the cofibre of a map between wedges of spheres. Furthermore, the cell-structure of a finite-dimensional complex X provides a restricted spherical cone decomposition of length $\leq$ the number of dimensions in which there are positive-dimensional cells.

We introduce a bit more notation before proving the main result of this section. Once again, Y is a complex of dimension N. We define

$$\mathcal{E}_k(Y) = \{f \in \mathcal{E}(Y) \mid f_* = 1\colon [X,Y] \to [X,Y], \text{ for every complex } X \text{ with } \dim X \leq N \text{ and } \mathrm{rscl}(X) \leq k\}.$$

In particular, we have $\mathcal{E}_1(Y) = \mathcal{E}_\#(Y)$. Also, there is a chain of subgroups

$$\mathcal{E}_\#(Y) = \mathcal{E}_1(Y) \supseteq \mathcal{E}_2(Y) \supseteq \cdots \supseteq \mathcal{E}_k(Y) \supseteq \cdots. \tag{1}$$

Clearly we have $\mathcal{E}_k(Y) \lhd \mathcal{E}_{k-1}(Y)$: For if $f \in \mathcal{E}_k(Y)$, $g \in \mathcal{E}_{k-1}(Y)$, $\dim X \leq N$ and $\mathrm{rscl}(X) \leq k$, then $f_*g_*^{-1} = g_*^{-1}\colon [X,Y] \to [X,Y]$. Hence

$$(gfg^{-1})_* = g_*f_*g_*^{-1} = g_*g_*^{-1} = 1,$$

and so $gfg^{-1} \in \mathcal{E}_k(Y)$. Therefore, the series (1) is a normal chain. Furthermore, if $\mathrm{rscl}(Y) \leq k$, then $\mathcal{E}_k(Y) = 1$. Then we have a normal chain

$$\mathcal{E}_\#(Y) \supseteq \mathcal{E}_2(Y) \supseteq \cdots \supseteq \mathcal{E}_k(Y) = \{1\}.$$

The following is the main result of this section.

THEOREM 3.3. *The series (1) is a solvability series, i.e., $[\mathcal{E}_i(Y), \mathcal{E}_i(Y)] \subseteq \mathcal{E}_{i+1}(Y)$. Consequently, we have*

$$\mathrm{solv}\, \mathcal{E}_\#(Y) \leq \mathrm{rscl}(Y) - 1.$$

PROOF. Let $f, g \in \mathcal{E}_i(Y)$ and X be a complex with $\mathrm{rscl}(X) = i+1$ and $\dim X \leq N$. It suffices to show that $f_*g_*(h) = g_*f_*(h)$ for every $h \in [X,Y]$. Consider the last cofibre sequence in a length-$(i+1)$ restricted spherical cone decomposition of X,

$$L_i \xrightarrow{\gamma} X_i \xrightarrow{j} X_{i+1},$$

where L_i is a wedge of spheres and $X_{i+1} \equiv X$. Now, since $f \in \mathcal{E}_i(Y)$, it follows that $j^*(fh) = f_*(hj) = j^*(h)$. Thus, from the properties of the coaction reviewed in Section 1, there is some $\alpha \in [\Sigma L_i, Y]$ such that $fh = h^\alpha$. Similarly, there is some $\beta \in [\Sigma L_i, Y]$ such that $g_*(h) = h^\beta$. Note also that $f\beta = \beta$ since $f \in \mathcal{E}_i(Y) \subseteq \mathcal{E}_1(Y)$, and ΣL_i is a wedge of spheres of dimension $\leq N$. Now we have

$$f_*g_*(h) = f(h^\beta) = (fh)^{f\beta} = (h^\alpha)^\beta = h^{\alpha+\beta}.$$

A similar computation yields $g_*f_*(h) = h^{\beta+\alpha}$. Since $[\Sigma L_i, Y]$ is abelian, the proof is complete. □

REMARK 3.4. A result analogous to Theorem 3.3 has been proved by Scheerer and Tanré in [**ST99**, Th.6]. We note the differences and similarities between these results. Theorem 6 in [**ST99**] is proved for the group of equivalences of a space Y relative to certain fixed classes of spaces (though our proof could be easily modified to hold for these classes). When the class consists of wedges of spheres, the corresponding group of equivalences is $\mathcal{E}_{\#\infty}(Y)$. The upper bound for the solvability of the group of equivalences $\mathcal{E}_{\#\infty}(Y)$ relative to the class of wedges of spheres given in [**ST99**] is then the so-called spherical category of Y, which is less than or equal to the spherical cone-length of Y minus one. On the other hand, the group of equivalences $\mathcal{E}_\#(Y)$ that we consider in Theorem 3.3 is larger than $\mathcal{E}_{\#\infty}(Y)$. Furthermore, the two proofs are similar, but the solvability series in Theorem 3.3 appears to be different from the one in [**ST99**, Th.6].

Theorem 3.3 easily gives the next two corollaries.

COROLLARY 3.5. *If* $\mathrm{rscl}(Y) \leq 2$, *that is,* Y *is the cofibre of a map between wedges of spheres, then* $\mathcal{E}_\#(Y)$ *is abelian.*

COROLLARY 3.6. *For* Y *any finite-dimensional complex,* $\mathcal{E}_\#(Y)$ *is solvable, with* $\mathrm{solv}\,\mathcal{E}_\#(Y) \leq k-1$, *where* k *is the number of dimensions in which there are positive-dimensional cells.*

PROOF. This follows by Remark 3.2. □

We can modify much of the previous material to deal with equivalences that fix all homotopy groups. This also allows us to deal with the case in which Y is an arbitrary space which is not necessarily a finite-dimensional complex. Define

$$\mathcal{E}'_k(Y) = \{f \in \mathcal{E}(Y) \mid f_* = 1\colon [X,Y] \to [X,Y], \text{ for all } X \text{ with } \mathrm{scl}(X) \leq k\}.$$

Then there is a normal chain

$$\mathcal{E}_{\#\infty}(Y) = \mathcal{E}'_1(Y) \supseteq \mathcal{E}'_2(Y) \supseteq \cdots \supseteq \mathcal{E}'_k(Y) \supseteq \cdots. \tag{2}$$

Now the proof of Theorem 3.3 yields the following analogous results.

THEOREM 3.7. [**ST99**, Th. 6] *The series (2) is a solvability series. Therefore,*

$$\mathrm{solv}\,\mathcal{E}_{\#\infty}(Y) \leq \mathrm{scl}(Y) - 1.$$

We note that from a spectral sequence of Didierjean [**Did85**], one can also obtain a different upper bound on the solvability of $\mathcal{E}_{\#\infty}(Y)$ in terms of the cohomology of Y with coefficients in the homotopy groups of Y.

COROLLARY 3.8. *If* $\mathrm{scl}(Y) \leq 2$, *then* $\mathcal{E}_{\#\infty}(Y)$ *is abelian.*

In view of the results in this section, together with the bounds found in [**AL96, FM97, FM98, ST99**], it is natural to believe that, for a finite-dimensional complex Y, the nilpotency class of $\mathcal{E}_{\#}(Y)$ is bounded above by $\mathrm{scl}(Y)-1$. We have not been able to prove this, and so we leave it as a conjecture.

CONJECTURE 3.9. *For a finite-dimensional complex Y,*

$$\mathrm{nil}\,\mathcal{E}_{\#}(Y) \leq \mathrm{rscl}(Y) - 1 \qquad \textit{and} \qquad \mathrm{nil}\,\mathcal{E}_{\#\infty}(Y) \leq \mathrm{scl}(Y) - 1.$$

We note that Scheerer and Tanré have conjectured that $\mathrm{nil}\,\mathcal{E}_{\#\infty}(Y)$ is bounded above by the spherical category of Y [**ST99**, §7, (6)].

Conjecture 3.9 would be established by showing that each of the series (1) and (2) is a nilpotency series. A direct proof of this would also give an independent proof of the Dror-Zabrodsky result on the nilpotency of $\mathcal{E}_{\#}(Y)$.

4. Equivalences that Fix Cohomology Groups

In this section, we dualize some of the ideas of the previous section. Although we did not use homotopy groups with coefficients there, we do use coefficients here, since this is more common with cohomology.

Let $\mathcal{G}$ be a collection of abelian groups and X be a space. Define

$$\mathcal{E}^*_{\mathcal{G}}(X) = \{f \in \mathcal{E}(X) \mid f^* = 1\colon H^i(X;G) \to H^i(X;G), \text{ for all } i \text{ and all } G \in \mathcal{G}\}.$$

The following cases are of special interest:

(1) $\mathcal{G} = \{\mathbb{Z}\}$. We write $\mathcal{E}^*_{\mathcal{G}}(X)$ as $\mathcal{E}^*(X)$ in this case.
(2) $\mathcal{G}$ = all cyclic groups. Then $f \in \mathcal{E}^*_{\mathcal{G}}(X)$ if and only if $f \in \mathcal{E}(X)$ and $f^* = 1\colon H^i(X;G) \to H^i(X;G)$ for every finitely-generated abelian group G. We write $\mathcal{E}^*_{\mathcal{G}}(X)$ as $\mathcal{E}^*_{\mathrm{fg}}(X)$ in this case. Note that $\mathcal{E}^*_{\mathrm{fg}}(X) \subseteq \mathcal{E}^*(X)$.

Next we define a topological invariant that plays a role dual to that of spherical cone-length in the previous section.

DEFINITION 4.1. For $\mathcal{G}$ a collection of abelian groups, call an Eilenberg-MacLane space $K(G,m)$, with $G \in \mathcal{G}$, a *$\mathcal{G}$-Eilenberg-MacLane space.* For any space X, a *$\mathcal{G}$-fibre decomposition of X of length n*, is a sequence of fibrations

$$X_{i+1} \xrightarrow{j_i} X_i \xrightarrow{p_i} K_i,$$

for $0 \leq i < n$, such that each K_i is a finite product of $\mathcal{G}$-Eilenberg-MacLane spaces, X_0 is contractible and $X_n \equiv X$. We define the *$\mathcal{G}$-fibre-length* of X, denoted by $\mathcal{G}\text{-fl}(X)$, by dualizing Definition 3.1 in a straightforward way.

Note that a space X with $\mathcal{G}\text{-fl}(X) = 1$ is homotopy equivalent to a product of $\mathcal{G}$-Eilenberg-MacLane spaces. A space X with $\mathcal{G}\text{-fl}(X) \leq 2$ is homotopy equivalent to the fibre of a map between products of $\mathcal{G}$-Eilenberg-MacLane spaces. Note also that when we mention a product of $\mathcal{G}$-Eilenberg-MacLane spaces, we allow factors with homotopy groups in different dimensions, so that a product of $\mathcal{G}$-Eilenberg-MacLane spaces is not itself a $\mathcal{G}$-Eilenberg-MacLane space in general.

Now define subgroups of $\mathcal{E}(X)$ as follows:

$$\mathcal{E}^*_{\mathcal{G},s}(X) = \{f \in \mathcal{E}(X) \mid f^* = 1\colon [X,Y] \to [X,Y], \text{ for all } Y \text{ with } \mathcal{G}\text{-fl}(Y) \leq s\}.$$

Then there is a normal chain of subgroups

$$(3) \qquad \mathcal{E}^*_{\mathcal{G}}(X) = \mathcal{E}^*_{\mathcal{G},1}(X) \supseteq \mathcal{E}^*_{\mathcal{G},2}(X) \supseteq \cdots \supseteq \mathcal{E}^*_{\mathcal{G},s}(X) \supseteq \cdots.$$

The proof of normality for (3) is similar to the proof of normality for (1) above.

A straightforward dualization of the proof of Theorem 3.3 yields the following result.

THEOREM 4.2. *The series (3) is a solvability series. Thus*

$$\operatorname{solv} \mathcal{E}^*_{\mathcal{G}}(X) \leq \mathcal{G}\text{-fl}(X) - 1.$$

In Corollary 3.6, we showed that the number of dimensions in which Y has cells may be used to estimate the spherical cone length. We now indicate briefly how the preceding notions can be modified for the dual result.

DEFINITION 4.3. A space X is called a *Postnikov piece* if there is some N such that $\pi_i(X) = 0$ for all $i > N$. The smallest such N is called the *homotopical dimension* of X, and is denoted $\text{h-dim}\, X$.

For a Postnikov piece X, we define another subgroup of $\mathcal{E}(X)$ as

$$\mathcal{E}^{*\prime}_{\mathcal{G}}(X) = \{f \in \mathcal{E}(X) \mid f^* = 1 \colon H^i(X;G) \to H^i(X;G), \text{ for all } i \leq \text{h-dim}\, X \text{ and all } G \in \mathcal{G}\}.$$

We then define a *restricted $\mathcal{G}$-fibre decomposition* of a Postnikov piece X, of length n, as above but with the additional condition that $\text{h-dim}\, K_i \leq \text{h-dim}\, X + 1$, for all i. This yields the *restricted $\mathcal{G}$-fibre-length* of X, denoted by $\text{r-}\mathcal{G}\text{-fl}(X)$.

REMARK 4.4. Let X be a 1-connected space and $X^{(N)}$ be the N'th Postnikov section of X. Set $\mathcal{G} =$ all cyclic groups and $s =$ the number of non-trivial homotopy groups of $X^{(N)}$. Then by taking the Postnikov decomposition of $X^{(N)}$, we see that $\text{r-}\mathcal{G}\text{-fl}(X^{(N)}) \leq s$.

Now if X is a Postnikov piece, define

$$\mathcal{E}^{*\prime}_{\mathcal{G},s}(X) = \{f \in \mathcal{E}(X) \mid f^* = 1 \colon [X, Y] \to [X, Y], \text{ for all Postnikov pieces } Y \text{ such that } \text{h-dim}\, Y \leq \text{h-dim}\, X \text{ and } \text{r-}\mathcal{G}\text{-fl}(Y) \leq s\}.$$

Then once again we have a normal chain of subgroups

$$\mathcal{E}^{*\prime}_{\mathcal{G}}(X) = \mathcal{E}^{*\prime}_{\mathcal{G},1}(X) \supseteq \mathcal{E}^{*\prime}_{\mathcal{G},2}(X) \supseteq \cdots \supseteq \mathcal{E}^{*\prime}_{\mathcal{G},s}(X) \supseteq \cdots. \tag{4}$$

Here also, the proof of normality is similar to the proof of normality for (1) above. A further adaptation of the proof of Theorem 3.3 gives the following result.

THEOREM 4.5. *The series (4) is a solvability series. Thus ,*

$$\operatorname{solv} \mathcal{E}^{*\prime}_{\mathcal{G}}(X) \leq \text{r-}\mathcal{G}\text{-fl}(X) - 1,$$

for a Postnikov piece X.

This leads to the dual of Corollary 3.6.

COROLLARY 4.6. *If X is a 1-connected finite complex, then $\mathcal{E}^*_{\text{fg}}(X)$ is solvable. In particular, if X has dimension N and there are s non-trivial homotopy groups in degrees $\leq N$, then $\operatorname{solv} \mathcal{E}^*_{\text{fg}}(X) \leq s - 1$.*

PROOF. Any map $f \colon X \to X$ induces a corresponding map $\theta(f) \colon X^{(N)} \to X^{(N)}$ of N'th Postnikov sections. This gives us a homomorphism $\theta \colon \mathcal{E}^*_{\mathcal{G}}(X) \to \mathcal{E}^{*\prime}_{\mathcal{G}}(X^{(N)})$ which is one-one. Therefore, $\operatorname{solv} \mathcal{E}^*_{\mathcal{G}}(X) \leq \operatorname{solv} \mathcal{E}^{*\prime}_{\mathcal{G}}(X^{(N)})$. But if $\mathcal{G}$ is the collection of all cyclic groups, then it is a consequence of Remark 4.4 and Theorem 4.5 that $\operatorname{solv} \mathcal{E}^{*\prime}_{\mathcal{G}}(X^{(N)}) \leq s - 1$. □

Next, we compare the subgroups $\mathcal{E}^*(X)$ and $\mathcal{E}^*_{\text{fg}}(X)$. Below, we give a simple example to illustrate that these two subgroups are distinct in general. First, however, we obtain conditions under which they agree.

PROPOSITION 4.7. *For any space X, if*

$$\mathrm{Hom}\big(\mathrm{Tor}(H^{i+1}(X),G),H^i(X)\otimes G\big)=0$$

for all i and all finitely-generated groups G, then $\mathcal{E}^*(X)=\mathcal{E}^*_{\mathrm{fg}}(X)$.

PROOF. If $f\in\mathcal{E}^*(X)$, then the universal coefficient theorem (cf. [**Spa66**, Thm. 5.10]) gives a commutative diagram with exact rows

$$\begin{array}{ccccccccc} 0 & \longrightarrow & H^i(X)\otimes G & \xrightarrow{\iota} & H^i(X;G) & \xrightarrow{\pi} & \mathrm{Tor}(H^{i+1}(X),G) & \longrightarrow & 0\\ & & \downarrow{\scriptstyle 1} & & \downarrow{\scriptstyle \phi} & & \downarrow{\scriptstyle 1} & & \\ 0 & \longrightarrow & H^i(X)\otimes G & \xrightarrow{\iota} & H^i(X;G) & \xrightarrow{\pi} & \mathrm{Tor}(H^{i+1}(X),G) & \longrightarrow & 0 \end{array}$$

where the middle homomorphism ϕ can be either f^* or the identity 1. Thus there is a homomorphism $\rho\colon\mathrm{Tor}(H^{i+1}(X),G)\to H^i(X)\otimes G$ such that $f^*-1=\iota\rho\pi$. By hypothesis, $\rho=0$ and so $f^*=1$. Therefore, $f\in\mathcal{E}^*_{\mathrm{fg}}(X)$. □

The following example illustrates that $\mathcal{E}^*(X)$ and $\mathcal{E}^*_{\mathrm{fg}}(X)$ may differ.

EXAMPLE 4.8. Let X be a Moore space $M(G,n)$, for $n\geq 2$ and G any infinite, finitely-generated abelian group with torsion. Then it follows from results of [**AM98**] that $\mathcal{E}^*(X)\neq\mathcal{E}^*_{\mathrm{fg}}(X)$

In Corollary 4.6, we showed that for a 1-connected, finite-dimensional complex X, $\mathcal{E}^*_{\mathrm{fg}}(X)$ is solvable. This raises the question of whether or not the group is nilpotent. We conclude the paper by showing that $\mathcal{E}^*(X)$, and therefore $\mathcal{E}^*_{\mathrm{fg}}(X)$, is a nilpotent group. We will use the following notation: Suppose that a group G acts on an abelian group A. We define inductively a decreasing sequence of subgroups of A by setting $\Gamma_1^G(A)=A$ and $\Gamma_i^G(A)$ is the subgroup generated by $\{ga-a\mid g\in G,\, a\in\Gamma_{i-1}^G(A)\}$. We say that the *action is nilpotent* if, for some i, $\Gamma_i^G(A)=\{0\}$.

PROPOSITION 4.9. *For any nilpotent finite complex X, $\mathcal{E}^*(X)$ is a nilpotent group.*

PROOF. We shall prove that $\mathcal{E}^*(X)$ acts nilpotently on $H_*(X)$. Then it follows from [**DZ79**] that $\mathcal{E}^*(X)$ is a nilpotent group.

Let $f\in\mathcal{E}^*(X)$ and consider the diagram with exact rows obtained from the universal coefficient theorem

$$\begin{array}{ccccccccc} 0 & \longrightarrow & \mathrm{Ext}(H_{*-1}(X),\mathbb{Z}) & \longrightarrow & H^*(X) & \longrightarrow & \mathrm{Hom}(H_*(X),\mathbb{Z}) & \longrightarrow & 0\\ & & \downarrow{\scriptstyle \mathrm{Ext}(f_*,\mathbb{Z})} & & \downarrow{\scriptstyle f^*=1} & & \downarrow{\scriptstyle \mathrm{Hom}(f_*,\mathbb{Z})} & & \\ 0 & \longrightarrow & \mathrm{Ext}(H_{*-1}(X),\mathbb{Z}) & \longrightarrow & H^*(X) & \longrightarrow & \mathrm{Hom}(H_*(X),\mathbb{Z}) & \longrightarrow & 0. \end{array}$$

Then both $\mathrm{Ext}(f_*,\mathbb{Z})$ and $\mathrm{Hom}(f_*,\mathbb{Z})$ are identity maps. Write $H_*(X)=F\oplus T$ as the sum of its free and torsion parts and let $p_T\colon F\oplus T\to T$ and $p_F\colon F\oplus T\to F$ be the projections. Since $\mathrm{Ext}(\mathbb{Z}/m,\mathbb{Z})=\mathbb{Z}/m$, it follows that $\mathrm{Ext}(H_*(X),\mathbb{Z})=\mathrm{Ext}(T,\mathbb{Z})=T$. Thus $\mathrm{Ext}(f_*,\mathbb{Z})=1$ implies that $p_T\circ f_*|_T\colon T\to T$ is the identity.

In the same way, since $\mathrm{Hom}(f_*,\mathbb{Z})=1$, we have that $p_F\circ f_*|_F\colon F\to F$ is also the identity. Therefore, $f_*\colon F\oplus T\to F\oplus T$ can be written as $f_*(x,y)=(x,y+\phi(x))$, with $\phi\colon F\to T$ a homomorphism that depends on f. Hence $\Gamma_1^{\mathcal{E}^*(X)}(H_*(X))$ is generated by elements of the form $f_*(x,y)-(x,y)=(0,\phi(x))$. On these elements, any $g \in \mathcal{E}^*(X)$ satisfies $g_*(0,\phi(x)) = (0,\phi(x))$. Hence $\Gamma_2^{\mathcal{E}^*(X)}(H_*(X)) = \{0\}$, and the result follows. □

References

[AL91] M. Aubry and J.-M. Lemaire, *Sur Certaines Equivalences d'Homotopies*, Ann. Inst. Fourier (Grenoble) **41** (1991), 173–187.

[AL96] M. Arkowitz and G. Lupton, *On the Nilpotency of Subgroups of Self-Homotopy Equivalences*, BCAT 94 Conference, Progress in Math., vol. 136, Birkhäuser, 1996, pp. 1–22.

[AM98] M. Arkowitz and K.-I. Maruyama, *Self-Homotopy Equivalences which Induce the Identity on Homology, Cohomology or Homotopy Groups*, Topology Appl. **87** (1998), 133–154.

[Cor94] O. Cornea, *There is Just One Rational Cone-Length*, Trans. A. M. S. **344** (1994), 835–848.

[Did85] G. Didierjean, *Homotopie de l'Espace des Equivalences d'Homotopie Fibrées*, Ann. Inst. Fourier (Grenoble) **35** (1985), 33–47.

[DZ79] E. Dror and A. Zabrodsky, *Unipotency and Nilpotency in Homotopy Equivalences*, Topology **18** (1979), 187–197.

[Fél89] Y. Félix, *La Dichotomie Elliptique-Hyperbolique en Homotopie Rationnelle*, Astérisque, no. 176, Soc. Math. France, 1989.

[FH82] Y. Félix and S. Halperin, *Rational LS Category and its Applications*, Trans. A. M. S. **273** (1982), 1–38.

[FM97] Y. Félix and A. Murillo, *A Note on the Nilpotency of a Subgroup of Self Homotopy Equivalences*, Bull. L. M. S. **29** (1997), 486–488.

[FM98] Y. Félix and A. Murillo, *A Bound for the Nilpotency of a Group of Self Homotopy Equivalences*, Proc. A. M. S. **126** (1998), 625–627.

[Got69] D. H. Gottlieb, *Evaluation Subgroups of Homotopy Groups*, Amer. J. Math. **91** (1969), 729–756.

[Hil65] P. J. Hilton, *Homotopy Theory and Duality*, Gordon and Breach, New York-London-Paris, 1965.

[HL87] S. Halperin and J.-M. Lemaire, *Suites Inertes dans les Algèbres de Lie Graduées*, Math. Scand. **61** (1987), 39–67.

[HMR75] P. J. Hilton, G. Mislin, and J. Roitberg, *Localization of Nilpotent Groups and Spaces*, North-Holland Publishing Co., Amsterdam, 1975, North-Holland Mathematics Studies, No. 15, Notas de Matemática, No. 55.

[Lan75] G. E. Lang, Jr., *Localizations and Evaluation Subgroups*, Proc. A. M. S. **50** (1975), 489–494.

[Mar89] K.-I. Maruyama, *Localization of a Certain Subgroup of Self-Homotopy Equivalences*, Pacific J. Math. **136** (1989), 293–301.

[Saw75] N. Sawashita, *On the Group of Self-Equivalences of the Product of Spheres*, Hiroshima Math. J. **5** (1975), 69–86.

[Spa66] E. H. Spanier, *Algebraic Topology*, McGraw-Hill, New York-Toronto-London, 1966.

[ST99] H. Scheerer and D. Tanré, *Variation zum Konzept der Lusternik-Schnirelmann-Kategorie*, Math. Nachr. **207** (1999), 183–194.

Department of Mathematics, Dartmouth College, Hanover NH 03755 U. S. A.
E-mail address: `Martin.Arkowitz@Dartmouth.edu`

Department of Mathematics, Cleveland State University, Cleveland OH 44115 U. S. A.
E-mail address: `Lupton@math.csuohio.edu`

Departmento de Algebra, Geometría y Topología, Universidad de Málaga, Ap. 59, 29080 Málaga, Spain
E-mail address: `Aniceto@agt.cie.uma.es`

Contemporary Mathematics
Volume **274**, 2001

The Space of Free Loops on a Real Projective Space

Sven Bauer, Michael Crabb, and Mauro Spreafico

ABSTRACT. Methods from fibrewise topology are used to give a stable splitting of the space of free loops on a real projective space as a wedge of Thom spaces. This splitting is equivariant with respect to the action of the isometry group.

1. Introduction

Suppose that $V \neq 0$ is a finite-dimensional real vector space with an inner product. Let $\mathcal{L}P(V)$ denote the space of continuous free loops on the projective space $P(V)$. The action of the orthogonal group $\mathrm{O}(V)$ on V induces an action of the projective orthogonal group $\mathrm{PO}(V)$ on $P(V)$. Our main result is a $\mathrm{PO}(V)$-equivariant stable splitting of $\mathcal{L}P(V)$.

To describe the stable summands let us write $\mathrm{O}(\mathbb{R}^2, V)$ for the Stiefel manifold of orthogonal 2-frames $b : \mathbb{R}^2 \to V$ in V and $\mathrm{PO}(\mathbb{R}^2, V)$ for the projective Stiefel manifold $\mathrm{O}(\mathbb{R}^2, V)/\{\pm 1\}$. The vector bundle over $\mathrm{PO}(\mathbb{R}^2, V)$ with fibre at $[b]$ the orthogonal complement in V of $b(\mathbb{R}^2)$ will be denoted by ζ.

THEOREM 1.1. *There is a* $\mathrm{PO}(V)$*-equivariant stable homotopy equivalence*

$$\big(\mathcal{L}P(V)\big)_+ \simeq P(V)_+ \vee \bigvee_{l=1}^{\infty} \mathrm{PO}(\mathbb{R}^2, V)^{(l-1)\zeta}.$$

Here, and throughout the paper, a subscript + stands for addition of a disjoint basepoint, and the Thom space of a vector bundle ξ over a base B is denoted by B^{ξ}. The stable splitting is understood as a splitting of equivariant spectra. We shall establish a more precise unstable splitting after one suspension in Proposition 3.1.

The methods of this paper are entirely homotopy-theoretic, but some geometric remarks are in order. The space of continuous loops $\mathcal{L}P(V)$ is homotopy equivalent to the manifold M of smooth loops $\omega : S^1 \to P(V)$. Consider the energy functional

$$E : M \to \mathbb{R}, \quad E(\omega) = \frac{1}{2}\int_0^{2\pi} ||\omega'(\theta)||^2 d\theta.$$

This is a Morse-Bott function with critical submanifolds C_l, $l \in \mathbb{N}$, where C_0 is the space of constant loops and C_l, for $l \geq 1$, is the space of closed geodesics of multiplicity l (so of length πl). One can easily identify C_l with $\mathrm{PO}(\mathbb{R}^2, V)$ for $l \geq 1$.

2000 *Mathematics Subject Classification.* Primary 55P35; Secondary 55P91.

SB partially supported by EPSRC.

The negative bundle of C_l is 0 if $l = 0$ and corresponds to $(l-1)\zeta$ if $l \geq 1$. See, for example, [**9**, Section 2] and [**3**].

The proof of Theorem 1.1 relies on an equivariant refinement of one of the main results in [**1**]. In Section 2 we review some of the constructions in [**1**] and show that they can be carried out equivariantly to obtain a stable splitting for the (based) loop space of the projective space of a real vector space (Proposition 2.2). In the next section we interpret $\mathcal{L}P(V)$ as a fibrewise (based) loop space and deduce the main theorem. The last section describes an $\mathrm{O}(V)$-equivariant stable splitting of the space $\mathcal{L}S(V)$ of continuous free loops on the unit sphere $S(V)$ in V.

2. An equivariant fibrewise stable splitting for projective bundles

Let W be a finite-dimensional real vector space with an inner product. Writing $[x]$ for the element of a projective space determined by a non-zero vector x, we choose the point $[1,0]$ as basepoint in the projective space $P(\mathbb{R} \oplus W)$. It is understood that $\mathrm{O}(W)$ acts trivially on the one-dimensional space $\mathbb{R}$. The space $P(\mathbb{R} \oplus W)$ then has an induced $\mathrm{O}(W)$-action.

Let $l \in \mathbb{N}$ and $l \geq 1$. We define a map

$$\tilde{\gamma}_l : S(W)^l \to \Omega P(\mathbb{R} \oplus W)$$

by setting

$$\tilde{\gamma}_l(x_1, x_2, \ldots, x_l)(t) = \big[\cos(l\pi t), \sin(l\pi t)x_j\big] \quad \text{for } (j-1)/l \leq t \leq j/l.$$

The loop $\tilde{\gamma}_1(x)$ is a closed geodesic lifting to a great semi-circle on the sphere $S(\mathbb{R} \oplus W)$ from $(1,0)$ through $(0,x)$ to $(-1,0)$. The map $\tilde{\gamma}_l$ assigns to an l-tuple of points $(x_1, x_2, \ldots, x_l)$ a piecewise smooth geodesic whose pieces are reparameterized closed paths $\tilde{\gamma}_1(x_j)$. In particular, if $x_1 = x_2 = \ldots = x_l$ this gives a closed geodesic of multiplicity l.

Let U be a codimension 1 subspace of W. Consider the embedding $S(\mathbb{R} \oplus U) \subseteq \mathbb{R} \oplus U$ with trivial normal bundle $\mathbb{R} \times S(\mathbb{R} \oplus U)$. The Pontrjagin-Thom construction gives a map

$$f : \Sigma U^+ = (\mathbb{R} \oplus U)^+ \to \big(\mathbb{R} \times S(\mathbb{R} \oplus U)\big)^+ = \Sigma S(\mathbb{R} \oplus U)_+, \tag{2.1}$$

where a superscript $+$ is used for one-point compactification. This map f splits the suspension of the stereographic projection $S(\mathbb{R} \oplus U)_+ \to U^+$. Making the appropriate identifications, we obtain maps

$$f_k : \Sigma(kU)^+ \to \Sigma(S(\mathbb{R} \oplus U)^k)_+$$

by defining $f_1 = f$ and $f_k = (f_{k-1} \wedge \mathrm{id}) \circ (\mathrm{id} \wedge f)$ for $k \geq 2$.

Viewing U as the tangent space at a point of the sphere $S(W)$, we now apply this construction fibrewise to the tangent bundle $TS(W)$. We write $T \to S$ for the tangent bundle $TS(W) \to S(W)$ for ease of notation and obtain fibrewise maps

$$\Sigma_S(kT)^+_S \to \Sigma_S(S(\mathbb{R} \oplus T)^k_S)_{+S}$$

over S, where $\mathbb{R}$ stands for the one-dimensional real trivial bundle and the subscript S indicates the appropriate fibrewise construction.

Collapsing the basepoint sections S to a point and identifying $\mathbb{R} \oplus T$ with the trivial bundle with fibre W gives maps

$$\alpha_k : \Sigma\big(S(W)^{kTS(W)}\big) \to \Sigma(S(W)^{k+1})_+$$

for $k \geq 1$. Additionally we define α_0 to be the identity on $\Sigma S(W)_+$.

Now define for $l \geq 1$ the map

$$\gamma_l : \Sigma\big(S(W)^l\big)_+ \to \Sigma\big(\Omega P(\mathbb{R} \oplus W)\big)_+$$

to be $(\Sigma\tilde{\gamma}_l)\alpha_{l-1}$. Let γ_0 be the suspension of the pointed map $S^0 \to \Omega P(\mathbb{R} \oplus W)_+$ that takes -1 to the constant loop $t \mapsto [1,0]$. Observe that the construction of the γ_l is $\mathrm{O}(W)$-equivariant.

PROPOSITION 2.2. *The map*

$$\gamma = \gamma_0 \vee \bigvee_{l=1}^{\infty} \gamma_l : \Sigma S^0 \vee \bigvee_{l=1}^{\infty} \Sigma\big(S(W)^{(l-1)TS(W)}\big) \to \Sigma\big(\Omega P(\mathbb{R} \oplus W)\big)_+$$

is an $\mathrm{O}(W)$*-equivariant homotopy equivalence.*

PROOF. We check first of all that the map γ is a non-equivariant homotopy equivalence. There is nothing to do if $W = 0$.

If $\dim W = 1$ the equivalence is clear. For the loop space is homotopically a disjoint union of points indexed by the degree. The lth component of the wedge corresponds to the points $\{l, -l\}$.

If $\dim W > 1$ the loop space has two components Ω_0 and Ω_1 indexed by the fundamental group $\pi_1(P(\mathbb{R} \oplus W)) = \mathbb{Z}/2$. The γ_l map into $\Sigma(\Omega_0)_+$ or $\Sigma(\Omega_1)_+$ according to whether l is even or odd. An integral homology calculation for each component shows that γ is an equivalence. We refer to **[1]** for details.

By **[6]** (or **[8]**), for any compact Lie group G an equivariant map between two G-CW-complexes (or G-ANRs respectively) is an equivariant homotopy equivalence if it induces (ordinary) homotopy equivalences on the fixed point sets of all closed subgroups of G. The spaces with which we are concerned all have the equivariant homotopy type of $\mathrm{O}(W)$-CW-complexes. Indeed, the sphere $S(W)$ and the projective space $P(\mathbb{R} \oplus W)$ are $\mathrm{O}(W)$-CW-complexes, and it follows from an equivariant generalization in **[11]** of Milnor's theorems on spaces having the homotopy type of a CW-complex that the loop space $\Omega P(\mathbb{R} \oplus W)$ has the equivariant homotopy type of an $\mathrm{O}(W)$-CW-complex. Alternatively, one can check that the spaces involved in this proof are of the equivariant homotopy type of $\mathrm{O}(W)$-ANRs.

It remains to show that γ induces equivalences on fixed point spaces of subgroups. Let K be a closed subgroup of $\mathrm{O}(W)$. Then the fixed point subspace $P(\mathbb{R} \oplus W)^K$ is a disjoint union of projective spaces

$$P(\mathbb{R} \oplus W)^K = P(\mathbb{R} \oplus W^K) \amalg \coprod_{\chi \neq 1} P(W_\chi)$$

indexed by the characters $\chi : K \to \{1, -1\}$. The spaces W_χ are the χ-isotypical summands of the K-representation W. Because a loop is fixed if and only if it is pointwise fixed we have $\big(\Omega P(\mathbb{R} \oplus W)\big)^K = \Omega\big(P(\mathbb{R} \oplus W^K)\big)$. Since $S(W)^K = S(W^K)$ and because the fixed point space of $TS(W)$ is $TS(W^K)$, the fixed subspace of the domain of γ is obtained by replacing W by W^K. As all our constructions are natural in W, the restriction of γ to K-fixed points is the corresponding map for W^K instead of W. So we have a homotopy equivalence on the K-fixed points, as required to complete the proof. □

We obtain a fibrewise version of this proposition by using the Borel construction. Let G be a compact Lie group, and let ξ be a G-vector bundle over a compact G-ENR B. Assume ξ has fibre W and let E be the associated G-principal

$\mathrm{O}(W)$-bundle. Then the fibrewise loop space $\Omega_B P(\mathbb{R}\oplus\xi)$ of the projective bundle $E\times_{\mathrm{O}(W)} P(\mathbb{R}\oplus W)$ is equal to

$$\Omega_B P\big(E\times_{\mathrm{O}(W)}(\mathbb{R}\oplus W)\big) = E\times_{\mathrm{O}(W)}\Omega P(\mathbb{R}\oplus W).$$

Abbreviating the $\mathrm{O}(W)$-equivariant homotopy equivalence γ from Proposition 2.2 to $\gamma : X\to Y$, we obtain a G-equivariant fibrewise homotopy equivalence

$$E\times_{\mathrm{O}(W)} X\to E\times_{\mathrm{O}(W)} Y.$$

Collapsing the basepoint sections yields a G-homotopy equivalence

$$(E\times_{\mathrm{O}(W)} X)/B\to(E\times_{\mathrm{O}(W)} Y)/B$$

and so establishes the following corollary.

COROLLARY 2.3. *Let G be a compact Lie group, and let ξ be a G-vector bundle over a compact G-ENR B. Then there is a G-equivalence*

$$\Sigma\big(\Omega_B P(\mathbb{R}\oplus\xi)\big)_+\simeq\Sigma B_+\vee\bigvee_{l=1}^{\infty}\Sigma\big(S(\xi)^{(l-1)T_B S(\xi)}\big),$$

where $T_B S(\xi)$ is the fibrewise tangent bundle of the unit sphere bundle $S(\xi)$ of ξ. □

3. Splitting the space of free loops on a projective space

Consider the trivial bundle $P(V)\times V\to P(V)$. There is a homeomorphism from $\mathcal{L}P(V)$ to the fibrewise loop space of the associated projective bundle:

$$\mathcal{L}P(V)\xrightarrow{\approx}\Omega_{P(V)}P\big(P(V)\times V\big) : \omega\mapsto(\omega(0),\omega).$$

Here, the basepoint in each fibre is given by the diagonal map. Thus the fibre over $L\in P(V)$ is just $P(V)$ with basepoint L. When we make the canonical identification

$$P(V)=P(L^*\otimes V)=P(L^*\otimes(L\oplus L^\perp))=P(\mathbb{R}\oplus(L^*\otimes L^\perp))$$

(given explicitly by the mapping $[y,z]\mapsto[x^*(y),x^*\otimes z]$, where $x^*\in L^*\setminus\{0\}$, $y\in L$, and $z\in L^\perp$), the basepoint $L\in P(V)$ corresponds to the usual basepoint $[1,0]$ in $P(\mathbb{R}\oplus(L^*\otimes L^\perp))$. Globally, this identifies $P(P(V)\times V)$ with $P(\mathbb{R}\oplus(H^*\otimes H^\perp))$, where H is the canonical line bundle over $P(V)$. Recognizing $H^*\otimes H^\perp$ as the tangent bundle $TP(V)$ we have a $\mathrm{PO}(V)$-equivariant homeomorphism $\mathcal{L}P(V)\approx\Omega_{P(V)}P(\mathbb{R}\oplus TP(V))$.

Setting $\xi=TP(V)$ and $G=\mathrm{PO}(V)$ in Corollary 2.3 gives:

PROPOSITION 3.1. *There is a $\mathrm{PO}(V)$-equivariant homotopy equivalence*

$$\Sigma\big(\mathcal{L}P(V)\big)_+\simeq\Sigma P(V)_+\vee\bigvee_{l=1}^{\infty}\Sigma\big(S(TP(V))^{(l-1)\tau}\big),$$

where τ is the fibrewise tangent bundle of $S\big(TP(V)\big)\to P(V)$. □

To complete the proof of Theorem 1.1 we need to see that τ over $S\big(TP(V)\big)$ is the same as ζ over $\mathrm{PO}(\mathbb{R}^2,V)$. The spaces $\mathrm{O}(\mathbb{R}^2,V)$ and $S\big(TS(V)\big)$ are easily identified via the map $b\mapsto\big(b(1,0),b(0,1)\big)$. Taking the quotient modulo the action of the group $\{\pm1\}$ gives the required identification. □

REMARK 3.2. Suppose V has a complex structure, and let $\mathrm{U}(V)$ be its unitary group. Then the lth wedge summand, for $l \geq 1$, in Proposition 3.1 has a stable $\mathrm{U}(V)$-equivariant decomposition

$$S(TP(V))^{(l-1)\tau} \simeq P(V)^{(l-1)\eta} \vee P(V)^{l\eta},$$

where η is the pull-back of the tangent bundle of the complex projective space of V. This can be seen by splitting $TP(V)$ as the sum of η and the trivial one-dimensional bundle and using f, (2.1), to split $S(\mathbb{R} \oplus \eta)$ over $P(V)$.

4. Free loops on spheres

Similar methods give a decomposition of the free loop space of the sphere. The map γ in Proposition 2.2 restricts to an $\mathrm{O}(W)$-equivalence

$$\Sigma S^0 \vee \bigvee_{m=1}^{\infty} \Sigma\big(S(W)^{(2m-1)TS(W)}\big) \to \Sigma\big(\Omega S(\mathbb{R} \oplus W)\big)_+,$$

and the argument of Section 3 gives:

THEOREM 4.1. *There is an $\mathrm{O}(V)$-equivariant stable homotopy equivalence*

$$\big(\mathcal{L}S(V)\big)_+ \simeq S(V)_+ \vee \bigvee_{m=1}^{\infty} \mathrm{O}(\mathbb{R}^2, V)^{(2m-1)\zeta}.$$

□

Non-equivariantly there is a finer decomposition. Suppose $V = \mathbb{R}^2 \oplus U$. As an easy special case of Miller's stable splitting of Stiefel manifolds (see [**10**]) we have

$$\mathrm{O}(\mathbb{R}^2, V)_+ \simeq S^0 \vee P(\mathbb{R}^2)^{H\otimes U} \vee \Sigma(2U)^+,$$

where H is the Hopf bundle over $P(\mathbb{R}^2)$. Since ζ is stably trivial, the mth summand for $m \geq 1$ in Theorem 4.1 is stably

$$\mathrm{O}(\mathbb{R}^2, V)^{(2m-1)\zeta} \simeq \big((2m-1)U\big)^+ \vee \big((2m-1)U\big)^+ \wedge P(\mathbb{R}^2)^{H\otimes U} \vee \Sigma\big((2m+1)U\big)^+.$$

The 0th summand is $S(V)_+ \simeq S^0 \vee \Sigma U^+$.

It is interesting to compare this decomposition with the Carlsson-Cohen splitting, [**4**],

$$\big(\mathcal{L}S(V)\big)_+ \simeq S^0 \vee \bigvee_{n=1}^{\infty} (S^1 \times_{\mathbb{Z}/n} nU)^+,$$

where $\mathbb{Z}/n$ permutes the summands in nU cyclically. (See also [**2**] and [**7**].) For $n = 2m$ even the nth summand is

$$\big((2m-1)U\big)^+ \wedge P(\mathbb{R}^2)^{H\otimes U}.$$

However, for $n = 2m-1$ odd it is a wedge

$$\Sigma\big((2(m-1)+1)U\big)^+ \vee \big((2m-1)U\big)^+$$

of two pieces coming from the $(m-1)$st and mth summands in Theorem 4.1.

References

1. S. Bauer, M. C. Crabb, and M. Spreafico, *The classifying space of the gauge group of an* SO(3)*-bundle over* S^2, Proc. Roy. Soc. Edinburgh, to appear.
2. C.-F. Bödigheimer and I. Madsen, *Homotopy quotients of mapping spaces and their stable splitting*, Quart. J. Math. Oxford **39** (1988), 401–409.
3. R. Bott, *Lectures on Morse theory, old and new*, Bull. Amer. Math. Soc. **7** (1982), 331–358.
4. G. E. Carlsson and R. L. Cohen, *The cyclic groups and the free loop space*, Comment. Math. Helv. **62** (1987), 423–449.
5. M. C. Crabb and I. M. James, *Fibrewise homotopy theory*, Springer, London, 1998.
6. T. tom Dieck, *Transformation groups*, Walter de Gruyter, Berlin, 1987.
7. N. Hingston, *An equivariant model for the free loop space of* S^N, Amer. J. Math. **114** (1992), 139–155.
8. I. M. James and G. B. Segal, *On equivariant homotopy type*, Topology **17** (1978), 267–272.
9. W. Klingenberg, *The space of closed curves on the sphere*, Topology **7** (1968), 395–415.
10. H. Miller, *Stable splittings of Stiefel manifolds*, Topology **24** (1985), 411–419.
11. S. Waner, *Equivariant homotopy theory and Milnor's theorem*, Trans. Amer. Math. Soc. **258** (1980), 351–368.

Department of Mathematical Sciences, King's College, University of Aberdeen, Aberdeen AB24 3UE, UK

E-mail address: `s.bauer@maths.abdn.ac.uk`

Department of Mathematical Sciences, King's College, University of Aberdeen, Aberdeen AB24 3UE, UK

E-mail address: `m.crabb@maths.abdn.ac.uk`

Dipartimento di Matematica e Applicazioni, Università di Milano / Bicocca, Via Bicocca degli Arcimboldi 8, 20126 Milano, Italy

E-mail address: `mauro@matapp.unimib.it`

Contemporary Mathematics
Volume **274**, 2001

Indecomposable Homotopy Types with at most two non–trivial Homology Groups

Hans–Joachim Baues and Yuri Drozd

ABSTRACT. We classify indecomposable spaces in the stable range with at most two non–trivial finitely generated homology groups.

It is a classical result of Brown–Copeland [**BC**] and Eckmann–Hilton [**EH**] that a 1–connected homotopy type X with at most two non trivial homology groups $H_m X = A$ and $H_n X = B, 2 \leq m < n$, is a mapping cone of a map

$$k_X : M(B, n-1) \to M(A, m) \tag{1}$$

Here $M(A, m)$ denotes the Moore space with homology A in degree m.

Let p be a prime and let $\mathbb{Z}_{(p)} \subset \mathbb{Q}$ be the smallest subring of $\mathbb{Q}$ containing $1/q$ for all primes $q \neq p$. We consider finitely generated $\mathbb{Z}_{(p)}$–modules A and B in the *stable range* $n < 2m - 2$. Hence X is a p–local space with at most two non–trivial homology groups in a stable range. Then the homotopy type of X admits a decomposition as a one point union

$$X \simeq X_1 \vee \ldots \vee X_j \tag{2}$$

where all X_i with $1 \leq i \leq j$ are indecomposable and this decomposition is unique up to permutation. We classify in this paper the indecomposable summands in (2).

THEOREM (A). *Let A and B be finitely generated free $\mathbb{Z}_{(p)}$–modules. Then the classification of indecomposable summands in* (2) *is*

$$\begin{cases} \textit{finite} & \textit{if } \pi_{n-1}(S^m) \otimes \mathbb{Z}_{(p)} \textit{ is cyclic,} \\ \textit{tame} & \textit{if } \pi_{n-1}(S^m) \otimes \mathbb{Z}_{(p)} \cong \mathbb{Z}/p \oplus \mathbb{Z}/p^k, k \geq 1, \textit{ and} \\ \textit{wild} & \textit{otherwise.} \end{cases}$$

Here $\pi_{n-1}(S^m) = \pi^s_{n-m-1}$ is given by stable homotopy group of spheres since we assume $n < 2m - 2$. For example the wild case appears for $n - m - 1 = 9$ since $\pi^s_9 = \mathbb{Z}/2 \oplus \mathbb{Z}/2 \oplus \mathbb{Z}/2$ and tame cases appear for $n - m - 1 = 8, 15$ since $\pi^s_8 = \mathbb{Z}/2 \oplus \mathbb{Z}/2$ and $\pi^s_{15} = \mathbb{Z}/2 \oplus \mathbb{Z}/480$. The proof of theorem (A) is based on the representation theory of matrices with entries in a fixed $\mathbb{Z}_{(p)}$–module M. For the finite and tame cases we describe the indecomposable summands explicitly.

2000 *Mathematics Subject Classification.* 55P15.

The space X in (2) is $(m-1)$–connected and we write $\mathsf{conn}(X) = m-1$ if $H_mX \neq 0$. Moreover if $H_nX \neq 0$ the *p–local dimension* of X is

$$\mathsf{dim}_{(p)}(X) = \begin{cases} n & \text{if } H_nX \text{ is a free } \mathbb{Z}_{(p)}\text{–module} \\ n+1 & \text{otherwise} \end{cases}$$

We say that X is trivial if X is a one point union of Moore spaces or equivalently if k_X in (1) is null homotopic.

THEOREM (B). *Let A and B be finitely generated $\mathbb{Z}_{(p)}$–modules and let $\mathsf{dim}_{(p)}(X) - m \leq 4p-5$. Then the classification of indecomposable summands in (2) is*

$$\begin{cases} \textit{tame} & \textit{if } n-m = 2p-2, \\ \textit{essentially finite} & \textit{if } n-m = 2p-3, 2p-1, \textit{ and} \\ \textit{trivial} & \textit{otherwise.} \end{cases}$$

For odd primes p this result can be deduced from the more general result of Henn [**H**]. For the prime $p=2$ and $\mathsf{dim}_{(p)}(X) - m \leq 2$ the result is due to J. H. C. Whitehead [**W1**] and Chang [**Ch**]. Therefore we give a proof only for the highly sophisticated case $p=2$ and $\mathsf{dim}_{(p)}(X) - m = 3$. This case was also treated by Baues–Hennes [**BH**]; see also [**B1**] and [**B2**]. Our approach here using representation theory yields a new proof and confirms the intricate computation in [**BH**].

THEOREM (C). *Let $p=2$ and let A and B be finitely generated $\mathbb{Z}/2$–vector spaces and let $n = m+3 \geq 9$. Then the classification of indecomposable summands in (2) is wild.*

This result solves an old question of homotopy theory: Let $\mathbf{A}^k = \mathbf{A}^k_m$ be the homotopy category of $(m-1)$–connected $(m+k)$–dimensional finite CW–spaces with $m \geq k+1$. Since the spaces of theorem (C) are objects in $\mathbf{A}^4$ we get:

COROLLARY (D). *$\mathbf{A}^k$ has wild representation type for $k \geq 4$.*

It was shown in Baues–Drozd [**BD1**] that $\mathbf{A}^k$ has wild representation type for $k \geq 6$. On the other hand J. H. C. Whitehead [**W1**] and Chang [**Ch**] computed the indecomposable objects of $\mathbf{A}^2$; compare also the books of Hilton [**H1**], [**H2**]. Moreover Baues–Hennes computed all indecomposable objects of $\mathbf{A}^3$. Hence since $\mathbf{A}^k$ is wild for $k \geq 4$ the representation type of $\mathbf{A}^k$ is now known for all k. This answers a classification problem started by J. H. C. Whitehead 50 years ago.

In this paper a space is a CW–complex. Let $\mathbf{Top}^*/\simeq$ be the homotopy category of pointed spaces and pointed maps. For pointed spaces X, Y let $[X,Y]$ be the set of homotopy classes of pointed maps $X \to Y$; this is the set of morphism $X \to Y$ in the category $\mathbf{Top}^*/\simeq$.

1. The torsion free case

A space X is *decomposable* if there exists a homotopy equivalence $X \simeq A \vee B$ where $A \vee B$ is the one point union of non–contractible spaces A and B; otherwise X is *indecomposable.*

DEFINITION 1.1. We say that X is a *p–local (m,n)–atom* if X is a 1–connected indecomposable space for which the homology groups $H_m(X)$ and $H_n(X)$ are non trivial finitely generated $\mathbb{Z}_{(p)}$–modules and $\widetilde{H}_i(X) = 0$ for $i \neq n, m$. We say that X is *torsion free* if the homology $H_m(X), H_n(X)$ are free $\mathbb{Z}_{(p)}$–modules.

The indecomposable summands of the space X in the introduction are either indecomposable Moore spaces or (m,n)–atoms as in (1.1).

For a prime p let $S^n_{(p)}$ be the p–local sphere. An element $\alpha \in \pi_{n-1}(S^m) \otimes \mathbb{Z}_{(p)}$ yields a map $\alpha : S^{n-1}_{(p)} \to S^m_{(p)}$ and the mapping cone of this map is denoted by

(1.1) $$C_\alpha = S^m_{(p)} \cup_\alpha e^n_{(p)}.$$

For $n \in \mathbb{Z}$ with $1/n \in \mathbb{Z}_{(p)} \subset \mathbb{Q}$ there is a homotopy equivalence $C_\alpha \simeq C_{n\alpha}$.

THEOREM 1.2. *If $\pi_{n-1}(S^m) \otimes \mathbb{Z}_{(p)} = \mathbb{Z}/p^k$ is a cyclic group with generator ξ then the spaces C_α with $\alpha = \xi, p\xi, \dots, p^{k-1}\xi$ form a complete list of torsion free p–local (n,m)–atoms.*

This result yields the finite case of theorem (A). The next result yields the tame case of theorem (A).

THEOREM 1.3. *If $\pi_{n-1}(S^m) \otimes \mathbb{Z}_{(p)} = \mathbb{Z}/p \oplus \mathbb{Z}/p^k$ with generators $\eta \in \mathbb{Z}/p$ and $\xi \in \mathbb{Z}/p^k$ then a complete list of torsion free p–local (n,m)–atoms $A(g)$ and $A(g,\varphi)$ is given as follows.*

We consider a finite connected non empty subword g of the infinite word $(i \in \mathbb{Z})$

(1) $$\dots \xi_i \eta \xi_{i+1} \eta \xi_{i+2} \eta \dots$$

where $\xi_i \in \{\xi, p\xi, \dots, p^{k-1}\xi\}$. Hence g corresponds to a connected subgraph of the infinite graph

(2)
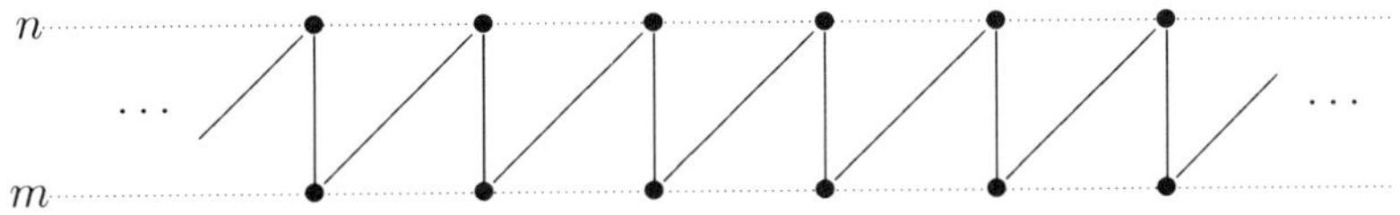

in which the vertical edges denote η and the diagonal edges denote elements in $\{\xi, p\xi, \dots, p^{k-1}\xi\}$. According to the graph g we attach p–local cells $e^n_{(p)}$ to a one point union of p–local spheres $S^n_{(p)}$. Here each vertex of level n in the graph g is a cell and each vertex of level m in g is a sphere and the cell is attached according to the edges of the graph g. More precisely, let B be the set of vertices of level m of g and let T be the set of vertices of level n of g and consider the one pont unions of p–local spheres

$$S^{n-1}_T = \bigvee_{e \in T} S^{n-1}_e \qquad (T\text{=top cells})$$

$$S^m_B = \bigvee_{e \in B} S^m_e \qquad (B\text{=bottom spheres})$$

with $S^{n-1}_e = S^n_{(p)}, S^m_e = S^m_{(p)}$. Then the *atom* $A(g)$ is the mapping cone of a map

(3) $$\alpha_g : S^{n-1}_T \to S^m_B$$

Here α_g is defined by the graph g, in particular, if $e \in T$ is the top vertex of $\xi_i\eta$ in g then the coordinate α^e_g is given by the sum $i_a\xi_i + i_b\eta : S^{n-1}_e \to S^m_a \vee S^m_b \subset S^m_B$ where a is the bottom vertex of ξ_i and b is the bottom vertex of η and i_a, i_b are the inclusions. Hence the space $A(g)$ is the p–local form of a "lightning flash space" defined in Baues–Drozd [**BD3**].

Next we define a *cyclic word* (g, φ) where g is a word as above of the form $g = \xi_1\eta \dots \xi_{c-1}\eta\xi_c$ with $c \geq 1$ and φ is an automorphism of a finite dimensional $\mathbb{Z}/p$–vector space $V = V(\varphi)$. Two cyclic words $(g, \varphi), (g', \varphi')$ are *equivalent* if g' is a cyclic permutation of g, that is $(1 \leq i \leq c)$

$$g' = \xi_i\eta\xi_{i+1}\eta \dots \xi_c\eta\xi_1\eta \dots \xi_{i-1}\eta,$$

and there is an isomorphism $\psi : V(\varphi) \to V(\varphi')$ with $\varphi = \psi^{-1}\varphi'\psi$. A cyclic word (g, φ) is a *special cyclic word* if g is not of the form $g = g'g' \dots g'$ where the right hand side is a k–fold power of a word g' with $k > 1$ and if φ is an indecomposable automorphism. Here we say that an automorphism of a homomorphism f between vector spaces is decomposable if f is a non–trivial direct sum of homomorphism.

For a special cyclic word $(g = \xi_1\eta \dots \eta\xi_c, \varphi)$ we define the p–local (m, n)*–atom* $A(g, \varphi)$ by the mapping cone of a map

$$\alpha_{g,\varphi} : \bigvee^d S^{n-1}_T \to \bigvee^d S^m_B \tag{4}$$

Here $d = \mathsf{dim}_{\mathbb{Z}/p}(V(\varphi))$ and $\bigvee^d Y$ denotes the d–fold connected sum of the space Y with inclusion $j_\delta : S \subset \bigvee^d S$ for $\delta = 1, \dots, d$. Let $b_0 \in B$ be the bottom vertex of ξ_1 and $t_0 \in T$ be the top vertex of ξ_c. Then the map $\alpha_{g,\varphi}$ on $j_\delta S^n_e$ with $e \in T - \{t_0\}$ is defined as α_g above compatible with the inclusion j_δ. The map $\alpha_{g,\varphi}$ restricted to $\bigvee^d S^{n-1}_{t_0}$, however, is a sum of the map $\bigvee^d_{\delta=1} j_\delta\xi_c$ and of the map

$$\bigvee^d S^{n-1}_{t_0} \xrightarrow{\varphi} \bigvee^d S^m_{b_0}$$

given by η and φ. We sketch the map $\alpha_{g,\varphi}$ for $d = 2$ and $g = \xi_1\eta\xi_2\eta\xi_3$ as follows

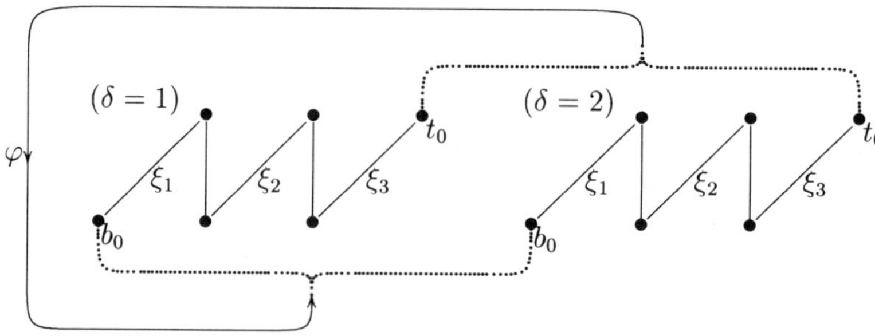

Now the atoms $A(g, \varphi)$ and $A(g', \varphi')$ given by special cyclic words are homotopy equivalent if and only if (g, φ) and (g', φ') are equivalent.

2. Proof of theorem (A)

For $1 < m < n$ and $R \subset \mathbb{Q}$ let

$$\mathbf{CW}(m, n)_R \subset \mathbf{Top}^*/\simeq \tag{2.1}$$

be the full subcategory consisting of 1–connected spaces X with at most two non trivial homology groups $H_mX = A$ and $H_nX = B$ which are finitely generated R–modules. Hence A and B are direct sums of the cyclic R–modules R and $\mathbb{Z}/p^k, k \geq$

$1, \frac{1}{p} \notin R$. We associate with X an element

$$k_X \in [M(B, n-1), M(A, m)] \tag{2.2}$$

obtained by the homology decomposition of X. Here $M(A, m)$ is the Moore space of A in degree m and X is homotopy equivalent to the mapping cone of a map representing k_X. Now let $\mathbf{M}(m, n)_R$ be the following category. Objects are triple $k_X = (A, B, k_X)$ with k_X as in (2.2) and morphisms $k_X \to k_Y$ are commutative diagrams

$$\begin{array}{ccc} M(B, n-1) & \xrightarrow{\beta} & M(B', n-1) \\ {\scriptstyle k_X}\downarrow & & \downarrow{\scriptstyle k_Y} \\ M(A, m) & \xrightarrow{\alpha} & M(A', m) \end{array} \tag{2.3}$$

in $\mathbf{Top}^*/\simeq$. Using the homology decomposition of spaces we get the following result.

LEMMA 2.1. *For $n \geq m + 3$ there is a functor*

$$k : \mathbf{CW}(m, n)_R \to \mathbf{M}(m, n)_R$$

which carries X to k_X in (2.2). *This functor reflects isomorphism, is full and representative.*

The lemma shows that k induces a bijection on equivalence classes of objects. For $n = m + 2$ there is slightly more delicate lemma which we describe in (3.1) below. For $n < 2m - 2$ the categories in (2.1) are additive categories with the direct sum defined by one point union of spaces and maps respectively. The functor k is additive since $k_{X \vee Y} = k_X \vee k_Y$.

Next we define for a finitely generated R–module M the following *category of matrices with entries* in M denoted by $\mathbf{Mat}_R(M)$. Objects are triple (A, B, k) where A and B are finitely generated free R–modules and $k \in \mathsf{Hom}(B, A \otimes M)$. Morphisms are pairs $\beta : B \to B', \alpha : A \to A'$ for which the diagram

$$\begin{array}{ccc} B & \xrightarrow{\beta} & B' \\ {\scriptstyle k}\downarrow & & \downarrow{\scriptstyle k'} \\ A \otimes M & \xrightarrow{\alpha \otimes 1} & A' \otimes M \end{array} \tag{2.4}$$

commutes. Let $\mathbf{M}(m, n)_R^{free}$ be the full subcategory of $\mathbf{M}(m, n)_R$ consisting of objects (A, B, k_X) for which A and B are finitely generated free R–modules. Then the following result is an easy application of standard facts of homotopy theory.

LEMMA 2.2. *Let $M = \pi_{n-1}(S^m) \otimes R$. Then for $n < 2m - 2$ one has an isomorphism of additive categories*

$$\mathbf{M}(m, n)_R^{free} \cong \mathbf{Mat}_R(M)$$

Proof. The isomorphism carries k_X to the induced morphism

$$k : B = \pi_{n-1} M(B, n-1) \xrightarrow{(k_X)_*} \pi_{n-1} M(A, m) = A \otimes M$$

and carries the morphism (α, β) in (2.3) to $(\pi_{n-1}(\beta), \pi_m(\alpha))$. q.e.d.

Using (2.2), (2.1) we see that theorem (A) and (1.2) and (1.3) are consequences of the representation theory in the category $\mathbf{Mat}_R(M)$.

Proof of (1.3). Let $R = \mathbb{Z}_{(p)}$ be given by a prime p. Let $M = \mathbb{Z}/p \oplus \mathbb{Z}/p^k$ be generated by elements $\eta \in \mathbb{Z}/p$ and $\xi \in \mathbb{Z}/p^k$. Then an object k in $\mathbf{Mat}_R(M)$. is given by

$$k = K\eta + L\xi$$

where K, L are $a \times b$–matrices with entries in $\mathbb{Z}/p$ and $\mathbb{Z}/p^k$ respectively. We may suppose that L is of block–diagonal form

$$L = \begin{pmatrix} I_1 & & & 0 \\ & pI_2 & & \\ & & p^2 I_3 & \\ 0 & & & \ddots \end{pmatrix}$$

where the identity matrices $I_1, I_2, I_3, \dots$ may have different sizes. Then for K one gets the known matrix problem

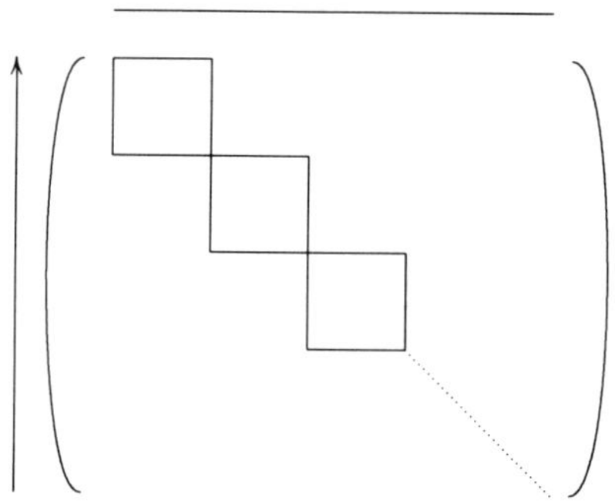

where the squares denote matrices which are transformed by conjugation and where the arrows indicate that we may add rows (columns) of the lower (right) stripes to those of the upper (left) ones. The answer of this problem is known by Bondarenko [**B**] or Drozd [**D**] and this yields the indecomposable objects described in (1.3). q.e.d.

We now recall some notation from representation theory. The *wild quiver* W consists of one vertex v and two arrows $a, b : v \to v$. Let **Vec** be the category of finite dimensional vector spaces over a field k and let **Wild** be the category of representations of W in **Vec** (i. e. objects are functors $A : W \to$ **Vec** and morphisms $A \to B$ are natural transformations). The direct sum of vector spaces yields a direct sum $A \oplus B$ in **Wild**. The *universal problem of representation theory* is the computation of all objects in **Wild** which are indecomposable with respect to direct sum. We call a classification problem *wild* if it requires the solution of the universal problem of representation theory.

PROPOSITION 2.3. *Let $R = \mathbb{Z}_{(p)}$ and let M be a finitely generated $\mathbb{Z}_{(p)}$–module which is not cyclic and not of the form $\mathbb{Z}/p \oplus \mathbb{Z}/p^l$. Then the classification of indecomposable objects in $\mathbf{Mat}_R(M)$ is a wild problem.*

This result proves by (2.2) and (2.1) the wild case of theorem (A).

Proof of (2.3). The proposition is well known if M has 3 cyclic summands. Hence one only has to prove wildness for $M = \mathbb{Z}/p^2 \oplus \mathbb{Z}/p^2$. Then an object in $\mathbf{Mat}(M)$ is a pair of $a \times b$–matrices (K, L) over $\mathbb{Z}/p^2$. We now consider special matrices K, L

of the following shape. We choose $r \times r$–matrices X, Y over $\mathbb{Z}/p$ and we choose $a = b = 7r$. Then let $K = I_a$ be the identity matrix and let $L = L(X, Y)$ be the matrix

$$L = \begin{bmatrix} 0 & 0 & 0 & 0 & 0 & 0 & 0 \\ 0 & 0 & 0 & I_r & 0 & 0 & 0 \\ 0 & 0 & 0 & 0 & I_r & 0 & 0 \\ 0 & 0 & 0 & 0 & 0 & I_r & 0 \\ 0 & 0 & 0 & 0 & 0 & 0 & I_r \\ pI_r & 0 & 0 & pX & pY & 0 & 0 \\ 0 & 0 & 0 & pI_r & 0 & 0 & 0 \end{bmatrix}$$

Then a simple though tedious calculation shows that $(K = I_a, L)$ is isomorphic in $\mathbf{Mat}_R(M)$ to $(K = I_a, L')$ with $L' = L(X', Y')$ if and only if there is an invertible $r \times r$–matrix R with $RX = X'R$ and $RY = Y'R$. This shows that the classification of indecomposable objects in $\mathbf{Mat}_R(M)$ for $M = \mathbb{Z}/p^2 \oplus \mathbb{Z}/p^2$ is wild. q.e.d.

3. Proof of theorem (B)

We consider the following diagram of functors $(m > 4)$

(3.1)
$$\begin{array}{ccc} \mathbf{CW}(m, m+2)_{\mathbb{Z}} & & \mathbf{M}(m, m+2)_{\mathbb{Z}} \\ & \searrow^{k} \quad \swarrow^{q} & \\ & \mathbf{M}(m, m+2)_{\mathbb{Z}}/\sim & \end{array}$$

Here $\mathbf{CW}(m, m+2)_{\mathbb{Z}}$ and $\mathbf{M}(m, m+2)_{\mathbb{Z}}$ are defined as in (2.2) and (2.3) for $R = \mathbb{Z}$. The functor k is defined as in (2.1) though this functor maps only to a quotient category of $\mathbf{M}(m, m+2)$. The functor q is the quotient functor for the following natural quivalence relation $\sim$ in $\mathbf{M}(m, m+2)_{\mathbb{Z}}$. Consider the diagram

$$\begin{array}{ccc} M(B, m+1) & \xrightarrow{\beta, \beta_1} & M(B', m+1) \\ {\scriptstyle k_X}\downarrow & \nearrow_{\xi} & \downarrow{\scriptstyle k_Y} \\ M(A, m) & \xrightarrow[\alpha, \alpha_1]{} & M(A', m) \end{array}$$

where $(\alpha, \beta), (\alpha_1, \beta_1)$ are morphisms $k_X \to k_Y$ in $\mathbf{M}(m, m+2)_{\mathbb{Z}}$. Then we set $(\alpha, \beta) \sim (\alpha_1, \beta_1)$ if there is

$$\xi \in [M(A, m), M(B', m+1)] = \mathsf{Ext}(A, B')$$

with $\beta_1 = \beta$ and $\alpha_1 = \alpha + k_Y \xi$. Since $m > 4$ we know that all categories and functors in (3.1) are additive.

Proposition 3.1. *The functor k in* (3.1) *is well defined. Moreover k reflects isomorphisms, is full and representative.*

This is a special case of the classification result 6.8.2 in Baues [**B1**]. Since also the quotient functor q reflects isomorphisms, is full and representative we see that the indecomposable objects in $\mathbf{CW}(m, m+2)$ coincide with the mapping cones of indecomposable objects in $\mathbf{M}(m, m+2)_{\mathbb{Z}}$.

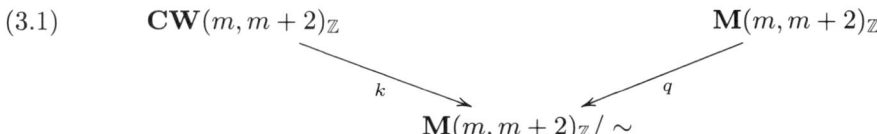

In order to obtain an algebraic description of the category $\mathbf{M}(m, m+2)_{\mathbb{Z}}$ we recall first some general notation on bimodule problems. Let $\mathcal{A}$, $\mathcal{B}$ be two categories (fully additive, i. e., such additive categories where every idempotent splits), U be an $\mathcal{A}$–$\mathcal{B}$–bimodule, i. e., a functor $\mathcal{A}^{op} \times \mathcal{B} \to \mathbf{Ab}$. Define the category $\mathbf{El}(U)$ of elements of this bimodule as follows:

(3.2) $$\mathrm{Ob}\mathbf{El}(U) = \coprod_{\substack{A \in \mathrm{Ob}\mathcal{A} \\ B \in \mathrm{Ob}\mathcal{B}}} U(A,B)$$

A morphism $u \to v$, where $u \in U(A,B), v \in U(A',B')$, is a pair (α,β), where $\alpha : A' \to A, \beta : B \to B'$, such that $\beta u = v\alpha$ (Here we write $v\alpha$ for $U(\alpha,1)v$ and βu for $U(1,\beta)u$).

We have the following example of a category of elements in a bimodule. Let $\mathbf{M}^m$ be the full subcategory of $\mathbf{Top}^*/\simeq$ consisting of Moore spaces $M(A,m)$ of finitely generated abelian groups. Then the suspension $\Sigma : \mathbf{M}^m \to \mathbf{M}^{m+1}$ is an equivalence of categories for $m \geq 3$ and $\mathbf{M}^m$ is an additive category for $m \geq 3$. We have a bimodule

(3.3) $$U : (\mathbf{M}^{m+1})^{op} \times \mathbf{M}^m \to \mathbf{Ab}$$

defined by the abelian group $[M(A,m+1), M(H,m)]$. Now the category $\mathbf{El}(U)$ coincides with $\mathbf{M}(m,m+2)_{\mathbb{Z}}$.

We describe $\mathbf{M}^m$ and the bifunctor U algebraically as follows. For an abelian group A, consider the short exact sequence

(1) $$A \otimes \mathbb{Z}/2 \overset{\eta^A}{\rightarrowtail} G(A) \twoheadrightarrow A * \mathbb{Z}/2$$

coresponding to the composition

$$A * \mathbb{Z}/2 \to A \to A \otimes \mathbb{Z}/2.$$

Applying $\mathsf{Hom}(\_,\mathbb{Z}/4)$, one gets the exact sequence

(2) $$\begin{array}{ccccc} \mathsf{Hom}(A * \mathbb{Z}/2, \mathbb{Z}/4) & \rightarrowtail & \mathsf{Hom}(G(A), \mathbb{Z}/4) & \twoheadrightarrow & \mathsf{Hom}(A \otimes \mathbb{Z}/2, \mathbb{Z}/4) \\ \wr\| & & & & \wr\| \\ \mathsf{Ext}(A, H \otimes \mathbb{Z}/2) & & & & \mathsf{Hom}(A, \mathbb{Z}/2) \end{array}$$

Now, applying $\_ \otimes H$, one gets

(3) $$\begin{array}{ccccccc} \mathsf{Ext}(A,\mathbb{Z}/2) \otimes H & \overset{\delta}{\to} & \mathsf{Hom}(G(A),\mathbb{Z}/4) \otimes H & \to & \mathsf{Hom}(A,\mathbb{Z}/2) \otimes H & \to & 0 \\ \wr\| & & & & \wr\| & & \\ \mathsf{Ext}(A, H \otimes \mathbb{Z}/2) & & & & \mathsf{Hom}(A, H \otimes \mathbb{Z}/2) & & \end{array}$$

(all groups are supposed to be finitely generated).

Now define $G(A,H)$ by the push–down diagram:

(4)

$$\begin{array}{ccccccc} \mathsf{Ext}(A, H \otimes \mathbb{Z}/2) & \xrightarrow{\Delta} & \mathsf{Hom}(G(A), \mathbb{Z}/4) \otimes H & \xrightarrow{\mu} & \mathsf{Hom}(A, H \otimes \mathbb{Z}/2) & \longrightarrow & 0 \\ \Big\downarrow{\scriptstyle \eta^H_*} & & \Big\downarrow & & \Big\| & & \\ \mathsf{Ext}(A, G(H)) & \xrightarrow[\Delta]{} & G(A,H) & \xrightarrow[\mu]{} & \mathsf{Hom}(A, H \otimes \mathbb{Z}/2) & \longrightarrow & 0 \end{array}$$

Define the category $\mathfrak{I}$ where objects are abelian groups (finitely generated) and morphisms are pairs $(\varphi, \widetilde{\varphi}), \varphi : A \to B, \widetilde{\varphi} : G(A) \to G(B)$ such that the following diagram commutes:

(5)
$$\begin{array}{ccccc} A\otimes\mathbb{Z}/2 & \longrightarrow & G(A) & \longrightarrow & A*\mathbb{Z}/2 \\ {\scriptstyle\varphi\otimes 1}\downarrow & & {\scriptstyle\widetilde{\varphi}}\downarrow & & \downarrow{\scriptstyle\varphi*1} \\ B\otimes\mathbb{Z}/2 & \longrightarrow & G(B) & \longrightarrow & B*\mathbb{Z}/2 \end{array}$$

Then $G(A,H)$ can be consider as $\mathfrak{I}$–$\mathfrak{I}$–bimodule, the (right) action of $(\varphi,\widetilde{\varphi}) : A \to A'$ induced by
$$(\varphi,\widetilde{\varphi})^* = \mathsf{Ext}(\varphi, G(H)) \oplus \mathsf{Hom}(\widetilde{\varphi},\mathbb{Z}/4)\otimes H$$
and the (left) action of $(\psi,\widetilde{\psi}) : H \to H'$ induced by
$$(\psi,\widetilde{\psi})_* = \mathsf{Ext}(A,\widetilde{\psi}) \oplus \mathsf{Hom}(G(A),\mathbb{Z}/4)\otimes \psi.$$
Hence, one can consider the category $\mathbf{El}(G)$ of the elements of this bimodule.

PROPOSITION 3.2. *For $m \geq 3$ there is an isomorphism of categories $\mathfrak{I} \cong \mathbf{M}^m$ and one has a natural isomorphism*
$$[M(B,m+1), M(H,m)] \cong G(B,H)$$
Hence the category $\mathbf{M}(m,m+2)$ is isomorphic to the category $\mathbf{El}(G)$.

This result is proved in (8.2.10) Baues [**B1**]. In order to obtain the indecomposable objects in $\mathbf{CW}(m,m+2)_{\mathbb{Z}}$ we thus have to classify the indecomposable objects in the category $\mathbf{El}(G)$.

Note that if $A = \bigoplus_j A_j$ and $H = \bigoplus_i H_i$, the elements of $G(A,H)$ can be naturally written as the matrices (g_{ij}) with $g_{ij} \in G(A_j,H_i)$. Hence, we only have to calculate $G(A,H)$ with indecomposable A,H and the action of the morphisms $A \to A'$ and $H \to H'$ for indecomposable $A,A';H,H'$. Thus, we are interested in the cases $A = \mathbb{Z}/2^a, H = \mathbb{Z}/2^h$ (putting $\mathbb{Z}/2^\infty = \mathbb{Z}$). First note that, if $A = \mathbb{Z}^a$ with $a > 1$, the sequence (1) splits and $A\otimes\mathbb{Z}/2 = A*\mathbb{Z}/2 = \mathbb{Z}/2$. Moreover, given $\varphi : A' \to A$, the homomorphisms $\widetilde{\varphi} : G(A') \to G(A)$ such that $(\varphi,\widetilde{\varphi})$ is a morphism from $\mathfrak{I}$, are given by the homomorphisms $\varphi' : A*\mathbb{Z}/2 \to A\otimes\mathbb{Z}/2$; namely, w. r. t. the decomposition $G(A) = A\otimes\mathbb{Z}/2 \oplus A*\mathbb{Z}/2$, $\widetilde{\varphi}$ is given by the matrix $\begin{pmatrix} \varphi\otimes 1 & \varphi' \\ 0 & \varphi*1 \end{pmatrix}$. Certainly, in this case (2) and (3) are also split sequences, hence, the lower row of the push–down (4) also splits, i. e.
$$G(A,H) = \mathsf{Ext}(A, G(H)) \oplus \mathsf{Hom}(A, H\otimes\mathbb{Z}/2).$$
If $h > 1$, one also has that $\mathsf{Ext}(A,G(H)) = \mathsf{Ext}(A,H*\mathbb{Z}/2)\oplus\mathsf{Ext}(A,H\otimes\mathbb{Z}/2)$, so $G(A,H) = \mathsf{Ext}(A,H*\mathbb{Z}/2)\oplus\mathsf{Ext}(A,H\otimes\mathbb{Z}/2)\oplus\mathsf{Hom}(A,H\otimes\mathbb{Z}/2)$ and all three direct summands here are $\mathbb{Z}/2$. We denote them, respectively, by $G(*^a,{}_h*), G(*^a,\otimes_h)$ and $G({}^a\otimes,\otimes_h)$. One easily sees that a homomorphism $(\varphi,\widetilde{\varphi})$, where $\widetilde{\varphi} = \begin{pmatrix} \varphi\otimes 1 & \varphi' \\ 0 & \varphi*1 \end{pmatrix}$ maps a triple (g_1,g_2,g_3) from this direct sum to $(\varphi^*g_1, \varphi^*g_2 + \hat{\varphi}'g_3, \varphi^*g_3)$, where $\hat{\varphi}'$ is the composition
$$\mathsf{Hom}(A,H\otimes\mathbb{Z}/2) \simeq \mathsf{Hom}(A\otimes\mathbb{Z}/2, H\otimes\mathbb{Z}/2) \xrightarrow{(\varphi')^*} \mathsf{Hom}(A'*\mathbb{Z}/2, H\otimes\mathbb{Z}/2) \simeq$$
$$\simeq \mathsf{Ext}(A'*\mathbb{Z}/2, H\otimes\mathbb{Z}/2) \simeq \mathsf{Ext}(A', H\otimes\mathbb{Z}/2).$$
If $(\psi,\psi') : H' \to H$, then it maps (g_1,g_2,g_3) to $(\psi_*g_1, \psi_*g_2 + (\psi')_*g_1, \psi_*g_3) \in G(A,H')$. If $h = 1$, the sequence (1) for H is indeed
$$\mathbb{Z}/2 \rightarrowtail \mathbb{Z}/4 \twoheadrightarrow \mathbb{Z}/2,$$

hence, for $a > 1$, one still has the exact sequence

$$\mathsf{Ext}(A, H \otimes \mathbb{Z}/2) \rightarrowtail \mathsf{Ext}(A, G(H)) \twoheadrightarrow \mathsf{Ext}(A, H * \mathbb{Z}/2);$$

which is indeed

$$\mathbb{Z}/2 \rightarrowtail \mathbb{Z}/4 \twoheadrightarrow \mathbb{Z}/2.$$

Again a homomorphism $(\varphi, \widetilde{\varphi}) : A' \to A$ maps a pair (g_1, g_3) with $g_1 \in \mathsf{Ext}(A, G(H)), g_2 \in G({}^a\otimes, \otimes_1) = \mathsf{Hom}(A, H \otimes \mathbb{Z}/2)$ to $(\varphi^* g_1 + \hat{\varphi}' g_3, \varphi^* g_3)$. On the other hand, to define a homomorphism $(\psi, \widetilde{\psi}) : H \to H'$, where $H = \mathbb{Z}/2$, one only has to define $\widetilde{\psi} : \mathbb{Z}/4 \to H'$: then $\psi = \widetilde{\psi} \otimes \mathbb{Z}/2 : \mathbb{Z}/2 \to H'$. Such a homomorphism maps a pair (g_1, g_3) as above to $(\widetilde{\psi}_* g_1, \psi_* g_3)$.

Suppose now that $a = 1$. Then (1) is indeed

$$\mathbb{Z}/2 \rightarrowtail \mathbb{Z}/4 \twoheadrightarrow \mathbb{Z}/2.$$

If $h > 1$, the tensor multiplication by H does not change this exact sequence; η_*^H in (4) is split monomorphism, hence, $G(A, H) \simeq \mathbb{Z}/4 \oplus \mathsf{Ext}(A, H * \mathbb{Z}/2)$, and we have an exact sequence

$$\mathbb{Z}/2 = G(*^1, \otimes_h) \to \mathbb{Z}/4 \to G({}^1\otimes, \otimes_h) = \mathbb{Z}/2.$$

A morphism $(\varphi, \widetilde{\varphi})$ is given by any homomorphism $\widetilde{\varphi} : G(A') \to \mathbb{Z}/4$: then $\varphi = \widetilde{\varphi}|_{A' \otimes \mathbb{Z}/2} \in \mathsf{Hom}(A' \otimes \mathbb{Z}/2, \mathbb{Z}/2) \simeq \mathsf{Hom}(A', \mathbb{Z}/2)$. It maps a pair $(g_1, g_2) \in \mathsf{Hom}(G(A), \mathbb{Z}/4) \otimes H \oplus \mathsf{Ext}(A, H * \mathbb{Z}/2)$ to $((\widetilde{\varphi}^* \otimes 1) g_1, \varphi^* g_2)$. A morphism $(\psi, \widetilde{\psi}) : H \to H'$, where $\widetilde{\psi} = \begin{pmatrix} \psi\otimes 1 & \psi' \\ 0 & \psi * 1 \end{pmatrix}$, maps (g_1, g_2) to $((1 \otimes \psi)_* g_1 + \psi'_* g_2, (\psi * 1)_* g_2)$. At last, if $a = 1 = h$, the sequence (3) is

$$\mathbb{Z}/2 \xrightarrow{0} \mathbb{Z}/2 \xrightarrow{1} \mathbb{Z}/2 \to 0 \text{ and } \eta_*^H = 0$$

Hence, $G(A, H) = \mathsf{Ext}(A, G(H)) \oplus \mathsf{Hom}(A, H \otimes \mathbb{Z}/2)$, both summand being $\mathbb{Z}/2$ (we denote them by $G({}_1*, *^1)$ and $G(\otimes_1, {}^1\otimes)$ respectively, putting $G({}_1*, {}^1\otimes) = 0$).

Denote by $\widetilde{G}$ the sub–bimodule of G such that, for $A = \mathbb{Z}/2^a$, $H = \mathbb{Z}/2^h$, $\widetilde{G}(A, H) = G({}_h*, {}^a\otimes)$. In the definition of the weak equivalence (6), one always has $\Delta(\mathrm{Im}\varphi^* \eta_*^H, \mu(\beta)_*) \subseteq \widetilde{G}(A, H)$. We will use this remark later to show that indeed the weak equivalence coincide with the isomorphism in $\mathbf{El}(G)$.

Now we are going to reformulate our bimodule problem in the matrix form.

This bimodule problem can obviously be rewritten in the matrix form as follws. We consider the "striped" matrices M with the horizontal stripes marked by the set $\mathcal{E} = \{\otimes_h, {}_h* | h \in \mathbb{N}\} \cup \{\otimes_\infty\}$ and the vertical stripes marked by the set $\mathcal{F} = \{{}^a\otimes, *^a | a \in \mathbb{N}\} \cup \{{}^\infty\otimes\}$. We denote by $M(x, y)$ the block placed at the intersection of the horizontal stripe x and the vertical stripe y. These matrices should satisfy the following conditions:

(1) The number of rows in the stripes $\otimes_h$ and ${}_h*$ is the same, as well as the number of columns in the stripes ${}^a\otimes$ and $*_a$.
(2) $M({}_h*, {}^a\otimes) = 0$; $M(\otimes_1, *^a) = 0$; $M(\otimes_h, *^1) = 0$. (We consider them as "matrices over the zero ring $\mathbb{Z}/1$").
(3) The matrices $M(\otimes_h, {}^1\otimes)$ with $h > 1$ and $M({}_1*, *^a)$ with $a > 1$ are with the entries from $\mathbb{Z}/4$, all other ones are with the entries from $\mathbb{Z}/2$.

Here is the picture describing such matrices (we always indicate the ring, where the entries of the blocks are from):

	${}^1\otimes$	${}^2\otimes$	${}^3\otimes$	$\dots$	${}^\infty\otimes$	$\dots$	$*^3$	$*^2$	$*^1$
$\otimes_1$	$\mathbb{Z}/2$	$\mathbb{Z}/2$	$\mathbb{Z}/2$	$\dots$	$\mathbb{Z}/2$	$\dots$	0	0	0
$\otimes_2$	$\mathbb{Z}/4$	$\mathbb{Z}/2$	$\mathbb{Z}/2$	$\dots$	$\mathbb{Z}/2$	$\dots$	$\mathbb{Z}/2$	$\mathbb{Z}/2$	0
$\otimes_1$	$\mathbb{Z}/4$	$\mathbb{Z}/2$	$\mathbb{Z}/2$	$\dots$	$\mathbb{Z}/2$	$\dots$	$\mathbb{Z}/2$	$\mathbb{Z}/2$	0
$\vdots$	$\dots$	$\dots$	$\dots$	$\dots$	$\dots$	$\dots$	$\dots$	$\dots$	$\dots$
$\otimes_\infty$	$\mathbb{Z}/4$	$\mathbb{Z}/2$	$\mathbb{Z}/2$	$\dots$	$\mathbb{Z}/2$	$\dots$	$\mathbb{Z}/2$	$\mathbb{Z}/2$	0
$\vdots$	$\dots$	$\dots$	$\dots$	$\dots$	$\dots$	$\dots$	$\dots$	$\dots$	$\dots$
${}_3*$	0	0	0	$\dots$	0	$\dots$	$\mathbb{Z}/2$	$\mathbb{Z}/2$	$\mathbb{Z}/2$
${}_2*$	0	0	0	$\dots$	0	$\dots$	$\mathbb{Z}/2$	$\mathbb{Z}/2$	$\mathbb{Z}/2$
${}_1*$	0	0	0	$\dots$	0	$\dots$	$\mathbb{Z}/4$	$\mathbb{Z}/4$	$\mathbb{Z}/2$

The following transformations of the matrix M are called "admissible transformations" (in what follows we denote by $M(x,\_)$ and by $M(\_,y)$, respectively, the horizontal stripe marked by x and the vertical stripe marked by y).

(a) Replacing the stripes $M(\otimes_h,\_)$ and $M({}_h*,\_)$ by $XM(\otimes_h)$ and $XM({}_h*)$.
(a') Replacing the stripes $M(\_,{}^a\otimes)$ and $M(\_,*^a)$ by $M(\_,{}^a\otimes)X$ and $M(\_,*^a)X$.
(b) Replacing $M(\otimes_h,\_)$ by $M(\otimes_h,\_)+XM(\otimes_{h'},\_)+YM({}_k*,\_)$, where $h'>h,k$ any.
(b') Replacing $M(\_,*^a)$ by $M(\_,*^a)+M(\_,*^{a'})X+M(\_,{}^b\otimes)$, where $a'>a,b$ any.
(c) Replacing $M({}_h*,\_)$ by $M({}_h*,\_)+XM({}_{h'}*,\_)$ and $M(\otimes_h,\_)Y$ by $M(\otimes_h,\_)+2^{h-h'}XM(\otimes_{h'},\_)$, where $h'<h$.
(c') Replacing $M(\_,{}^a\otimes)$ by $M(\_,{}^a\otimes)+M(\_,{}^{a'}\otimes)X$ and $M(\_,*^a)$ by $M(\_,*^a)+2^{a-a'}M(\_,*^{a'})X$, where $a'<a$.
(d) Replacing $M({}_1*,\_)$ by $M_1(*,\_)+2XM({}_h*,\_)+2YM(\otimes_k,\_)$ for arbitrary h,k.
(d') Replacing $M(\_,{}^1\otimes)$ by $M(\_,{}^1\otimes)+2M(\_,{}^a\otimes)X+2M(\_,*^b)Y$ for arbitrary a,b.
(e) Replacing $M(\otimes_h,{}^1\otimes), h>1$, by $M(\otimes_h,{}^1\otimes)+2XM({}_{h'}*,*^1)$, h' any.
(e') Replacing $M({}_1*,*^a), a>1$, by $M({}_1*,*^a)+2M(\otimes_1,{}^{a'}\otimes)X$, a' any.

Here X,Y always denote arbitrary matrices of the appropriate size with the entries from $\mathbb{Z}/4$; in the transformations of type (a) and (a') the matrix X should be invertible. Certainly, if an original block was with the entries from $\mathbb{Z}/2$, the resulting one should also be calculated modulo 2; if an original block was over zero ring, so is the resulting one.

Two matrices, M,M' (of the same size) are called *equivalent* if they can be transformed to each other by a sequence of admissible transformations. Call M,M' *equivalent modulo 2* if M can be transformed by a sequence of admissible transformations to a matrix M'' such that $M''\equiv M' \mod 2$. Of course, considering the equivalence modulo 2, we may reduce the stripes $M(\_,{}^1\otimes)$ and $M({}_1*,\_)$ modulo 2 (thus forgetting the transformations of types (d), (d')), as well as always suppose X,Y being over $\mathbb{Z}/2$.

One can easily see that for the equivalence modulo 2 we get a sort of representations of a bunch of chains in the sense of [**D**]. Therefore, one can write down all indecomposable matrices. Namely, we have the chain $\mathcal{E}$ for rows, with the order:

$$\otimes_1 > \otimes_2 > \ldots > \otimes_\infty > \ldots > {}_3* > {}_2* > {}_1*,$$

and the chain $\mathcal{F}$ for columns, with the order:

$$^1\otimes < {}^2\otimes < \ldots < {}^\infty\otimes < \ldots < *^3 < *^2 < *^1.$$

The equivalence relation $\sim$ is given by

$$\otimes_h \sim_h *(h \in \mathbb{N}); {}^a\otimes \sim *^a(a \in \mathbb{N}).$$

Remind the corresponding combinatorics. Put $\mathcal{X} = \mathcal{E} \cup \mathcal{F}$ and write $x - y$ if $x \in \mathcal{E}, y \in \mathcal{F}$ or vice versa. Then $\mathcal{X}$–word is a sequence $w = x_1 r_2 x_2 r_3 \ldots r_n x_n$, where $x_i \in \mathcal{X}, r_i \in \{\sim, -\}$, such that $r_i \neq r_{i+1} (i = 2 \ldots, n-1), x_i r_{i+1} x_{i+1}$ in $\mathcal{X}(i = 1, \ldots, n-1)$. Such a word is said to be *full* if:
either $r_2 =\sim$ or $x_1 \nsim y$ for any $y \neq x_1$;
either $r_n =\sim$ or $x_n \nsim y$ for any $y \neq x_n$.
w is called *cyclic* if $r_2 = r_n = -$ and $x_n \sim x_1$; it is called *aperiodic* if it is not of the type $v \sim v \sim \ldots \sim v$ for a shorter word v.

Call a polynomial $\pi(t) \in \mathbb{Z}/2[t]$ *primitive* if it is a power of an irreducible polynomial (with the leading coefficient 1). Then the indecomposable representations of this bunch of chains are in 1–1 correspondence with the set $\mathcal{S} \cup \mathcal{B}$, where $\mathcal{S}$ consists of all full words and $\mathcal{B}$ consists of the pairs $(w, \pi(t))$, where w is an aperiodic cyclic word and $\pi(t) \neq t^d$ is a primitive polynomial. More precisely, we should identify any word w with its inverse and any cycle with its cyclic shift!

We call the representations corresponding to $\mathcal{S}$ "strings" and those corresponding to $\mathcal{B}$ "bands".

The condition (2) from the definition of the matrix M imposes some restrictions on the representations, which can occur as equivalence classes modulo 2 for such matrices. In terms of the $\mathcal{X}$–words, they mean that the corresponding word contains no subword ${}^a\otimes -{}_h*, \otimes_1 - *^a, *^1 - \otimes_h$ or their inverse. (Call such words admissible.) One can easily check that such a word can only contain at most one subword of the form $\otimes_h - *^a$ (or its inverse). To simplify the notations, we replace the subword ${}^a\otimes - \otimes_h$ by ${}^a\otimes_h$, $\otimes_h - {}^a\otimes$ by ${}_h\otimes^a$; ${}_h* - *^a$ by ${}_h*^a$, $*^a - {}_h*$ by ${}^a*_h$; $\otimes_h - *^a$ by ${}_h\theta^a$. We also omit all $\sim$ signs and alway replace a superscript ^{aa} by ^{a} as well as a subscript ${}_{hh}$ by ${}_h$. At last, we omit x_1 if $r_2 =\sim$ and x_n if $r_n =\sim$. Certainly, one can always restore the original $\mathcal{X}$–word having the result of all these simplifications.

Now, any admissible word (or its inverse) can be written as a subword of

(i) $${}^{a_1}\otimes_{h_1} *^{a_2} \otimes_{h_2} * \ldots {}^{a_n} \otimes_{h_n} \quad (\text{"usual word"})$$

or

(ii) $${}_{h_{-m}}\otimes \ldots {}^{a_{-2}} *_{h_{-2}} \otimes^{a_{-1}} *_{h_{-1}} \theta^{a_1} \otimes_{h_1} *^{a_2} \otimes_{h_2} \ldots *^{a_n} \quad (\text{"}\theta\text{–word"})$$

The strings correponding to θ–words are called "θ–strings", all other ones are called "usual string". Any admissible cyclic word (or its shift) can be written in the form

$${}^{a_1}\otimes_{h_1} *^{a_2} \otimes_{h_2} * \ldots {}^{a_n} \otimes_{h_n} *^{a_1}$$

Moreover, the following conditions hold:

(1) $a_i = \infty$ or $h_j = \infty$ can only occur at the end of a word as ${}^\infty\otimes$ $(\otimes^\infty)$ or $\otimes_\infty ({}_\infty\otimes)$;

(2) In any θ–word, $h_{-1} \neq 1$ and $a_1 \notin \{1, \infty\}$.

Note that the description of the representations of a bunch of chains [**D**] implies:

- Any row (column) of a string contains at most 1 non–zero element.
- There are at most 2 zero rows or columns in a string, namely, they are in the following stripes:
  - $M({}_h*, {}_-)$ if w has an end $\otimes_h$ (or ${}_h\otimes$), $h \neq \infty$;
  - $M({}_-, *^a)$ if w has an end ${}^a\otimes$, $a \neq \infty$;
  - $M(\otimes_h, {}_-)$ if w has an end ${}_h*$;
  - $M({}_-,{}^a \otimes)$ if w has an end $*^a$ (or ${}^a*$).
- The horizontal and vertical stripes of a band can be subdivided in such a way that every new horizontal or vertical sub–stripe contains exactly 1 non–zero block, which is invertible.

COROLLARY 3.3. *Let $\overline{M}$ be the reduction modulo 2 of an admissible matrix M, $\overline{N}$ be its indecomposable direct summand. If $\overline{N}$ is a band or a 0–string, then $M \simeq N \oplus M'$ such that $\overline{N} \equiv N \mod 2$.*

So, from now on, we may only consider such M that $\overline{M}$ is a direct sum of usual strings. Denote $\widetilde{M}$ the matrix obtained from M by replacing all invertible entries by zeroes (in particular, all its elements are of the form $2c$). Let $\overline{M} = \bigoplus_{i=1}^r \overline{M_i}$, where $\overline{M_i}$ is the usual string corresponding to a word w_i.

COROLLARY 3.4. *$M \simeq M'$, where $\overline{M'} = \overline{M}$ and the only non–zero rows and columns of $\widetilde{M}'$ may be the following:*

(1) *the columns of $\widetilde{M}'({}_-,{}^1 \otimes)$ corresponding to the ends ${}^1\otimes$ occurring in the word w_i;*
(2) *the rows of $\widetilde{M}'({}_1*, {}_-)$ corresponding to the ends ${}_1*$ occurring in the words w_i;*
(3) *the rows of $\widetilde{M}'({}_-,{}^1 \otimes)$ corresponding to the ends ${}_h*$ $(h > 1)$;*
(4) *the columns of $\widetilde{M}'({}_1*, {}_-)$ corresponding to the ends ${}^a\otimes$ $(1 < a < \infty)$.*

We call all these ends the "distinguished ends".

In what follows, we always suppose that already $M = M'$.

To prove Corollaries 3.3, 3.4, one only has to use the transformations (d), (d') and (e), (e').

The following Lemma is now decisive.

LEMMA 3.5. *Suppose that two words w_i, w_j have a common distinguished end. There is a sequence of admissible transformations of $M = M'$ which does not change the matrix $\overline{M}$ and the resulting transformation of $\widetilde{M}$ adds the row (column) corresponding to the end of w_i to that corresponding to the end of w_j or vice versa.*

COROLLARY 3.6. *There is a sequence of admissible transformations of $M = M'$, which does not change $\overline{M}$ and transform $\widetilde{M}$ to a matrix having at most one non–zero element in every row and every column.*

Proof of Lemma. We consider the case of the end ${}^1\otimes$, all other cases being quite analogous. So let $w_i = {}^1 \otimes_{h_1} *^{a_2} \otimes_{h_2} \dots$, while $w_j = {}^1 \otimes_{k_1} *^{b_2} \otimes_{k_2} \dots$. Suppose that $h_1 \leq k_1$. If indeed $h_1 < k_1$, one can add the column corresponding to the end ${}^1\otimes_{h_1}$ of w_i to that corresponding to the end ${}^1\otimes_{k_1}$ of w_j and then subtract the row corresponding to the latter one from that corresponding to the first one to restore $\overline{M}$. If $h_1 = k_1$, compare a_2 and b_2. If $a_2 < b_2$, one can perform the same transformations as above and afterwards subtract the column corresponding to ${}_{h_1}*^{b_2}$ from that corresponding to ${}_{h_1}*^{a_2}$, to restore $\overline{M}$. Continuing these considerations, one sees that we can add the column corresponding to the end of w_i to that corresponding to the end of w_j if $(h_1, a_2, h_2, \dots) < (k_1, b_2, k_2, \dots)$ lexicographically. As the lexicographical order is linear, it proves the lemma.

COROLLARY 3.7. *Suppose M an indecomposable matrix such that $\widetilde{M'} \neq 0$ for any matrix M' equivalent to M. Let $\overline{M} = \bigoplus_{i=1}^{r} \overline{M_i}$, where $\overline{M_i}$ are usual strings. Then, up to equivalence, there are but the following possibilities:*

(1) *$r = 1, \overline{M} = \overline{M_1}$ corresponds to a word $w_1 =^1 \otimes_{h_1} *^{a_2} \otimes_{h_2} \dots$, $\widetilde{M}$ has one non-zero element in the matrix $M(\otimes_h, {}^1\otimes)$. We denote this case by the word $w =_h \theta^1 \otimes_{h_1} *^{a_2} \otimes_{h_2} \dots$*
(2) *$r = 1, \overline{M} = \overline{M_1}$ corresponds to a word ${}_1 *^{a_1} \otimes_{h_1} *^{a_2} \dots$; $\widetilde{M}$ has one non-zero element in the matrix $M({}_1*, *^a)$. We denote this case by the word $w = \dots{}^{a_2} *_{h_1} \otimes^{a_1} *_1 \theta^a$.*
(3) *$r = 1, \overline{M} = \overline{M_1}$ corresponds to a word ${}^{a_1} \otimes_{h_1} *^{a_2} \dots$; $\widetilde{M}$ has one non-zero element in the matrix $M({}_1*, *^{a_1})$. We denote this case by the word $w =_1 \theta^{a_1} \otimes_{h_1} *^{a_2} \dots$*
(4) *$r = 1, \overline{M} = \overline{M_1}$ corresponds to a word ${}_{h_1} *^{a_1} \otimes_{h_2} * \dots$; $\widetilde{M}$ has one non-zero element in the matrix $M(\otimes_{h_1}, {}^1\otimes)$. We denote this case by the word $w = \dots *_{h_2} \otimes^{a_1} *_{h_1} \theta^1$.*
(5) *$r = 2, \overline{M_1}$ corresponds to a word ${}^1 \otimes_{h_1} *^{a_1} \otimes_{h_2} \dots$, $\overline{M_2}$ corresponds to a word ${}_{h_{-1}} *^{a_{-1}} \otimes_{h_{-2}} * \dots$; the unique non-zero element of $\widetilde{M}$ is in the matrix $M(\otimes_{h_{-1}}, {}^1\otimes)$. We denote this case by the word $w = \dots *_{h_{-1}} \otimes^{a_{-1}} *_{h_{-1}} \theta^1 \otimes_{h_1} *^{a_1} \otimes_{h_2} \dots$*
(6) *$r = 2, \overline{M_1}$ corresponds to a word ${}^{a_1} \otimes_{h_1} *^{a_2} \otimes_{h_2} \dots$; $\overline{M_2}$ corresponds to a word ${}_1 *^{a_{-1}} \otimes_{h_{-1}} *^{a_{-2}} \dots$; the unique non-zero element of $\widetilde{M}$ is in the matrix $M({}_1*, *^{a_1})$. We denote this case by the word $w = \dots{}^{a_{-2}} \otimes_{h_{-1}} \otimes^{a_{-1}} *_1 \theta^{a_1} \otimes_{h_1} *^{a_2} \otimes_{h_2} \dots$*

Obviously, the cases (1)–(4) always give indecomposable representations. On the other hand, in the case (5), if $h_1 \geq h_{-1}$, the representation M evidently decomposes as $M_1 \oplus M_2$; the same is with the case (6) and $a_{-1} \geq a_1$. If $h_1 < h_{-1} - 1$ in the case (5) or $a_{-1} < a_1 - 1$ in the case (6), the corresponding representation is evidently indecomposable. Suppose, in the case (5), $h_1 = h_{-1} - 1$. Then we can delete the non-zero element of $\widetilde{M}(\otimes_{h_{-1}}, {}^1\otimes)$ using the transformation (c), but it changes the zero entry in $M({}_{h_{-1}}*, *^{a_1})$. To restore it, we need that $a_{-1} \geq a_1$, moreover, if $a_{-1} = a_1$, we also need that $h_2 \geq h_{-2}$, etc. Thus, such a representation remains indecomposable if and only if $(h_1 + 1, a_{-1}, h_2, a_{-2}, \dots) < (h_{-1}, a_1, h_{-1}, a_2, \dots)$ lexicographically. Just in the same way, the representation of type (6) is indecomposable if and only if $(a_{-1} + 1, h_1, a_{-2}, h_2, \dots) < (a_1, h_{-1}, a_2, h_{-2}, \dots)$ lexicographically.

Therefore, the complete list of indecomposable matrices looks like follows.

THEOREM 3.8. *Indecomposable objects in $\mathbf{El}(G)$ are in 1-1 correspondence with the following types of data:*

(1) *usual words, i. e. subword of the words ${}^{a_1} \otimes_{h_1} *^{a_2} \otimes_{h_2} * \dots{}^{a_n} \otimes_{h_n}$, where $a_i, h_j \in \mathbb{N} \cup \{\infty\}, \infty$ being only possible for a_1 or h_n.*
(2) *θ-words, i. e. subwords containing θ of the words ${}_{h_{-m}} \otimes \dots{}^{a_{-2}} *_{h_{-2}} \otimes^{a_{-1}} *_{h_{-1}} \theta^{a_1} \otimes_{h_1} *^{a_2} \otimes_{h_2} \dots *^{a_n} \otimes_{h_n}$, where $a_i, h_j \in \mathbb{N} \cup \{\infty\}, \infty$ being only possible for h_{-m} or h_n, $(a_1, h_{-1}) \neq (1, 1)$ and,*
if $h_{-1} = 1$,

then $(a_{-1}+1, h_1, a_{-2}, h_2, \dots) < (a_1, h_{-2}, a_2, h_{-3}, \dots)$ *lexicographically, and, if* $a_1 = 1$,
then $(h_1+1, a_{-2}, h_2, a_{-3}, \dots) < (h_{-1}, a_2, h_{-2}, a_3, \dots)$ *lexicographically. Moreover the* θ*–word does not coincide with* ${}_\infty\theta^1\otimes_\infty$.

(3) *Pairs* $(w, \pi(t))$, *where* w *is an aperiodic cyclic word:* $w = {}^{a_1}\otimes_{h_1} *^{a_2}\otimes_{h_2} \dots {}^{a_n}\otimes_{h_n} *^{a_1}$ *(up to shift), and* $\pi(t) \neq t^d$ *is a primitive polynomial over* $\mathbb{Z}/2$.

COROLLARY 3.9. *Let* α *be an indecomposable element of the bimodule* G. *Denote by* $\widetilde{\alpha}$ *its component belonging to the sub–bimodule* $\widetilde{G}$ *and by* $\widetilde{G_\alpha}$ *the sub-bimodule of* $\widetilde{G}$ *generated by* $\widetilde{\alpha}$ *Then* $\widetilde{G_\alpha} \cap \operatorname{Im}\eta_*\mu(\alpha)_* = 0$.

Proof. Follows immediately from the description above.

Here are examples of indecomposable representations.

(1) Usual string corresponding to the word ${}_2*^3\otimes_2*^4\otimes_1*^3\otimes_4$

	$^3\otimes$		$^4\otimes$	$*^4$	$*^3$	
$\otimes_1$	0	0	1	0		
$\otimes_2$	0	0	0			
$\otimes_2$	1	0	0			
$\otimes_4$	0	1	0			
$_4*$				0	0	0
$_2*$				0	1	0
$_2*$				1	0	0
$_1*$				0	0	1

$$A = 2\cdot\mathbb{Z}/8 \oplus \mathbb{Z}/16$$
$$H = \mathbb{Z}/2 \oplus 2\cdot\mathbb{Z}/4 \oplus \mathbb{Z}/16$$

(2) θ–string corresponding to the word ${}_\infty\otimes^4*_3\otimes^2*_2\theta^4\otimes_3*^3$

	$^2\otimes$	$^3\otimes$	$^4\otimes$		$*^4$		$*^3$	$*^2$
$\otimes_2$	0	0	0	0	0	1	0	0
$\otimes_3$	1	0	0	0	0	0	0	0
$\otimes_3$	0	0	0	1	0	0	0	0
$\otimes_\infty$	0	0	1	0	0	0	0	0
$_3*$					1	0	0	0
$_3*$					0	0	1	0
$_2*$					0	0	0	1

$$A = \mathbb{Z}/4 \oplus \mathbb{Z}/8 \oplus 2\cdot\mathbb{Z}/16$$
$$H = \mathbb{Z}/4 \oplus 2\cdot\mathbb{Z}/8 \oplus \mathbb{Z}$$

(3) θ–string corresponding to the word ${}^2*_4\otimes^1*_3\theta^1\otimes_2*^3\otimes_1$

	$^1\otimes$		$^2\otimes$	$^3\otimes$	$*^3$	$*^2$	$*^1$	
$\otimes_1$	0	0	0	1	0			
$\otimes_2$	0	1	0	0				
$\otimes_3$	0	2	0	0				
$\otimes_4$	1	0	0	0				
$_1*$					0	0	0	0
$_2*$					1	0	0	0
$_3*$					0	0	1	0
$_4*$					0	1	0	0

$$A = 2 \cdot \mathbb{Z}/2 \oplus \mathbb{Z}/4 \oplus \mathbb{Z}/8$$
$$H = \mathbb{Z}/2 \oplus \mathbb{Z}/4 \oplus \mathbb{Z}/8 \oplus \mathbb{Z}/16$$

(4) band corresponding to the word ${}^2\otimes_4 *^3 \otimes_1 *^3 \otimes_2 *^2$ and the polynomial $\pi(t) = t^2 + t + 1$

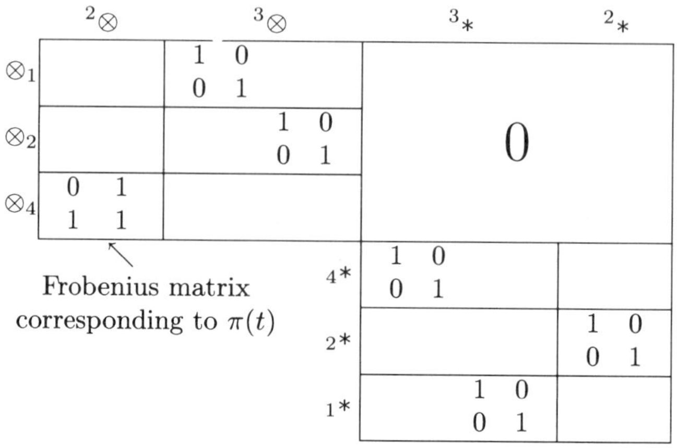

$$A = 2 \cdot \mathbb{Z}/4 \oplus 4 \cdot \mathbb{Z}/8$$
$$H = 2 \cdot \mathbb{Z}/2 \oplus 2 \cdot \mathbb{Z}/4 \oplus 2 \cdot \mathbb{Z}/16$$

4. Proof of theorem (C)

Let $\mathbf{M}^m(\mathbb{Z}/2)$ be the homotopy category of Moore spaces $M(A,m)$ where A is a finitely generated $\mathbb{Z}/2$–vector space. Then we know by Baues [**B1**] 1.6.7 that $\mathbf{M}^m(\mathbb{Z}/2)$ for $m \geq 3$ is isomorphic to the category $\mathbf{mod}(\mathbb{Z}/4)$ of finitely generated free $\mathbb{Z}/4$–modules. Using this result we show:

THEOREM 4.1. *Let $m \geq 6$. Then the bimodule $\mathbf{M}^{m+3}(\mathbb{Z}/2)^{op} \times \mathbf{M}^m(\mathbb{Z}/2) \to$ $\mathbf{Ab}$ given by $[M(a,m+3), M(B,m)]$ is natural isomorphic to the bimodule*

$$\mathbf{mod}(\mathbb{Z}/4)^{op} \times \mathbf{mod}(\mathbb{Z}/4) \to \mathbf{Ab}$$

which carries $(\overline{A}, \overline{B})$ to $\mathsf{Hom}(\overline{A} \otimes \mathbb{Z}/2, \overline{B} \otimes (\mathbb{Z}/2 \oplus \mathbb{Z}/2 \oplus \mathbb{Z}/2))$.

Since the quiver

$$\bullet \overset{\longrightarrow}{\underset{\longrightarrow}{\longrightarrow}} \bullet$$

is wild we see that theorem (4.1) implies theorem (C) in the introduction.

Proof of (4.1). Since the bimodule is biadditive it suffices to compute for $A = B = \mathbb{Z}/2$ the $\mathbb{Z}/4$–bimodule $\pi = [M(A,m+3), M(B,m)]$. Here the $\mathbb{Z}/4$–module structure is already determined by the structure of π as an abelian group. We know that

$$\pi_{m+2}M(B,m) = \mathbb{Z}/4$$
$$\pi_{m+3}M(B,m) = \mathbb{Z}/2 \oplus \mathbb{Z}/2$$

and that the Hopf map $\eta : S^{m+3} \to S^{m+2}$ induces the map

$$\eta^* : \pi_{m+2}M(B,m) \to \pi_{m+3}M(B,m)$$

corresponding to the composite

$$\eta^* : \mathbb{Z}/4 \twoheadrightarrow \mathbb{Z}/2 \xrightarrow{i_1} \mathbb{Z}/2 \oplus \mathbb{Z}/2$$

of the inclusion i_1 and the quotient map. This follows for example from Baues–Goerss [**BG**] 5.2. Moreover using Baues [**B1**] 1.6.11 we have the push out diagram of abelian groups

$$\begin{array}{ccccc}
\mathbb{Z}/2 \oplus \mathbb{Z}/2 & = & \mathsf{Ext}(A, \pi_{m+3}M(B,m)) & \longrightarrow & \pi \\
 & & \uparrow{\scriptstyle (\eta^*)_*} & & \uparrow \\
\mathbb{Z}/2 & = & \mathsf{Ext}(A, \pi_{m+2}(B,m) \otimes \mathbb{Z}/2) & \xrightarrow{j} & \mathbf{G}(A, \pi_{m+2}M(B,m))
\end{array}$$

Here $(\eta^*)_*$ is induced by η^* above so that $(\eta^*)_*$ is split injective with cokernel $\mathbb{Z}/2$. Moreover we have by definition of the category $\mathbf{G}$ in Baues [**B1**] the isomorphism

$$\mathbf{G}(A, \pi_{m+2}(B,m)) = \mathbf{G}(\mathbb{Z}/2, \mathbb{Z}/4) = \mathbb{Z}/2 \oplus \mathbb{Z}/2$$

with j above corresponding to the inclusion $i_1 : \mathbb{Z}/2 \subset \mathbb{Z}/2 \oplus \mathbb{Z}/2$. This shows $\pi = \mathbb{Z}/2 \oplus \mathbb{Z}/2 \oplus \mathbb{Z}/2$. q.e.d.

References

[B] V. Bondarenko, *Representations of bundles of semichained sets and their applications.* St. Petersburg Math. J. **3** (1992) 973–996.

[B1] Baues, H.–J., *Homotopy type and homology*, Oxford Math. Monographs, Oxford University Press, 1996, 596 pages.

[B2] Baues, H.–J., *Homotopy types*, Handbood of algebraic topology, chapter I (1995), Edited by I. M. James, Elsevier Science, 1–71.

[BC] Brown, E. H. and Copeland, A. H., *Homology analogue of Postnikov Systems*, Mich. Math. Journ. **6** (1959) 313–330.

[BD1] Baues, H.–J. and Drozd, Y., *Representation theory of homotopy types with at most two non trivial homotopy groups*, Math. Proc. Camp. Phil. Soc. **128** (2000) 283–300.

[BD2] Baues, H.–J. and Drozd, Y., *The homotopy classification of $(n-1)$–connected $(n+1)$–dimensional polyhedra with torsion free homology, $n \geq 5$*, Expositiones Mathematicae, **17** (1999) 161–180.

[BD3] Baues, H.–J. and Drozd, Y., *Classification of stable homotopy types with torsion free homology*, to appear in Topology.

[BG] Baues, H.–J. and Goerss, P., *A homotopy operation spectral sequence for the computation of homotopy groups*, Topology **39** (2000) 161–192.

[BH] Baues, H.–J. and Hennes, M., *The homotopy classification of $(n-1)$–connected $(n+3)$–dimensional polyhedra, $n \geq 4$*, Topology **30** (1991) 373–408.

[Ch] Chang, S. C., *Homotopy invariants and continous mappings*, Proc. R. Soc. London Ser. A **202** (1950) 253–263.

[D] Y. Drozd, *Finitely generated quadratic modules.* Preprint, MPI, 1999.

[EH] Eckmann, B. and Hilton, P. J., *On the homology and homotopy decompositions of continuous maps.* Pro. Nat. Acad. Science **45** (1959) 372–375.

[H] Henn, H.–W., *Classification of p–local low dimensional spectra.* J. Pure Appl. Algebra **19** (1980) 159–169.

[H1] Hilton, P. J., *An introduction to homotopy theory*, Cambridge University Press, 1953.

[H2] Hilton, P. J., *Homotopy theory and duality*, Nelson Gordon and Breach, 1965.

[W1] Whitehead, J. H. C., *The homotopy type of a special kind of polyhedron*, Ann. Soc. Polon. Math. **21** (1948) 176–186.

Hans–Joachim Baues
Max–Planck–Institut für Mathematik
P.O.Box: 7280
D–53072 Bonn
Germany
E-mail address: baues@mpim-bonn.mpg.de

Yuri A. Drozd
Faculty of Mechanics and Mathematics
Kiev Taras Shevchenko University
RUS–252033 Kiev
Ukraine

Contemporary Mathematics
Volume **274**, 2001

Square Rings Associated to Elements in Homotopy Groups of Spheres

Hans–Joachim Baues and Norio Iwase

Abstract. In this paper we compute for $\alpha \in \pi_{n-2}(S^{m-1})$ with $n < 3m-3$ the full homotopy category consisting of finite one point unions $\Sigma C_\alpha \vee \ldots \vee \Sigma C_\alpha$ with $\Sigma C_\alpha = S^m \cup_{\Sigma\alpha} e^n$. For this we describe the square ring $\mathrm{End}(\Sigma C_\alpha)$ only in terms of primary homotopy operations on spheres. In low dimensions with $n - m \leq 19$ these homotopy operations are computed in the book of Toda [**T**], so that we get this way many explicit examples of square rings. In particular we shall describe algebraically the square rings $\mathrm{End}(\Sigma\mathbb{C}P_2)$, $\mathrm{End}(\Sigma\mathbb{H}P_2)$ and $\mathrm{End}(\Sigma\mathbb{C}a)$ where $\mathbb{C}P_2$ and $\mathbb{H}P_2$ are the complex and quaternionic projective plane respectively and where $\mathbb{C}a$ is the Cayley plane. The structure of $\mathrm{End}(\Sigma C_\alpha)$ leads to a theory of extensions for square rings.

1. Introduction

For pointed spaces X, Y let $[X, Y]$ be the set of homotopy classes of pointed maps $X \to Y$. Hence $[X, Y]$ is the set of morphisms in the homotopy category $\mathrm{Top}^*/_{\simeq}$. In this paper spaces are CW-complexes.

We consider a suspended space ΣX which is $(m-1)$–connected and of dimension $< 3m-3$ with $m \geq 2$ (that is ΣX is *metastable*) and we consider the full subcategory

$$\underline{\underline{\mathrm{Add}}}(\Sigma X) \subset \mathrm{Top}^*/_{\simeq} \tag{1.1}$$

consisting of one point unions $\bigvee^k \Sigma X$ of k-copies of the space ΣX with $k \geq 0$. On the other hand we associate with ΣX the diagram

$$\begin{aligned} \mathrm{End}(\Sigma X) = Q &= (Q_e \xrightarrow{\bar{H}} Q_{ee} \xrightarrow{\bar{P}} Q_e), \\ Q_e = [\Sigma X, \Sigma X], &\quad Q_{ee} = [\Sigma X, \Sigma X \wedge X] \end{aligned} \tag{1.2}$$

where $\bar{H}$ is the Hopf invariant and $\bar{P}$ is induced by the Whitehead product $[1_{\Sigma X}, 1_{\Sigma X}]$: $\Sigma X \wedge X \to X$. Here Q_e and Q_{ee} are groups by the co–H–structure of ΣX and Q_e is also a monoid by composition of maps. Moreover since ΣX is metastable the group Q_{ee} is abelian. The next lemma is shown in [**BHP**].

1991 *Mathematics Subject Classification.* 55Q05.

Key words and phrases. 2–cell complex, Hopf invariant, Whitehead product, square ring, metastable range.

The authors thank Kaoru Morisugi for his helpful comments on many parts of this paper.

LEMMA 1.1. *The diagram $Q = \mathrm{End}(\Sigma X)$ has the structure of a square ring (see Section 3 below) and the algebraic biproduct completion $\underline{\underline{\mathrm{Add}}}(Q)$ of the square ring Q is isomorphic to the category $\underline{\underline{\mathrm{Add}}}(\Sigma X)$.*

The lemma shows that the computation of the square ring $Q = \mathrm{End}(\Sigma X)$ yields an algebraic characterization of the category $\underline{\underline{\mathrm{Add}}}(\Sigma X)$ by the category $\underline{\underline{\mathrm{Add}}}(Q)$ in which the object corresponding to $\bigvee^k \Sigma X$ is denoted by $\coprod^k Q$ and in which the morphisms are certain matrices defined in (2.3) below. If ΣX is *stable* (that is, if ΣX is of dimension $\leq 2m-2$), then $Q_{ee} = 0$ and $Q_e = \mathrm{End}(\Sigma X)$ is a ring, i.e. the endomorphism ring of an object in an additive category. In this case the lemma states the well known fact that $\underline{\underline{\mathrm{Add}}}(\Sigma X)$ is isomorphic to the category of free Q_e-modules $\coprod^k Q = \bigoplus^k Q_e$.

The k-th *general linear group* of the square ring Q is the group

(1.3)
$$GL(Q,k) = \mathrm{Aut}(\coprod^k Q)$$

of automorphisms of $\coprod^k Q$ in the category $\underline{\underline{\mathrm{Add}}}(Q)$. Hence we obtain by (1.1) the following computation of the group of homotopy equivalences $\mathrm{Aut}(\bigvee^k \Sigma X)$ in $\mathrm{Top}^*/_{\simeq}$.

COROLLARY 1.2. *For $k \geq 0$ one has a canonical isomorphism of groups*

$$\mathrm{Aut}(\bigvee^k \Sigma X) \cong GL(Q,k)$$

where the right hand side is algebraically determined by the square ring $Q = \mathrm{End}(\Sigma X)$ in (1.2).

The purpose of this paper is the computation of the square ring $\mathrm{End}(\Sigma X)$ if ΣX is a 2-cell complex. Let $\pi_k(S^m)$ be the k-th homotopy group of the m-sphere S^m. We consider an element

(1.4)
$$\alpha \in \pi_{n-2}(S^{m-1}) \quad \text{with} \quad n-2 \geq m-1 \geq 1, n < 3m-3$$

which yields the mapping cone C_α and its suspension ΣC_α which are CW-complexes of the form

$$\begin{cases} C_\alpha = S^{m-1} \cup_\alpha e^{n-1} \\ \Sigma C_\alpha = S^m \cup_{\Sigma\alpha} e^n \end{cases}$$

The assumptions on m, n show that ΣC_α is in the meta-stable range so that the endomorphism square ring $\mathrm{End}(\Sigma C_\alpha)$ is defined. For example, if $\alpha = \eta_2 \in \pi_3(S^2)$ is the Hopf map then $C_\alpha = \mathbb{C}P_2$ is the complex projective plane and the square ring $\mathrm{End}(\Sigma\mathbb{C}P_2)$ was computed in (8.6) of [**BHP**]. In this paper we describe more generally $\mathrm{End}(\Sigma C_\alpha)$ only in terms of primary homotopy operations on spheres. In low dimensions with $n - m \leq 19$ these homotopy operations are computed in the book of Toda [**T**], so that we get this way many explicit examples of square rings. In particular we shall describe algebraically the square rings $\mathrm{End}(\Sigma\mathbb{H}P_2)$ and $\mathrm{End}(\Sigma\mathcal{C}a)$ where $\mathbb{H}P_2$ is the quaternionic projective plane and where $\mathcal{C}a$ is the Cayley plane.

2. The biproduct completion of a square ring

We recall the definition of square group and square ring from [**BHP**].

DEFINITION 2.1. A *square group* $Q = (Q_e \xrightarrow{\bar{H}} Q_{ee} \xrightarrow{\bar{P}} Q_e)$ is given by a group Q_e and an abelian group Q_{ee}. Both groups are written additively. Moreover $\bar{P}$ is a homomorphism and $\bar{H}$ is a quadratic function, that is, the cross effect

$$\langle x|y\rangle_{\bar{H}} = \bar{H}(x+y) - \bar{H}(x) - \bar{H}(y)$$

is linear in $x, y \in Q_e$. In addition the following properties are satisfied for $x, y \in Q_e$ and $u, v \in Q_{ee}$.

(1) $\langle \bar{P}(u)|y\rangle_{\bar{H}} = 0$ and $\langle x|\bar{P}(v)\rangle_{\bar{H}} = 0$
(2) $\bar{P}(\langle x|y\rangle_{\bar{H}}) = x + y - x - y$
(3) $\bar{P}\bar{H}\bar{P}(u) = \bar{P}(u) + \bar{P}(u)$

DEFINITION 2.2. A *square ring* $Q = (Q_e \xrightarrow{\bar{H}} Q_{ee} \xrightarrow{\bar{P}} Q_e)$ is given by a square group $(Q_e \xrightarrow{\bar{H}} Q_{ee} \xrightarrow{\bar{P}} Q_e)$ for which Q_e has the additional structure of a monoid with unit $1 \in Q_e$ and the multiplication is denoted by $x \circ y \in Q_e$. This monoid structure induces a ring structure on the abelian group $\bar{R} = \operatorname{cok}(\bar{P})$ through the canonical projection $Q_e \xrightarrow{\bar{\epsilon}} \bar{R}$. We write $\bar{\epsilon}(a) = \bar{a}$. Moreover the abelian group Q_{ee} is an $\bar{R}\otimes\bar{R}\otimes\bar{R}^{op}$-module with action denoted by $(\bar{t}\otimes\bar{s})\cdot u\cdot\bar{r} \in Q_{ee}$ for $\bar{t}, \bar{s}, \bar{r} \in \bar{R}, u \in Q_{ee}$. In addition the following properties are satisfied where $\bar{H}(2) = \bar{H}(1+1)$.

(1) $\langle x|y\rangle_{\bar{H}} = (\bar{y}\otimes\bar{x})\cdot\bar{H}(2)$
(2) $T = \bar{H}\bar{P} - 1$ is an isomorphism of abelian groups satisfying $T((\bar{t}\otimes\bar{s})\cdot u\cdot\bar{r}) = (\bar{s}\otimes\bar{t})\cdot T(u)\cdot\bar{r}$.
(3) $\bar{P}(u)\circ x = \bar{P}(u\cdot\bar{x})$
(4) $x\circ\bar{P}(u) = \bar{P}((\bar{x}\otimes\bar{x})\cdot u)$
(5) $\bar{H}(x\circ y) = (\bar{x}\otimes\bar{x})\cdot\bar{H}(y) + \bar{H}(x)\cdot\bar{y}$
(6) $(x+y)\circ z = x\circ z + y\circ z + \bar{P}((\bar{x}\otimes\bar{y})\cdot\bar{H}(z))$
(7) $x\circ(y+z) = x\circ y + x\circ z$

By [**BP**], we know that the category of square groups is the same as the category of quadratic functors $\underline{\underline{\mathrm{Gr}}} \to \underline{\underline{\mathrm{Gr}}}$ where $\underline{\underline{\mathrm{Gr}}}$ is the category of groups. With respect to the monoidal structure in this category a square ring is also a monoid in the category of square groups; see [**BP**].

We remark that the definition of a square ring in [**BHP**] or [**BP**] uses also the equation

(8) $\bar{H}\bar{P}\bar{H}(x) + \bar{H}(x+x) - 4\bar{H}(x) = \bar{H}(2)\cdot\bar{x}$.

which is redundant. In fact, by the condition (6) of Definition 2.2, we have

$$2\circ x = (1+1)\circ x = 1\circ x + 1\circ x + \bar{P}((\bar{1}\otimes\bar{1})\cdot\bar{H}(x)) = x + x + \bar{P}\bar{H}(x).$$

Applying $\bar{H}$ using the condition (1) of Definition 2.1 we get

$$\bar{H}(2\circ x) = \bar{H}(x+x) + \bar{H}(\bar{P}\bar{H}(x)) + \langle x+x|\bar{P}\bar{H}(x)\rangle_{\bar{H}} = \bar{H}(x+x) + \bar{H}\bar{P}\bar{H}(x).$$

On the other hand by condition (5) we have

$$\bar{H}(2\circ x) = (\bar{2}\otimes\bar{2})\cdot\bar{H}(x) + \bar{H}(2)\cdot\bar{x} = 4\bar{H}(x) + \bar{H}(2)\cdot\bar{x}.$$

Comparing the equations we obtain (8).

DEFINITION 2.3. Given a square ring Q as above we define the *biproduct completion* $\underline{\underline{\mathrm{Add}}}(Q)$. We obtain the category $\underline{\underline{\mathrm{Add}}}(Q)$ in terms of matrices as follows. Objects in $\underline{\underline{\mathrm{Add}}}(Q)$ are denoted by $\coprod^x Q$ will $x \in \{0,1,2,\cdots\}$. For $x=0$ this is the initial object and for $x=1$ we write $Q = \coprod^1 Q$. Sets of morphisms are defined by product sets

$$\mathrm{Mor}(Q, \coprod^x Q) = (\prod_{i=1}^x Q_e) \times (\prod_{1 \le i < j \le x} Q_{ee})$$

$$f \in \mathrm{Mor}(\coprod^y Q, \coprod^x Q) = \prod_{k=1}^y \mathrm{Mor}(Q, \coprod^x Q)$$

where we write $f = (f_i^k, f_{ij}^k)$. Now let $g = (g_k^s, g_{k\ell}^s)$ be an element in $\mathrm{Mor}(\coprod^z Q, \coprod^y Q)$. Then the *composition*

$$fg = ((fg)_i^s, (fg)_{ij}^s)$$

is given by the coordinates

$$(fg)_i^s = f_i^1 \circ g_1^s + f_i^2 \circ g_2^s + \cdots + f_i^y \circ g_y^s + \sum_{k<\ell} \bar{P}((\overline{f_i^k} \otimes \overline{f_i^\ell}) \cdot g_{k\ell}^s)$$

$$(fg)_{ij}^s = \sum_k (f_{ij}^k \cdot \overline{g_k^s}) + \sum_{k<\ell} ((\overline{f_i^k} \otimes \overline{f_j^\ell}) \cdot g_{k\ell}^s + (\overline{f_i^\ell} \otimes \overline{f_j^k}) \cdot T g_{k\ell}^s + \overline{(f_i^\ell \cdot g_\ell^s)} \otimes \overline{(f_j^k \cdot g_k^s)} \cdot \bar{H}(2))$$

3. The main result

We associate with α in (1.4) the following data determined by α.

DEFINITION 3.1. Given $\alpha \in \pi_{n-2}(S^{m-1})$ with $n < 3m-3$. Let $U_\alpha \subset \pi_n(S^m)$ be the subgroup generated by $\eta_m(\Sigma^2\alpha)$ and $(\Sigma\alpha)\eta_{n-1}$ where η_t is the Hopf element, $t \ge 2$. Then the quotient group $\pi_n(S^m)/U_\alpha$ is part of the diagram

(3.1)
$$M = (M_e \xrightarrow{H} M_{ee} \xrightarrow{P} M_e)$$
$$M_e = \pi_n(S^m)/U_\alpha, \quad M_{ee} = \pi_n(S^{2m-1})$$

where H is given by the Hopf invariant and P is induced by the Whitehead square $[\iota_m, \iota_m] : S^{2m-1} \to S^m$, that is, $P(u) = [\iota_m, \iota_m]_* u$. The element $\Sigma\alpha$ is a torsion element of *order* k in $\pi_{n-1}(S^m)$ so that $\Sigma(k\alpha) = 0$ and hence by the exactness of the EHP-sequence with $n \le 3m-3$ (see [**B1**] A.6.7)

$$\pi_n(S^{2m-1}) \xrightarrow{P_0} \pi_{n-2}(S^{m-1}) \xrightarrow{E_0} \pi_{n-1}(S^m) \xrightarrow{H_0} \pi_{n-1}(S^{2m-1}) \to \cdots,$$

there exists

(3.2)
$$\begin{cases} \mu \in \pi_n(S^{2m-1}) \quad \text{with} \\ P_0(\mu) = [\iota_{m-1}, \iota_{m-1}]_* (\Sigma^2)^{-1}\mu = -k\alpha \end{cases}$$

Here we use the inverse $(\Sigma^2)^{-1}$ of the double suspension $\Sigma^2 : \pi_{n-2}(S^{2m-3}) \cong \pi_n(S^{2m-1})$. Moreover let

(3.3)
$$\lambda = \Sigma^2 H(\alpha)$$

be given by the Hopf invariant $H : \pi_{n-2}(S^{m-1}) \to \pi_{n-2}(S^{2m-3})$.

THEOREM 3.2. *In terms of the data (M, λ, μ, k) associated to α we define below a square ring $Q(M, \lambda, \mu, k)$ together with an isomorphism*

$$\mathrm{End}(\Sigma C_\alpha) \cong Q(M, \lambda, \mu, k)$$

of square rings.

For $k \geq 1$ let $R = \mathbb{Z}\times_k\mathbb{Z}$ be the subring of $\mathbb{Z}\times\mathbb{Z}$ consisting of all pairs $a = (a_0, a_1)$ with $a_0 - a_1 \equiv 0 \mod k$. This is the pull back ring of $\mathbb{Z} \longrightarrow \mathbb{Z}/k \longleftarrow \mathbb{Z}$. Then $1 = \eta(1) = (1,1) \in R$ is the unit and we have an augmentation $\epsilon : R \to \mathbb{Z}$ with $\epsilon(a) = a_0$ for $a = (a_0, a_1)$. The kernel of ϵ is generated by $\bar{k} = (0, k)$ so that 1 and $\bar{k}$ form a $\mathbb{Z}$-basis of the free abelian group $\mathbb{Z}\times_k\mathbb{Z}$. We have a surjection map

$$\deg : [\Sigma C_\alpha, \Sigma C_\alpha] \twoheadrightarrow \mathbb{Z}\times_k\mathbb{Z} = R \tag{3.4}$$

which carries $u : \Sigma C_\alpha \to \Sigma C_\alpha$ to the pair $\deg(u) = (a_0, a_1)$ where a_0 is the degree of $H_m(u)$ on $H_m(\Sigma C_\alpha) = \mathbb{Z}$ and a_1 is the degree of $H_n(u)$ on $H_n(\Sigma C_\alpha) = \mathbb{Z}$. We shall prove the following crucial lemma.

LEMMA 3.3. *For the square ring End(ΣC_α) given by diagram (1.2) and for the data (M, λ, μ, k) in definition 3.1 one gets a commutative diagram*

$$\begin{array}{ccccc}
M_e & \xrightarrow{\quad H \quad} & M_{ee} & \xrightarrow{\quad P \quad} & M_e \\
{\scriptstyle i}\downarrow & & \| & & \downarrow{\scriptstyle i} \\
[\Sigma C_\alpha, \Sigma C_\alpha] & \xrightarrow{\bar{H}} & [\Sigma C_\alpha, \Sigma C_\alpha \wedge C_\alpha] & \xrightarrow{\bar{P}} & [\Sigma C_\alpha, \Sigma C_\alpha] \\
{\scriptstyle s}\uparrow\downarrow{\scriptstyle \deg} & \nearrow{\scriptstyle h} & & & \\
R & & & &
\end{array}$$

where the column $M_e \overset{i}{\hookrightarrow} [\Sigma C_\alpha, \Sigma C_\alpha] \overset{\deg}{\to} R$ is a split short exact sequence of abelian groups. For a splitting s of deg *let $h = \bar{H}s$. We can choose s such that $s(1) = \iota_\alpha$ is the identity of ΣC_α and $s(\bar{k}) = \mu_0$ such that $\bar{H}(\mu_0) = h(\bar{k}) = \mu$ and $\bar{H}(1+1) = h(1+1) = \lambda$.*

DEFINITION 3.4. A quadratic $\mathbb{Z}$-module [**B2**]

$$M = (M_e \xrightarrow{H} M_{ee} \xrightarrow{P} M_e) \tag{3.5}$$

consists of abelian groups M_e and M_{ee} and homomorphisms H and P satisfying $HPH = 2H$ and $PHP = 2P$. We consider $k \in \mathbb{N}$ and $\lambda, \mu \in M_{ee}$ with relations

$$\begin{cases} P(\lambda) = 0 \\ \\ HP(\mu) = 2\mu + k\lambda. \end{cases} \tag{3.6}$$

One can easily check that (M, λ, μ, k) in (3.1) satisfies these relations. Then the *square ring* $Q(M, \lambda, \mu, k)$ with

$$Q = (R\oplus M_e \xrightarrow{\bar{H}} M_{ee} \xrightarrow{\bar{P}} R\oplus M_e) \tag{3.7}$$

is defined as follows with $R = \mathbb{Z} \times_k \mathbb{Z}$. Let $h : R \to M_{ee}$ be the unique function satisfying

$$\begin{cases} h(1) = 0, \quad h(\bar{k}) = \mu, \\ \\ \langle a, b\rangle_h = h(a+b) - h(a) - h(b) = a_0 b_0 \lambda. \end{cases} \tag{3.8}$$

Then $\bar{H}$ is the function given by

$$\bar{H}(a,x) = h(a) + H(x). \tag{3.9}$$

Moreover $\bar{P}$ is defined by

$$\bar{P}(y) = (0, P(y)) \in R \oplus M_e. \tag{3.10}$$

As a group $R \oplus M_e$ is the direct sum of abelian groups and the monoid structure of $R \oplus M_e$ is given by the product formula

$$(a,x)\circ(b,y) = (a{\cdot}b, x{\cdot}b_1 + a_0 {*} y + \Delta(a,b)) \tag{3.11}$$

where

$$a_0 {*} y = a_0 y + \frac{a_0(a_0-1)}{2} PH(y) \tag{3.12}$$

and

$$\Delta(a,b) = \frac{a_0(a_0-1)(b_1-b_0)}{2k} P(\mu). \tag{3.13}$$

Now $\bar{R} = \operatorname{cok}(\bar{P}) = R \oplus \operatorname{cok}(P)$ is a ring and the projection $\bar{R} \to R$ is a ring homomorphism. Moreover M_{ee} as $\bar{R} \otimes \bar{R} \otimes \bar{R}^{op}$-module is defined by

$$(\overline{(a,x)} \otimes \overline{(b,y)}){\cdot}u{\cdot}\overline{(c,z)} = (a_0 b_0){\cdot}u{\cdot}c_1 \tag{3.14}$$

for $\overline{(a,x)}, \overline{(b,y)}, \overline{(c,z)} \in \bar{R}$ and $u \in M_{ee}$.

We shall prove that $Q = Q(M, \lambda, \mu, k)$ given by 3.5 through 3.14 above is a well-defined square ring. Using the section s in Lemma 3.3 one obtains the isomorphism

$$R \oplus M_e = [\Sigma C_\alpha, \Sigma C_\alpha] \tag{3.15}$$

as in the proof of Corollary 8.5 carrying (a,x) to $s(a)+i(x)$. This is an isomorphism of abelian groups and of monoids and this isomorphism yields the isomorphism $Q \cong \operatorname{End}(\Sigma C_\alpha)$ of square rings in Theorem 3.2. We point out that in general the group $[\Sigma C_\alpha, W]$ for some space W needs not to be abelian, e.g, for $W = \Sigma C_\alpha \vee \Sigma C_\alpha$ the group $[\Sigma C_\alpha, W]$ is abelian if and only if $H(\alpha) = 0$, in other words, C_α is itself a co-H-space.

4. Examples and applications

We consider the special case of the square ring $Q = Q(M, \lambda, \mu, k)$ for a quadratic $\mathbb{Z}$-module M with $M_e = 0$. In this case we obtain for the ring $R = \mathbb{Z} \times_k \mathbb{Z}$ and for $\lambda, \mu \in M_{ee}$ with $2\mu + \lambda = 0$ the *square ring*

$$Q(M_{ee}, \lambda, \mu, k) = Q = (R \xrightarrow{h} M_{ee} \xrightarrow{0} R) \tag{4.1}$$

with $h(1) = 0, h(\bar{k}) = \mu$ and $\langle a,b \rangle_h = h(a+b) - h(a) - h(b) = a_0 b_0 \lambda$. Moreover M_{ee} is an $R \otimes R \otimes R^{op}$-module by $(a \otimes b){\cdot}u \otimes c = (a_0 b_0){\cdot}u{\cdot}c_1$.

For the complex projective plane $\mathbb{C}P_2 = C_{\eta_2}$ where $\alpha = \eta_2$ is the Hopf map we get as a special case of (4.1):

EXAMPLE 4.1. For $\alpha = \eta_2 \in \pi_3(S^2)$ one has

$$\operatorname{End}(\Sigma \mathbb{C}P_2) \cong Q(M_{ee}, \lambda, \mu, k) \quad \text{with}$$

$$M_{ee} = \mathbb{Z}, \quad \lambda = 1, \quad \mu = -1, \quad k = 2.$$

This example was also computed in (8.6)(2) of [**BHP**]. Moreover the category $\underline{\underline{\mathrm{Add}}}(\Sigma\mathbb{C}P_2)$ was computed by [**U**] and [**Y**]. Using the computation of the square ring $Q = \mathrm{End}(\Sigma\mathbb{C}P_2)$ above we know that $\underline{\underline{\mathrm{Add}}}(\Sigma\mathbb{C}P_2) = \underline{\underline{\mathrm{Add}}}(Q)$ is algebraically determined by Q. These results can be generalised as follows.

We describe the square rings for the Hopf maps ν_4, σ_8 for which the mapping cones

$$C_{\nu_4} = \mathbb{H}P_2 \quad \text{and} \quad C_{\sigma_8} = \mathbb{C}a$$

are the *quaternionic projective space* and the *Cayley plane* respectively. By inspection of Toda's book [**T**] we get the following square rings:

EXAMPLE 4.2. For $\alpha = \nu_4 \in \pi_7(S^4)$ one has

$$\mathrm{End}(\Sigma\mathbb{H}P_2) \cong Q(M_{ee}, \lambda, \mu, k) \quad \text{with}$$

$$M_{ee} = \mathbb{Z}, \quad \lambda = 1, \quad \mu = -12, \quad k = 24.$$

EXAMPLE 4.3. For $\alpha = \sigma_8 \in \pi_{15}(S^8)$ one has

$$\mathrm{End}(\Sigma\mathbb{C}a) \cong Q(M_{ee}, \lambda, \mu, k) \quad \text{with}$$

$$M_{ee} = \mathbb{Z}, \quad \lambda = 1, \quad \mu = -120, \quad k = 240.$$

Hence the examples $\mathrm{End}(\Sigma\mathbb{C}P_2)$, $\mathrm{End}(\Sigma\mathbb{H}P_2)$ and $\mathrm{End}(\Sigma\mathbb{C}a)$ are special cases of the square ring Q in (4.1). The endomorphism square rings of $\Sigma\mathbb{C}P_2$, $\Sigma\mathbb{H}P_2$, $\Sigma\mathbb{C}a$ satisfy $P = 0$. The next examples satisfy $P \neq 0$.

EXAMPLE 4.4. For the double Hopf map $\alpha = \eta_3^2 \in \pi_5(S^3)$ one has

$$\mathrm{End}(\Sigma C_\alpha) \cong Q(M, \lambda, \mu, k) \quad \text{with}$$

$$M = (\mathbb{Z} \oplus \mathbb{Z}/6 \xrightarrow{(1,0)} \mathbb{Z} \xrightarrow{(2,1)} \mathbb{Z} \oplus \mathbb{Z}/6), \quad \lambda = 0, \quad \mu = 0, \quad k = 2.$$

EXAMPLE 4.5. For the Whitehead square $\alpha = [\iota_5, \iota_5] \neq 0$ in $\pi_9(S^5) = \mathbb{Z}/2$ one has

$$\mathrm{End}(\Sigma C_\alpha) \cong Q(M, \lambda, \mu, k) \quad \text{with}$$

$$M = (\mathbb{Z} \xrightarrow{2} \mathbb{Z} \xrightarrow{1} \mathbb{Z}), \quad \lambda = 0, \quad \mu = 1, \quad k = 1.$$

EXAMPLE 4.6. For the Whitehead square $\alpha = [\iota_8, \iota_8] \in \pi_{15}(S^8)$ one has

$$\mathrm{End}(\Sigma C_\alpha) \cong Q(M, \lambda, \mu, k) \quad \text{with}$$

$$M = (\mathbb{Z}/2 \oplus \mathbb{Z}/2 \oplus \mathbb{Z}/2 \xrightarrow{0} \mathbb{Z} \xrightarrow{(1,1,1)} \mathbb{Z}/2 \oplus \mathbb{Z}/2 \oplus \mathbb{Z}/2), \quad \lambda = 2, \quad \mu = -1, \quad k = 1.$$

In Theorem 3.13 of Oka, Sawashita and Sugawara [**OSS**] extending results of Oka [**O**], the group $\mathrm{Aut}(S^n \cup_f e^m)$ is computed up to an extension problem if $f = \Sigma\alpha$ is a suspension. Also in Theorem A of Yamaguchi [**Y**] and in Section 2 of Unsöld [**U**], the group $\mathrm{Aut}(\bigvee^k \Sigma^r \mathbb{C}P^2)$ is determined for $r \geq 1$. By our result we get:

PROPOSITION 4.7. *Let $\alpha \in \pi_{n-2}(S^{m-1})$ be with $n < 3m-3$. Then the group $\mathrm{Aut}(S^m \cup_f e^m)$ with $f = \Sigma\alpha$ is the group of units in the monoid Q_e determined by the square ring $Q = Q(M, \lambda, \mu, k)$ given by α. In fact, the monoid of self maps of $\Sigma C_\alpha = S^m \cup_f e^n$ coincides with the monoid Q_e. In addition the group*

$$\mathrm{Aut}(\bigvee^k \Sigma C_\alpha) \cong GL(Q, k).$$

is algebraically determined by the square ring $Q = \mathrm{End}(\Sigma C_\alpha)$ for $k \geq 1$.

We now consider for a metastable space ΣX the groups $M = [\Sigma X, W]$ where W is a pointed space. For $Q = \mathrm{End}(\Sigma X)$ in (4.1) we get the *operations*

$$(4.2) \qquad \begin{cases} [\Sigma X, W] \times [\Sigma X, \Sigma X] \longrightarrow [\Sigma X, W] \\ \\ [\Sigma X, W] \times [\Sigma X, W] \times [\Sigma X, \Sigma X \wedge X] \longrightarrow [\Sigma X, W] \end{cases}$$

which carry (m, a) and (m, n, x) to the composites $m \circ a$ and $[m, n] \circ x$ respectively where $[m, n] \in [\Sigma X \wedge X, W]$ is the Whitehead product of $m, n \in [\Sigma X, W]$. These operations give $M = [\Sigma X \wedge X, W]$ the following structure of a Q-module. The structure of $[\Sigma X \wedge X, W]$ as a Q-module determines completely the functor

$$\underline{\underline{\mathrm{Add}}}(\Sigma X)^{op} \longrightarrow \underline{\underline{\mathrm{Set}}}$$

which carries $\bigvee^k \Sigma X$ to the set of homotopy classes $[\bigvee^k \Sigma X, W]$; see [**BHP**].

DEFINITION 4.8. A *Q–module* M is given by a group M which we write additively and by Q-operations which are functions

$$M \times Q_e \longrightarrow M, \quad (m, a) \longmapsto m{\cdot}a,$$
$$M \times M \times Q_{ee} \longrightarrow M, \quad (m, n, x) \longmapsto [m, n]{\cdot}x.$$

For $a, b \in Q_e$, $x, y \in Q_{ee}$, $m, n \in M$ the following relations hold where $[M] = \{[m, n]{\cdot}x\,;\, m, n \in M,\, x \in Q_{ee}\} \subset M$:

$m{\cdot}1 = m$, $(m{\cdot}a){\cdot}b = m{\cdot}(a{\cdot}b)$, $m{\cdot}(a+b) = m{\cdot}a + m{\cdot}b$,
$(m+n){\cdot}a = m{\cdot}a + n{\cdot}a + [m, n]{\cdot}H(a)$,
$m{\cdot}P(x) = [m, m]{\cdot}x$,
$[m, n]{\cdot}T(x) = [n, m]{\cdot}x$,
$[m{\cdot}a, n{\cdot}b]{\cdot}x = [m, n]{\cdot}(a \otimes b){\cdot}x$ and $([m, n]{\cdot}x){\cdot}a = [m, n]{\cdot}(x{\cdot}a)$,
$[m, n]{\cdot}x$ is linear in m, n and x,
$[m, n]{\cdot}x = 0$ for $m \in [M]$.

These equations imply that the commutator in M satisfies

$$n + m - n - m = -n - m + n + m = [m, n]{\cdot}H(2).$$

Hence M is a group of nilpotency degree 2 and $[M]$ is central in M. Morphisms in the category $\underline{\underline{\mathrm{Mod}}}(Q)$ of Q-modules are homomorphisms $M \to M'$ which are compatible with the Q-operations.

Since we computed $\mathrm{End}(\Sigma C_\alpha)$ we then get:

PROPOSITION 4.9. *For $\alpha \in \pi_{n-2}(S^{m-1})$ with $n < 3m-3$ the group $[\Sigma C_\alpha, W]$ is a Q-module where $Q = Q(M, \lambda, \mu, k) \cong \mathrm{End}(\Sigma C_\alpha)$ is given by α as in section 3.*

REMARK 4.10. Let ΣX be a metastable space and let G be a connected topological group. Then the set $[X, B]$ of homotopy classes of maps from X to G

has the structure of a Q-module where $Q = \mathrm{End}(\Sigma X)$ is the endomorphism square ring of ΣX. This follows since we have natural isomorphism of groups

$$[X, G] = [X, \Omega BG] = [\Sigma X, BG]$$

where BG is the classifying space of G. In particular the groups

$$[\mathbb{R}P_2, G], \quad [\mathbb{C}P_2, G], \quad [\mathbb{H}P_2, G], \quad [\mathcal{C}a, G]$$

have the structure of a Q-module where Q is the endomorphism square ring for $\Sigma\mathbb{R}P_2$, $\Sigma\mathbb{C}P_2$, $\Sigma\mathbb{H}P_2$, $\Sigma\mathcal{C}a$ respectively; in fact, algebraic descriptions of these square rings are given in (8.2) of [**BHP**], (4.1), (4.2) and (4.3).

5. A quadratic action

The following three sections are purely algebraic. We study first a quadratic action denoted by $*$ which will be used in the next section for the computation of certain square rings. This way we show that the square ring $Q(M, \lambda, \mu, k)$ used in our main result is in fact a well defined square ring satisfying all properties in Definition 2.2.

Let R be an augmented ring with unit, i.e. two ring homomorphisms $\eta : \mathbb{Z} \to R$ and $\epsilon : R \to \mathbb{Z}$ are given to satisfy $\epsilon\eta = 1_{\mathbb{Z}}$. We write $\eta(\ell) = \ell$ for $\ell \in \mathbb{Z}$ and $\epsilon(a) = \tilde{a}$ for $a \in R$. For example the ring $R = \mathbb{Z} \times_k \mathbb{Z}$ is augmented by $\varepsilon(a) = \tilde{a} = a_0 \in \mathbb{Z}$ with $a = (a_0, a_1)$ and the unit is $1 = (1, 1)$.

A quadratic R-module $M = (M_e \xrightarrow{H} M_{ee} \xrightarrow{P} M_e)$ is given by right R-modules M_e, M_{ee} and R-linear homomorphisms H, P with $HPH = 2H$ and $PHP = 2P$.

For any quadratic R-module $M = (M_e \xrightarrow{H} M_{ee} \xrightarrow{P} M_e)$, we define a left action of R on M_e by

$$\tilde{a}{*}x = \tilde{a}x + \frac{\tilde{a}(\tilde{a}-1)}{2} PH(x),$$

$\tilde{a} \in \mathbb{Z}$ for $a \in R, x \in M_e$. Then the following proposition holds.

PROPOSITION 5.1. *The action $*$ satisfies following formulas.*

(1) $\tilde{a}{*}(\tilde{b}{*}x) = (\widetilde{ab}){*}x$
(2) $(\widetilde{a+b}){*}x = \tilde{a}{*}x + \tilde{b}{*}x + \tilde{a}\tilde{b}PH(x)$
(3) $\tilde{a}{*}(x+y) = \tilde{a}{*}x + \tilde{a}{*}y$
(4) $\tilde{a}{*}(x{\cdot}b) = (\tilde{a}{*}x){\cdot}b$
(5) $H(\tilde{a}{*}x) = \tilde{a}\tilde{a}H(x)$
(6) $\tilde{a}{*}P(x) = \tilde{a}\tilde{a}P(x)$

6. Square extension

We study square extensions which are motivated by the commutative diagram in Lemma 3.3. Let R be an augmented ring and let $M = (M_e \xrightarrow{H} M_{ee} \xrightarrow{P} M_e)$ be a quadratic R–module. For the group $Q_e = R \oplus M_e$ (given by the direct sum of the abelian groups R and M_e) we consider the following extension diagram

$$\begin{array}{ccccc} M_e & \xrightarrow{H} & M_{ee} & \xrightarrow{P} & M_e \\ {\scriptstyle i}\downarrow & & \| & & \downarrow \\ Q_e & \xrightarrow{\bar{H}} & M_{ee} & \xrightarrow{\bar{P}} & Q_e \\ {\scriptstyle s}\uparrow\downarrow{\scriptstyle \pi} & \nearrow_{h=\bar{H}s} & & & \\ R & & & & \end{array}$$

This is a generalization of the diagram in Lemma 3.3. Here i and s are the inclusions for $Q_e = R \oplus M_e$ and π is the projection. We now consider conditions on (H, P, h) which yield a square group and a square ring respectively.

PROPOSITION 6.1. *Given (H, P, h) as above, the following two conditions are equivalent.*

(1) $Q = (Q_e \xrightarrow{\bar{H}} M_{ee} \xrightarrow{\bar{P}} Q_e)$ *is a square group with*

$$\bar{H}(a, x) = h(a) + H(x) \quad \text{and} \quad \bar{P}(u) = iP(u) = (0, P(u))$$

for $a \in R$, $x \in M_e$ and $u \in M_{ee}$.

(2) *The data (H,P,h) satisfies the following conditions.*
  i) *A cross effect $\langle a|b\rangle_h = h(a+b) - h(a) - h(b)$ is linear in $a, b \in R$.*
  ii) *$P(\langle a|b\rangle_h) = 0$, in other words, $Ph(a)$ is linear in a.*

Proof: By the definitions of $\bar{H}$ and cross effects, we have

$$\begin{aligned} \langle (a,x)|(b,y)\rangle_{\bar{H}} &= \bar{H}((a,x)+(b,y)) - \bar{H}(a,x) - \bar{H}(b,y) \\ &= h(a+b) + H(x+y) - h(a) - H(x) - h(b) - H(y) = h(a+b) - h(a) - h(b). \end{aligned}$$

Thus we have

$$\langle (a,x)|(b,y)\rangle_{\bar{H}} = \langle a|b\rangle_h. \tag{6.1}$$

Suppose (1). The condition (1) of Definition 2.1 implies the condition (2i) by (6.1). The condition (2) of Definition 2.1 implies $\bar{P}(\langle a,b\rangle_h) = a + b - a - b = 0$, and hence we have (2ii).

Conversely suppose (2). By (6.1), the condition (2i) implies the condition (2) of Definition 2.1. The condition (1) of Definition 2.1 is a direct consequence of $\operatorname{im}\bar{P} = 0 \oplus \operatorname{im} P \subset 0 \oplus M_e = \ker \pi$ and (6.1). The condition (2) of Definition 2.1 is obtained by (6.1) and the condition (2ii). The condition (3) of Definition 2.1 is automatically satisfied since M is a quadratic R-module. *qed.*

For a given quadratic R-module $M = (M_e \xrightarrow{H} M_{ee} \xrightarrow{P} M_e)$, let $\bar{R} = R \oplus (M_e/\operatorname{im}(P))$ and let

$$\bar{\epsilon} : R \oplus M_e \to R \oplus (M_e/\operatorname{im}(P)) = \bar{R} \text{ and}$$

$$p : \bar{R} = R \oplus (M_e/\operatorname{im}(P)) \to R$$

be the canonical projections. We write $\bar{\epsilon}(a, x) = \overline{(a, x)}$. There is an action of $R \otimes R \otimes R^{op}$ on M_{ee} given by

$$(t \otimes s)\cdot u \cdot r = \widetilde{ts}(u \cdot r), \tag{6.2}$$

which makes M_{ee} an $R \otimes R \otimes R^{op}$-module. We show the following theorem.

THEOREM 6.2. *Given (H, P, h) as above and a fuction $\Delta : R\times R \to M_e$ with $\Delta(0,b) = \Delta(1,b) = \Delta(a,0) = \Delta(a,1) = 0$ the following two statements are equivalent.*

(1) $Q = (Q_e \xrightarrow{\bar{H}} M_{ee} \xrightarrow{\bar{P}} Q_e)$ *is a square ring with*

$$\bar{H}(a,x) = h(a) + H(x) \quad \text{and} \quad \bar{P}(u) = (0, P(u))$$

for $a \in R$, $x \in M_e$ and $u \in M_{ee}$ and multiplication $\circ$ of the monoid Q_e given by

$$(a,x)\circ(b,y) = (ab, x\cdot b + \tilde{a}*y + \Delta(a,b))$$

for $a,b \in R$, $x,y \in M_e$, which yields a ring structure on $\bar{R} = R\oplus(M_e/\operatorname{im}(P)) = Q_e/\operatorname{im}(\bar{P})$ as an extension of R with the action of $\bar{R} \otimes \bar{R} \otimes \bar{R}^{op}$ on M_{ee} through $p \otimes p \otimes p$.

(2) *The data (H,P,h,Δ) satisfy the following conditions.*
   i) $Ph(2) = 0$,
   ii) $\langle a|b\rangle_h = \tilde{a}\tilde{b}h(2)$,
   iii) $h(a\cdot b) + H(\Delta(a,b)) = \tilde{a}\tilde{a}h(b) + h(a)\cdot b$,
   iv) $\Delta(a,b)\cdot c + \Delta(ab,c) = \tilde{a}*\Delta(b,c) + \Delta(a,bc)$,
   v) $\Delta(a,b+c) = \Delta(a,b) + \Delta(a,c)$,
   vi) $\Delta(a+b,c) = \Delta(a,c) + \Delta(b,c) + \tilde{a}\tilde{b}Ph(c)$.

We call the data (H, P, h, Δ) with the properties in the theorem a *square extension.*

Proof: Firstly, we observe the properties of the multiplication $\circ$ on $Q_e = R\oplus M_e$.

$$(a,x)\circ(1,0) = (a, x\cdot 1 + \Delta(a,1)) = (a,x),$$
$$(1,0)\circ(b,y) = (b, \tilde{1}*y + \Delta(1,b)) = (b,y).$$

Thus $(1,0)$ gives the two-sided unit for $\circ$. The following equations illustrates the conditions for $\circ$ to satisfy the associativity law in Q_e.

$$\begin{aligned}
(a,x)\circ((b,y)\circ(c,z)) &= (a,x)\circ(bc, \tilde{b}*z + y\cdot c + \Delta(b,c)) \\
&= (abc, \tilde{a}*(\tilde{b}*z + y*c + \Delta(b,c)) + x\cdot(bc) + \Delta(a,bc)) \\
&= (abc, (\tilde{a}\tilde{b})*z + (\tilde{a}*y)\cdot c + x\cdot(bc) + \tilde{a}*\Delta(b,c) + \Delta(a,bc)), \\
((a,x)\circ(b,y))\circ(c,z) &= (ab, \tilde{a}*y + x\cdot b + \Delta(a,b))\circ(c,z) \\
&= (abc, (\tilde{a}\tilde{b})*z + (\tilde{a}*y)\cdot c + (x\cdot b)\cdot c + \Delta(a,b)\cdot c + \Delta(ab,c)).
\end{aligned}$$

Thus $\circ$ gives a monoid structure on Q_e if and only if the condition (2iv) is satisfied. Next we see $Q_e\circ\operatorname{im}(\bar{P}) \subset \operatorname{im}(\bar{P})$ and $\operatorname{im}(\bar{P})\circ Q_e \subset \operatorname{im}(\bar{P})$. For $(a,x),(b,y) \in Q_e$ and $u \in M_{ee}$, we have the following equations by the definition of $\circ$:

$$(a,x)\circ(0,P(u)) = (0, \tilde{a}*P(u) + x\cdot 0 + \Delta(a,0)) = (0, \tilde{a}\tilde{a}P(u)) = \bar{P}(\tilde{a}\tilde{a}u),$$
$$(0,P(u))\circ(b,y) = (0, \tilde{0}*x + P(u)\cdot b + \Delta(0,b)) = (0, P(u)\cdot b) = \bar{P}(u\cdot b).$$

Thus $\circ$ induces a monoid structure also on $\bar{R}$ such that the canonical projection $\bar{\epsilon} : Q_e \to \bar{R} = Q_e/\operatorname{im}(\bar{P})$ preserves the monoid structures. Moreover the following equations illustrates the conditions for $\circ$ to satisfy the distributive laws in Q_e.

$$
\begin{aligned}
((a,x)+(b,y))\circ(c,z) &= (a+b,x+y)\circ(c,z)\\
&= ((a+b)c, \widetilde{(a+b)}*z + (x+y)\cdot c + \Delta(a+b,c))\\
&= (ac+bc, (\tilde{a}+\tilde{b})*z + x\cdot c + y\cdot c + \Delta(a+b,c))\\
&= (ac+bc, \tilde{a}*z + x\cdot c + \tilde{b}*z + y\cdot c + \Delta(a+b,c) + \tilde{a}\tilde{b}PH(z)),\\
(a,x)\circ(c,z)+(b,y)\circ(c,z) &= (ac, \tilde{a}*z + x\cdot c + \Delta(a,c)) + (bc, \tilde{b}*z + y\cdot c + \Delta(b,c))\\
&= (ac+bc, \tilde{a}*z + x\cdot c + \tilde{b}*z + y\cdot c + \Delta(a,c) + \Delta(b,c)),\\
(a,x)\circ((b,y)+(c,z)) &= (a,x)\circ(b+c,y+z)\\
&= (a(b+c), \tilde{a}*(y+z) + x\cdot(b+c) + \Delta(a,b+c))\\
&= (ab+ac, \tilde{a}*y + x\cdot b + \tilde{a}*z + x\cdot c + \Delta(a,b+c))\\
(a,x)\circ(b,y)+(a,x)\circ(c,z) &= (ab, \tilde{a}*y + x\cdot b + \Delta(a,b)) + (ac, \tilde{a}*z + x\cdot c + \Delta(a,c))\\
&= (ab+ac, \tilde{a}*y + x\cdot b + \tilde{a}*z + x\cdot c + \Delta(a,b) + \Delta(a,c)).
\end{aligned}
$$

Thus the multiplication $\circ$ gives a monoid structure in Q_e with conditions (6) and (7) of Definition 2.2 if and only if the conditions (2v) and (2vi) are satisfied. Hence the conditions (2iv), (2v) and (2vi) imply that the multiplication $\circ$ induces a ring structure on $\bar{R}$ such that the canonical projection $p : \bar{R} \to R$ as well as $\bar{\epsilon} : Q_e \to \bar{R}$ preserves the ring structures, which induces an action of $\bar{R}\otimes\bar{R}\otimes\bar{R}^{op}$ on M_{ee} through $p\otimes p\otimes p : \bar{R}\otimes\bar{R}\otimes\bar{R}^{op} \to R\otimes R\otimes R^{op}$: For any $\overline{(a,x)}, \overline{(b,y)} \in \bar{R}$, we have

$$p(\overline{(a,x)}\circ\overline{(b,y)}) = p(\overline{(a,x)\circ(b,y)}) = p(\overline{(ab, x\cdot b + \tilde{a}*y + \Delta(a,b))}) = ab.$$

Also the conditions (2iv), (2v) and (2vi) imply conditions (2), (3) and (4) of Definition 2.2:

$$
\begin{aligned}
TT(u) &= HP(T(u)) - T(u)\\
&= \bar{H}\bar{P}(HP(u)-u) - (HP(u)-u) = HP(u) - HP(u) + u = u,\\
T((\overline{(a,x)}\otimes\overline{(b,y)})\cdot u\cdot\overline{(c,z)}) &= T(\tilde{a}\tilde{b}u\cdot c) = HP(\tilde{a}\tilde{b}u\cdot c) - \tilde{a}\tilde{b}u\cdot c\\
&= \tilde{a}\tilde{b}HP(u)\cdot c - \tilde{a}\tilde{b}u\cdot c = \tilde{a}\tilde{b}(HP(u)-u)\cdot c = \tilde{b}\tilde{a}T(u)\cdot c\\
&= (\overline{(b,y)}\otimes\overline{(a,x)})\cdot\bar{T}(u)\cdot\overline{(c,z)},
\end{aligned}
$$

$$
\begin{aligned}
&\bar{P}(u\cdot\overline{(a,x)}) = (0, P(u\cdot a)) = (0, P(u)\cdot a) = \bar{P}(u)\cdot\overline{(a,x)},\\
&\bar{P}((\overline{(a,x)}\otimes\overline{(a,x)})\cdot u) = (0, P(\tilde{a}\tilde{a}u)) = (0, \tilde{a}\tilde{a}P(u)) = (0, \tilde{a}*P(u)) = (a,x)\circ\bar{P}(u),
\end{aligned}
$$

where $T = \bar{H}\bar{P} - 1 = HP - 1$. Thus the conditions (2), (3), (4), (6) and (7) of Definition 2.2 are satisfied with inducing a ring structure on $\bar{R}$ with an action on M_{ee} via $p\otimes p\otimes p$ if and only if $\circ$ gives a multiplication with the conditions (2iv) (2v) and (2vi).

Secondly, Proposition 6.1 shows that the conditions (2i) and (2ii) are necessary and sufficient conditions for Q_e to be square group satisfying the condition (1) of Definition 2.2, since we have

$$
\begin{aligned}
\langle (a,x)|(b,y)\rangle_{\bar{H}} &= \bar{H}((a,x)+(b,y)) - \bar{H}(a,x) - \bar{H}(b,y)\\
&= \bar{H}(a+b,x+y) - \bar{H}(a,x) - \bar{H}(b,y)\\
&= h(a+b) + H(x+y) - H(x) - h(a) - H(y) - h(b)\\
&= \langle a|b\rangle_h = \tilde{a}\tilde{b}h(2) = \tilde{a}\tilde{b}\bar{H}(2),
\end{aligned}
$$

if (2ii) is satisfied.

Finally, the condition (2iii) is equivalent to the condition (5) of Definition 2.2, since

$$\begin{aligned}
\bar{H}((a,x)\circ(b,y)) &= \bar{H}(ab, x\cdot b + \tilde{a}*y + \Delta(a,b)) \\
&= h(ab) + H(x)\cdot b + H(\tilde{a}*y) + H(\Delta(a,b)) \\
&= h(ab) + H(\Delta(a,b)) + \tilde{a}\tilde{a}H(y) + H(x)\cdot b \\
(\overline{(a,x)}\otimes\overline{(a,x)})\cdot\bar{H}(b,y) + \bar{H}(a,x)\cdot\overline{(b,y)} &= \tilde{a}\tilde{a}(h(b)+H(y)) + (h(a)+H(x))\cdot\overline{(b,y)} \\
&= \tilde{a}\tilde{a}h(b) + h(a)\cdot b + \tilde{a}\tilde{a}H(y) + H(x)\cdot b
\end{aligned}$$

This completes the proof of the theorem. *qed.*

7. The square ring $Q(M,\lambda,\mu,k)$

For $k \geq 1$ let $R = \mathbb{Z}\times_k\mathbb{Z}$ be the subring of $\mathbb{Z}\times\mathbb{Z}$ consisting of all pairs $a = (a_0, a_1)$ with $a_0 - a_1 \equiv 0 \mod k$. This is the pull back ring of $\mathbb{Z} \longrightarrow \mathbb{Z}/k \longleftarrow \mathbb{Z}$. Then $\eta(\ell) = (\ell,\ell) \in \mathbb{Z}\times_k\mathbb{Z} = R$ gives the unit $1 = \eta(1) = (1,1)$. The augmentation $\epsilon : R = \mathbb{Z}\times_k\mathbb{Z} \longrightarrow \mathbb{Z}$ is defined by $\epsilon(a) = a_0$ for $a = (a_0, a_1)$. A free $\mathbb{Z}$-basis of $R = \mathbb{Z}\times_k\mathbb{Z} \cong \mathbb{Z}\oplus\mathbb{Z}$ is given by 1 and $\bar{k}$ where $\bar{k} = (0,k)$ a generator of $\ker(\epsilon)$.

PROPOSITION 7.1. *Let $M = (M_e \xrightarrow{H} M_{ee} \xrightarrow{P} M_e)$ be a quadratic $\mathbb{Z}$-module. Also let $k \in \mathbb{N}$ and $\lambda, \mu \in M_{ee}$ with relations $P(\lambda) = 0$ and $HP(\mu) = 2\mu + k\lambda$ be given. Then we obtain a square extension (H, P, h, Δ) as follows. Let*

(1) *the right action of R on M_e and M_{ee} be the multiplication given as $x\cdot(a_0, a_1) = a_1 x$ so that the homomorphisms H and P are R-linear and*

(2) *$h : \mathbb{Z}\times_k\mathbb{Z} \longrightarrow M_{ee}$ be the unique quadratic function satisfying*

$$\begin{cases} h(1) = 0, \quad h(\bar{k}) = \mu, \\ \langle a|b\rangle_h = a_0 b_0 \lambda \end{cases}$$

for $a, b \in \mathbb{Z}\times_k\mathbb{Z}$. Moreover let

(3) *$\Delta : (\mathbb{Z}\times_k\mathbb{Z}) \times (\mathbb{Z}\times_k\mathbb{Z}) \longrightarrow M_e$ be defined by*

$$\Delta((a_0,a_1),(b_0,b_1)) = \frac{a_0(a_0-1)(b_1-b_0)}{2k}P(\mu). \tag{7.1}$$

Since (H, P, h, Δ) is a square extension we thus obtain by theorem 6.2 the well defined square ring

$$Q(M,\lambda,\mu,k) = ((\mathbb{Z}\times_k\mathbb{Z})\oplus M_e \xrightarrow{\bar{H}} M_{ee} \xrightarrow{\bar{P}} (\mathbb{Z}\times_k\mathbb{Z})\oplus M_e)$$

which coincides with Definition 3.4.

Proof: One can easily check all the necessary conditions as follows. Firstly we give the following relations which makes our computations easy.

$$h(a_0,a_1) = \frac{a_0(a_0-1)}{2}\lambda + \frac{a_1-a_0}{k}\mu, \quad \text{and}$$

$$h(2) = h(2,2) = \lambda \quad \text{and} \quad Ph(a_0,a_1) = \frac{a_1-a_0}{k}P(\mu).$$

Then Definition 6.2(2i) and 6.2(2ii) are obtained by the equations $P(h(2)) = P(\lambda) = 0$ and $\langle a|b\rangle_h = a_0b_0\lambda = a_0b_0h(2)$.

Definition 6.2(2iii) is obtained as follows.

$$\begin{aligned}
&\widetilde{(a_0,a_1)}\widetilde{(a_0,a_1)}h(b_0,b_1)+h(a_0,a_1)\cdot(b_0,b_1)={a_0}^2h(b_0,b_1)+b_1h(a_0,a_1)\\
&={a_0}^2\frac{b_0(b_0-1)}{2}\lambda+{a_0}^2\frac{b_1-b_0}{k}\mu+\frac{a_0(a_0-1)}{2}b_1\lambda+\frac{a_1-a_0}{k}b_1\mu\\
&=\frac{{a_0}^2b_0(b_0-1)}{2}\lambda+\frac{a_0(a_0-1)}{2}b_0\lambda+\frac{a_0(a_0-1)(b_1-b_0)}{2k}k\lambda+\frac{{a_0}^2(b_1-b_0)+(a_1-a_0)b_1}{k}\mu\\
&=\frac{{a_0}^2{b_0}^2-a_0b_0}{2}\lambda+\frac{a_0(a_0-1)(b_1-b_0)}{2k}(HP(\mu)-2\mu)+\frac{{a_0}^2(b_1-b_0)+(a_1-a_0)b_1}{k}\mu\\
&=\frac{(a_0b_0)^2-a_0b_0}{2}\lambda+\frac{a_1b_1-a_0b_0}{k}\mu+H(\frac{a_0(a_0-1)(b_1-b_0)}{2k}P(\mu))\\
&=h((a_0b_0,a_1b_1))+H(\Delta((a_0,a_1),(b_0,b_1)))=h((a_0,a_1)\cdot(b_0,b_1))+H(\Delta((a_0,a_1),(b_0,b_1))).
\end{aligned}$$

Definition 6.2(2iv) is obtained as follows.

$$\begin{aligned}
&\Delta((a_0,a_1),(b_0,b_1))\cdot(c_0,c_1)+\Delta((a_0,a_1)(b_0,b_1),(c_0,c_1))\\
&\quad=c_1\frac{a_0(a_0-1)(b_1-b_0)}{2k}P(\mu)+\frac{a_0b_0(a_0b_0-1)(c_1-c_0)}{2k}P(\mu)\\
&\quad=\frac{a_0(a_0-1)(b_1-b_0)c_1+a_0b_0(a_0b_0-1)(c_1-c_0)}{2k}P(\mu),
\end{aligned}$$

and hence

$$\begin{aligned}
&\widetilde{(a_0,a_1)}*\Delta((b_0,b_1),(c_0,c_1))+\Delta((a_0,a_1),(b_0,b_1)(c_0,c_1))\\
&\quad=a_0*(\frac{b_0(b_0-1)(c_1-c_0)}{2k}P(\mu))+\frac{a_0(a_0-1)(b_1c_1-b_0c_0)}{2k}P(\mu)\\
&\quad=\frac{a_0a_0b_0(b_0-1)(c_1-c_0)+a_0(a_0-1)(b_1c_1-b_0c_0)}{2k}P(\mu)\\
&\quad=\frac{a_0b_0(a_0b_0-1)(c_1-c_0)+a_0(a_0-1)(b_1-b_0)c_1}{2k}P(\mu)\\
&\quad=\Delta((a_0,a_1),(b_0,b_1))\cdot(c_0,c_1)+\Delta((a_0,a_1)(b_0,b_1),(c_0,c_1)).
\end{aligned}$$

Definition 6.2(2v) is obtained as follows.

$$\Delta((a_0,a_1),(b_0,b_1)+(c_0,c_1))=\frac{a_0(a_0-1)(b_1+c_1-b_0-c_0)}{2k}P(\mu),$$

and hence

$$\begin{aligned}
&\Delta((a_0,a_1),(b_0,b_1))+\Delta((a_0,a_1),(c_0,c_1))=\frac{a_0(a_0-1)(b_1-b_0)}{2k}P(\mu)+\frac{a_0(a_0-1)(c_1-c_0)}{2k}P(\mu)\\
&=\frac{a_0(a_0-1)(b_1-b_0)+a_0(a_0-1)(c_1-c_0)}{2k}P(\mu)=\frac{a_0(a_0-1)(b_1+c_1-b_0-c_0)}{2k}P(\mu)\\
&=\Delta((a_0,a_1),(b_0,b_1)+(c_0,c_1)).
\end{aligned}$$

Definition 6.2(2vi) is obtained as follows.

$$\begin{aligned}
\Delta((a_0, a_1) + (b_0, b_1), (c_0, c_1)) &= \Delta((a_0 + b_0, a_1 + b_1), (c_0, c_1)) \\
&= \frac{(a_0 + b_0)(a_0 + b_0 - 1)(c_1 - c_0)}{2k} P(\mu) \\
&= \frac{a_0(a_0 - 1)(c_1 - c_0)}{2k} P(\mu) + \frac{b_0(b_0 - 1)(c_1 - c_0)}{2k} P(\mu) + \frac{2a_0 b_0 (c_1 - c_0)}{2k} P(\mu) \\
&= \Delta((a_0, a_1), (c_0, c_1)) + \Delta((b_0, b_1), (c_0, c_1)) + a_0 b_0 \frac{(c_1 - c_0)}{k} P(\mu) \\
&= \Delta((a_0, a_1), (c_0, c_1)) + \Delta((b_0, b_1), (c_0, c_1)) + a_0 b_0 P(h(c_0, c_1))
\end{aligned}$$

qed.

8. Proof of Lemma 3.3

Now let $\alpha \in \pi_{n-2}(S^{m-1})$ be an element as in (1.4) which induces the following cofibration sequence.

$$S^{n-2} \xrightarrow{\alpha} S^{m-1} \xrightarrow{i} C_\alpha \xrightarrow{j} S^{n-1}.$$

We give here a picture of related maps and Hopf invariants.

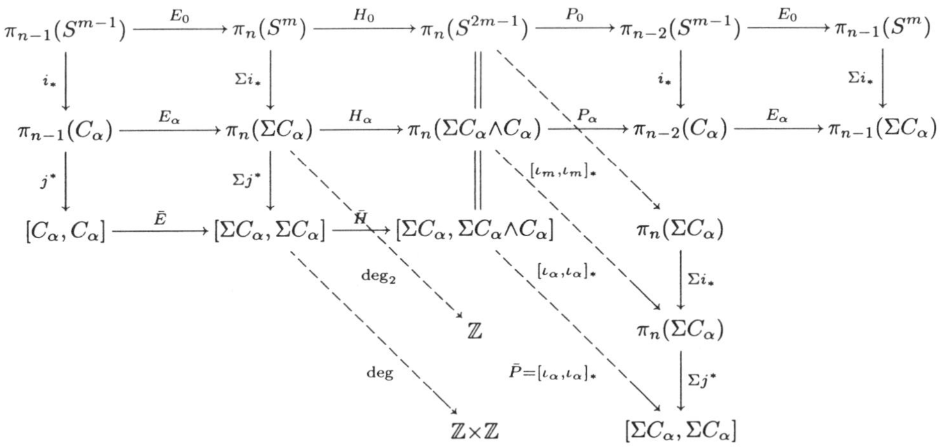

where $\deg : [\Sigma C_\alpha, \Sigma C_\alpha] \to \mathrm{End}(H^m(\Sigma C_\alpha)) \times \mathrm{End}(H^n(\Sigma C_\alpha)) = \mathbb{Z} \times \mathbb{Z}$ and $\deg_2 : \pi_n(\Sigma C_\alpha) \to \mathrm{Hom}(H^n(S^n), H^n(\Sigma C_\alpha)) = \mathbb{Z}$ is taking the degree of a map. We remark that the EHP sequences as rows are exact, when $n < 3m - 3$ by Toda [**T**] with H_0 and H_α the Hilton-Hopf invariants.

PROPOSITION 8.1. *The homomorphism* deg *has its image in the pull back ring* $R = \mathbb{Z} \times_k \mathbb{Z}$, *where* k *is the order of the suspension element* $\Sigma\alpha$ *in the group* $\pi_{n-1}(S^m)$.

Proof: A map $f : \Sigma C_\alpha \to \Sigma C_\alpha$ with $\deg(f) = (d_0, d_1)$ induces the commutative diagram

$$
(8.1)\qquad
\begin{array}{ccccccc}
S^{n-1} & \xrightarrow{\Sigma\alpha} & S^m & \xrightarrow{\Sigma i} & \Sigma C_\alpha & \xrightarrow{\Sigma j} & S^n \\
\downarrow{\scriptstyle d_1\iota_{n-1}} & & \downarrow{\scriptstyle d_0\iota_m} & & \downarrow{\scriptstyle f} & & \downarrow{\scriptstyle d_1\iota_n} \\
S^{n-1} & \xrightarrow{\Sigma\alpha} & S^m & \xrightarrow{\Sigma i} & \Sigma C_\alpha & \xrightarrow{\Sigma j} & S^n.
\end{array}
$$

Thus we have $d_1\Sigma\alpha = \Sigma\alpha\circ d_1\iota_{n-1} = d_0\iota_m\circ\Sigma\alpha = d_0\Sigma\alpha$, and hence $(d_1 - d_0)\Sigma\alpha = 0$, $d_1 - d_0 \equiv 0 \mod k$. *qed.*

LEMMA 8.2. *Let $\alpha \in \pi_{n-2}(S^{m-1})$. If $k\Sigma\alpha = 0$, then for any choice of an element $\mu_1 \in \pi_{n-2}(S^{2m-3})$ with $[\iota_{m-1}, \iota_{m-1}]_*\mu_1 = -k\alpha$, we can find out an element $\mu_0 \in \pi_n(\Sigma C_\alpha)$ with $H_\alpha(\mu_0) = \Sigma^2\mu_1$ and $\deg_2(\mu_0) = k$, i.e. the degree of $\mu_{0*} : H_n(S^n; \mathbb{Z}) \to H_n(\Sigma C_\alpha; \mathbb{Z})$ is k.*

Proof: We use the exact sequences of homotopy groups associated to the pairs $(\Omega S^m, S^{m-1})$ and $(\Omega\Sigma C_\alpha, C_\alpha)$, since there is the following commutative diagram (see **[W]** and **[B1]**).

$$
\begin{array}{ccccccc}
\pi_n(S^m) & \xrightarrow{H_0} & \pi_n(S^{2m-1}) & \xrightarrow{P_0} & \pi_{n-2}(S^{m-1}) & \xrightarrow{E_0} & \pi_{n-1}(S^m) \\
\| & & \uparrow{\scriptstyle J_2(S^{m-1})_*} & & \| & & \| \\
\pi_{n-1}(\Omega S^m) & \xrightarrow{\hat{H}_0} & \pi_{n-1}(\Omega S^m, S^{m-1}) & \xrightarrow{\hat{P}_0} & \pi_{n-2}(S^{m-1}) & \xrightarrow{\hat{E}_0} & \pi_{n-2}(\Omega S^m) \\
\downarrow{\scriptstyle \Omega\Sigma i_*} & & \|{\scriptstyle \Omega\Sigma i_*} & & \downarrow{\scriptstyle i_*} & & \downarrow{\scriptstyle \Omega\Sigma i_*} \\
\pi_{n-1}(\Omega\Sigma C_\alpha) & \xrightarrow{\hat{H}_\alpha} & \pi_{n-1}(\Omega\Sigma C_\alpha, C_\alpha) & \xrightarrow{\hat{P}_\alpha} & \pi_{n-2}(C_\alpha) & \xrightarrow{\hat{E}_\alpha} & \pi_{n-2}(\Omega\Sigma C_\alpha) \\
\| & & \downarrow{\scriptstyle J_2(C_\alpha)_*} & & \| & & \| \\
\pi_n(\Sigma C_\alpha) & \xrightarrow{H_\alpha} & \pi_n(\Sigma C_\alpha \wedge C_\alpha) & \xrightarrow{P_\alpha} & \pi_{n-2}(C_\alpha) & \xrightarrow{E_\alpha} & \pi_{n-1}(\Sigma C_\alpha)
\end{array}
$$

where $J_2(S^{m-1})_*$ and $J_2(C_\alpha)_*$ are surjective. Then we see that it is sufficient to show the existence of elements $\mu_1 \in \pi_{n-2}(S^{2m-3})$ and $\mu_0 \in \pi_n(\Sigma C_\alpha)$ with $[\iota_{m-1}, \iota_{m-1}]_*\mu_1 = -k\alpha$, $H_\alpha(\mu_0) = \Sigma^2\mu_1$ and $\deg_2(\mu_0) = k$. In fact, for any other element $\mu_1' \in \pi_{n-2}(S^{2m-3})$ with $[\iota_{m-1}, \iota_{m-1}]_*\mu_1' = -k\alpha$, there is an element $\mu_2' \in \pi_{n-1}(\Omega S^m, S^{m-1})$ with $J_2(S^{m-1})_*\mu_2' = \Sigma^2\mu_1'$, and hence $\hat{P}_0(\mu_2') = -k\alpha = \hat{P}_0\hat{H}_\alpha(\mu_0)$. Then we can take an element $\gamma \in \pi_n(S^m)$ with $\mu_2' = \hat{H}_\alpha(\mu_0) + H_0(\gamma) = \hat{H}_\alpha(\mu_0 + \gamma)$. By putting $\mu_0' = \mu_0 + \gamma \in \pi_{n-1}(C_\alpha)$, we get $H_\alpha(\mu_0') = \Sigma^2(\mu_1')$ with $\deg_2(\mu_0') = k$.

Since the order of $\Sigma\alpha$ is $k \geq 1$, we have $\hat{E}_0(k\alpha) = 0$, and hence there is an extension $\widehat{k\alpha} : (C(S^{n-2}), S^{n-2}) \to (\Omega S^m, S^{m-1})$ of $k\alpha$. Let $H : (C(S^{n-2}), S^{n-2}) \to (C_\alpha, S^{m-1})$ be a relative homeomorphism giving a null-homotopy of α. By adding k-copies of H, we obtain a null-homotopy $kH : (C(S^{n-2}), S^{n-2}) \to (C_\alpha, S^{m-1})$ of $k\alpha$ which gives the commutative diagram

$$
\begin{array}{ccc}
C(S^{n-2}) & \xrightarrow{kH} & C_\alpha \\
\downarrow & & \downarrow \\
S^{n-1} & \xrightarrow{k\iota_{n-1}} & S^{n-1},
\end{array}
$$

where the two columns are the canonical collapsions. Since the two maps $\widehat{k\alpha}$ and kH coincide on S^{n-2}, by gluing $-\widehat{k\alpha}$ with the direction altered to $kH : 0 \to k\alpha$ we get a new map

$$\mu_0 = kH - \widehat{k\alpha} : S^{n-1} \to \Omega\Sigma S^{m-1} \cup C_\alpha \subset \Omega\Sigma C_\alpha$$

which gives the following diagram commutative up to homotopy.

$$\begin{array}{ccccc} S^{n-1} & \xrightarrow{\mu_0} & \Omega\Sigma S^{m-1}\cup C_\alpha & \hookrightarrow & \Omega\Sigma C_\alpha \\ \Vert & & \downarrow & & \downarrow \\ S^{n-1} & \xrightarrow{k\iota_{n-1}} & S^{n-1} & = & S^{n-1}. \end{array}$$

By taking the adjoint we get $\mu_0 \in \pi_{n-1}(\Omega\Sigma C_\alpha)$ as an element in $\pi_n(\Sigma C_\alpha)$. Then by the definition of Hilton-Hopf invariant H_0, $H_0(\mu_0) = \mu_0$. *qed.*

Next we show the following Propositions.

PROPOSITION 8.3. *There is a central extension*

$$0 \longrightarrow \pi_n(S^m)/U_\alpha \longrightarrow [\Sigma C_\alpha, \Sigma C_\alpha] \xrightarrow{\deg} \mathbb{Z} \times_k \mathbb{Z} \longrightarrow 0$$

where the subgroup U_α of $\pi_n(S^m)$ is generated by the elements $\eta_m(\Sigma^2\alpha)$ and $(\Sigma\alpha)\eta_{n-1}$ with $\eta_t = \Sigma^{t-2}\eta \in \pi_{t+1}(S^t)$ the Hopf element, $t \geq 2$.

Proof: Since $\mathbb{Z} \times_k \mathbb{Z} \cong \mathbb{Z}\{(1,1)\} \oplus \mathbb{Z}\{(0,k)\}$ as modules, deg is surjective by Lemma 8.2.

Let $f : \Sigma C_\alpha \to \Sigma C_\alpha$ be an element in $\ker \deg$. Then $f \circ \Sigma i : S^m \to S^m$ must be trivial, since $\deg(f) = (0,0)$. Hence there exists a map $f_1 : S^n \to \Sigma C_\alpha$ such that $f \sim f_1 \circ \Sigma j$. We also see that $\Sigma j \circ f : S^n \to S^n$ is trivial. Hence there exists a map $f_0 : S^n \to S^m$ such that $f_1 \sim \Sigma i \circ f_0$ and $f \sim \Sigma i \circ f_0 \circ \Sigma j$. Conversely, an element $f_0 : S^n \to S^m$ induces $\Sigma i \circ f_0 \circ \Sigma j : \Sigma C_\alpha \to \Sigma C_\alpha$ which is in $\ker \deg$. Thus we have a short exact sequence

$$0 \longrightarrow \operatorname{im}(\Sigma j^* \circ \Sigma i_*) \longrightarrow [\Sigma C_\alpha, \Sigma C_\alpha] \xrightarrow{\deg} \mathbb{Z} \times_k \mathbb{Z} \longrightarrow 0 \tag{8.2}$$

and an isomorphism $\operatorname{im}(\Sigma j^* \circ \Sigma i_*) \cong \pi_n(S^m)/\ker \Sigma j^* \circ \Sigma i_*$. Since $S^n = \Sigma C_\alpha / S^m$ co-acts on ΣC_α, the image $\operatorname{im} \Sigma j^*$ is in the center of $[\Sigma C_\alpha, \Sigma C_\alpha]$, and hence so is $\operatorname{im}(\Sigma j^* \circ \Sigma i_*)$. Thus the short exact sequence (8.2) is a central extension.

So we are left to show that $\ker \Sigma j^* \circ \Sigma i_* = U_\alpha$. Let g be an element in $\ker \Sigma j^* \circ \Sigma i_*$, i.e, $\Sigma i \circ g \circ \Sigma j \sim 0$. Let us consider the diagram

$$\begin{array}{ccccc} \Sigma C_\alpha & \xrightarrow{\Sigma j} & S^n & \xrightarrow{\Sigma^2\alpha} & S^{m+1} \\ & & & \searrow^{g} & \\ S^{n-1} & \xrightarrow{\Sigma\alpha} & S^m & \xrightarrow{\Sigma i} & \Sigma C_\alpha \end{array}$$

with cofibration rows. Let $F_{\Sigma i} \xrightarrow{\widehat{\Sigma i}} S^m$ denote the homotopy fibre of Σi with CW decomposition

$$F_{\Sigma i} \simeq S^{n-1} \cup (\text{cells in dimension} \geq n + m - 2 > n),$$

where $\widehat{\Sigma i}|_{S^{n-1}} = \Sigma\alpha$. Since $\Sigma i \circ g \circ \Sigma j \sim 0$, there is a map $g_0 : \Sigma C_\alpha \to F_{\Sigma i}$ whose image is in $S^{n-1} \subset F_{\Sigma i}$ such that $g \circ \Sigma j \sim \widehat{\Sigma i} \circ g_0$. For the dimensional reasons, we can take an element $x\eta_{n-1} \in \pi_n(S^{n-1})$ such that $g_0 \sim x\eta_{n-1}\circ\Sigma j$, and hence $g\circ\Sigma j \sim x\Sigma\alpha\circ\eta_{n-1}\circ\Sigma j$. Thus we get that $(g - x\Sigma\alpha\circ\eta_{n-1})\circ\Sigma j \sim 0$. Then there exists an element $y\eta_m \in \pi_{m+1}(S^m)$ such that $g - x\eta_{n-1} \sim y\eta_m$, and hence g is in U_α. This implies that $\ker \Sigma j^*\circ\Sigma i_* \subseteq U_\alpha$. The converse is clear. *qed.*

PROPOSITION 8.4. *The epimorphism* deg *has a splitting homomorphism* $s : \mathbb{Z} \times_k \mathbb{Z} \to [\Sigma C_\alpha, \Sigma C_\alpha]$ *given by the following formula.*

$$s(a_0, a_1) = a_0\iota_\alpha + \frac{(a_1 - a_0)}{k}\mu_0\circ\Sigma j$$

Proof: Since $s(1,1) = \iota_\alpha$ and $s(0,k) = \mu_0\circ\Sigma j$, it is sufficient to show that ι_α and $\mu_0\circ\Sigma j$ commutes up to homotopy. To see this, we take adjoint maps of them. The adjoint of the identity ι_α is the canonical inclusion $\widehat{\iota_\alpha} : C_\alpha \hookrightarrow \Omega\Sigma C_\alpha$ and the adjoint of $\mu_0\circ\Sigma j$ is described as a composition $\widehat{\mu_0}\circ j$ where $\widehat{\mu_0} : S^{n-1} \to \Omega\Sigma C_\alpha$ is the adjoint of μ_0. Also the adjoint of the commutator of ι_α and $\mu_0\circ\Sigma j$ is given, up to sign, by the composition

$$\langle \widehat{\iota_\alpha}, \widehat{\mu_0}\circ j\rangle : C_\alpha \xrightarrow{\bar{\Delta}} C_\alpha \wedge C_\alpha \xrightarrow{1\wedge j} C_\alpha \wedge S^{n-1} \xrightarrow{\widehat{\iota_\alpha}\wedge\widehat{\mu_0}} \Omega\Sigma C_\alpha \wedge \Omega\Sigma C_\alpha \xrightarrow{c} \Omega\Sigma C_\alpha,$$

where $\bar{\Delta} : C_\alpha \to C_\alpha \wedge C_\alpha$ is the reduced diagonal map and c denotes the commutator of the first and second projections $\Omega\Sigma C_\alpha \times \Omega\Sigma C_\alpha \to \Omega\Sigma C_\alpha$. Since C_α is of dimension $n - 1 < n + m - 2$, we can compress $\bar{\Delta}$ into a subspace $S^{m-1}\wedge S^{m-1}$ which is collapsed in $C_\alpha \wedge S^{n-1}$. Thus we have $\langle \widehat{\iota_\alpha}, \widehat{\mu_0}\circ j\rangle \sim 0$, and hence ι_α and $\mu_0\circ\Sigma j$ are commutative up to homotopy. *qed.*

COROLLARY 8.5. *The group* $[\Sigma C_\alpha, \Sigma C_\alpha]$ *is an abelian group isomorphic with* $(\mathbb{Z} \times_k \mathbb{Z}) \times (\pi_n(S^m)/U_\alpha)$.

Proof: Let $\phi : (\mathbb{Z} \times_k \mathbb{Z}) \times (\pi_n(S^m)/U_\alpha) \to [\Sigma C_\alpha, \Sigma C_\alpha]$ be the homomorphism given by

$$\phi(a, x) = s(a) + \Sigma j^*\circ\Sigma i_*(x).$$

Then by Propositions 8.3 and 8.4, one can easily see that ϕ is an isomorphism of groups. Thus the group $[\Sigma C_\alpha, \Sigma C_\alpha]$ is an abelian group isomorphic with $(\mathbb{Z} \times_k \mathbb{Z}) \times (\pi_n(S^m)/U_\alpha)$. *qed.*

Thus for any choice of the data λ, μ in (3.1), the homomorphism s given in Proposition 8.4 gives a splitting of the homomorphism deg : $[\Sigma C_\alpha, \Sigma C_\alpha] \to R$ of abelian groups. We summarise the results obtained in this section as follows.

Lemma 3.3 *For the square ring* $End(\Sigma C_\alpha)$ *given by (1.2) and for the data* (M, λ, μ, k) *in (3.1) one gets a commutative diagram*

$$\begin{array}{ccccc}
M_e & \xrightarrow{H} & M_{ee} & \xrightarrow{P} & M_e \\
{\scriptstyle i}\downarrow & & \| & & \downarrow{\scriptstyle i} \\
[\Sigma C_\alpha, \Sigma C_\alpha] & \xrightarrow{\bar{H}} & [\Sigma C_\alpha, \Sigma C_\alpha \wedge C_\alpha] & \xrightarrow{\bar{P}} & [\Sigma C_\alpha, \Sigma C_\alpha] \\
{\scriptstyle s}\uparrow\downarrow{\scriptstyle \deg} & \nearrow{\scriptstyle h} & & & \\
R & & & &
\end{array}$$

where the column $M_e \overset{i}{\hookrightarrow} [\Sigma C_\alpha, \Sigma C_\alpha] \overset{\deg}{\to} R$ *is a split short exact sequence of abelian groups. For a splitting* s *of* deg, *let* $h = \bar{H}s$. *We can choose* s *such that* $s(1) = \iota_\alpha$

is the identity of ΣC_α *and* $s(\bar{k}) = \mu_0$ *such that* $\bar{H}(\mu_0) = h(\bar{k}) = \mu$ *and* $\bar{H}(1+1) = h(1+1) = \lambda$.

9. Proof of Theorem 3.2

Using Lemma 3.3 with notations in (3.1), we show Theorem 3.2

PROPOSITION 9.1. *We have the following formulae for any* $\ell \in \mathbb{Z}$, $f, f' \in [\Sigma C_\alpha, \Sigma C_\alpha]$ *and* $g, g' \in \pi_n(\Sigma C_\alpha)$.

(1) $(f + (g\circ\Sigma j))\circ f' = f\circ f' + (g\circ\Sigma j)\circ f'$.

(2) $f\circ s(\ell,\ell) = \ell f$.

(3) $s(\ell,\ell)\circ(g'\circ\Sigma j) = \ell g'\circ\Sigma j + \frac{\ell(\ell-1)}{2}\bar{P}\bar{H}(g'\circ\Sigma j)$.

(4) $(g\circ\Sigma j)\circ(g'\circ\Sigma j) = (\deg_2 g')g\circ\Sigma j$.

Proof: Firstly we show the formula (1). By a Hilton-Milnor theorem with $n \le 3m-3$ (see [**W**]), we have

$$(f + (g\circ\Sigma j))\circ f' = f\circ f' + (g\circ\Sigma j)\circ f' + [f, g\circ\Sigma j]\circ H_\alpha(f'),$$
$$[f, g\circ\Sigma j]\circ H_\alpha(f') = [f,g]\circ(\Sigma\iota_\alpha\wedge j)\circ(\Sigma i\wedge i)\circ H_\alpha(f') = 0.$$

Secondly we show the formula (2). For a map $f : \Sigma C_\alpha \to \Sigma C_\alpha$, we have $f\circ s(\ell,\ell) = f\circ(\ell\iota_\alpha) = f\circ(\iota_\alpha + \cdots \iota_\alpha) = f + \cdots + f = \ell f$.

Thirdly we show the formula (3). For $\ell = 2$, a Hilton-Milnor theorem with $n \le 3m-3$ (see [**W**]) implies

$$2\iota_\alpha\circ(g'\circ\Sigma j) = 2g'\circ\Sigma j + [\iota_m, \iota_m]\circ H_\alpha(g'\circ\Sigma j) = 2g\circ\Sigma j + \bar{P}\bar{H}(g'\circ\Sigma j),$$

By the induction on $\ell \ge 2$, we get the desired formula (3).

We show the last formula (4). We have $(g\circ\Sigma j)\circ(g'\circ\Sigma j) = g\circ(\Sigma j\circ g'\circ\Sigma j) = g\circ((\deg_2 g')\iota_n) = (\deg_2 g')f$. *qed.*

COROLLARY 9.2. *The splitting* $s : \mathbb{Z}\times_k\mathbb{Z} \to [\Sigma C_\alpha, \Sigma C_\alpha]$ *satisfies the following formulae.*

(1) $(g\circ\Sigma j)\circ s(a_0,a_1) = a_1(g\circ\Sigma j)$, *where* g *is in* $\pi_n(\Sigma C_\alpha)$ *or* $\pi_n(\Sigma C_\alpha\wedge C_\alpha)$.

(2) $s(a_0,a_1)\circ s(b_0,b_1) = s(a_0b_0, a_1b_1) + \frac{a_0(a_0-1)(b_1-b_0)}{2k}\bar{P}(H_\alpha(\mu_0)\circ\Sigma j)$.

Proof: The formula (1) is clear by the proof of Proposition 9.1 (4). So we show the formula (2) using Proposition 9.1 (1) through (4).

$$\begin{aligned}
s(a_0,a_1)\circ s(b_0,b_1) &= s(a_0,a_1)\circ(s(b_0,b_0) + s(0,b_1-b_0)) \\
&= s(a_0,a_1)\circ s(b_0,b_0) + s(a_0,a_1)\circ s(0,b_1-b_0) \\
&= b_0 s(a_0,a_1) + (s(a_0,a_0) + s(0,a_1-a_0))\circ s(0,b_1-b_0) \\
&= b_0 s(a_0,a_1) + s(a_0,a_0)\circ s(0,b_1-b_0) + s(0,a_1-a_0)\circ s(0,b_1-b_0) \\
&= s(a_0b_0,a_1b_0) + a_0 s(0,b_1-b_0) + \frac{b_1-b_0}{k}s(a_0,a_0)\circ s(0,k) + (b_1-b_0)s(0,a_1-a_0) \\
&= s(a_0b_0,a_1b_1) + \frac{(b_1-b_0)}{k}(\frac{a_0(a_0-1)}{2}\bar{P}(H_\alpha(\mu_0)\circ\Sigma j)) \\
&= s(a_0b_0,a_1b_1) + \frac{a_0(a_0-1)(b_1-b_0)}{2k}\bar{P}(H_\alpha(\mu_0)\circ\Sigma j)
\end{aligned}$$

qed.

For the two elements $\bar{H}(2\iota_\alpha), H_\alpha(\mu_0)\circ\Sigma j \in [\Sigma C_\alpha, \Sigma C_\alpha\wedge C_\alpha]$, the following holds.

PROPOSITION 9.3.

(1) $\bar{H}(2\iota_\alpha) = \Sigma^2 H(\alpha)$ *where* $\iota_\alpha : C_\alpha \to C_\alpha$ *denotes the identity.*

(2) $\bar{P}\bar{H}(2\iota_\alpha) = 0$.

(3) $\bar{H}\bar{P}(H_\alpha(\mu_0)\circ\Sigma j) = 2H_\alpha(\mu_0)\circ\Sigma j + k\bar{H}(2\iota_\alpha)$.

i) $\bar{H}\bar{P}(H_\alpha(\mu_0)\circ\Sigma j) = 0$ *and* $2H_\alpha(\mu_0)\circ\Sigma j + k\bar{H}(2\iota_\alpha) = 0$ *when* m *is odd.*

ii) $\bar{H}\bar{P}(H_\alpha(\mu_0)\circ\Sigma j) = 2\bar{H}(\mu_0)\circ\Sigma j$ *and* $k\bar{H}(2\iota_\alpha) = 0$ *when* m *is even.*

Proof: Firstly we show (1). By a Hilton-Milnor theorem with $n \leq 3m-3$ (see [**W**]), we have

$$(\iota_1 + \iota_2)\circ(2\iota_\alpha) = \iota_1\circ(2\iota_\alpha) + \iota_2\circ(2\iota_\alpha) + [\iota_1, \iota_2]\circ\bar{H}(2\iota_\alpha),$$

where $\iota_t : \Sigma C_\alpha \to \Sigma C_\alpha \vee \Sigma C_\alpha$ is the inclusion to the t-th factor. Since $[\iota_1, \iota_2]\circ\bar{H}(2\iota_\alpha)$ is in the center of the group $[\Sigma C_\alpha, \Sigma C_\alpha \vee \Sigma C_\alpha]$ for dimensional reasons, we have the relation

$$\iota_1 + \iota_2 - \iota_1 - \iota_2 = [\iota_2, \iota_1]\circ\bar{H}(2\iota_\alpha).$$

The adjoint of $[\iota_2, \iota_1]$ is given by a Samelson product (commutator) $c_\alpha : C_\alpha \wedge C_\alpha \to \Omega\Sigma C_\alpha$ of the adjoints of ι_1 and ι_2. Also the adjoint of $\iota_1 + \iota_2 - \iota_1 - \iota_2$ is given by the composition of the commutator c_α with reduced diagonal map $\hat{\Delta}_2 : C_\alpha \to C_\alpha \wedge C_\alpha$ which is given by the suspension of the Hilton-Hopf invariant $H_0(\alpha)$ (see Theorem 5.14 of Boardmann and Steer [**BS**]). Thus we have $[\iota_2, \iota_1]\circ\Sigma^2 H_0(\alpha) = [\iota_2, \iota_1]\circ\bar{H}(2\iota_\alpha)$, and hence $\bar{H}(2\iota_\alpha) = \Sigma^2 H_0(\alpha)$.

Secondly we show (2). By a Hilton-Milnor theorem with $n \leq 3m-3$ (see [**W**]), we have

$$(2\iota_\alpha)f = 2f + [\iota_\alpha, \iota_\alpha]\circ\bar{H}(f).$$

For $f = \ell\iota_\alpha$ with $\ell \in \mathbb{Z}$, we then have $2\ell\iota_\alpha = 2\ell\iota_\alpha + [\iota_\alpha, \iota_\alpha]\circ\bar{H}(\ell\iota_\alpha)$, and hence we have $\bar{P}\bar{H}(\ell\iota_\alpha) = [\iota_\alpha, \iota_\alpha]\circ\bar{H}(\ell\iota_\alpha) = 0$.

So we are left to show (3). For the dimensional reasons, Hilton-Hopf invariant satisfies the following derivation formula.

$$\bar{H}(\ell\iota_\alpha\circ(\mu_0\circ\Sigma j)) = \bar{H}(\ell\iota_\alpha)\circ(\mu_0\circ\Sigma j) + (\Sigma(\ell\iota_m)\wedge(\ell\iota_m))\circ(H_\alpha(\mu_0)\circ\Sigma j)$$

Since the image of $\bar{H}(\ell\iota_\alpha)$ is in S^{2m-1}, it factors through $\Sigma j : \Sigma C_\alpha \to S^n$, and $\Sigma j\circ\mu_0\circ\Sigma j = (k\iota_n)\circ\Sigma j = k\Sigma j$. It then follows that $\bar{H}(\ell\iota_\alpha)\circ(\mu_0\circ\Sigma j) = k\bar{H}(\ell\iota_\alpha)$, and hence $\bar{H}((\ell\iota_\alpha)\circ(\mu_0\circ\Sigma j)) = k\bar{H}(\ell\iota_\alpha) + \ell^2 H_\alpha(\mu_0)\circ\Sigma j$. By putting $\ell = 2$, we get

$$\bar{H}((2\iota_\alpha)\circ(\mu_0\circ\Sigma j)) = k\bar{H}(2\iota_\alpha) + 4H_\alpha(\mu_0)\circ\Sigma j.$$

On the other hand, by a Hilton-Milnor theorem with $n \leq 3m-3$ (see [**W**]), we have

$$(2\iota_\alpha)\circ\mu_0 = 2\mu_0 + [\iota_\alpha, \iota_\alpha]\circ H_\alpha(\mu_0) = 2\mu_0 + [\iota_m, \iota_m]\circ H_\alpha(\mu_0)$$

and hence

$$\begin{aligned}\bar{H}((2\iota_\alpha)\circ(\mu_0\circ\Sigma j)) &= H_\alpha((2\iota_\alpha)\circ\mu_0)\circ\Sigma j \\ &= 2H_\alpha(\mu_0)\circ\Sigma j + \Sigma i\circ H_0([\iota_m, \iota_m]\circ H_\alpha(\mu_0))\circ\Sigma j = 2H_\alpha(\mu_0)\circ\Sigma j + \bar{H}\bar{P}(H_\alpha(\mu_0)\circ\Sigma j)\end{aligned}$$

where $\bar{H}\bar{P}(H_\alpha(\mu_0)\circ\Sigma j) = \Sigma i\circ H_0([\iota_m, \iota_m])\circ H_\alpha(\mu_0)\circ\Sigma j$. Thus we get the relation $2H_\alpha(\mu_0)\circ\Sigma j + \bar{H}\bar{P}(H_\alpha(\mu_0)\circ\Sigma j) = k\bar{H}(2\iota_\alpha) + 4H_\alpha(\mu_0)\circ\Sigma j$, and hence $\bar{H}\bar{P}(H_\alpha(\mu_0)) = k\bar{H}(2\iota_\alpha) + 2H_\alpha(\mu_0)\circ\Sigma j$.

In case when m is odd, the Whitehead square $[\iota_m, \iota_m]$ has order 2, an¡d hence its Hilton-Hopf invariant is trivial. Thus $\bar{H}\bar{P}(H_\alpha(\mu_0)\circ\Sigma j) = 0$ and $k\bar{H}(2\iota_\alpha) + 2H_\alpha(\mu_0)\circ\Sigma j = 0$.

In case when m is even, the Whitehead square $[\iota_m, \iota_m]$ has order ∞ and its Hilton-Hopf invariant is 2. Thus $\bar{H}\bar{P}(H_\alpha(\mu_0)\circ\Sigma j) = 2H_\alpha(\mu_0)\circ\Sigma j$ and $k\bar{H}(2\iota_\alpha) = 0$. *qed.*

Then we define the quadratic $\mathbb{Z}$-module

(9.1)
$$M = (\pi_n(S^m)/U_\alpha \xrightarrow{H} \pi_n(S^{2m-1}) \xrightarrow{P} \pi_n(S^m)/U_\alpha) \quad \text{with elements}$$
$$\lambda \in \pi_n(S^{2m-1}), \quad \mu \in \pi_n(S^{2m-1}), \quad k \in \mathbb{N}$$

as in (3.1) above. With the notations in (1.4), Proposition 7.1 and (9.1) we get: *In terms of the data (M, λ, μ, k) associated to α in (9.1) we obtain by 7.1 the square ring $Q(M, \lambda, \mu, k)$ together with an isomorphism*

(9.2)
$$\mathrm{End}(\Sigma C_\alpha) \cong Q(M, \lambda, \mu, k)$$

of square rings.

Proof: We show that (M, λ, μ, k) satisfies the hypothesis in Proposition 7.1. Since a higher homotopy groups are abelian, M_e and M_{ee} are abelian groups. Also from the fact that H_0 and $[\iota_m, \iota_m]_*$ is a homomorphism, it follows that H and P are homomorphisms. For $x \in M_{ee} = \pi_n(S^{2m-1})$, since x is in stable range, $PHP(x) = P(H([\iota_m, \iota_m])\circ x)$. If m is even, $H([\iota_m, \iota_m]) = 2$ and $PHP(x) = 2P(x)$. But if m is odd, $H([\iota_m, \iota_m]) = 0$ and $PHP(x) = 0$ while the order of $P(x) = [\iota_m, \iota_m]\circ x$ is 2, and hence $PHP(x) = 0 = 2P(x)$. For $x \in M_e$, $H(x)$ is in $M_{ee} = \pi_n(S^{2m-1})$, and hence $H(x)$ is in stable range, $HPH(x) = H([\iota_m, \iota_m])\circ H(x)$. If m is even, $H([\iota_m, \iota_m]) = 2$ and $HPH(x) = 2H(x)$. But if m is odd, $H([\iota_m, \iota_m]) = 0$ and $HPH(x) = 0$. For dimensional reasons, there is an element $x_0 \in \pi_{n-2}(S^{2m-3})$ with $\Sigma^2 x_0 = H(x) = H_0(x)$ and $[\iota_{m-1}, \iota_{m-1}]_* x_0 = P_0(H_0(x)) = 0$. Taking its Hilton-Hopf invariant, we get $2x_0 = 0$ and $2H_0(x) = \Sigma^2(2x_0) = 0$, and hence $HPH(x) = 0 = 2H(x)$. By Proposition 9.3, it follows that (M, λ, μ, k) satisfy the required conditions to define a square extension.

So we are left to show that the isomorphism $\phi : (\mathbb{Z} \times_k \mathbb{Z}) \times (\pi_n(S^m)/U_\alpha) \to [\Sigma C_\alpha, \Sigma C_\alpha]$ in the proof of Corollary 8.5 carries the product given in Theorem 6.2 to the composition. For $a = (a_0, a_1), b = (b_0, b_1) \in \mathbb{Z}\times_k\mathbb{Z}$ and $[x], [y] \in \pi_n(S^m)/U_\alpha$, we obtain by using Proposition 9.1

$$\begin{aligned}
\phi(a,[x])\circ\phi(b,[y]) &= (s(a)+\Sigma j^*(\Sigma i_*(x)))\circ(s(b)+\Sigma j^*(\Sigma i_*(y)))\\
&= (s(a)+\Sigma i\circ x\circ\Sigma j)\circ s(b)+(s(a_0,a_1)+\Sigma i\circ x\circ\Sigma j)\circ(\Sigma i\circ y\circ\Sigma j)\\
&= s(a)\circ s(b)+(\Sigma i\circ x\circ\Sigma j)\circ s(b)+(a_0\iota_\alpha+(\frac{a_1-a_0}{k}\mu_0+\Sigma i\circ x)\circ\Sigma j)\circ(\Sigma i\circ y\circ\Sigma j)\\
&= s(ab)+b_1\Sigma i\circ x\circ\Sigma j+\frac{a_0(a_0-1)(b_1-b_0)}{2k}\Sigma i\circ[\iota_m,\iota_m]\circ H_\alpha(\mu_0)\circ\Sigma j\\
&\quad +(a_0\iota_\alpha)\circ\Sigma i\circ y\circ\Sigma j+(\frac{a_1-a_0}{k}\mu_0+\Sigma i\circ x)\circ\Sigma j\circ\Sigma i\circ y\circ\Sigma j\\
&= s(ab)++b_1\Sigma i\circ x\circ\Sigma j+a_0\Sigma i\circ y\circ\Sigma j+\frac{a_0(a_0-1)}{2}\Sigma i\circ[\iota_m,\iota_m]\circ H_\alpha(y)\circ\Sigma j\\
&\quad +\frac{a_0(a_0-1)(b_1-b_0)}{2k}\Sigma i\circ([\iota_m,\iota_m]\circ H_\alpha(\mu_0))\circ\Sigma j\\
&= s(ab)+\Sigma i\circ(b_1x+a_0y+\frac{a_0(a_0-1)}{2}P_\alpha H_\alpha(y)+\frac{a_0(a_0-1)(b_1-b_0)}{2k}P_\alpha H_\alpha(\mu_0))\circ\Sigma j\\
&= \phi(ab,b_1[x]+a_0[y]+\frac{a_0(a_0-1)}{2}P_\alpha H_\alpha(y)+\Delta(a,b))=\phi((a,[x])\circ(b,[y]))
\end{aligned}$$

qed.

References

[B1] H. Baues, *"Homotopy Type and Homology"*, Oxford Math. Monograph, Clarendon Press, Oxford (1996), 489 pages.

[B2] H. Baues, Quadratic functors and metastable homotopy, J. Pure Appl. Alg. **91** (1994), 49–107.

[BHP] H. Baues, M. Hartl, T. Pirashvili, Quadratic categories and square rings, J. Pure Appl. Alg. **122** (1997), 1–40.

[BP] H. Baues, T. Pirashvili, Quadratic endofunctors of the category of groups, Adv. Math. **141** (1999), 167–206.

[BS] J. M. Boardman and B. Steer, On Hopf invariants, Comment. Math. Helv. **42** (1967), 180–221.

[O] S Oka, Groups of self-equivalences of certain complexes, Hiroshima Math. J. **2** (1972), 285–293.

[OSS] S Oka, N. Sawashita and M. Sugawara, On the group of self-equivalences of a mapping cone, Hiroshima Math. J. **4** (1974), 9–28.

[U] H. M. Unsöld, A_n^4-polyhedra with free homology, Manuscripta Mathematica **65** (1989), 123–145.

[T] H. Toda, *"Composition methods in Homotopy groups of spheres"*, Princeton Univ. Press, Princeton N.Y., Ann. of math. studies **49** (1962).

[W] G. W. Whitehead, *"Elements of Homotopy Theory"*, Springer Verlag, Berlin, GTM series **61** (1978).

[Y] K. Yamaguchi, Self-Homotopy Equivalence and Highly Connected Poincaré Complex, Springer, Berlin, Lecture Notes in Math. **1425**, (1990), 157–169.

Hans–Joachim Baues, Max-Planck-Institut für Mathematik, Bonn, Germany.
E-mail address: `baues@mpim-bonn.mpg.de`

Norio Iwase, Graduate School of Mathematics, Kyushu University, Japan.
Current address: Max-Planck-Institut für Mathematik, Bonn, Germany.
E-mail address: `iwase@math.kyushu-u.ac.jp`

Contemporary Mathematics
Volume **274**, 2001

Fibrations with Product of Eilenberg-MacLane Space Fibres I

Peter I. Booth

ABSTRACT. We consider the problem of classifying 3-stage Postnikov towers, i.e. towers constructed from a base space and two Eilenberg-MacLane spaces. Appropriately defined spaces of partial maps act as classifying spaces for certain 3-towers, i.e. those for which the underlying fibration has the product of the two Eilenberg-MacLane spaces as its distinguished fibre. The point set topology of these functional classifying spaces is investigated, and they are shown to classify the aforementioned class of 3-towers up to a suitable form of fibre homotopy equivalence. Alternative classification results and computations will be given in a sequel to this paper.

1. Introduction

The results of this paper, and its sequel [B6], can be viewed in two different ways. On one level, they discuss a straightforward topic in homotopy theory, i.e. the problem of classifying the class of Hurewicz fibrations whose fibres are the product of a pair of *Eilenberg-MacLane Spaces (= EM-spaces)*. On a second level, they constitute an investigation that is relevant to a fundamental—and difficult—question, i.e. the *Homotopy Decomposition Problem.* We refer here to the *Postnikov decomposition* procedure for constructing models of homotopy types of topological spaces using "unit blocks" (EM-spaces), and the problem of finding a satisfactory algebraic description of the homotopy types of the "compound blocks" (= spaces) that can be constructed in this way.

Moore-Postnikov decomposition is a procedure that allows us to build models of maps, i.e. *Moore-Postnikov towers.* The construction technique involves starting with an initial block and attaching (some may prefer the term coattaching) a unit block, thereby producing a 2-stage compound block. This new block is then treated as if it were an initial block, and a further unit block may be attached, producing a 3-stage compound block. Thus we have the first steps in an inductive procedure that can produce models of fibre homotopy types of fibrations and, when the initial block is a point, models of homotopy types of spaces.

Let $\mathcal{W}$ denote the class of spaces with the homotopy types of CW-complexes. The term *EM-space* will be taken to include the assumption that the space in $\mathcal{W}$.

2000 *Mathematics Subject Classification.* Primary 55R15, 55S45, 55P15, Secondary 55P20.
The author thanks NSERC of Canada for its support.

Our initial block (or base space) is simply a topological space. The procedure of attaching an EM-space $K(G,m)$ to a base space B involves selecting a principal $K(G,m)$-fibration over B, and designating its total space as the manufactured compound block.

We will be concerned, in this paper and [B6], with the following question.

Problem: Informal Description. *Classify the possible outcomes when two unit blocks are attached, one at a time, to a given initial block, thereby producing a 3-stage compound block.*

We will restate the question more precisely. Let G and H be Abelian groups and m and n be integers, with $0 < m < n$. Our initial block is a base space B, the first unit block is the EM-space $K(G,m)$ and the second unit block is the EM-space $K(H,n)$.

We are interested in classifying fibrations up to *fibre homotopy equivalence* ($= FHE$), which is also known as *fibrewise homotopy equivalence*. When a pair of fibrations are fibre homotopy equivalent, they will be said to have the same *fibre homotopy type* ($= FHT$).

Problem: Formal Description. *Classify the underlying fibrations of 3-stage Moore-Postnikov towers (or 3-towers) over B. Thus we wish to classify all composite fibrations $p_1 \circ p_2 : E_2 \to B$ up to FHE, where $p_1 : E_1 \to B$ is a principal $K(G,m)$-fibration and $p_2 : E_2 \to E_1$ is a principal $K(H,n)$-fibration.*

These composite fibrations occur when one carries out a Moore-Postnikov decomposition of a fibration whose fibres have just two non-zero homotopy groups, i.e. G in dimension m and H in dimension n. Conversely, let us consider starting with a principal $K(G,m)$-fibration p_1 and a principal $K(H,n)$-fibration p_2 such that $p_1 \circ p_2$ is defined. Then it is standard—and easily verified using the exact homotopy sequences of the fibrations involved—that the fibres of $p_1 \circ p_2$ have just two non-zero homotopy groups, as specified previously.

It is a common practice, when investigating a mathematical topic, to first consider an important special case. We follow such an approach in this paper and [B6], by looking at 3-towers with distinguished fibres that are either $K(G,m) \times K(H,n)$, or at least have the homotopy type of that space. Those for which this fibre is $K(G,m) \times K(H,n)$ will be referred to as *$K(G,m) \times K(H,n)$-towers*.

Our method of investigation can be explained by backtracking and reviewing the methods used in classifying 2-stage compound blocks, i.e. 2-stage Postnikov towers over a base space B. Thus we refer to the classification of principal $K(G,m)$-fibrations over a simply connected CW-complex B.

This longstanding result is implicit in [St, section 23]. The procedure involved can be broken down into four steps as follows.

(I). Specify an appropriate classifying space for the fibrations under consideration, i.e. $K(G,m+1)$.

(II). Show that $[B, K(G,m+1)]$, the set of free (or based) homotopy classes of maps from B to $K(G,m+1)$, classifies the fibrations in question up to a strong form of FHE.

(III). Show that $[B, K(G,m+1)]/Aut(G)$, the orbit set determined by the action of $Aut(G)$ on $K(G,m+1)$, classifies the fibrations in question up to FHE.

(IV). Use the well-known homotopy classification thereom, i.e. that there is a group isomorphism $[B, K(G, m+1)] \approx H^{m+1}(B; G)$ in cases where $B \in \mathcal{W}$, for computations of the group $[B, K(G, m+1)]$ and the set $[B, K(G, m+1)]/Aut(G)$.

We will now describe the equivalence relation amongst fibrations that occurs in (II). Let B be a pointed space and $p_X : X \to B$ and $p_Y : Y \to B$ be Hurewicz fibrations, both with distinguished fibres F. Then p_X will be said to be *identity fibre homotopy equivalent (1FHE) to* p_Y if there is an FHE $f : X \to Y$ such that $f|F : F \to F$ is homotopic to 1_F. It is easily seen that *1FHE* is an equivalence relation amongst fibrations over B whose distinguished fibre is F.

In this paper we will produce arguments and results analagous to those of (I) and (II), but for $K(G, m) \times K(H, n)$-towers. In particular we will (I) define a type of functional classifying space and investigate some of its topological properties, and (II) use this space to classify such fibrations up to *1FHE*.

In [B6], the sequel to this paper, we will establish the analogue of (III), classifying towers up to FHE, and then give some computational methods, comparable to (IV), all for 3-towers whose fibres have the homotopy type of $K(G, m) \times K(H, n)$.

If X and Y are spaces, then a *partial map* from X to Y is a map from a subspace of X to Y. Partial maps from the real line to itself are pervasive in elementary calculus, and much of modern mathematics grew out of calculus. Nevertheless, spaces of partial maps are a neglected topic in present-day mathematics (but see [BB1] and [HZ]). Our technique for dealing with (I) involves using classifying spaces that are spaces of partial maps. If P denotes the space of paths in $K(G, m+1)$ that start at the base point, then our classifying spaces $M_\infty = M_\infty(P, K(H, n))$ are appropriately defined spaces of partial maps from P to $K(H, n)$.

It is well known, of course, that classifying spaces exist for classes of fibrations with any given homotopy type of fibre. Such spaces are usually produced either via the Brown Representability Theorem or by a bar construction procedure [B5]. Sometimes, however, there are simpler alternative classifying spaces available.

(i) If we are working with Hurewicz fibrations with fibres of the homotopy type of $K(G, m)$, for G an Abelian group and $m > 0$, then for many purposes it is simpler to work with a classifying space $K(G, m+1)$ rather than one produced by the aforementioned "heavyweight" procedures.

(ii) If we are working with Hurewicz fibrations with fibres of the homotopy type of $K(G, m) \times K(H, n)$, where G and H are Abelian groups and $n > m > 0$, then our **functional classifying spaces have similar advantages over those produced by the above standard procedures.**

In particular we note the following points.

(a) Our alternative approach facilitates a strategy that is often successful with problems in this area, i.e. consideration of the case with simply connected base spaces before the more difficult general case is tackled.
(b) Our classifying spaces are explicitly defined function spaces. No infinite limiting processes, simplicial sets or geometric realizations are used in

their definition. As a result, **our discussion of maps into and out of our functional classifying spaces can be framed in the language of elementary homotopy theory. Thus we are frequently able to specify such maps by writing down explicit formulae that define them.** This is apparent for the maps p_∞, j_∞ and s_∞ that are defined in Section 3, as well as in the rule, relating 3-towers $\tau(f,g)$ and their classifying maps $g^\bullet$, that is used in (M6) and Theorem 7.5. This aspect will be important when, at a later stage, we come to consider applications.

After a review of our terminologies and some known or easy results, in Section 2, our functional classifying spaces are defined and discussed in Section 3. Some point-set topological and homotopy properties of these spaces are presented there.

The properties of functional classifying spaces can, of course, be derived from first principles. It is, however, most economical to view them as particular examples of fibred (or fibrewise) mapping spaces and to derive the properties of functional classifying spaces from known properties of fibred mapping spaces. So in Sections 4 and 5 we review some properties of the latter spaces, and use them in Section 6 to prove our previously stated functional classifying spaces properties. Then, in Section 7, we prove a classification theorem for $K(G,m) \times K(H,n)$-towers over a space B. This tells us that, if B is simply connected and in $\mathcal{W}$, then the set of *1FHE* classes of such towers is in bijective correspondence with the set of based homotopy classes of based maps from B into the appropriate functional classsifying space M_∞.

If X is a space, then $\mathcal{E}(X)$ will denote the group of free homotopy classes of self-homotopy equivalences of X. In the sequel [B6] to this paper we will prove that $\mathcal{E}(K(G,m) \times K(H,n))$ acts on the set of based homotopy classes of based maps $[B, M_\infty]^0$. It will be shown that, under light conditions, the corresponding set of orbits, $[B, M_\infty]^0/\mathcal{E}(K(G,m) \times K(H,n))$ classifies the set of FHE classes of 3-towers over B whose fibres have the homotopy type of $K(G,m) \times K(H,n)$. Let us now assume that, for example, B is a simply connected space and has the homotopy type of a CW-complex. Then Hurewicz fibrations over B, with fibres of the homotopy type of $K(G,m) \times K(H,n)$, are classified by that same orbit set. Some basic computations, for example in a stable range or where B is a co-H-space, will also be presented in that paper.

More sophisticated applications, where M_∞ is approximated by an appropriately defined pullback space, will be given elsewhere. Further developments of this work will be motivated by their relevance to the homotopy decomposition problem. For example, our arguments will be extended to cover 3-stage towers with fibres that have just two non-zero homotopy groups but are not necessarily of the homotopy type of a product space.

2. Preliminaries

We will always work in the context of the category of *compactly generated spaces (= cg-spaces)* and maps, as described in [V, section 5, example (ii)].

(2.1) Mapping Spaces

If Y and Z are spaces, then $M(Y,Z)$ will denote *the space of all maps from Y to Z*. If $f:Y \to Z$ is a given map, then $M(Y,Z;f)$ will represent the path component of $M(Y,Z)$ that contains f. If Y_0 and Z_0 are subspaces of Y and Z, respectively, then $M(Y,Y_0;Z,Z_0)$ will represent the space of maps f, from Y to Z, such that $f(Y_0) \subset Z_0$. In particular, for pointed spaces (Y,y_0) and (Z,z_0), either $M(Y,y_0;Z,z_0)$ or $M^0(Y,Z)$ will denote *the space of pointed maps from Y to Z*. All of our spaces of maps will carry the appropriate form of the *cg-ified compact-open topology*.

We will make use of the following results concerning mapping spaces.

(2.1.1) The Exponential Law of Spaces = *[V, Thm.3.6]. If X, Y and Z are spaces, then maps $f^> : X \times Y \to Z$ are in bijective correspondence with maps $f^< : X \to M(Y,Z)$, via the rule $f^>(x,y) = f^<(x)(y)$, where $x \in X$ and $y \in Y$.*

Further, there is a homeomorphism $M(X \times Y, Z) \to M(X, M(Y,Z))$ determined by the rule $f^> \rightsquigarrow f^<$.

Lemma 2.1.2. *Let G and H be Abelian groups, and j, m and n be positive integers. Then there are group isomorphisms:*

$$H^{n-j}(K(G,m);H) \;\approx\; \pi_j(M^0(K(G,m),\, K(H,n);c_0)$$

Proof. $H^{n-j}(K(G,m);H) \approx [K(G,m),K(H,n-j)]^0 = [K(G,m),\Omega^j K(H,n)]^0$
$= [K(G,m),M^0(S^j,K(H,n))]^0 \approx [S^j,M^0(K(G,m),K(L,n))]^0$
$= \pi_j(M^0(K(G,m),K(H,n);c_0))$.

(2.2) Maps

Certain notations for maps will be used repeatedly in this work.

Let Y and Z be spaces, and $z \in Z$. Then $c_z : Y \to Z$ will denote the constant map with value z.

Projections such as the map $Y \times Z \to Y$, $(y,z) \rightsquigarrow y$, where $y \in Y$ and $z \in Z$, will be denoted either by π or some slight variation on π.

Evaluation maps such as $M(Y,Z) \times Y \to Z, (f,y) \rightsquigarrow f(y)$, where $f \in M(Y,Z)$ and $y \in Y$, and "fibred" analogues of this map, will be denoted by either e, some slight variation on e or, in one case, v.

(2.3) Homotopy

The symbols $\simeq$, $\simeq^A$, $\simeq_B$ and $\simeq^A_B$ will represent homotopies in the *free, under A, over B* and *under A and over B* senses, respectively. In the case where A is a point, then $\simeq^A$ and $\simeq^A_B$ reduce to $\simeq^0$ and $\simeq^0_B$, i.e. *pointed homotopy* or *based homotopy*, and *pointed homotopy over B* or *based homotopy over B*, respectively.

(2.4) Fibrations

(2.4.1) We work with fibrations in the sense of *Hurewicz*, i.e. that have the covering homotopy property (CHP). The principal fibrations used, i.e. fibrations induced from the path fibrations of (2.6), are also Hurewicz fibrations.

(2.4.2) Let $r : Z \to C$ be a fibration and A be a subspace of C. Then the *subspace of* Z *over* A, i.e. $r^{-1}(A)$, will be denoted by $Z|A$. In particular, if $c \in C$, then the *fibre of* r *over* c will be denoted by $Z|c$.

(2.4.3) *Let* C *be a simply connected space and* $r : Z \to C$ *be a Hurewicz fibration. If* $c \in C$ *and* $i : Z|c \to Z$ *denotes the inclusion, then the associated induced function* $\pi_0(i) : \pi_0(Z|c) \to \pi_0(Z)$ *is injective.*

Proof. Let z_1 and z_2 be points of $Z|c$, and l be a path from z_1 to z_2 in Z. Then $r \circ l$, a loop at c in C, shrinks down to the constant loop at c. It follows, from the CHP, that l can be deformed into a path from z_1 to z_2 in $Z|c$.

(2.4.4) *Let* $r : Z \to C$ *be a Hurewicz fibration and* A *be a closed subspace of* C. *If* (C, A) *has the homotopy extension property (HEP), then* $(Z, Z|A)$ *also has the HEP.*

This is proved, in the context of all spaces, in [Str2, Thm.12]. It is proved in the compactly generated context in [B3, Corollary 3], with the additional assumption that B is weak Hausdorff. The latter case is adequate for our purposes.

(2.4.5): The Relative Covering Homotopy Property= *Thm.2 of [Str1]. Suppose that* $r : Z \to C$ *is a Hurewicz fibration and that* A *is a closed subspace of* B *such that the pair* (B, A) *has the HEP. If there is a map* $f : (B \times 0) \cup (A \times I) \to Z$ *and a homotopy* $F : B \times I \to C$ *such that* $r \circ f = F|((B \times 0) \cup (A \times I))$, *then there is a homotopy* $H : B \times I \to Z$ *such that* $r \circ H = F$ *and* $H|(B \times 0) \cup (A \times I) = f$.

(2.4.6) Let B be a pointed space and $p_X : X \to B$ and $p_Y : Y \to B$ be Hurewicz fibrations, each with distinguished fibre F. In the introduction we introduced the concept of a *1FHE*, i.e. an FHE $f : X \to Y$ such that $f|F \simeq 1_F$. If a *1FHE* has $f|F = 1_F$ then it will be said to be a *strict 1FHE*.

Let B be a pointed space and $p_X : X \to B$ and $p_Y : Y \to B$ be Hurewicz fibrations, each with a distinguished fibre F. Then *1FHE(X, Y)* and *S1FHE(X, Y)* will denote the set of *1FHEs* and the set of *strict 1FHEs* from X to Y, respectively. We notice that there is an inclusion map ι:*S1FHE(X, Y)*$\to$*1FHE(X, Y)*, and an associated function $\pi_0(\iota) : \pi_0$*(S1FHE(X, Y))* $\to \pi_0$*(1FHE(X, Y))*.

(2.4.7) *If* B *has a non-degenerate base point, i.e. if* $(B, *)$ *has the HEP, then* $\pi_0(\iota)$ *is surjective. If* $\pi_1(M(F, F), 1_F) = 0$, *then* $\pi_0(\iota)$ *is injective.*

Proof. Let us consider a *1FHE* f from X to Y. We then use a homotopy between $f|F : F \to F$ and 1_F, along with (2.4.4) and (2.4.5), to obtain an *S1FHE* from X to Y. So $\pi_0(\iota)$ is surjective.

Let us assume that f and g are a pair of *strict 1FHEs* from X to Y, that are homotopic via a family of *1FHEs*. Thus the restriction of this homotopy to $F \times I$ is a self-homotopy of 1_F. It follows from our data that this homotopy can be deformed, relative to $F \times \{0, 1\}$, to the projection $F \times I \to F$. Using this deformation, and the results (2.4.4) and (2.4.5), we obtain a homotopy from f to g that is a family of strict *1FHEs*.

The following remark, whilst it does not refer directly to fibrations, is relevant to (2.4.4), (2.4.5) and (2.4.7).

(2.4.8) If B is a CW-complex and $*$ is a base point and 0-cell of B, then $*$ is a subcomplex of B and so $*$ is a non-degenerate base point of B. If $*$ is not a 0-cell, then $*$ belongs to an open cell. This cell may be decomposed (= partitioned) into a set of cells, in such a way that $*$ becomes a 0-cell. Hence any base point in a CW-complex is non-degenerate.

(2.5) Fibred Product Spaces

(2.5.1) Let $g : B \to C$ and $r : Z \to C$ be maps. Then $B \sqcap Z$ will denote the associated *fibred product* or pullback space, and $g^*r : B \sqcap Z \to B$ the corresponding *induced projection*. In cases where there is some possibility of confusion, $B \sqcap Z$ will be written $B \sqcap_g Z$. If the above spaces and maps are pointed, then $B \sqcap Z$ will have distinguished point $(*, *')$, where $*$ and $*'$ are the distinguished points of B and Z, respectively. We will list some well known or easily verified properties of fibred product spaces, that we wish to refer to later.

(2.5.2) If r is either a Hurewicz or a principal fibration then it is well known, and easily verified, that g^*r will be a fibration in the same sense

(2.5.3) If B and C are pointed spaces and $g \in M^0(B, C)$, then the distinguished fibre $(B \sqcap Z)|* = \{*\} \times (Z|*)$ of g^*r may be identified, in an obvious way, with the distinguished fibre $Z|*$ of r.

(2.5.4) Let D be a space and $\pi : C \times D \to C$ the projection. Then we may identify $\pi^*r : (C \times D) \sqcap Z \to C \times D$ with $r \times 1_D : Z \times D \to C \times D$, via the homeomorphism $Z \times D \to (C \times D) \sqcap Z, (z, d) \rightsquigarrow (r(z), d, z)$, where $z \in Z$ and $d \in D$. In that case the projection $(C \times D) \sqcap Z \to Z$ is identified with the projection $Z \times C \to Z$. The inverse homeomorphism $(C \times D) \sqcap Z \to Z \times D$ is simply the projection.

If $\pi_Z : Z \times D \to Z$ denotes the projection, then the fibred product map $\pi \sqcap r : (C \times D) \sqcap Z \to C$ will be identified with $r \circ \pi_Z : Z \times D \to C$.

(2.5.5) Let us now assume that $g : B \to C$ is an embedding, i.e. a homeomorphism into. Then $g^*r : B \sqcap Z \to B$ can be identified with $r|r^{-1}(g(B)) : r^{-1}(g(B)) \to B$, via the homeomorphisms $g : B \to g(B)$ and $B \sqcap Z \to r^{-1}(g(B)), (b, z) \rightsquigarrow z$, where $(b, z) \in B \sqcap Z$. So $B \sqcap Z$ embeds in Z and we may regard g^*r as being *embedded* in r, or r as being an extension of g^*r.

In particular, if q is the identity map $C \to C$, then the projection $C \sqcap Z \to Z$ is a homeomorphism over C, so g^*r can be identified with r.

(2.5.6) Let us now assume that $f : A \to B$ and $g : B \to C$ are arbitrary maps. Then the induced projections $(g \circ f)^*r : A \sqcap Z \to A$ and $f^*(g^*r) : A \sqcap (B \sqcap Z) \to A$ may be identified together, using the homeomorphism

$$A \sqcap Z \to A \sqcap (B \sqcap Z), \ (a, z) \rightsquigarrow (a, (f(a), z)),$$

where $(a, z) \in A \sqcap Z$. In that case the projection $A \sqcap (B \sqcap Z) \to B \sqcap Z$ is identified with the map $f \sqcap 1_Z : A \sqcap Z \to B \sqcap Z$. The inverse of the above homeomorphism is simply the projection $A \sqcap (B \sqcap Z) \to A \sqcap Z$.

(2.5.7) Let $f, g \in M(B, C)$ and $r : Z \to B$ be a Hurewicz fibration. If $f \simeq g$ then it is standard that f^*r is FHE to g^*r (see, for example, [B4, Prop.6.2]).

(2.5.8) Given maps $g : (B, B_0) \to (C, *)$ and $r : Z \to C$, then $g^*r : B \sqcap Z \to B$ extends the projection $B_0 \times Z \to B_0$.

Let $r : Z \to C$ and $f : B \to C$ be maps. Then $f^*r : B \sqcap Z \to B$ allows us to define a space $M_B(B \sqcap Z, B \sqcap Z)$. The maps $f \sqcap r : B \sqcap Z \to C$ and $r : Z \to C$ allow us to define a space $M_C(B \sqcap Z, Z)$. Further, composition with the projection $B \sqcap Z \to Z$, defines a map $\psi : M_B(B \sqcap Z, B \sqcap Z) \to M_C(B \sqcap Z, Z)$.

(2.5.9) *The map ψ is a homeomorphism*

Proof. It can be seen, via the universal property of pullbacks, that ψ is a bijection. Consider now an arbitrary space W, a function from W to $M_B(B \sqcap Z, B \sqcap Z)$, and the function from W to $M_C(B \sqcap Z, Z)$ obtained by composing ψ with the original function. It can be shown, using (2.1.1), that one of these functions is continuous if and only if the other is continuous.

(2.6) Eilenberg-MacLane Spaces and Path Fibrations

(2.6.1) Let G be an Abelian group, m be a positive integer and K denote the EM-space $K(G, m+1)$. Then ΩK, the space of loops at the base point of K, is the EM-space $K(G, m)$. The group or H-group operations on these spaces will be denoted by $+$ and their identities by 0. When base points are required the 0 points will, of course, be used.

We will use P to denote the space of paths, that start at 0, in K. Then $p : P \to K$ will denote the *path fibration over* K, i.e. the map that evaluates at 1. For $0 \in K$, $P|0 = \Omega K$.

Let H be an Abelian group, n be a positive integer and L denote the EM-space $K(H, n+1)$. Then $\Omega L = K(H, n)$. The path fibration over L will be denoted by $q : Q \to L$.

(2.6.2) Given l and m in P, we will use the group structure on K to define a path $l + m \in P$. Thus we let $(l+m)(t)$ be $l(2t)$ if $t \leq 1/2$, and $l(1) + m(2t-1)$ if $t \geq 1/2$. So $l + m$ is a path in P that starts at $l(0) + m(0) = 0$ and ends at $l(1) + m(1)$.

Clearly this extends the usual operation $\Omega K \times \Omega K \to \Omega K$.

(2.6.3) We will sometimes use the map $1^+ : \Omega K \to \Omega K$, $1^+(l) = c_0 + l$, where $l \in \Omega K$, and $c_0 \in \Omega K$ is the constant loop value 0.

(2.6.4) *There is a weak homotopy equivalence*

$$\chi : K \to M(K, K; 1_K), \quad \chi(x)(y) = x + y,$$

where $x, y \in K$.

Proof. The map $K \times K \to K$, $(x,y) \to x+y$, where x and $y \in K$, determines a map $K \to M(K,K)$, $x \rightsquigarrow (y \rightsquigarrow x+y)$ via (2.1.1). Now K is path connected and this last map clearly takes $0 \in K$ into $M(K,K;1_K)$, so χ is a well defined map into $M(K,K;1_K)$.

Let X and Y be spaces, X having a non-degenerate base point. It is standard, and easily verified using (2.1.1) and the HEP for the inclusion $* \to X$, that the evaluation map $e : M(X,Y) \to Y$, $e(f) = f(*)$, where $f \in M(X,Y)$, is a Hurewicz fibration. It follows that $e : M(K,K) \to K$ and restriction $e' : M(K,K;1_K) \to K$ are Hurewicz fibrations. We see from (2.4.3) that the distinguished fibre of e' is $M^0(K,K;1_K)$.

We know that $M^0(K,K;1_K)$ is path connected and that $\pi_j(M^0(K,K;1_K)) = \pi_j(M^0(K,K)) \approx [K,\Omega^j K]^0 \approx [K(G,m),K(G,m-j)]^0 \approx \tilde{H}^{m-j}(K(G,m);G) = 0$, for all positive integers j. So $\pi_j(M^0(K,K;1_K)) = 0$ for all non-negative integers j.

It follows, from the exact homotopy sequence of e', that e' is a weak homotopy equivalence. Now $e' \circ \chi = 1_K$, so for all non-negative integers j, $\pi_j(e') \circ \pi_j(\chi)$ is the identity on $\pi_j(K)$. The right inverse of a bijection is always a bijection, so χ is a weak homotopy equivalence.

(2.7) 3-Stage Postnikov Towers

Let us assume that B has a basepoint, $f \in M^0(B,K), B \sqcap P$ is the pullback space formed using f and p, and $g \in M^0(B \sqcap P, L)$. Then there are induced principal fibrations $f^*p : B \sqcap P \to B$ and $g^*q : (B \sqcap P) \sqcap Q \to B \sqcap P$.

(2.7.1) The pair of fibrations (f^*p, g^*q) will be called a *3-stage Postnikov tower over* B, or simply a *3-tower over* B, and denoted by $\tau(f,g)$. The Hurewicz fibration $f^*p \circ g^*q$ will be called the *underlying fibration of* $\tau(f,g)$ and denoted by $t(f,g)$. The *distinguished fibre of* $\tau(f,g)$ will be the space $((B \sqcap P) \sqcap Q)|* = t(f,g)^{-1}(*)$, where $*$ is the base point of B.

(2.7.2) Let $\tau(f,g)$ and $\tau(h,k)$ be 3-towers over A and B, respectively. A *morphism* $< \epsilon, \eta, \lambda >: \tau(f,g) \to \tau(h,k)$ will consist of a maps $\epsilon : A \to B$, $\eta : A \sqcap_f P \to B \sqcap_h P$, and $\lambda : (A \sqcap_f P) \sqcap_g Q \to (B \sqcap_h P) \sqcap_k Q$ such that $(h^*p) \circ \eta = \epsilon \circ f^*p$ and $(k^*q) \circ \lambda = \eta \circ (g^*q)$.

We now assume that $A = B$. An *over* B *morphism* $< \eta, \lambda >: \tau(f,g) \to \tau(h,k)$ refers to a morphism $< 1_B, \eta, \lambda >$. We notice that λ is then a map over B, from the fibration $t(f,g)$ to the fibration $t(h,k)$.

If η and λ are homotopy equivalences, then $< \eta, \lambda >$ will be said to be an FHE *from* $\tau(f,g)$ *to* $\tau(h,k)$. If η and λ are homotopy equivalences whose respective restrictions to the fibres $\{*\} \times \Omega K$ and $(\{*\} \times \Omega K) \times \Omega L$ are homotopic to the identity maps on those fibres, then $< \eta, \lambda >$ will be said to be a *1FHE from* $\tau(f,g)$ *to* $\tau(h,k)$. If the aforementioned restrictions are equal to the identity map on the fibres, then $< \eta, \lambda >$ will be said to be a *strict 1FHE from* $\tau(f,g)$ *to* $\tau(h,k)$.

We notice that if $< \eta, \lambda >: \tau(f,g) \to \tau(h,k)$ is an *FHE, 1FHE* or *strict 1FHE*, then λ is an FHE, in the same sense, from $t(f,g)$ to $t(h,k)$. The converse of these statwements is also true. If λ is an FHE of fibrations from $t(f,g)$ to $t(h,k)$, then it follows by [K, Thm.2.2] that there is a homotopy equivalence η such that $< \eta, \lambda >$

is a FHE from $\tau(f,g)$ to $\tau(h,k)$. Similar results apply in the *1FHE* and strict *1FHE* cases.

(2.7.3) *Let $\tau(f,g)$ and $\tau(h,k)$ be 3-towers over a space B with a non-degenerate base point. If $<\eta,\lambda>: \tau(f,g) \to \tau(h,k)$ is a 1FHE, then there is also a strict 1FHE between the same towers.*

Proof. We can use the surjection portion of (2.4.7) to, if necessary, replace η by a homotopic map that is a strict *1FHE* from f^*p to h^*p. The CHP allows us to replace λ with a homotopic map that covers our possibly new η. A further application of (2.4.7) allows a further modification of λ so that our final new $<\eta,\lambda>$ is the required strict *1FHE.*

(2.7.4) Let A and B be pointed spaces, $\tau(f,g)$ be a 3-tower over B and take $\mu \in M^0(A,B)$. Pulling $f^*p : B \sqcap P \to B$ back over μ, we obtain the principal fibration $(f\circ\mu)^*p : A \sqcap P \to A$, and a "pairwise" map $<\mu, \mu\sqcap 1_P>$ from $(f\circ\mu)^*p$ to f^*p (see (2.5.6)). Pulling g^*q back over $\mu \sqcap 1_P : A \sqcap P \to B \sqcap P$, we obtain a principal fibration $(g\circ(\mu \sqcap 1))^*q : (A\sqcap P)\sqcap Q \to A \sqcap P$, and a pairwise map $<\mu\sqcap 1_P, (\mu\sqcap 1_P)\sqcap 1_Q>$ from $(g\circ(\mu\sqcap 1))^*q$ to g^*q.

Thus we have an *induced 3-tower* $\mu^*\tau(f,g) = \tau(f\circ\mu, g\circ(\mu\sqcap 1_P))$, and a morphism of 3-towers $<\mu, \mu\sqcap 1_P, (\mu\sqcap 1_P)\sqcap 1_Q>$ from $\mu^*\tau(f,g)$ to $\tau(f,g)$.

(2.7.5) We now reconsider the situation of (2.7.4), but with the additional assumption that $\mu : A \to B$ is an embedding. Then the morphism $<\mu, \mu\sqcap 1_P, (\mu\sqcap 1_P)\sqcap 1_Q>$ consists of three embeddings, and $\mu^*\tau(f,g)$ will be said to be *embedded* in $\tau(f,g)$. Further, the pairwise map $<\mu, (\mu\sqcap 1_P)\sqcap 1_Q>$ is a pair of embeddings from $\mu^*t(f,g)$ to $t(f,g)$, so the fibration $\mu^*t(f,g)$ will be *embedded* in the fibration $t(f,g)$.

(2.7.6) Let $\tau(f,g)$ be a 3-tower and $t(f,g)$ have distinguished fibre $\Omega K\times\Omega L$; then $\tau(f,g)$ will be said to be a *$\Omega K\times\Omega L$-tower.* To be more precise, the distinguished fibre in such cases is actually $(\{*\}\times\Omega K)\times\Omega L$. However, it is usually convenient to identify this last space with $\Omega K\times\Omega L$.

(2.7.7) *A 3-tower $\tau(f,g)$ is an $\Omega K\times\Omega L$-tower if and only if $g|\{*\}\times\Omega K = c_0$.*

Proof. Let $\tau(f,g)$ be a 3-tower for which $g|\{*\}\times\Omega K = c_0 : \{*\}\times\Omega K \to L$. It follows, from (2.7.5), taking A to be the distinguished point of B and μ the inclusion of that point in B, that $\tau(f,g)$ is a $\Omega K\times\Omega L$-tower.

Conversely, let us assume that $\tau(f,g)$ is a $\Omega K\times\Omega L$-tower. This fibre is the pullback space obtained from $c_0 : \{*\}\times\Omega K \to L$ and $q : Q \to L$, so it follows from (2.7.5) that $g|\{*\}\times\Omega K = c_0$.

3. The Topology of Functional Classifying Spaces

We will now define our classifying spaces M_∞ and present some of their major properties. The proofs are removed to section 6. Our basic data, throughout this section, consists of the path fibration $p : P \to K$ and the space L.

The reader will recall that, if X and Y are spaces, then a *partial map from X to Y* is a map from a subspace of X to Y.

We define the set $M_\infty = M_\infty(P, L)$ to be a set of partial maps from P to L, i.e. the set of all null-homotopic maps from fibres of p to L. Thus

$$M_\infty = \bigcup_{x \in K} M(P|x, L; c_0).$$

The function $p_\infty : M_\infty \to K$ will be defined by $p_\infty(f) = x$, where $P|x$ is the domain of f.

We will now topologize M_∞. Let $\pi : K \times L \to L$ denote the projection, V be a subset of $K \times L$ and $x \in K$. Let us define V_x to be the subset of L determined by $V_x = \pi(V \cap (\{x\} \times L))$. Then we give M_∞ the topology determined by the subbasis that consists of the following two types of set:

$U_\infty = \bigcup_{x \in U} M(P|x, L; c_0)$, for all choices of an open subset U of K, and

$W_\infty(A, V) = \bigcup_{x \in K} \{f \in M(P|x, L; c_0) \mid f(A \cap (P|x)) \subset V_x\}$, for all choices of a compact subset A of P and an open subset V of $K \times L$.

However, we wish to work in the context of the category of cg-spaces. So the space M_∞, as described above, must be cg-ified, i.e. given the final (= weak) topology relative to all incoming maps from compact Hausdorff spaces. We note that $p_\infty : M_\infty \to K$ is then continuous.
The map $c_0 : P|0 \to L$, i.e. $c_0 : \Omega K \to L$, will be taken to be the base point of M_∞. We notice that p_∞ is then a base point preserving map.

Our definition of M_∞ is unsymmetrical relative to K and L. Nevertheless, two of the properties that follow are to some extent symmetrical in that respect. We start with those properties.

(M1) *There is an embedding, i.e. a homeomorphism into,*

$$j_\infty : K \times L \to M_\infty, (x, y) \rightsquigarrow c_y : P|x \to L,$$

where $x \in K$ and $y \in L$.
We note that the composite map $p_\infty \circ j_\infty : K \times L \to K$ is the projection.

So $K \times L$ may be embedded as a subspace of M_∞ in such a way that $p_\infty : M_\infty \to K$ (essentially) extends the projection $K \times L \to K$.

(M2) *Let us define $\hat{K}$ and $\hat{L}$ to be subspaces of M_∞, as follows:*

$$\hat{K} = j_\infty(K \times \{0\}) = \{c_0 : P|x \to L| \ \ x \in K\}$$

$$\hat{L} = j_\infty(\{0\} \times L) = \{c_y : \Omega K \to L| \ \ y \in L\}.$$

Thus $\hat{K} \cong K$ and $\hat{L} \cong L$ in the obvious canonical fashions.

If $G = 0$, then $K = K(G, m+1) = 0$. In that case M_∞ collapses to $\hat{L}$, i.e. $M_\infty = \hat{L} \cong L$.

If $H = 0$, then $L = K(H, n+1) = 0$. In that case M_∞ collapses to $\hat{K}$, i.e. $M_\infty = \hat{K} \cong K$.

We now move on to a group of fibration related conditions that are quite unsymmetrical relative to K and L.

(M3) *The map $p_\infty : M_\infty \to K$ is a Hurewicz fibration.*

(M4) *The fibre of p_∞ over $x \in K$ is $M(P|x, L; c_0)$; in particular the distinguished fibre over $x = 0$ is $M(\Omega K, L; c_0)$.*

(M5) *The fibration p_∞ has a cross section $s_\infty : K \to M_\infty$, $s_\infty(x) = c_0 : P|x \to L$, where $x \in K$, i.e. $s_\infty(x)(z) = 0$, where $z \in P|x$.*

We note that $\hat{K} = s_\infty(K)$ and $\hat{L} \subseteq M(\Omega K, L; c_0)$.

The next property - the most important of this list of results - is a precursor to our main classification theorems. It tells us that $\Omega K \times \Omega L$-towers over B are classified by the based maps from B to M_∞.

(M6) We will consider $\Omega K \times \Omega L$-towers $\tau(f, g)$. Thus we allow f and g to vary, but insist that the fibre is always $\Omega K \times \Omega L$. This condition requires, as we saw in (2.7.7), that $g|\{*\} \times \Omega K$ is in all cases $c_0 : \{*\} \times \Omega K \to L$.

Then there is a bijective correspondence between

(i) *$\Omega K \times \Omega L$-towers $\tau(f, g)$, and*

(ii) *maps $g^\bullet \in M^0(B, M_\infty)$.*

This correspondence is determined by the rule $g(b, z) = g^\bullet(b)(z)$, for $(b, z) \in B \sqcap_f P$, where $f = p_\infty \circ g^\bullet$. The map $g^\bullet$ will be referred to as the *classifying map* for $\tau(f, g)$.

Let $F : B \times I \to K$ be a pointed homotopy, with induced principal fibration $F^*p : (B \times I) \sqcap P \to B \times I$. We note that the subspace $((B \times I) \sqcap P)|(\{*\} \times I)$ is $\{*\} \times I \times \Omega K$.

(M7) *Let $\tau(f, g)$ and $\tau(h, k)$ be $\Omega K \times \Omega L$-towers over B with classifying maps $g^\bullet$ and $k^\bullet \in M^0(B, M_\infty)$.*

Then $g^\bullet \simeq^0 k^\bullet$ if and only if $f \simeq^0 h$ via a pointed homotopy F and there is a "generalized homotopy" H from g to k. Thus we are here referring to a map $H : ((B \times I) \sqcap_F P, \{*\} \times I \times \Omega K) \to (L, 0)$ such that $H(b, 0, l) = g(b, l)$ and $H(b, 1, l) = k(b, l)$, where $(b, 0, l)$ and $(b, 1, l)$ are in $(B \times I) \sqcap_F P$.

(M8) Let $M_\infty \sqcap P$ be the pullback space obtained using $p_\infty : M_\infty \to K$ and $p : P \to K$. Then there is an evaluation map $e_\infty : M_\infty \sqcap P \to L, (f, z) \rightsquigarrow f(z)$, where $p_\infty(f) = p(z)$, and an associated $\Omega K \times \Omega L$-tower $\tau(p_\infty, e_\infty)$ over M_∞.

(M9) *All $\Omega K \times \Omega L$-towers are induced from $\tau(p_\infty, e_\infty)$ by their classifying maps. Thus if $\tau(f, g)$ is a $\Omega K \times \Omega L$-tower with associated classifying map $g^\bullet$, then the tower $\tau(f, g)$ is $(g^\bullet)^*\tau(p_\infty, e_\infty)$*

On the basis of this result it is reasonable to refer to $\tau(p_\infty, e_\infty)$ as the *universal $\Omega K \times \Omega L$-tower*. Its underlying fibration

$$t(p_\infty, e_\infty) : (M_\infty \sqcap P) \sqcap Q \to M_\infty$$

will be referred to as the *underlying universal* $\Omega K \times \Omega L$*-fibration.* We will later prove that they are universal in other more applicable senses.

(M10) *The universal principal* $\Omega K \times \Omega L$*-fibration,* $p \times q : P \times Q \to K \times L$, *can be embedded in the underlying universal* $\Omega K \times \Omega L$*-fibration* $t(p_\infty, e_\infty)$, *via the embeddings* $j_\infty : K \times L \to M_\infty$ *and*

$$z_\infty : P \times Q \to (M_\infty \sqcap P) \sqcap Q,\ (l,m) \rightsquigarrow ((c_{q(m)} : M_\infty|(p(l)) \to L),\ l,\ m),$$

where $l \in P$ and $m \in Q$.

4. Functional Fibrations in Fibrewise Topology

The reader will recall that fibrewise topology applies to the category of spaces and maps over a fixed space B. Thus the objects are *spaces over* B, i.e. maps such as $p_X : X \to B$ and $p_Y : Y \to B$ of spaces to B. The morphisms are *maps over* B, so a morphism $f : p_X \to p_Y$ is a map $f : X \to Y$ such that $p_Y \circ f = p_X$.

Let $p_X : X \to B$ and $p_Y : Y \to B$ be spaces over B. Their product in the category of spaces and maps over B is the *fibred product space over* B, i.e. the map $p_X \sqcap p_Y : X \sqcap Y \to B$, where $(p_X \sqcap p_Y)(x,y) = p_X(x) = p_Y(y)$, for $(x,y) \in X \sqcap Y$.

We will describe the corresponding *fibred mapping spaces.* If $p_Y : Y \to B$ and $p_Z : Z \to B$ are spaces over B, then we define the set $(Y;Z)$ as $\bigcup_{b \in B} M(Y|b, Z|b)$. The function $(p_Y;p_Z)$ is defined to be the obvious projection $(Y;Z) \to B$.

If A is a subset of Y and V a subset of Z, we define

$$W(A,V) = \{f \in (Y;Z)|\ f(A) \subseteq V\}.$$

We note that $f(A)$, for a map $f : X|b \to Y|b$, is $f(A \cap dom(f))$ in the usual sense. Then $(Y;Z)$ can be given a *modified compact-open topology* with subbasis consisting of all sets $(p_Y : p_Z)^{-1}(U)$ and $W(A,V)$, where U is open in B, A is compact in Y and V is open in Z. However, we wish to work in the context of the category of *cg*-spaces, so the space $(Y;Z)$ must be *cg*-ified, i.e. retopologized with the final (= weak) topology relative to all incoming maps from compact Hausdorff spaces. Then $(p_Y;p_Z) : (Y;Z) \to B$ is, of course, continuous.

It will be assumed, from this point on, that B *is weak Hausdorff,* i.e. that the diagonal of $B \times B$ is closed in the *cg*-ified product topology on $B \times B$. This includes, of course, the case where B is Hausdorff.

Theorem 4.1: Fibred Exponential Law. *Let* $p_X : X \to B$, $p_Y : Y \to B$ *and* $p_Z : Z \to B$ *be spaces over* B*. Then there is a bijective correspondence between:*

(i) *maps* $f^> : X \sqcap Y \to Z$ *over* B*, and*
(ii) *maps* $f^< : X \to (Y;Z)$ *over* B*,*

determined by the rule $f^>(x,y) = f^<(x)(y)$ *where* $p_X(x) = p_Y(y)$.

Further there is a homeomorphism $M_B(X \sqcap Y, Z) \to M_B(X,(Y;Z))$ *determined by the rule* $f^> \rightsquigarrow f^<$.

We notice that when B is a point, $X \sqcap Y = X \times Y, (Y;Z) = M(Y,Z)$ and the last theorem reduces to the exponential law of spaces (2.1.1).

Proofs of this result are given in [B1, Thm.1.1] and [BB1, Thm.7.3] (but see also Remark 7.4(1) of [BB1]).

Corollary 4.2. *There is a bijective correspondence between:*

(i) *maps* $g : Y \to Z$ *over* B, *and*
(ii) *sections to* $(p_Y; p_Z)$, *i.e. maps* $s : B \to (Y; Z)$ *with* $(p_Y; p_Z) \circ s = 1_B$.

determined by the rule $s(b)(y) = g(y)$, *where* $p_Y(y) = b$.

Proof. This is obtained by taking $p_X = 1_B$ in the previous result, and identifying X with $X \sqcap B$ as in (2.5.5).

Theorem 4.3. *Let us assume that* A *is a weak Hausdorff space (as is* B*), and that* $f : A \to B, p_Y : Y \to B$ *and* $p_Z : Z \to B$ *are maps. Then* $A \sqcap (Y; Z)$ *can be identified with* $(A \sqcap Y; A \sqcap Z)$, *in which case the projections* $f^*(p_Y; p_Z) : A \sqcap (Y; Z) \to A$ *and* $(f^*(p_Y); f^*(p_Z)) : (A \sqcap Y; A \sqcap Z) \to A$ *coincide.*

Thus, there is a homeomorphism $\xi : A \sqcap (Y; Z) \to (A \sqcap Y; A \sqcap Z)$ *such that, for each* $a \in A$ *and* $h \in M(Y|f(a), Z|f(a)), \xi(a, h)$ *is the corresponding map*

$$1_{\{a\}} \times h \in M(\{a\} \times Y|f(a), \{a\} \times Z|f(a)).$$

Further, ξ *is a map over* A, *i.e.* $(f^*(p_Y); f^*(p_Z)) \circ \xi = f^*(p_Y; p_Z)$.

Proof. This is explained in (8.2) of [BB2].

We now consider the case where B is a point, so $f : A \to \{*\}, p_Y : Y \to \{*\}$ and $p_Z : Z \to \{*\}$.

Corollary 4.4. *Let* π_1, π_2 *and* π_3 *denote the projections* $A \times Y \to A, A \times Z \to A$ *and* $A \times M(Y, Z) \to A$, *respectively. Then* $(\pi_1; \pi_2)$ *can be identified with* π_3. *To be precise there is a homeomorphism over* A

$$\xi : A \times M(Y, Z) \to (A \times Y; A \times Z),\ \xi(a, f) = 1_{\{a\}} \times f : \{a\} \times Y \to \{a\} \times Z,$$

where $a \in A$ *and* $f \in M(Y, Z)$. *Further* $(\pi_1; \pi_2) \circ \xi = \pi_3$.

Corollary 4.5. *Let* A *be a subspace of* B. *Then* $(Y|A; Z|A)$ *and* $(Y; Z)|A$ *are identical, and so* $(Y|A; Z|A)$ *is a subspace of* $(Y; Z)$.

In particular, the fibre of $(p_Y; p_Z)$ *over* $b \in B$ *is* $M(Y|b, Z|b)$.

Proof. We note that the underlying sets of $(Y|A; Z|A)$ and $(Y; Z)|A$ are identical.

Let $f : A \to B$ be the inclusion map. We know, from (2.5.5), that $(Y; Z)|A \cong A \sqcap (Y; Z)$, $A \sqcap Y \cong Y|A$, and $A \sqcap Z \cong Z|A$. These homeomorphisms are all over A, so $(A \sqcap Y; A \sqcap Z) \cong (Y|A; Z|A)$.

Now Theorem 4.3 tells us that $A \sqcap (Y; Z) \cong (A \sqcap Y; A \sqcap Z)$, so $(Y; Z)|A \cong A \sqcap (Y; Z) \cong (A \sqcap Y; A \sqcap Z) \cong (Y|A; Z|A)$. The reader, who checks the definitions of these last three homeomorphisms, will find that the composite homeomorphism $(Y; Z)|A \cong (Y|A; Z|A)$ has the identity as underlying function. Hence the spaces $(Y; Z)|A$ and $(Y|A; Z|A)$ are identical.

The last part follows since $(Y|b; Z|b) = M(Y|b, Z|b)$.

Theorem 4.6. *If* $p_Y : Y \to B$ *and* $p_Z : Z \to B$ *are Hurewicz fibrations, then* $(p_Y; p_Z) : (Y; Z) \to B$ *is also a Hurewicz fibration.*

Proof. The argument of [B1, Thm.(6.1)(i)] applies in the cg-space context.

Theorem 4.7: Homotopy Fibred Exponential Law. *Let $p_X : X \to B, p_Y : Y \to B$ and $p_Z : Z \to B$ be spaces over B, and $f^>$ and $g^>$ be maps of $X \sqcap Y$ to Z over B. We will use $f^<$ and $g^<$ to denote the maps of X to $(Y;Z)$ over B that correspond to $f^>$ and $g^>$, respectively. Then $f^> \simeq_B g^>$ if and only if $f^< \simeq_B g^<$.*

Proof. We simply apply the functor π_0 to the last part of Theorem 4.1.

The following application of fibred mapping spaces was originally pointed out to the author by Ronnie Brown.

Proposition 4.8. *If $f : A \to B$ is a map and $r : Z \to B$ is an identification, then $f^*r : A \sqcap Z \to A$ is an identification.*

Proof. Let W be a space, $g : A \sqcap Z \to W$ be a map and $h : A \to W$ be a function such that $h \circ (f^*r) = g$. So we want to prove that h is continuous.

We will now consider the function $h_b = h|(A|b) : A|b \to W$, the projection $\pi_b : (A|b) \times (Z|b) \to A|b$, and $g_b = g|(A|b) \times (Z|b) : (A|b) \times (Z|b) \to W$, where $b \in B$. Then $g_b = h_b \circ \pi_b$ and π_b is an identification, so h_b is continuous.

We will define $g^> : A \sqcap Z \to W \times B$ by $g^>(a, z) = (g(a, z), f(a))$, where $f(a) = r(z)$. Then $g^>$ is a map over B, i.e. $g^> : f \sqcap r \to \pi$, where $\pi : W \times B \to B$ is the projection. Let $k : A \to W \times B$ be defined by $k(a) = (h(a), f(a))$, where $a \in A$. We note that $k \circ (f^*r) = g^>$, and $k_b = k|(A|b) : A|b \to W \times \{b\}, a \rightsquigarrow (h(a), f(a)) = (h_b(a), b)$ is continuous.

It follows from Theorem 4.1 that there is a map $g^< : Z \to (A\,; W \times B)$ over B, i.e. a map $g^< : r \to (f; \pi)$ defined by $g^<(z)(a) = g^>(a, z)$.

Let $s : B \to (A;\ W \times B)$ be the function defined by $s(b) = k_b$, hence we have $s(b) \in M(A|b, W \times \{b\})$ with $s(b)(a) = (h(a), f(a))$ and $f(a) = b$. We now see the point of our previous discussion of h_b and k_b — since it tells us that s is a well defined function.

Now if $r(z) = f(a) = b$, then $(sr(z))(a) = s(b)(a) = (h(a), f(a)) = g^>(z, a) = g^<(z)(a)$ and so $s \circ r = g^<$. Now $g^<$ is continuous and r is an identification, so s is continuous.

We see, from the definitions involved, that s is a section to $(f; \pi)$, i.e. $(f; \pi) \circ s = 1_B$. It is a consequence of Corollary 4.2 that the associated function $k : A \to W \times B$ is continuous. Then $\pi \circ k = h$, so h is continuous.

Example 4.9. *(i) If $p_Y : Y \to B$ and $p_Z : Z \to B$ are spaces over B, then there is an injective map over B, $i : Z \to (Y; Z)$, $i(z) = c_z$, where $z \in Z$.*
(ii) If $p_Y : Y \to B$ is an identification, then i is an embedding.

Proof. (i) The function i is clearly an injection. The projection $Z \sqcap Y \to Z$ is a map over B, i.e. from $p_Z \sqcap p_Y$ to p_Z. It follows, via Theorem 4.1, that there is an associated map $i : p_Z \to (p_Y; p_Z)$, i.e. the map i as specified.

(ii) Let $h : A \to Z$ be a function such that $i \circ h = g^< : A \to (Y; Z)$ is continuous. We just have to show that h is continuous.

Let $f = p_Z \circ h = (p_Y; p_Z) \circ i \circ h = (p_Y; p_Z) \circ g^< : A \to B$. Hence we have a map $g^< : f \to (p_Y; p_Z)$ and there is a map $g^> : f \sqcap p_Y \to p_Z, g^>(a, y) = g^<(a)(y) = (i \circ h)(a)(y) = (c_{h(a)})(y) = h(a) = h((f^*(p_Y))(a, y))$, so $g^> = h \circ (f^*(p_Y))$. Now $f^*(p_Y)$ is an identification (Proposition 4.8) and $g^>$ is continuous, so h is continuous.

5. Restricted Functional Fibrations

We will develop modified versions of the results of the previous section, i.e. that apply to functional fibrations whose domain has been restricted to being a single path component of their original domain. The properties of our functional classifying spaces are easily derived, in section 6, since they are essentially just particular cases of these modified results.

We assume, from this point on, that B is a simply connected pointed space, X is a path connected space, $p_X : X \to B$ is a map, and $p_Y : Y \to B$ and $p_Z : Z \to B$ are Hurewicz fibrations.

Let $\beta \in M(Y|*, Z|*)$. *We will define $(Y; Z; \beta)$ to be the path component of $(Y; Z)$ that contains β, and $(p_Y; p_Z; \beta) = (p_Y; p_Z)|(Y; Z; \beta) : (Y; Z; \beta) \to B$.*

We note that $(Y; Z; \beta)$ has the cg-ification of the topology that has subbasis consisting of all sets of the form $((p_Y; p_Z)^{-1}(U)) \cap (Y; Z; \beta) = (p_Y; p_Z; \beta)^{-1}(U)$ and $W(A, V) \cap (Y; Z; \beta)$, where U is an open subset of B, A is a compact subset of Y and V is an open subset of Z.

We recall that $(p_Y; p_Z)$ is a Hurewicz fibration (Theorem 4.6), with fibre $M(Y|b, Z|b)$ over $b \in B$ (Corollary 4.5). It follows from (2.4.3) that the the inclusion $M(Y|b, Z|b) \to (Y; Z)$ induces an injection $\pi_0(M(Y|b, Z|b)) \to \pi_0(Y; Z)$. Now the restriction of a Hurewicz fibration to a path component of its domain is itself a Hurewicz fibration, so we have established the following result.

Theorem 5.1. *The map $(p_Y; p_Z; \beta) : (Y; Z; \beta) \to B$ is a fibration with path connected fibres. Its distinguished fibre is $M(Y|*, Z|*, \beta)$.*

Theorem 5.2: Fibred Exponential Law. *Let $x_0 \in X|*$. We assume that $\alpha : \{x_0\} \times Y|* \to Z|*$ is a map, and that $\beta : Y|* \to Z|*$ is the composite of α with the canonical homeomorphism from $Y|*$ to $\{x_0\} \times Y|*$.*

Then there is a bijective correspondence between

(i) *maps $f^> : X \sqcap Y \to Z$ over B that extend α, and*
(ii) *maps $f^< : X \to (Y; Z; \beta)$ over B such that $f^<(x_0) = \beta$,*

determined by the rule $f^>(x, y) = f^<(x)(y)$, where $p_X(x) = p_Y(y)$.

Further, the rule $f^> \rightsquigarrow f^<$ defines a homeomorphism from the subspace of $M_B(X \sqcap Y, Z)$ of maps that extend α to the space $M_B(X, x_0; (Y; Z; \beta), \beta)$.

Proof. This is an easy consequence of the definition of $(Y; Z; \beta)$, the path-connectivity of X and Theorem 4.1.

Corollary 5.3. *Let $\beta \in M(Y|*, Z|*)$. There is a bijective correspondence between*

(i) *maps $g : Y \to Z$ over B that extend β, and*
(ii) *sections s to $(p_Y; p_Z; \beta)$ such that $s(*) = \beta$.*

determined by the rule $s(b)(y) = g(y)$, where $p_Y(y) = b$.

Proof. This follows either from the previous result, or from the definition of $(Y; Z; \beta)$ and corollary 4.2.

Example 5.4. *Let Y and Z be spaces, A be a pointed weak Hausdorff space and $\beta \in M(Y, Z)$. We will use π_1, π_2 and π_4 to denote the projections $A \times Y \to$*

A, $A \times Z \to A$ and $A \times M(Y;Z;\beta) \to A$ (to be consistent with the notation of Corollary 4.4). Then $(\pi_1;\pi_2;\beta)$ can be identified with π_4. To be precise, there is a homeomorphism

$$\kappa : A \times M(Y,Z:\beta) \to (A \times Y; A \times Z; 1_{\{*\}} \times \beta)$$

defined by $\kappa(a,f) = 1_{\{a\}} \times f$, such that $(\pi_1;\pi_2;\beta) \circ \kappa = \pi_4$

Proof. This follows from Corollary 4.4, and the definitions of $M(Y,Z;\beta)$ and $(Y;Z;\beta)$ and

Examples 5.5. *(i) Let Z be path connected and $c \in M(Y|*, Z|*)$ be the constant map to some point of $Z|*$. Then $(Y;Z;c)$ is independent of our choice of b and c, and there is an injective map over B*

$$j : Z \to (Y;Z;c),\ j(z) = c_z \in M(Y|p_Z(z), Z|p_Z(z)),$$

where $z \in Z$. If B is a CW-complex, then j is an embedding.

(ii) Let $\beta \in M(Y|*, Z|*)$. Then there is an evaluation map

$$e : (Y;Z;\beta) \sqcap Y \to Z,\ (f,y) \rightsquigarrow f(y),$$

where $(p_Y;p_Z;\beta)(f) = p_Y(y)$.

(iii) Let 1 denote the identity on ΩK and $v : M(S^1,K) \to K$ be the evaluation map determined by $v(f) = f(*)$, where $f \in M(S^1,K)$ and $* = (1,0) \in S^1$. Then there is an over K weak homotopy equivalence

$$\rho : M(S^1,K) \to (P;P;1),\ \rho(f)(l) = l + f,$$

where $f \in M(S^1,K)$ and $l \in P|f(*)$.

(iv) If $f,g \in M^0(B,K)$, then $f+g \in M^0(B,K)$ is defined by pointwise addition. Then there is an over B map and weak homotopy equivalence

$$\omega : B \sqcap_f P \to (B \sqcap_g P; B \sqcap_{f+g} P; 1),\ \omega(b,l)(b,m) = (b, l+m),$$

$f(b) = p(l)$, $g(b) = p(m)$. In this example the map 1 denotes the identity on $\{*\} \times \Omega K$.

Proof. (i) If $u,v \in Z$, then the continuity of i of Example 4.9 tells us that there is a path in $(Y;Z)$ from c_u to c_v. So the path component $(Y;Z;c)$ of $(Y;Z)$ is defined independently of our choice of b and c.

Now $i(Z)$ consists of constant maps, i.e. $i(Z) \subset (Y;Z;c)$. Hence j, which is just i with its range restricted, is well defined.

It follows from [M, Prop.2.1] that q_Y is an identification, so the last part follows from (ii) of Example 4.9.

(ii) The map e corresponds, via Theorem 5.2, to the identity map on $(Y;Z;c)$.

(iii) There is a map $M(S^1,K) \sqcap P \to P$, $(f,l) \rightsquigarrow l+f$, where $f(*) = v(f) = p(l) = l(1)$, i.e. a map $v \sqcap p \to p$. The continuity and over K property of ρ follow via Theorem 5.2. The Hurewicz fibrations v and $(p;p;1)$ have exact homotopy sequences; the map $\rho : v \to (p;p;1)$ determines a corresponding exact homotopy ladder. Restricting ρ to the distinguished fibre over $0 \in K$, we obtain the weak homotopy equivalence $\chi^\circ : \Omega K \to M(\Omega K, \Omega K; 1), \chi^\circ(f)(l) = l + f$, where f and l are in ΩK. The homotopy commutativity of loop addition on K and (2.1.1) ensure that $\chi^\circ \simeq \chi$, where χ is the weak homotopy equivalence described in (2.6.4). So χ° is a weak homotopy equivalence.

The result follows via the 5-Lemma.

(iv) There is a map over B

$$(B \sqcap_f P) \sqcap (B \sqcap_g P) \to B \sqcap_{f+g} P,\ ((b,l),(b,m)) \rightsquigarrow (b, l+m),$$

where $f(b) = p(l)$ and $g(b) = p(m)$. The continuity and over B property of ω follow via Theorem 5.2.

The rest of the proof is now similar to that of (iii), but replacing ρ by the map $\omega : f^*p \to (g^*p;\ (f+g)^*p; 1)$. The argument is easier this time, since χ° is now replaced by χ.

Proposition 5.6. *If Z is path connected, then the underlying set of $(Y;Z;c)|b$ is $\{f \in M(Y|b, Z|b) \mid f \text{ is null homotopic}\}$, and the underlying set of $(Y;Z;c)$ is $\bigcup_{b \in B} \{f \in M(Y|b, Z|b) \mid f \text{ is null homotopic}\}$.*

Proof. Let us assume that $f \in M(Y|b, Z|b)$ and that f is null homotopic, i.e. $f \simeq c'$, where $c' : Y|b \to Z|b$ is a constant map. So there is a path from f to c' in $M(Y|b, Z|b)$, and hence a path from f to c' in $(Y;Z)$. Now $c' \in (Y;Z;c)$, so $f \in (Y;Z;c)$.

Conversely, p_Y and p_Z are fibrations over a path connected space B, so $Y|b$ and $Z|b$ are non-empty. Hence there is a constant map $Y|b \to Z|b$, and $(Y;Z;c)|b$ contains at least one constant map. Now we know from Theorem 5.1 that $(Y;Z;c)|b$ is a single path component of $M(Y|b, Z|b)$, so $f \in (Y;Z;c)|b$ implies that f is homotopic to the aforementioned constant map, i.e. f is null homotopic.

Proposition 5.7. *Let $* \in A \subseteq B$ and $\beta \in M(Y|*, Z|*)$. Then $(Y|A; Z|A; \beta)$ is a subspace of $(Y;Z;\beta)$, in the sense that $(Y|A; Z|A; \beta) = (Y;Z;\beta)|A$.*

Proof. This follows from the definition of $(Y;Z;\beta)$ and Corollary 4.5.

Examples 5.8. *(i) Given maps f and $g : (B, B_0) \to (K, 0)$, there is an embedding*

$$\phi : B_0 \times M(\Omega K, \Omega K; 1) \to (B \sqcap_f P; B \sqcap_g P; 1),\ \phi(b,l) = 1_{\{b\}} \times l.$$

where $b \in B_0$ and $l \in M(\Omega K, \Omega K; 1)$. Then $(f^*p; g^*p; 1)$ (essentially) extends the projection $B_0 \times M(\Omega K, \Omega K; 1) \to B_0$, i.e. $(f^*p; g^*p; 1) \circ \phi = \pi_4$ where π_4 is as in Example 5.4.

(ii) There is a map $\omega : B \sqcap_f P \to (B \sqcap_g P; B \sqcap_{f+g} P; 1)$ as described in Example 5.5(iv), and a map $\chi : K \to M(K, K; 1_K)$ as described in (2.6.1). Then $\omega|(B_0 \times \Omega K)$ is essentially $1 \times \chi$, where 1 refers to the identity on B_0. More precisely $\omega|(B_0 \times \Omega K) = \phi \circ (1 \times \chi)$.

Proof. Part (i) follows by composing κ of Example 5.4 with the inclusion of Proposition 5.7; part (ii) is an immediate consequence of the definitions of the functions involved.

6. Functional Classifying Space Proofs

We will now show that our functional classifying spaces $M_\infty = M_\infty(P, L)$ can be viewed as fibred mapping spaces, of the type discussed in section 5, and use the results of that section to verify the properties of M_∞ that are stated in section 3.

The order of the results of section 3 was based on topic: first were results that were (almost) symmetric in K and L, then fibration related results and finally exponential law style results. The order of the proofs in this section - a logical order - is quite different.

We will use $\pi : K \times L \to K$ to denote the obvious projection, and will take $c_{00} : \Omega K \to \{0\} \times L$ to denote the constant function value $(0, 0)$.

Proposition 6.1. *The functional classifying space $M_\infty = M_\infty(P, L)$ can be identified with the fibred mapping space $(P; K \times L; c_{00})$, and the map $p_\infty : M_\infty \to K$ with the functional fibration $(p; \pi; c_{00}) : (P; K \times L; c_{00}) \to K$.*

To be precise, there is a homeomorphism $\theta : M_\infty(P, L) \to (P; K \times L; c_{00})$ such that if $f \in M(P|x, L) \subset M_\infty(P, L)$, then $\theta(f) \in M(P|x, \{x\} \times L)$ is obtained by composing f with the canonical homeomorphism $L \to \{x\} \times L$. Further we have $(p; \pi; c_{00}) \circ \theta = p_\infty$.

Proof. It is clear that θ is a well defined injection; it is a surjection by Proposition 5.6 and hence is a bijection.

If U is a subset of K, then $\theta(U_\infty) = (p; \pi; c_{00})^{-1}(U)$. If A is a subset of P and V is a subset of $K \times L$, then $\theta(W_\infty(A, V)) = W(A, V) \cap (P; K \times L; c_{00})$. So the previously discussed subbases for $M_\infty(P, L)$ and $(P; K \times L; c_{00})$ correspond, and θ induces a homeomorphism between the modified compact-open topologies on these spaces, and hence between the associated cg-ified topologies.

Notes. (i) $M_\infty(P, L)$ has basepoint $c_0 : \Omega K \to L$ and $(P; K \times L; c_{00})$ has basepoint c_{00}. We note that $\theta(c_0) = c_{00}$.
(ii) If $f \in (P; K \times L; c_{00})$, i.e. $f \in M(P|x, \{x\} \times L)$ for some $x \in K$, then $\theta^{-1}(f)$ is obtained by composing f with the projection and homeomorphism $\{x\} \times L \to L$.

Proof of (M1), (M3), (M4) and (M8). These follows from Proposition 6.1 and the results of section 5, with $p_Y = p, p_Z = \pi$ and $\beta = c_{00} : \Omega K \to \{0\} \times L$. Thus (M1) uses Example 5.5(i), (M3) uses Theorem 5.1, (M4) uses Proposition 5.6, and (M8) uses Example 5.5(ii).

Proof of (M2). Composing the two embeddings, $K \to K \times L, x \rightsquigarrow (x, 0)$ and $L \to K \times L, y \rightsquigarrow (0, y)$, with j_∞, we obtain the required embeddings of K and L as the subspaces $\hat{K}$ and $\hat{L}$ of M_∞, respectively.

If $G = 0$, then $K = \{0\}$ and $M_\infty = M_\infty(P, L) = M_\infty(\{0\}, L) = \hat{L}$.

If $H = 0$, then $L = \{0\}$ and $M_\infty = M_\infty(P, L) = M_\infty(P, \{0\})$
$= \bigcup_{x \in K} M(P|x, \{0\}) = \hat{K}$

Proof of (M5). There is a map $f : P \to K \times L, f(l) = (l(1), 0)$, where $l \in P$, that extends c_{00}. It follows, from corollary 5.3, that there is a section s to $(P; K \times L; c_{00})$ such that $s(x) : P|x \to \{x\} \times L$ is the constant function with value $(x, 0)$, where

$x \in K$. We note that s_∞, as defined in (M5), is $\theta^{-1} \circ s$. Hence s_∞ is continuous. It is clear, from the definitions of p_∞ and s_∞, that $p_\infty \circ s_\infty = 1_K$.

Proof of (M6). Let $\tau(f, g)$ be a $\Omega K \times \Omega L$-tower. So $f \in M^0(B, K)$ and $g \in M(B \sqcap_f P, \{*\} \times \Omega K; L, \{0\})$. Then there is a map

$$g^> : B \sqcap P \to K \times L,\ g^>(b, l) = (l(1), g(b, l)),$$

where $f(b) = l(1)$. We notice that $g^>$ extends $c_{00} : \{*\} \times \Omega K \to \{0\} \times L$ and is a map over K, i.e. $g^> : f \sqcap p \to \pi$ where $\pi : K \times L \to K$.

The fibred exponential law (Theorem 5.2) then determines a map

$$g^< : B \to (P; K \times L; c_{00}),\ g^<(b)(l) = g^>(b, l),$$

i.e. $g^<(b)(l) = (l(1), g(b, l))$, where $f(b) = p(l)$. Then $g^<$ is based in that we have $g^<(*) = c_{00}$.

Composing $g^<$ with the map $\theta^{-1} : (P; K \times L; c_{00}) \to M_\infty$, we obtain a map $g^\bullet = \theta^{-1} \circ g^< : B \to M_\infty$. Now $g^<$ is the map $b \rightsquigarrow (l \rightsquigarrow (l(1), g(b, l))$, so $g^\bullet$ is $b \rightsquigarrow (l \rightsquigarrow g(b, l))$, where $f(b) = l(1)$. Hence $g^\bullet(b)(l) = g(b, l)$. We also notice that

$$g^\bullet(*) = \theta^{-1}(g^<(*)) = \theta^{-1}(c_{00}) = c_0 : \Omega K \to L,$$

so $g^\bullet \in M(B, *; M_\infty, c_0)$.

Conversely, let $g^\bullet : B \to M_\infty$ be a based map. We define $f = p_\infty \circ g^\bullet : B \to K$ and $g^< = \theta \circ g^\bullet : B \to (P; K \times L; c_{00})$, i.e. $b \rightsquigarrow (l \rightsquigarrow (l(1), g^\bullet(b)(l)))$, where $f(b) = l(1)$. We notice that

$$g^<(*) = \theta(g^\bullet(*)) = \theta(c_0) = c_{00},$$

i.e. that $g^<$ is a based map. Also $(p; \pi; c_{00}) \circ g^< = (p; \pi, c_{00}) \circ \theta \circ g^\bullet = p_\infty \circ g^\bullet = f$, so $g^<$ is a map over K in the sense that $g^< : f \to (p; \pi; c_{00})$.

It follows, via the fibred exponential law (Theorem 5.2), that there is a map $g^> : B \sqcap_f P \to K \times L$ determined by the rule $g^>(b, l) = g^<(b)(l)$, where $f(b) = l(1)$. Thus $g^>(b, l) = (l(1), g^\bullet(b)(l))$. Also $g^>$ extends $c_{00} : \{*\} \times \Omega K \to \{0\} \times L$ and $g^>$ is over K, i.e. $g^> : f \sqcap p \to \pi$.

Let us define $g = \pi' \circ g^> : B \sqcap_f P \to L$, where $\pi' : K \times L \to L$ denotes the projection. So $g(b, l) = \pi'(l(1), g^\bullet(b)(l)) = g^\bullet(b)(l)$, where $f(b) = g(l)$. Further, g extends $\pi' \circ c_{00} = c_0 : \{*\} \times \Omega K \to L$ and so $\tau(f, g)$ is a $\Omega K \times \Omega L$-tower.

The procedures in the two halves of this proof are inverse to each other, so our argument is complete.

Proof of (M7). This proof is similar to that of (M6), except that we use the over B homotopies of Theorem 4.7, instead of the over B maps of Theorem 4.1, and utilize (2.5.4) in conjunction with the projection $B \times I \to B$.

Proof of (M9). We recall from (M6) that $p_\infty \circ g^\bullet = f$. So it follows from (2.5.6) that the principal ΩK-fibration of the tower $(g^\bullet)^* \tau(p_\infty, e_\infty)$ is

$$(g^\bullet)^*(p_\infty)^* p = (p_\infty \circ g^\bullet)^* p = f^* p : B \sqcap_f P \to B.$$

It follows, from (2.5.6), that the pullback space $B \sqcap (M_\infty \sqcap P)$, and its projections $B \sqcap (M_\infty \sqcap P) \to B$ and $B \sqcap (M_\infty \sqcap P) \to M_\infty \sqcap P$, can be identified with the pullback space $B \sqcap_f P$, the projection $B \sqcap_f P \to B$, and the map $g^\bullet \sqcap 1_P : B \sqcap_f P \to M_\infty \sqcap P$, respectively.

Then we have $e_\infty \circ (g^\bullet \sqcap 1_P)(b,l) = e_\infty(g^\bullet(b),l) = g^\bullet(b)(l) = g(b,l)$, where $(b,l) \in B \sqcap_f P$, so $e_\infty \circ (g^\bullet \sqcap 1_P) = g$. It follows from (2.5.6) that the principal ΩL-fibration of the tower $(g^\bullet)^*\tau(p_\infty, e_\infty)$ may be taken to be

$$(g^\bullet \sqcap 1_P)^*(e_\infty)^* q = (e_\infty \circ (g^\bullet \sqcap 1_P))^* q = g^* q.$$

So, in more detail, the pullback space $(B \sqcap_f P) \sqcap ((M_\infty \sqcap P) \sqcap Q)$ and its projections on $(B \sqcap_f P)$ and $(M_\infty \sqcap P) \sqcap Q)$ can be identified with the space $(B \sqcap_f P) \sqcap_g Q$, its projection on $B \sqcap_f P$ and the map $(g^\bullet \sqcap 1_P) \sqcap 1_Q : (B \sqcap_f P) \sqcap_g Q \to (M_\infty \sqcap P) \sqcap Q$, respectively.

Hence, to within the aforementioned canonically defined homeomorphisms, the $\Omega K \times \Omega L$-tower $(g^\bullet)^*\tau(p_\infty, e_\infty)$ consists of the principal fibrations f^*p and g^*q, i.e. it is $\tau(f,g)$.

Proof of (M10). We have already seen that $j_\infty : K \times L \to M_\infty$ is an embedding (M1), so it follows from (2.7.5) that the induced tower $(j_\infty)^*\tau(p_\infty, e_\infty)$ embeds in $\tau(p_\infty, e_\infty)$. We will now determine the nature of this embedded 3-tower.

It follows from (2.5.6) that the principal ΩK-fibration of the embedded 3-tower is induced from p via $p_\infty \circ j_\infty : K \times L \to K$. Now $p_\infty \circ j_\infty$ is the projection $\pi : K \times L \to K$ (M1), so it follows from (2.5.4) and (2.5.6) that this principal ΩK-fibration can be taken to be $p \times 1_L : P \times L \to K \times L$. The embedding of $P \times L$ in $M_\infty \sqcap P$ is then the composite of a homeomorphism described in (2.5.4), i.e. $t_\infty : P \times L \to (K \times L) \sqcap P, (l,y) \rightsquigarrow (p(l)), y, l)$, where $l \in P$ and $y \in L$, and an embedding described in (2.5.6), i.e. $j_\infty \sqcap 1 : (K \times L) \sqcap P \to M_\infty \sqcap P$. So this embedding is the map

$$u_\infty : P \times L \to M_\infty \sqcap P,\ (l,y) \rightsquigarrow (c_y : M_\infty|(p(l) \to L,\ l),$$

where $l \in P$ and $y \in L$.

It also follows from (2.5.6) that the principal ΩL-fibration of the embedded 3-tower is induced from q by the map $e_\infty \circ u_\infty = \pi'$, i.e. by the projection $P \times L \to L$. Hence, using (2.5.4), we see that the principal ΩL-fibration of the embedded 3-tower can be taken to be $1_P \times q : P \times Q \to P \times L$. The embedding of $P \times Q$ into $(M_\infty \sqcap P) \sqcap Q$ is the composite of $v_\infty : P \times Q \to (P \times L) \sqcap Q, (l,m) \rightsquigarrow ((l, q(m)), m)$, where $l \in P$ and $m \in Q$, as described in (2.5.4), and the map $u_\infty \sqcap 1_Q : (P \times L) \sqcap Q \to (M_\infty \sqcap P) \sqcap Q$, as described in (2.5.6). So this embedding is the map

$$z_\infty : P \times Q \to (M_\infty \sqcap P) \sqcap Q,\ (l,m) \rightsquigarrow ((c_{q(m)} : M_\infty|(p(l)) \to L),\ l,\ m),$$

where $l \in P$ and $m \in Q$.

Hence we have found an embedding of the fibration $(p \times 1_L) \circ (1_K \times q) = p \times q : P \times Q \to K \times L$ into the universal 3-tower fibration and projection $t(p_\infty, e_\infty)$, using the embeddings $j_\infty : K \times L \to M_\infty$ and $z_\infty : P \times Q \to (M_\infty \sqcap P) \sqcap Q$.

7. The 1FHE Classification of $\Omega K \times \Omega L$-Towers

We assume, throughout this section, that B is a simply connected CW-complex. The base point $* \in B$ is necessarily non-degenerate (see (2.4.8)).

If A is a subspace of B and $f : (B, A) \to (K, 0)$ is a map, then we note that the restriction of $f^*p : B \sqcap_f P \to B$ over A is the projection $A \times \Omega K \to A$.

Lemma 7.1. *Let A be a closed subspace of B such that (B, A) has the HEP and f and g be maps of (B, A) to $(K, 0)$. Then $f \simeq^A g$ is and only if there is an FHE from $B \sqcap_f P$ to $B \sqcap_g P$ that extends the identity on $A \times \Omega K$.*

Proof. ($\Rightarrow$) Let us assume that $f \simeq^A g$. Now K is a topological group, so we can pointwise subtract f from the above homotopy. Thus we obtain a homotopy $c_0 \simeq^A g - f$, i.e. a homotopy $H^> : B \times I \to K$ such that $H^>(b,0) = 0$ and $H^>(b,1) = (g-f)(b) = g(b)f(b)$, for all $b \in B$, and $H^>(a,t) = 0$, for all $a \in A$ and $t \in I$.

Applying (2.1.1), we see that there is a corresponding map $H^< : B \to M(I,K)$ defined by $H^<(b)(t) = H^>(b,t)$. Now $H^<(b)(0) = H^>(b,0) = 0$ and, since PK has underlying set $\{f \in M(I,K) | f(0) = 0\}$, $H^<$ can be taken to be a map $H^< : B \to PK$. We notice that $H^<(b)(1) = H^>(b,1) = (g-f)(b)$, for $b \in B$, so $p \circ H^< = g - f$. Also $g(a)(t) = H^>(a,t) = 0$ for all $a \in A$ and $t \in I$, so $H^<(a) = c_0 \in PK$. It follows, by the universal property of pullbacks, that $(g-f)^*(p) : B \sqcap_{g-f} P \to B$ has a section $\sigma : B \to B \sqcap_{g-f} P$, with $\sigma(b) = (b, H^<(b))$, where $b \in B$. Then $\sigma(a) = (a, c_0) \in A \times \Omega K \subseteq B \sqcap_{g-f} P$, where $a \in A$.

We know, by examples 5.5(iv) and 5.8(ii), that there is a weak homotopy equivalence

$$\omega : B \sqcap_{g-f} P \to (B \sqcap_f P; B \sqcap_g P; 1)$$

that is over B, and such that $\omega | A \times \Omega K = 1_A \times \chi : A \times \Omega K \to A \times M(\Omega K, \Omega K; 1)$. Composing, $\omega \circ \sigma$ is a section to $(f^*p; g^*p; 1)$ and $\omega \circ \sigma(a)$ is the map $\omega(a, c_0) = (1_A \times \chi)(a, c_0) = (a, 1^+)$ (see 2.6.3).

Applying Corollary 5.3, we see that there is a map $\eta : f^*p \to g^*p$ that restricts over A to $1_A \times 1^+ : A \times \Omega K \to A \times \Omega K$. Now $1^+ \simeq 1_{\Omega K}$, so $\eta | A \times \Omega K \simeq 1_{A \times \Omega K}$. It follows by (2.4.4) and (2.4.5) that there is an FHE of the required type.

($\Leftarrow$) Most of the first half of this proof is directly reversible; the only exception is the argument involving the composition of the section σ with ω, to obtain a section to $(f^*p; g^*p; 1)$.

So we need to show that, given a section s_1 to $(f^*p; g^*p; 1)$ that extends

$$A \to (B \sqcap_f P; B \sqcap_g P; 1), a \rightsquigarrow (a, 1^+),$$

there must exist a section s to $(g-f)^*(p)$ that extends

$$A \to B \sqcap_{g-f} P, a \rightsquigarrow (a, c_0).$$

We know by [Sp, Cor.7.6.23] that there is a map $s_2 : B \to B \sqcap_{g-f} P$ such that $\omega \circ s_2 \simeq s_1$. Then $((g-f)^*p) \circ s_2 = (f^*p; g^*p; 1) \circ \omega \circ s_2 \simeq (f^*p; g^*p; 1) \circ s_1 = 1_B$.

It follows, by the CHP for $(g-f)^*p$, that there is a map $s_3 : B \to B \sqcap_{g-f} P$ with $s_3 \simeq s_2$ and such that $((g-f)^*p) \circ s_3 = 1_B$, i.e. s_3 is a section to $(g-f)^*p$. Then $(\omega \circ s_3)|A \simeq (\omega \circ s_2)|A \simeq s_1|A = (1_A, c_{1^+})$. Now $(1_A, c_0)$ is a section to the projection $A \times \Omega K \to A$ and $\omega \circ (1, c_0) = (1_A, c_{1^+})$, so we see, via [Sp, Cor.7.6.23], that $s_3|A \simeq (1_A, c_0)$. Now homotopic sections are vertically homotopic [JT, Lem.1.1], so $s_3|A \simeq_A (1_A, c_0) : A \times I \to B \sqcap_{g-f} P$.

Identifying B with $B \times \{0\}$ and combining s_3 with the last mentioned homotopy, we determine a map $\nu : (B \times 0) \cup (A \times I) \to B \sqcap_{g-f} P$. Now $((g-f)^*(p)) \circ \nu$ is the restriction of the projection $\pi : B \times I \to B$, so it follows by (2.4.5), that there is a homotopy $N : B \times I \to B \sqcap_{g-f} P$ that extends ν and is such that $((g-f)^*p) \circ N = \pi$.

Restricting N to $B \times \{1\}$, we necessarily obtain a section s to $(g-f)^*p$. Then if $a \in A$, $s(a) = N(a,1) = (1_A, c_0)(a,1) = (a, c_0) \in A \times \Omega K \subseteq B \sqcap_{g-f} P$. So the proof is complete.

Let f be the base point of $M^0(B, K; f)$. If $\zeta \in M^0(S^1, M^0(B, K; f))$, then we will see that it determines a pair of maps ζ° and $\zeta^{\circ\circ}$, and look at a few of their properties.

(i) We define $\zeta^\circ : B \to M(S^1, K)$ by the rule $\zeta^\circ(b)(t) = \zeta(t)(b)$, where $b \in B$ and $t \in S^1$. The continuity of ζ° follows after two applications of (2.1.1). We notice that, for each $b \in B, \zeta^\circ(b)$ is a loop in K at $f(b)$. Taking $e : M(S^1, K) \to K$ to be the evaluation map at $* \in S^1$, and noticing that $\zeta(*)(b) = f(b)$, we see that $e(\zeta^\circ(b)) = \zeta^\circ(b)(*) = \zeta(*)(b) = f(b)$. So $e \circ \zeta^\circ = f$, i.e. ζ° is a lifting of f over e. Also $\zeta^\circ(*)(t) = \zeta(t)(*) = 0$, so $\zeta^\circ(*) = c_0 : S^1 \to K$. Hence $\zeta^\circ \in M^0_K(B, M(S^1, K; c_0))$.

It is easily verified that the above argument is reversible, so $\zeta \rightsquigarrow \zeta^\circ$ defines a bijective correspondence $M^0(S^1, M^0(B, K; f)) \to M^0_K(B, M(S^1, K))$.

(ii) We define a homotopy $\zeta^{\circ\circ} : B \times I \to K, \zeta^{\circ\circ}(b, t) = (\zeta \circ \eta)(t)(b)$, where $b \in B, t \in I$ and $\eta : I \to S^1$ is the obvious identification involving multiplication by 2π. The continuity of $\zeta^{\circ\circ}$ is an application of (2.1.1). If $t = 0$ or 1, then $\zeta^{\circ\circ}(b, t) = \zeta(b)(*) = f(b)$, and $\zeta^{\circ\circ}(*, t) = (\zeta \circ \eta)(t)(*) = 0$. Hence $\zeta^{\circ\circ}$ is a based self-homotopy of f.

In the following result we contiue to assume that $f \in M^0(B, K)$, and use $B \sqcap P$ to denote the pullback space $B \sqcap_f P$.

Lemma 7.2. *(i) Let $\delta : M^0(S^1, M^0(B, K; f)) \to 1FHE(B \sqcap P, B \sqcap P)$ be the map defined by $\delta(\zeta)(b, l) = (b, l + \zeta^\circ(b))$, where $\zeta \in M^0(S^1, M^0(B, K; f))$, $(b, l) \in B \sqcap P$, and ζ° is as defined previously.*

(ii) The function $\pi_0(\delta) : \pi_1(M^0(B, K; f)) \to \pi_0(1FHE(B \sqcap P, B \sqcap P))$ is a bijection.

Proof (i) It is easily seen via (2.1.1) that, given $\zeta \in M^0(S^1, M(B, K; f))$, the corresponding $\delta(\zeta)$ is a well defined and continuous function over B. We know that $\zeta^\circ(b)$ is a loop in K at $f(b)$, so there is an inverse loop $-\zeta^\circ(b)$ in K at $f(b)$. The rule $(b, l) \rightsquigarrow (b, l - \zeta^\circ(b))$ then defines a fibre homotopy inverse $\gamma(\zeta)$ to $\delta(\zeta)$. Hence $\delta(\zeta)$ is a self-FHE of $B \sqcap P$ over B.

Restricting $\delta(\zeta)$ to the distinguished fibre $\{*\} \times \Omega K$, we obtain the map $\{*\} \times \Omega K \to \{*\} \times \Omega K, (*, l) \rightsquigarrow (*, l + \zeta^\circ(*))$, where $l \in \Omega K$. Now $\zeta^\circ(*) = c_0 : I \to K$. Hence this self-map of $\{*\} \times \Omega K$ simply adds the constant loop c_0, i.e. after identifying $\{*\} \times \Omega K$ with ΩK, it is the map $1^+ : \Omega K \to \Omega K$ of (2.6.3). Now 1^+ is homotopic to the identity on ΩK, so δ is a self-*1FHE* of $B \sqcap P$ over B.

(ii) We notice that $\delta = \delta_3 \circ \delta_2 \circ \delta_1$, where:

$\delta_1 : M^0(S^1, M^0(B, K; f)) \to M^0_K(B, M(S^1, K; c_0)), \delta_1(\zeta) = \zeta^\circ$ with, as explained previously, $\zeta \in M^0(S^1, M^0(B, K; f))$,

$\delta_2 : M^0_K(B, M(S^1, K; c_0)) \to M^0_K(B, (P; P; 1^+)), \delta_2(\upsilon) = \rho \circ \upsilon$, where ρ is the weak homotopy equivalence $M(S^1, K) \to (P; P; 1^+)$ of Example 5.5(iii) and $\upsilon \in M^0_K(B, M(S^1, K; c_0))$, and

$\delta_3 : M^0_K(B, (P; P; 1^+)) \to 1FHE(B \sqcap P, B \sqcap P), \delta_3(\nu)(b, l) = (b, \nu(b)(l))$, for $\nu \in M_K(B, (P; P; 1^+)$ and $(b, l) \in B \sqcap P$.

We know from (2.1.1) that $M(S^1, M(B, K)) \cong M(S^1 \times B, K)$, and $M(S^1 \times B, K) \cong M(B, M(S^1, K))$. So $M(S^1, M(B, K)) \cong M(B, M(S^1, K)$. Then δ_1 is a restriction of the latter homeomorphism that maps a subspace of its domain to a subspace of its range. So δ_1 is a homeomorphism and $\pi_0(\delta_1)$ a bijection.

It was shown in Ex 5.5(ii) that ρ is a weak homotopy equivalence, so it follows from [Sp, Cor.7.6.23] that $\pi_0(\delta_2)$ is a bijection.

There are homeomorphisms $M_K(B,(P;P)) \cong M_K(B \sqcap P, P)$ (see Theorem 5.2), and $M_K(B \sqcap P, P) \cong M_B(P \sqcap P, B \sqcap P)$ (see (2.5.9)). Then, restricting the composite homeomorphism $M_K(B,(P;P)) \cong M_B(B \sqcap P, B \sqcap P)$ to $M_K^0(B,(P;P;1^+))$, we obtain a homeomorphism from the latter space to $F(1^+)$. Here $F(1^+)$ is the subspace of $M_B(B \sqcap P, B \sqcap P)$ that consists of maps that restrict to $1 \times 1^+$ on the distinguished fibre $\{*\} \times \Omega K$.

If the identity map on ΩK is replaced by 1^+, throughout the argument of (2.4.7), then that argument is still applicable. It follows that π_0 of the inclusion $F(1^+) \to 1FHE(B \sqcap P, B \sqcap P)$ is a bijection. Hence $\pi_0(\delta_3)$ is a bijection.

So $\pi_0(\delta)$ is the composite of three bijections, and is therefore itself a bijection.

Lemma 7.3. *Let $\tau(f,k)$ be a $\Omega K \times \Omega L$-tower over B and $\zeta \in M^0(S^1, M^0(B,K;f))$. There is a map $R : (B \times I) \sqcap_{\zeta^{oo}} P \longrightarrow L$,*

$$R(b,s,l) = \begin{cases} k(b, t \rightsquigarrow l\left(\frac{2t}{2-s}\right)) & s + 2t \le 2 \\ k(b, t \rightsquigarrow \zeta^{oo}\left(b, \frac{2-2t}{s}\right)) & s + 2t \ge 2 \end{cases}$$

such that $R|(B \times \{0\}) \sqcap P = k$ and $R|(B \times \{1\}) \sqcap P = k \circ \gamma(\zeta)$. Further the space $((B \times I) \sqcap_{\zeta^{oo}} P)|(\{*\} \times I) = (* \times I) \sqcap_{\zeta^{oo}} P = \{*\} \times I \times \Omega K$, and the map $R|\{*\} \times I \times \Omega K = c_0 : \{*\} \times I \times \Omega K \to L$.

The proof is routine and left to the reader.

Lemma 7.4. *Let $\tau(f,g)$ and $\tau(h,k)$ be $\Omega K \times \Omega L$-towers over B with $f = h$. If $\tau(f,g)$ is 1FHE to $\tau(h,k)$, then $g^\bullet \simeq^0 k^\bullet$.*

Proof. We see via (2.7.3) that there is a strict *1FHE* $<\eta, \lambda>$: $\tau(f,g) \to \tau(h,k)$. Applying the universal property of pullbacks and (2.5.6), we see that there is an *FHE* $g^*q \to (k \circ \eta)^*q$, i.e. a homotopy equivalence

$$(B \sqcap_f P) \sqcap_g Q \to (B \sqcap_f P) \sqcap_{k\eta} Q$$

over $B \sqcap_f P$, defined by

$$(b,l,m) \rightsquigarrow (b,l,\pi_Q(\lambda(b,l,m)))$$

where $(b,l,m) \in (B \sqcap_f P) \sqcap_g Q$ and π_Q is the projection $(B \sqcap_f P) \sqcap_k Q \to Q$.

We see via (2.7.5) that $(\{*\} \times \Omega K) \times \Omega L$ is the subspace of both $(B \sqcap_f P) \sqcap_g Q$ and $(B \sqcap_f P) \sqcap_{k\eta} Q$ that is over the subspace $\{*\} \times \Omega K$ of $B \sqcap_f P$. Now λ is the identity on $(\{*\} \times \Omega K) \times L$, hence so also is our given FHE $g^*q \to (k \circ \eta)^*q$. It follows from Lemma 7.1 that $k \circ \eta \simeq^{\Omega K} g$.

Further, Lemma 7.2 tells us that there is a $\zeta \in M^0(S^1, M^0(B,K;f))$, such that $\eta \simeq_B^{\Omega K} \delta(\zeta)$. So $k \circ \delta(\zeta) \simeq^{\Omega K} k \circ \eta \simeq^{\Omega K} g$.

Let $\gamma(\zeta)$ be the previously specified fibre homotopy inverse of $\delta(\zeta)$. Then $k \simeq^{\Omega K} k \circ \delta(\zeta) \circ \gamma(\zeta) \simeq^{\Omega K} g \circ \gamma(\zeta)$. These homotopies take $\{*\} \times \Omega K$ to $0 \in L$, so it follows from the (M7) that $k^\bullet \simeq^0 (g \circ \gamma(\zeta))^\bullet$ ... (1)

The map R satisfies the criteria of (M7); it follows that
$g^\bullet \simeq^0 (g \circ \gamma(\zeta))^\bullet$ ... (2)

Combining (1) and (2), we see that $g^\bullet \simeq^0 k^\bullet$.

Let B be a given pointed space. We will take F to be a space constructed by a finite Postnikov decomposition process, thus it is the total space of a Postnikov tower with base space a single point. Then the set of *1-fibre homotopy types*, i.e. the set of 1FHE-classes, of Moore-Postnikov towers over B with distinguished fibre F will be denoted by *1FHT(F: B)*.

Theorem 7.5. *(i) Let B be a simply connected CW-complex with a (necessarily non-degenerate) base point. Let $\tau(f,g)$ and $\tau(h,k)$ be 3-stage towers over B, with associated classifying maps $g^\bullet$ and $k^\bullet$ respectively. Then $\tau(f,g)$ is 1FHE to $\tau(h,k)$ if and only if $g^\bullet \simeq^0 k^\bullet$.*
(ii) The rule that the 1FHT of $\tau(f,g)$ corresponds to the based homotopy class of $g^\bullet$ determines a bijective correspondence

$$1FHT(\Omega K \times \Omega L : B) \approx [B, M_\infty]^0 .$$

Proof. In reading the following argument one should recall (2.7.2) that a pair of 3-towers are *1FHE* if and only if the corresponding underlying fibrations are *1FHE*.

((i) $\Leftarrow$) We recall (M9) that $\tau(f,g) = (f^\bullet)^*\tau(p_\infty, e_\infty)$. We notice that our present concept of $\tau(f,g)$ and $\tau(h,k)$ being *1FHE* agrees with the [B4, p.142] concept of $\tau(f,g)$ being $\mathcal{E}$*1FHE* to $\tau(h,k)$, where $\mathcal{E}$ is the class of all spaces. Then $f^\bullet \simeq^0 h^\bullet$ implies that $(f^\bullet)^*\tau(p_\infty, e_\infty)$ is *1FHE* to $(h^\bullet)^*\tau(p_\infty, e_\infty)$ [B4, Prop.6.1], and so $\tau(f,g)$ is $1FHE$ to $\tau(h,k)$.

((i) $\Rightarrow$) Let us assume that there is a $1FHE$ $<\eta, \lambda>$ from $\tau(f,g)$ to $\tau(h,k)$. We know, by (2.7.3), that $<\eta, \lambda>$ can be taken to be a strict *1FHE*. Then η is a strict *1FHE* from $B \sqcap_f P$ to $B \sqcap_h P$, and so $f \simeq^0 h$ (Lemma 7.1). We know (M6) that $p_\infty : M_\infty \to K$ satisfies the CHP, and that $g^\bullet$ lifts f over p_∞. It follows by (2.4.5) that there is a based map $g^{\bullet\prime} : B \to M_\infty$ that lifts h and is such that $g^{\bullet\prime} \simeq^0 g^\bullet$.

Then $(g^{\bullet\prime})^*t(p_\infty, e_\infty)$	*is 1FHE to*	$(g^\bullet)^*t(p_\infty, e_\infty)$	by [B4, Prop.6.1]
	$=$	$t(f,g)$	by (M9)
	is 1FHE to	$t(h,k)$	by our data
	$=$	$(k^\bullet)^*t(p_\infty, e_\infty)$	by (M9)

Hence $(g^{\bullet\prime})^*t(p_\infty, e_\infty)$ is *1FHE* to $(k^\bullet)^*t(p_\infty, e_\infty)$. Now $(g^{\bullet\prime})^*t(p_\infty, e_\infty)$ and $(k^\bullet)^*t(p_\infty, e_\infty)$ both have first k-invariant h. So it follows, by lemma 7.4, that $g^{\bullet\prime} \simeq^0 k^\bullet$. We already know that $g^\bullet \simeq^0 g^{\bullet\prime}$, so $g^\bullet \simeq k^\bullet$.

(ii) This follows from (M9) and (i) of this present result.

References

B1. P. I. BOOTH, The Exponential Law of Maps I, Proc. Lon. Math. Soc.(3) 20 (1970), 179-192.

B2. PETER. I. BOOTH, The Exponential Law of Maps II, Math. Z. 121, 311-319 (1971).

B3. P. BOOTH, Fibrations and Cofibred Pairs, Math. Scand. 35 (1974), 145-148.

B4. PETER I. BOOTH, Local to Global Properties in the Theory of Fibrations, Cahiers de Topologie et Géométrie Différentielle Catégoriques, XXXIV-2 (1993), 127-151.

B5. PETER I. BOOTH, Fibrations and Classifying Spaces: Overview and the Classical Examples, to appear in Cahiers de Topologie et Géométrie Différentielle Catégoriques.

B6. PETER I. BOOTH, Fibrations with Product of Eilenberg-MacLane Space Fibres II, to appear.

BB1. PETER I. BOOTH and RONALD BROWN, Spaces of Partial Maps, Fibred Mapping Spaces and the Compact-Open Topology, Gen. Top. and its Applics. 8 (1978) 181-195.

BB2. PETER I. BOOTH and RONALD BROWN, On the Application of Fibred Mapping Spaces to Expontential Laws for Bundles, Ex-Spaces and other Categories of Maps, Gen. Top. and its Applics. 8 (1978), 165-179.

D. ALBRECHT DOLD, Partitions of Unity in the Theory of Fibrations, Ann. of Math.78 (1963), 223-255.

HZ. L'UBICA HOLÁ and LÁSZLÓ ZSILINSZKY, Completeness Properties of the Generalized Compact-Open Topology on Partial Functions with Closed Domains, to appear.

JT. IOAN JAMES and EMERY THOMAS, Note on the Classification of Cross-Sections, Topology 4 (1966), 351-359.

K. DONALD. W. KAHN, Induced Maps for Postnikov Systems, Trans. Amer. Math. Soc. 107 (1963), 432 - 450.

M. M. RYUJI MAEHARA, An Obstruction Theory for Fibre Preserving Maps, Ph.D. Thesis, Iowa State University, Ames, Iowa, 1972.

Sp. EDWIN. H. SPANIER, Algebraic Topology, McGraw-Hill, New York, 1966.

St. NORMAN. E. STEENROD, Cohomology Operations, and Obstructions to Extending Continuous Functions, Advances in Math. 8 (1972), 371 - 416.

Str1 ARNE STRØM, Note on Cofibrations, Math. Scand. 19 (1966), 11-14.

Str2 ARNE STRØM, Note on Cofibrations II, Math. Scand. 22 (1968), 130-142.

V. RAINER. M. VOGT, Convenient Categories of Topological Spaces for Homotopy Theory, Arch. Math. XXII (1971), 545 - 555.

DEPARTMENT OF MATHEMATICS AND STATISTICS, MEMORIAL UNIVERSITY, ELIZABETH AVENUE, ST JOHN'S, NEWFOUNDLAND, CANADA A1C5S7

E-mail address: pbooth@math.mun.ca

Contemporary Mathematics
Volume **274**, 2001

Self Homotopy Equivalences of Equivariant Spheres

Davide L. Ferrario

ABSTRACT. The aim of the paper is to compute the groups of self-homotopy equivalences of equivariant spheres, on which a finite group acts. A theorem is proved, and tools or algorithms are provided as to be used in a computer-algebra environment (GAP). Actual computation has been performed on some examples, and for the 213 regular G-spheres for non-abelian groups of order less than 63.

1. Introduction

The problem itself is quite simple: let G be a finite group, V an orthogonal real representation of G and $S(V)$ the unit sphere in V with respect to a G-invariant metric. We want to compute $\mathcal{E}_G(S(V))$, i.e. the group of units (homotopy equivalences) in the monoid $[S(V), S(V)]_G$ of free equivariant homotopy classes of self-maps of $S(V)$. If the representation V is large enough (cf. tom Dieck [**tD78**], [**tD79**], [**tD87**] or Rubinsztein [**Rub76**]) then $[S(V), S(V)]_G$ is isomorphic to the Burnside ring of G (with respect to the smash-product of maps and composition), so that the problem is related to finding units in the Burnside ring and classifying equivariant maps up to homotopy. Unless otherwise stated, by representation we always mean *real* representation.

First results about self homotopy equivalences came out by Matsuda [**Mat78**], [**Mat79**], [**Mat82**], [**MM83**], [**Mat86**] and Yoshida [**Yos83**], [**Yos90**]. Other important results are due to Dress [**Dre69**], tom Dieck [**tD75**], [**tD78**], Rubinsztein [**Rub76**], Waner [**Wan87**], Izydorek and Marzantowicz [**IM95**] and Komiya [**Kom89**], [**Kom95**], [**Kom98**]. About the equivariant homotopy classification of maps problem, well-known results are of tom Dieck [**tD87**], Rubinsztein [**Rub76**] and Tornehave [**Tor82**].

In this paper we prove the following theorem.

THEOREM 1.1. *Let G be a finite group, V an orthogonal finite-dimensional representation of G and $X := S(V)$ the unit sphere in V. Let*

$$d_G : \mathcal{E}_G(X) \to \prod_{(H)\in \mathrm{Iso}(X)} \mathbb{Z}$$

2000 *Mathematics Subject Classification.* Primary: 55P91 (Equivariant Homotopy theory); Secondary: 55P10 (Homotopy Equivalences).

Key words and phrases. Equivariant Spheres, Homotopy Equivalences.

be the equivariant degree homomorphism. Then

- *The kernel* $K := \ker d_G$ *is torsion-free, finitely generated and solvable. If for each isotropy subgroup* H *of* G *we denote by* γ_H *the number of components of* X_H/W_GH *and by* $\nu(X)$ *the number of isotropy subgroups with* $\gamma_H > 1$ *and* $\dim X_H > 0$*, then the number of generators of* K *is given by the sum*
$$\sum_{(H)} (\gamma_H - 1),$$
where (H) *ranges over the isotropy types of* X *such that* $\dim X_H > 0$*. The derived length of* K *is* $\leq \nu(X)$*.*
- *The image* $\operatorname{Im} d_G$ *is an elementary abelian* 2*-group. Its order is equal to the number of solutions in* $\mathbb{Z}^N$ *of*
$$y \cdot T \in \{0, 2\}^N,$$
where T *is the* $N \times N$ *Table of Isotropy Marks matrix, and* $N = \operatorname{niso}(X)$ *is the number of distinct isotropy types of* X*. As a consequence,* $\operatorname{Im} d_G$ *has at most* $\kappa(X)$ *generators,*
$$\#\operatorname{Im} d_G \leq 2^{\kappa(X)},$$
where $\kappa(X)$ *denotes the number of isotropy types* $(H) \in \operatorname{Iso}(X)$ *such that* $|W_GH| \in \{1, 2\}$*.*

The structure of the paper resembles the structure of the proof of the theorem. First, in Section 2, we recall some preliminaries in equivariant topology. The Section is splitted into some subsequences, for the convenience of the reader. Some preliminaries are well-known, others not, so in some subsections we give more details. In Section 3 we compute the kernel of the degree homomorphism, using all the properties proved or quoted from the preliminaries Section. The same happens in Section 4, where we compute the image of the degree homomorphism. In Section 5 we prove some easy but interesting consequences of the main theorem, while in Section 6 we apply the theorem and the GAP tools, introduced throughout the paper, to some actual computations. The short Section 7 is just a list of notes or remarks on the paper. At the end of the paper, there are some tables in which the sizes of the groups of self homotopy equivalences are listed, for equivariant spheres of regular representations (regular G-spheres).

I wish to express my sincere thanks to the Organizers of the Gargnano Workshop for the invitation, their kindness and for the fine and stimulating environment that they were able to create during the workshop. Also, I couldn't perform the amount of GAP computations without having access to the clusters of PC's of the Dipartimento di Matematica e Applicazioni, Università di Milano-Bicocca. I wish to thank them for their hospitality.

2. Preliminaries

Let G be a group, and X a G-space. If $x \in X$, then denote with G_x the isotropy group of x (also known as the stabilizer of x, $G_x := \{g \in G \mid gx = x\}$). With this notation, we can define the set of fixed points of a subgroup H of G as
$$\begin{aligned} X^H &:= \{x \in X \mid G_x \supset H\} \\ &= \{x \in X \mid Hx = x\} \end{aligned}$$

and in the same way the *singular set*

$$X_s^H := \{x \in X \mid G_x \supsetneq H\}$$

and the other subsets

$$\begin{aligned} X_H &:= \{x \in X \mid G_x = H\} \\ &= X^H \setminus X_s^H \end{aligned}$$

$$\begin{aligned} X_{(H)} &:= \{x \in X \mid (G_x) = (H)\} \\ &= G(X^H \setminus X_s^H) \end{aligned}$$

For every subgroup H, an equivariant map $f : X \to X$ induces a self-map $f^H : X^H \to X^H$ by restriction.

If H is a subgroup of G, then let (H) denote the conjugacy class of H in G, or, equivalently, the orbit type of G/H. If (H) and (K) are two classes, we use the symbol $(H) \leq (K)$ or $(H) < (K)$ to denote that H is subconjugated in K (that is, there is a conjugate of H contained in K). We recall that $N_G H$ denotes the normalizer of H in G, $N_G H := \{g \in G \mid g^{-1}Hg = H\}$, and $W_G H := N_G H/H$ denotes the Weyl group of H. Let $\operatorname{Iso}(X)$ denote the poset of all the isotropy types in X (with respect to the G-action).

If Y is a space, then let $[Y, Y]$ denote the monoid (with respect of composition of maps) of the (free) homotopy classes of self-maps of Y; if A is a subspace of Y, then $[Y, Y]^A$ denotes the monoid of homotopy classes of self-maps of Y which are the identity on A, where the homotopy is relative to A. If X is a G-space, then let $[X, X]_G$ denote the monoid of the free G-homotopy classes of G-self-maps of X. If G is the trivial group, then of course $[X, X]_G = [X, X]$. Again, if $A \subset X$ is a G-subspace, then let $[X, X]_G^A$ denote the monoid of the G-homotopy classes relative to A of G-self-maps of X which are the identity on A.

The equivariant self homotopy equivalences of X are just the units of the monoid $[X, X]_G$. In particular, if the representation V of G such that $X = S(V)$ is big enough (in the sense of [**Rub76, tD87**]), then $[X, X]_G$ is isomorphic to the Burnside ring $A(G)$ (as multiplicative monoids; in fact, the additive structure of $A(G)$ corresponds to the smash product of G-self-maps of X under this correspondence; see [**tD87**]).

If X is a G-sphere, then we can easily define the equivariant degree homomorphism

$$d_G : [X, X]_G \to \prod_{(H) \in \operatorname{Iso}(X)} \mathbb{Z}$$

as follows. For every isotropy type (H), and every G-homotopy class $[f] \in [X, X]_G$, the (H)-component of $d_G([f])$ is

$$d_G([f])(H) = \deg(f^H),$$

where $\deg(f^H)$ is the degree of the map $f^H : X^H \to X^H$. In case the dimension of X^H is 0, the degree $\deg(f^H)$ is set to be 0 if f^H is not a bijection, 1 if it is the identity and -1 if it is the antipodal map.

Note that the definition of d_G does not depend upon the choice of the representative f in $[f]$, nor upon the choice of H in its conjugacy class (H). Also, if $[f]$ is an equivalence, then its G-degree is actually an element of

$$\prod_{(H)\in \mathrm{Iso}(X)} \mathbb{Z}^*$$

where $\mathbb{Z}^* = \{-1,+1\}$ is the group of units in the ring $\mathbb{Z}$.

2.1. Isotropy subgroups and induction on isotropy types. The poset $\mathrm{Iso}(X)$ is finite, hence we can assume that the isotropy types are indexed as $\{(H_i)\}_{i=1\ldots N}$, such that $(H_i) \leq (H_j) \implies i \geq j$ (N is the cardinality of $\mathrm{Iso}(X)$). This implies that (H_1) is a maximal isotropy type. For $i = 1\ldots N$ let

$$X_i := \bigcup_{j=1}^{i} X_{(H_i)},$$

that is

$$X_i = \{x \in X \mid (G_x) \in \{(H_1),(H_2),\ldots,(H_i)\}\}.$$

Note that because of the choice of the ordering of the indexes, for each $i \leq j$, $X_j^{H_i} = X^{H_i}$: therefore $X_i^{H_i} = X^{H_i}$, and $X_{i-1}^{H_i} = X_s^{H_i}$.

The key step in using induction over isotropy types is that the restriction map $f \mapsto f^{H_i}$ induces an isomorphism

$$[X_i, X_i]_G^{X_{i-1}} \cong [X^{H_i}, X^{H_i}]_{W_G H_i}^{X_s^{H_i}}.$$

For details we refer to [**tD87**], Proposition I.7.4, page 52.

2.2. Computing isotropy subgroups. The first step is to compute the isotropy subgroups, given the representation V of G, such that $X = S(V)$. As in the standard software GAP, we assume to have access to the following data of representations and groups: characters, subgroups, fusion maps and scalar products.

Hence, let $\chi : G \to \mathbb{C}$ the character of a representation V of G, and $H_1,\ldots,H_n$ a list of representatives of conjugacy classes of subgroups of G. The algorithm that allows to detect which of them are isotropy subgroups is the following: for every i, take $H = H_i$ and compute the dimension $\dim V^H$; this can be done simply by computing the scalar product of the trivial character of H with the character χ restricted to H:

$$\dim V^H = \frac{1}{|H|}\sum_{h\in H} \chi(h) = < 1_H, \chi >_H$$

Now consider a subgroup H of G with the property (*) that if there is a subgroup K such that $K \supsetneq H$ then $\dim V^K < \dim V^H$: since

$$V^H = \bigcup_{K\supset H} V_K$$

there is always an isotropy subgroup $K \supset H$ with $\dim V_K = \dim V^H$: because of the assumption, it must be that $H = K$ and therefore H must be an isotropy subgroup. On the other hand it is easy to see that every isotropy subgroup has the property (*).

Thus, in order to detect the isotropy subgroups from the set of conjugacy classes $\{H_1, \dots, H_n\}$ of subgroups of G, it suffices to compute the dimensions $\dim V^{H_i}$ for $i = 1 \dots n$ and then consider those subgroups with the property above.

An efficient way to do this computation is done through the Table of Marks of the group G.

2.3. The Table of Marks. The Table of Marks (first introduced by Burnside in the 20's) has become now an interesting tool to explore group properties, because many informations on the structure of the group can be deduced by its entries. Details on how to compute the Table of Marks of a group can be found e.g. in [**Pfe97**]. As a matter of fact, Tables of Marks of thousands of finite groups are stored (or computable from those stored) in the standard libraries of GAP [**S$^+$95**], and are readily available to the programmer, exactly as the Character Tables of the ATLAS of finite groups [**CCN$^+$85**].

Consider two subgroups H and K of G: then the set G/H of *left* cosets of H is endowed with an action of K, by left multiplication. The fixed subspace $(G/K)^H$ has cardinality $|(G/K)^H|$: the table of marks is simply defined as the matrix T_G whose (i,j)-entry is $|(G/H_i)^{H_j}|$, for $i, j \in 1 \dots n$.

$$T_G := \begin{pmatrix} |(G/H_1)^{H_1}| & |(G/H_1)^{H_2}| & |(G/H_1)^{H_3}| & \dots & |(G/H_1)^{H_n}| \\ |(G/H_2)^{H_1}| & |(G/H_2)^{H_2}| & |(G/H_2)^{H_3}| & \dots & |(G/H_2)^{H_n}| \\ |(G/H_3)^{H_1}| & |(G/H_3)^{H_2}| & |(G/H_3)^{H_3}| & \dots & |(G/H_3)^{H_n}| \\ \vdots & \vdots & \vdots & \ddots & \vdots \\ |(G/H_n)^{H_1}| & |(G/H_n)^{H_2}| & |(G/H_n)^{H_3}| & \dots & |(G/H_n)^{H_n}| \end{pmatrix}$$

REMARK 2.1. Since there is always an ordering $H_1 \dots H_n$ of the set of conjugacy classes of subgroups of G such that $(H_i) \le (H_j)$ implies $i \le j$, we can assume the subgroups ordered in this way. This order is the opposite order than the one we defined for isotropy types. The reason for this double notation is that in usual induction over orbit types that order is needed, while in GAP the order is already stored in this way. Let us call this ordering the *lattice* order.

Moreover, it is easy to see that $(G/K)^H \neq \emptyset$ implies $(K) \supseteq (H)$, hence

$$|(G/H_i)^{H_j}| \neq 0 \implies (H_i) \ge (H_j) \implies i \ge j$$

and therefore the Matrix of the Table of Marks with this choice of ordering is lower triangular.

REMARK 2.2. It is equivalent to compute the table of marks considering the *right* cosets $H \backslash G$.

EXAMPLE 2.3. Let A_5 denote the alternating group of order 60. A_5 has 9 conjugacy classes of subgroups:

$$\{1\}, \mathbb{Z}_2, \mathbb{Z}_3, \mathbb{Z}_2 \oplus \mathbb{Z}_2, \mathbb{Z}_5, S_3 = D_6, D_{10}, A_4, A_5$$

where $\mathbb{Z}_k$ denotes the cyclic group of order k, S_3 the symmetric group of order 6, and D_k denotes the dihedral group of order k. With this ordering, the table of marks is as follows.

$$\begin{pmatrix} 60 & 0 & 0 & 0 & 0 & 0 & 0 & 0 & 0 \\ 30 & 2 & 0 & 0 & 0 & 0 & 0 & 0 & 0 \\ 20 & 0 & 2 & 0 & 0 & 0 & 0 & 0 & 0 \\ 15 & 3 & 0 & 3 & 0 & 0 & 0 & 0 & 0 \\ 12 & 0 & 0 & 0 & 2 & 0 & 0 & 0 & 0 \\ 10 & 2 & 1 & 0 & 0 & 1 & 0 & 0 & 0 \\ 6 & 2 & 0 & 0 & 1 & 0 & 1 & 0 & 0 \\ 5 & 1 & 2 & 1 & 0 & 0 & 0 & 1 & 0 \\ 1 & 1 & 1 & 1 & 1 & 1 & 1 & 1 & 1 \end{pmatrix}$$

This matrix can be also easily computed by GAP as follows:

```
gap> G:=AlternatingGroup(5);
Alt( [ 1 .. 5 ] )
gap> Display(TableOfMarks(G));
1:  60
2:  30 2
3:  20 . 2
4:  15 3 . 3
5:  12 . . . 2
6:  10 2 1 . . 1
7:   6 2 . . 1 . 1
8:   5 1 2 1 . . . 1
9:   1 1 1 1 1 1 1 1 1
```

The algorithm that computes the isotropy groups, relying on the data of the table of marks $(T_G)_{ij} := T_G$ of G, is as follows: (H) is an isotropy type if and only if $\dim V^H > \dim V^K$ whenever $(H) \leq (K)$; hence first compute the vector of all the fixed spaces dimensions

$$v := (\dim V^{H_1}, \dim V^{H_1}, \dots, \dim V^{H_n})$$

and then see that H_i is an isotropy subgroup if and only if for all the j's such that $(T_G)_{ij} \neq 0$, it happens that $v_i > v_j$. Thus in a finite number of steps the subgroups of isotropy can be determined.

An example of GAP code that can perform this task on the alternating group A_5 is the following.

```
G:=AlternatingGroup(5);
tm:=TableOfMarks(G);
TOMMatrix:=MatTom(tm);
l:=Length(SubsTom(tm));
AllTheSubgroups:=List([1..l],k -> RepresentativeTom(tm,k));
tg:=CharacterTable(G);
chars:=Irr(tg);

DimFixed:=function(H,chi)
local th;
if IsTrivial(H)
then return Degree(chi);
else
```

```
th:=CharacterTable(H);
FusionConjugacyClasses(H,G);;
return  ScalarProduct(
        th,
        TrivialCharacter(th),
        RestrictedClassFunction(tg,chi,H)
);
fi;
end;

dimsfixed:= List(chars, chi ->
        List(AllTheSubgroups, H ->
                DimFixed(H,chi)
));

Isotropies:= List([1..Length(chars)], p ->
Filtered([1..l], j ->
Length(
    Filtered([1..l], i -> ( TOMMatrix[i][j]>0
                            and
                            dimsfixed[p][i]=dimsfixed[p][j]
                            and
                            dimsfixed[p][j]>0
                          )
              )
    )=1

)
);
```

An important property of the table of marks is the following: if $m(H,K)$ denotes the number of conjugated copies of K containing H, then $|(G/K)^H| = |W_GK|m(H,K)$. In fact,

$$\begin{aligned}
|(G/K)^H| &= \frac{1}{|K|}\#\{g \in G \mid HgK = gK\} \\
&= \frac{1}{|K|}\#\{g \in G \mid g^{-1}Hg \subseteq K\} \\
&= \frac{1}{|K|}\#\{g \in G \mid H \subseteq gKg^{-1}\} \\
&= \frac{1}{|K|}m(H,K)|N_GK| \\
&= m(H,K)|W_GK|
\end{aligned}$$

2.4. The fixed point index and the classification of maps. Let Y be an ENR, and $U \subset Y$ an open subset. We denote by $\operatorname{ind}(f)$ the fixed point index of f in U (see [**Dol65, Dol80**]), where $f : U \to Y$ is a compactly fixed map.

The following Lemma was first proved by Komiya [**Kom87**] and Wilczyński [**Wil84**] in a more general situation.

LEMMA 2.4. *Let $f : X \to X$ be a G-map. Then there is a G-map h G-homotopic to f such that for every isotropy subgroup H of G, there is an equivariant W_GH-neighbourhood N_H of X_s^H in X^H and a W_GH-equivariant retraction $r_H : N_H \to X_s^H$ such that $h^H|N_H = h^H r_H$.*

PROOF. In our settings the proof is almost trivial, because there is an equivariant mapping cylinder neighbourhood of X_s^H in X^H. □

It is not difficult to prove that if h and h' are two G-homotopic maps with the property of Lemma 2.4 (they are called *taut maps*), then for every isotropy H and every component U_H of X_H, $\operatorname{ind}(h^H|U_H) = \operatorname{ind}(h'^H|U_H)$.

Consider the projection over the quotient

$$p_H : X_H \to X_H/W_GH$$

and let $C_1, \ldots, C_{\gamma_H}$ the components of X_H/W_GH (where γ_H is the number of such components).

There is a function

$$\alpha : [X, X]_G \to \prod_{(H)} \mathbb{Z}^{\gamma_H}$$

(where (H) ranges over the isotropy types) defined by $[f] \in [X, X]_G \mapsto \alpha([f])$, where for every isotropy type (H)

$$\begin{aligned} \alpha([f])(H) \quad &:= \quad \frac{1}{|W_GH|}(\operatorname{ind}(h^H|p_{H'}^{-1}C_1), \operatorname{ind}(h^H|p_H^{-1}C_2), \ldots \\ &\quad \ldots, \operatorname{ind}(h^H|p_H^{-1}C_{\gamma_H})) \in \mathbb{Z}^{\gamma_H} \end{aligned}$$

and h is the taut approximation of f as in Lemma 2.4.

As we noted above, since for every H, $p_H^{-1}C_i$ is a union of components of X_H, the values of α do not depend upon the choice of the approximation h, and the indexes $\operatorname{ind}(h^H|p_H^{-1}C_i)$ are well-defined since h is compactly-fixed in X_H. Also, because the action of W_GH on X_H is free, the values are actually integers. Let us define

$$A'(X) := \{l \in \prod_{(H)} \mathbb{Z}^{\gamma_H} \mid \text{ if } \dim X_H = 0, \text{ then } l(H) \in \{0,1\}^{\gamma_H} \subset \mathbb{Z}^{\gamma_H}\}.$$

Note that if $\dim X_H = 0$, then γ_H can be only 2 or 1, according to the case whether W_GH is trivial or not.

The following result is due to Wilczyński [**Wil84**] (Theorem 2.7, page 51):

THEOREM 2.5. *The above defined function*

$$\alpha : [X, X]_G \to A'(X)$$

is a bijection.

2.5. The induction formula for fixed points. The following Lemma is the main step to compute the image of the degree homomorphism d_G.

LEMMA 2.6. *Consider a map h with the property stated in Lemma 2.4: for every isotropy subgroup H,*

$$\operatorname{ind}(f^H) = \sum_{K \supseteq H} \operatorname{ind}(f^K|X_K).$$

PROOF. First note that

$$\mathrm{Fix}(f^H) = \bigcup_{K \supseteq H} \mathrm{Fix}(f^K | X_K)$$

(where the union is disjoint). Because of the Retraction Property, the index $\mathrm{ind}(f^K|X_K)$ coincides with the index of f^H in a neighbourhood of $\mathrm{Fix}(f^K|X_K)$, thus the formula. □

2.6. The Table of Isotropy Marks. Consider the matrix of the Table of Marks T_G, and delete the rows and columns not corresponding to isotropy groups H. This matrix (called the Table of Isotropy Marks) gives rise to an homomorphism

$$T : \prod_{(H)} \mathbb{Z} \to \prod_{(H)} \mathbb{Z}$$

by right multiplication of row-vectors in $\prod_{(H)} \mathbb{Z}$

$$T(v) := v \cdot T,$$

i.e.

$$T(v_1, v_2, \ldots, v_N)_j = \sum_{i=1}^{N} v_i T_{ij},$$

where $T_{ij} = |(G/H_i)^{H_j}|$ and the isotropy subgroups are ordered according to the lattice order. It is a monomorphism, because the Table of Isotropy Marks is a triangular matrix.

Now we describe its main property. Let $L : [X,X]_G \to \prod_{(H)} \mathbb{Z}$ be the homomorphism defined as

$$L([f])(H) := \frac{1}{|W_G H|} \mathrm{ind}(f^H | X_H)$$

for every isotropy H, where f is a taut representative in the G-homotopy class $[f] \in [X,X]_G$. This definition does not depend upon the choice of f.

Moreover, let $\beta : \prod_{(H)} \mathbb{Z} \to \prod_{(H)} \mathbb{Z}$ defined as

$$\beta(y)(H) := 1 + (-1)^{n(H)} y(H)$$

for every $y \in \prod_{(H)} \mathbb{Z}$, where (H) range over the isotropy types of X. It is a bijection.

We can arrange the previous functions in a diagram.

$$\begin{array}{ccc} [X,X]_G & \xrightarrow{d_G} & \prod_{(H)} \mathbb{Z} \\ {\scriptstyle L}\downarrow & & \downarrow{\scriptstyle \beta}\,\cong \\ \prod_{(H)} \mathbb{Z} & \overset{T}{\dashrightarrow} & \prod_{(H)} \mathbb{Z} \end{array} \tag{2.1}$$

LEMMA 2.7. *If T, L, β and d_G are as above, then*

$$TL = \beta d_G,$$

i.e. the diagram is commutative.

PROOF. Consider an element $[f] \in [X,X]_G$, where $f : X \to X$ is a taut representative (cf. Lemma 2.4). By definition,

$$L([f])(H) = \frac{1}{|W_G H|}\text{ind}(f^H|X_H),$$

and hence

$$TL([f])(K) = \sum_{(H)} L([f])(H)|(G/H)^K|$$

$$TL([f])(K) = \sum_{(H)} \frac{1}{|W_G H|}\text{ind}(f^H|X_H)|(G/H)^K| =$$

$$\sum_{(H)} m(K,H)\text{ind}(f^H|X_H)$$

where $m(K,H)$ as before denotes the number of conjugated copies of H containing K. The latter sum is nothing but

$$\sum_{H \supset K} \text{ind}(f^H|X_H) = \text{ind}(f^K).$$

That is, we have proved that for each isotropy type (K) $TL([f])(K) = \text{ind}(f^K)$. It is now straightforward to conclude also that

$$\beta d_G([f])(K) = 1 + (-1)^{n(K)}\deg(f^K) = \text{ind}(f^K).$$

Thus the diagram is commutative. □

3. The kernel K of d_G

We now prove the first half of the Theorem.

PROPOSITION 3.1. *Let $K := \ker d_G$, where $d_G : \mathcal{E}_G(X) \to \prod_{(H)} \mathbb{Z}^*$ is the degree homomorphism. Then there is an abelian series*

$$K = K_0 \supset K_1 \supset \dots K_{N-1} \supset K_N = \{1\}$$

whose factors $\frac{K_{i-1}}{K_i}$ are torsion-free abelian groups with $\gamma_{H_i} - 1$ generators, in case $n(H_i) > 0$, or 0 if $n(H_i) = 0$, where $n(H_i) = \dim(X^{H_i})$ and $\gamma_{H_i} = \#\pi_0(X_{H_i}/W_G H_i)$. By this it is meant that if either $\dim(X^{H_i}) = 0$ or $\gamma_{H_i} = 1$, then $K_i = K_{i-1}$.

PROOF. Let us define the subgroups K_i. For every $i = 1 \dots N$ (order the isotropies in the usual way, as said above, to get induction over orbit types), let $P_i : \mathcal{E}_G(X) \to [X_i, X_i]_G$ be defined as the restriction homomorphism, and let K_i be the kernel $\ker P_i$. For $i = 0$, let $K_0 = K$. Because $X_N = X$, we have that P_N is mono, hence K_N is the trivial subgroup of $\mathcal{E}_G(X)$. On the other hand, it is easy to see that for each $i = 1 \dots N$, $H_i \subset H_{i-1}$.

An element of K_{i-1} is a G-homotopy class $[f]$ (assume it taut) such that its restriction to $[X_{i-1}, X_{i-1}]_G$ is homotopic to the identity map. Without loss of generality we can hence assume that f is the identity on X_{i-1} and taut. Now consider its restriction $f^{H_i} : X^{X_i} \to X^{H_i}$: because of the assumptions it is the identity on $X_s^{H_i}$. Actually, because of the classification theorem 2.5, we can deform the map until it is the identity outside a finite number $\{D_1, \dots, D_{\gamma_{H_i}}\}$ of $W_G H_i$-disks contained in X_{H_i}, one for each component C_l of $X_{H_i}/W_G H_i$: because the Weyl action is free in X_{H_i}, just take the pre-image of some small disks in X_{H_i}/W_G, one in each component (we recall that by a $W_G H$-disk in this case we mean the disjoint union of $|W_G H|$ copies under the Weyl action of a disk). If the dimension of X_{H_i} is 0, this means that $X^{H_i} = X_{H_i}$ and the disks are actually points. This map now is not taut.

For $i = 1 \dots \gamma_{H_i}$ let $\operatorname{ind}(h^{H_i}|p_{H_i}^{-1} C_l)$ be the fixed point index of a taut $W_G H_i$-approximation h^{H_i} of f^{H_i}, where $p_{H_i} : X_{H_i} \to X_{H_i}/W_G H_i$ denotes the projection over the quotient. It is easy to see that for every $l = 1 \dots \gamma_{H_i}$,

$$(-1)^{n(H_i)} \deg_P(f^{H_i}|D_l) = \frac{1}{|W_G H_i|} \operatorname{ind}(h^{H_i}|p_{H_i}^{-1} C_l),$$

where $\deg_P(f^{H_i}|D_l)$ is the local degree at a point P inside D_l of the map restricted to the $W_G H_i$-disk D_l. Note also that the local degree of $f^{H_i}|D_l$ at any point Q outside the $W_G H_i$-disk D_l is simply $\deg_P(f^{H_i}|D_l) - 1$. Now we can define a function

$$\delta : K_{i-1} \to \mathbb{Z}^{\gamma_{H_i}}$$

as follows. For every $l = 1 \dots \gamma_{H_i}$, let

$$\delta([f])(l) := \deg_{P_l}(f^{H_i}|D_l) - 1,$$

where P_l is as above a point inside D_l (i.e. $\delta([f])(l) = \deg_{P_l'}(f^{H_i}|D_l)$ for a point P_l' not in D_l.

PROPOSITION 3.2. *If $[f], [g] \in K_{i-1}$, then for each $l = 1 \dots \gamma_{H_i}$,*

$$\delta([f \circ g])(l) = \delta([f])(l) + \delta([g])(l) \cdot \left(1 + |W_G H_i| \sum_{s=1}^{\gamma_{H_i}} \delta([f])(s) \right).$$

PROOF. Consider two representatives f and g of $[f]$ and $[g]$ such that f^{H_i} and g^{H_i} are the identity outside the $|W_G H_i|$ small disks as above, and they are regular in the interior of the disks. Then their composition $f \circ g$ has the same property, and therefore $\delta([f \circ g])(l)$ is the degree at a (regular) point P_l' not in D_l of $f^{H_i} \circ g^{H_i}$. Choose a regular point P' not in D_l for each l. The set of pre-images of P' is

$$(f \circ g)^{-1} P' = \bigcup_{P \in f^{-1} P'} g^{-1} P;$$

first, note that $f^{-1}P'$ contains P' (because $f(P') = P'$ with multiplicity 1 (f is the identity on a neighbourhood of P'), and then $|W_G H_i| \cdot \gamma_{H_i}$ disjoint subsets of points, contained in $D_1 \cup \dots \cup D_{\gamma_{H_i}}$. Hence, because the subsets contained in D_s with $s \neq l$ contribute with a term

$$|W_G H_i| \deg_{P'}(f^{H_i}|D_s) \cdot \deg_{P'}(g^{H_i}|D_l),$$

while D_l (i.e. when $s = l$) contributes with a term

$$|W_G H_i| \deg_{P'}(f^{H_i}|D_l) \cdot \deg_{P'}(g^{H_i}|D_l) + \deg_{P'}(f^{H_i}|D_l),$$

the following equality holds:

$$\begin{aligned} \deg_{P'}(f^{H_i} \circ g^{H_i}|D_l) &= \deg_{P'}(g^{H_i}|D_l) + \\ &+ \sum_{s=1}^{\gamma_{H_i}} \left(|W_G H_i| \deg_{P'}(f^{H_i}|D_s) \cdot \deg_{P'}(g^{H_i}|D_l) \right) + \\ &+ \deg_{P'}(f^{H_i}|D_l). \end{aligned}$$

But this is what we wanted to prove, therefore the proof is complete. □

COROLLARY 3.3. *The above defined function* $\delta : K_{i-1} \to \mathbb{Z}^{\gamma_{H_i}}$ *is a homomorphism, with respect to the sum in* $\mathbb{Z}^{\gamma_{H_i}}$. *Moreover, the composition*

$$K_{i-1} \to^{\delta} \mathbb{Z}^{\gamma_{H_i}} \to^{\epsilon} \mathbb{Z}$$

is trivial, where $\epsilon(y)$ *is defined as the sum of the coordinates of* y.

PROOF. We need to prove that given $[f]$ and $[g]$ in K_i, then $\delta([f \circ g])) = \delta([f]) + \delta([g])$. Because of Proposition 3.2, this is equivalent to prove that if $[f]$ and $[g]$ are in K_i then

$$\deg_{P'}(g^{H_i}|D_l) \cdot \sum_{s=1}^{\gamma_{H_i}} \left(|W_G H_i| \deg_{P'}(f^{H_i}|D_s)\right) = 0$$

Since $[f]$ and $[g]$ are in K_{i-i}, which is contained in $\ker d_G$, it is necessary that $\deg(f^{H_i}) = 1$. This simply means, by the additivity of the degree, that

$$1 + \sum_{s=1}^{\gamma_{H_i}} \deg_{P'}(f^{H_i}|D_s) = \deg(f^{H_i}) = 1$$

This implies that

$$\sum_{s=1}^{\gamma_{H_i}} \deg_{P'}(f^{H_i}|D_s) = 0$$

and therefore our claim. Furthermore, we implicitly proved also that the homomorphism $\epsilon \circ \delta$, as defined above, is zero.

□

The next step in the proof of Proposition 3.1 is the following Lemma:

LEMMA 3.4. *If* $n(H_i) > 0$, *then the following sequence is exact:*

$$0 \longrightarrow K_i \longrightarrow K_{i-1} \xrightarrow{\delta} \mathbb{Z}^{\gamma_{H_i}} \xrightarrow{\epsilon} \mathbb{Z} \longrightarrow 0$$

PROOF. The arrow $K_i \to K_{i-1}$ is simply the inclusion, so that the sequence is exact in K_i.

It is clear by the definition of δ that if a map f is in K_i, then it can be deformed G-homotopically to be the identity on X_i, and therefore $\delta([f])(l) = 0$ for every $l = 1 \dots \gamma_{H_i}$. On the other hand, because of the classification theorem 2.5, it is not difficult to see that $\ker \delta \subset K_i$, and therefore the sequence is exact in K_{i-1}.

Now the exactness in $\mathbb{Z}^{\gamma_{H_i}}$: we already proved that $\ker \epsilon \supset \operatorname{Im} \delta$: we need to show that for every γ_{H_i}-uple y of integers whose sum is zero, there is an element $[f]$ in K_{i-1} such that $\delta([f]) = y$: again, this follows easily from the classification theorem 2.5.

□

Now we can prove Proposition 3.1: in fact, we have exhibited an abelian series $\{K_i\}$ whose factors are isomorphic to the kernels of the homomorphisms $\epsilon : \mathbb{Z}^{\gamma_{H_i}} \to \mathbb{Z}$, and these of course are torsion-free abelian groups with $\gamma_{H_i} - 1$ generators, whenever $n(H_i) > 0$. If $n(H_i) = 0$, then $X_s^{H_i}$ is empty, and therefore $K_{i-1} = K_i$, so that there is nothing to say.

□

COROLLARY 3.5. *The degree homomorphism d_G is mono if and only if*

$$\gamma_H > 1 \implies \dim X^H = 0.$$

PROOF. It is an immediate corollary of Proposition 3.1. □

4. The image $\operatorname{Im} d_G$

Now we are going to prove the second half of the main Theorem.

PROPOSITION 4.1. *The image $\operatorname{Im} d_G$ is an elementary abelian 2-group, whose order is equal to the number of solutions in $\mathbb{Z}^N$ of*

$$y \cdot T \in \{0,2\}^h,$$

where T is the $N \times N$ Table of Isotropy Marks matrix, and $N = \operatorname{niso}(X)$ is the number of isotropy types of X.

PROOF. It is trivial to see that $\operatorname{Im} d_G$ is an elementary abelian finite 2-group. Consider the homomorphisms T, d_G, β and L, as in diagram 2.1. Take an element $[f] \in [X,X]_G$: because of James-Siegal [**JS78, tD87**], $[f]$ is an equivalence if and only if $d_G([f])$ is a (multiplicative) unit, that is if and only if $d_G([f])$ is in $\prod_{(H)} \mathbb{Z}^*$. Thus,

$$\mathcal{E}_G(X) = d_G^{-1}(\prod_{(H)} \mathbb{Z}^*).$$

Moreover, because of the definition of β,

$$\beta(\prod_{(H)} \mathbb{Z}^*) = \{0,2\}^N \subset \mathbb{Z}^N = \prod_{(H)} \mathbb{Z},$$

and thus

$$\mathcal{E}_G(X) = d_G^{-1}\beta^{-1}(\{0,2\}^N).$$

Now consider $L(\mathcal{E}_G(X)) \subset \prod_{(H)} \mathbb{Z}$; the diagram is commutative, therefore $L(\mathcal{E}_G(X)) \subset T^{-1}\{0,2\}^N$. We want to show that the inclusion is actually an equality. If $y = (y_1, \ldots, y_N) \in \prod_{(H)} \mathbb{Z}$ is a solution of

$$y \cdot T \in \{0,2\}^N,$$

then consider the following commutative diagram:

$$(4.1) \qquad \begin{array}{ccc} [X,X]_G & \xrightarrow[\cong]{\alpha} & A'(X) \\ \Big\downarrow L & & \Big\downarrow j \\ \prod_{(H)} \mathbb{Z} & \xleftarrow{\epsilon} & \prod_{(H)} \mathbb{Z}^{\gamma_H} \end{array}$$

where ϵ is, with an abuse of notation, the direct product of all the $\epsilon : \mathbb{Z}^{\gamma_H} \to \mathbb{Z}$, as in Corollary 3.3, and j is simply the inclusion. The homomorphism ϵ is of course onto, so that we can select a $\tilde{y} \in \prod_{(H)} \mathbb{Z}^{\gamma_H}$ such that $\epsilon(\tilde{y}) = y$. Now, consider the (H)-coordinate $y(H)$ of y: if the dimension $\dim X^H$ is zero, then the isotropy group H is maximal, and therefore in the (H)-column of T there is just the element in the diagonal, namely $|W_G H|$. Also, because by assumption $y \cdot T \in \{0,2\}^N$, reading the (H)-coordinates of both sides of this equation implies

$$y(H) \cdot |W_G H| \in \{0,2\}.$$

This means that either $|W_G H| = 1$ and $y(H)$ is 0or 2, or $|W_G H| = 2$ and $y(H)$ is 0 or 1. In both cases, it is easy to see that there is a $W_G H$-homotopy equivalence $\phi : X^H \to X^H$ such that $\operatorname{ind}(\phi) = 2$ or $\operatorname{ind}(\phi) = 0$ (just take the identity on $X^H = S^0$, or the antipodal map). The same argument applies to all the orbit types (K) such that $\dim X^K = 0$. Now, this means we have defined a G-map $\phi' : Y \to Y$, where Y is the union of all the 0-dimensional fixed spheres in X. This map can be easily extended equivariantly to the whole X, so that actually we can modify $\tilde{y}$ in a way that $\epsilon(\tilde{y}) = y$ and $\tilde{y} \in j(A'(X))$. Again, because of the James-Siegal theorem, actually $\tilde{y} \in \alpha(\mathcal{E}_G(X)) \subset A'(X)$, and therefore $y \in L(\mathcal{E}_G(X))$.

So far, we have proved that $L(\mathcal{E}_G(X)) = T^{-1}\{0,2\}^N$. Now, because the matrix of T is triangular, as noticed before, it induces a bijection

$$T : T^{-1}\{0,2\}^N \to \operatorname{Im}(\beta d_G) \cong \operatorname{Im} d_G,$$

and this concludes the proof. □

Corollary 4.2. *The image* $\operatorname{Im} d_G$ *has at most* $\kappa(X)$ *generators, so that its order*

$$\#\operatorname{Im} d_G \leq 2^{\kappa(X)},$$

where $\kappa(X)$ *denotes the number of isotropy types* $(H) \in \operatorname{Iso}(X)$ *such that*

$$|W_G H| \in \{1,2\}.$$

Proof. The elements of $T^{-1}\{0,2\}$ are less then 2^k, where k denotes the number of diagonal entries of the matrix T equal to 1 or 2. But we already know that the diagonal entries are just the orders of the Weyl groups $W_G H$, therefore the thesis. □

Corollary 4.3 ((Matsuda [**Mat79**])). *Let* G *be a finite abelian group, and* V *an orthogonal representation of* G *such that* $\dim V^G \geq 2$. *Then we have*

$$|\mathcal{E}_G(X)(S(V))| = 2^{k+1},$$

where k *is the number of isotropy subgroups of* $S(V)$ *of index* 2 *in* G.

Proof. Let X denote the sphere $S(V)$. In this case G itself is an isotropy subgroup of X, with trivial Weyl group, and the isotropy subgroups with $|W_G H| = 2$ are simply those of index 2, because $W_G H = G/H$ for every H in G. There are no self-normalizing subgroups in G other than G. Therefore the number $\kappa(X)$ is equal to $k+1$. Also, because G is abelian, $|(G/K)^H| = |G/K|$ if K contains H and $|(G/K)^H| = 0$ otherwise. Another key point is that if an isotropy group has index 2 in G, then it cannot contain other isotropy groups with index 2, and so the solutions of

$$y \cdot T \in \{0,2\}^N$$

are exactly $2^{\kappa(x)}$. □

5. Consequences

COROLLARY 5.1. *Let G be a finite 2-split group, V an orthogonal real representation of G, and $X = S(V)$ the unit sphere in V. Then*

$$\mathcal{E}_G(X) \cong \mathbb{Z}_2^k,$$

where $\mathbb{Z}_2^k$ denotes the elementary abelian 2-group of order 2^k, and the order of $\mathcal{E}_G(X)$ is equal to the number of integer solutions y of

$$y \cdot T \in \{0, 2\}^h,$$

where T denotes the Table of Isotropy Marks of V, and h the number of isotropy types in V. Furthermore, $k \leq \kappa(X)$, where $\kappa(X)$ denotes the number of isotropy types in V whose Weyl group $W_G H$ has order either 1 or 2.

PROOF. I have proved in [**Fer00**] that if the group G is finite, then for every isotropy subgroup H of G the quotient space $X_H/W_G H$ is connected. That is, for every H, $\gamma_H = 1$. This means that the Kernel $\ker d_G$ is trivial, and therefore

$$d_G : E_G(X) \to \prod_{(H)} \mathbb{Z}$$

is mono. Moreover, because of the main theorem, $\operatorname{Im} d_G$ is an elementary abelian 2-group with that order. □

COROLLARY 5.2. *If G is an odd-order nilpotent group and X is an orthogonal G-sphere, then*

$$E_G(X) = \begin{cases} 1 & \text{if } X^G = \emptyset \\ \mathbb{Z}_2 & \text{if } X^G \neq \emptyset \end{cases}$$

PROOF. Because G is nilpotent, there are no self-normalizing subgroups of G other than G itself. Because the order of G is odd, there are no subgroups with the Weyl group of order 2. Thus, if $X^G = \emptyset$, then $\kappa(X) = 0$, and d_G is the trivial homomorphism. Because every nilpotent group is 2-split, the kernel of d_G is trivial, and this concludes the proof in the first case. On the other hand, if $X^G \neq \emptyset$, then it is easy to see that the equation $y \cdot T \in \{0, 2\}^h$ has at least the two solutions $y_1 = (0, 0, \ldots, 0)$ and $y_2 = (0, 0, \ldots, 2)$. Thus $|\mathcal{E}_G(X)| \geq 2$. But in this case $\kappa(X) = 1$, and therefore $|\mathcal{E}_G(X)| \leq 2^{\kappa(X)} = 2$, i.e. $\mathcal{E}_G(X) \cong \mathbb{Z}_2$. □

COROLLARY 5.3. *Let G be a finite group. If G is not 2-split, then there is an orthogonal G-sphere X such that $\ker d_G$ is not trivial; hence $E_G(X)$ contains at least one copy of $\mathbb{Z}$ and is infinite.*

PROOF. This is also a consequence of the main theorem of [**Fer00**]: if G is not 2-split, then there is a representation V and a self-normalizing subgroup H of G such that H is an isotropy subgroup of G, $\dim V^H = 1$ and $V^G = 0$. Thus, if we add to V a copy of the field of the real numbers $\mathbb{R}$, with trivial G-action, and take the unit sphere in the G-module that we obtain, we get a G-sphere with the property that $X_H/W_G H = X_H$ (because $W_G H$ is trivial by assumption), and X_H is homeomorphic to a 1-sphere minus a 0-sphere. Hence $\gamma_H = 2$, and $\dim X_H > 0$: we can apply the main theorem to get that the kernel $\ker d_G$ has at least 1 (free) generator. That is, it contains a copy of $\mathbb{Z}$. □

COROLLARY 5.4. *If G is a finite group, and X the unit sphere in the real regular representation of G, then $\ker d_G$ is trivial.*

PROOF. In case of a regular representation, if K and H are two isotropy subgroups such that $K \supset H$, then the codimension of X^K in X^H is at least 2. Therefore X_H itself is connected, if $H \neq G$. Thus $\gamma_H = 1$, whenever $H \neq G$. On the other hand, $\dim X^G = 0$, and thus because of the main theorem the kernel is trivial. □

6. Examples

We have seen that, because of Matsuda's Theorem, if G is abelian, then the order of $\mathcal{E}_G(X)$ is equal to the number $2^{\kappa(X)}$, where $\kappa(X)$ is the number of isotropy types of X whose Weyl group has order 1 or 2. This means that the inequality given in Corollary 4.2

$$\#\mathrm{Im}\, d_G \leq 2^{\kappa(X)}$$

is actually an equality for abelian groups. It is natural to ask whether the equality holds in a wider class of finite groups. In the following example we show that, at least without additional hypotheses, this is not the case.

EXAMPLE 6.1. We now consider the 2 non-isomorphic non-abelian 2-groups of order 8. Let G be the 2-group of order 8 with the presentation:

$$G := < a, b, c | a^2 = b^2 = c^2 = acac = bcbc = cabab = 1 > \cong$$

$$< a, b | a^2 = b^2 = abababab = 1 >$$

and let L be the one presented as follows:

$$L := < a, b | ab^{-2}a = a^{-1}bab = 1 > .$$

Consider now the G-sphere X contained in the regular real representation of G, and the L-sphere Y contained in the regular real L-representation of L. Because G and L are 2-groups, and hence 2-split groups, the kernel of d_G and d_L is trivial (this follows also from Corollary 5.4), and hence the order of $\mathcal{E}_G(X)$ and $\mathcal{E}_G(Y)$ is given by the number of solutions of $y \cdot T \in \{0, 2\}^h$. Thus we have to compute the Table of (Isotropy) Marks.

The Tables of Isotropy Marks are exactly the Tables of Marks, because in a regular representation every subgroup is of isotropy: this is the Table of G

```
1:  8
2:  4 4
3:  4 . 2
4:  4 . . 2
5:  2 2 2 . 2
6:  2 2 . 2 . 2
7:  2 2 . . . . 2
8:  1 1 1 1 1 1 1 1
```

and this is the Table of L:

```
1:  8
2:  4 4
3:  2 2 2
4:  2 2 . 2
5:  2 2 . . 2
6:  1 1 1 1 1 1
```

It is clear that $\kappa(X) = 6$ and $\kappa(Y) = 4$, and so as a first result we conclude that

$$|\mathcal{E}_G(X)| \leq 2^6 = 64$$

$$|\mathcal{E}_G(Y)| \leq 2^4 = 16.$$

To get the exact order of $\mathcal{E}_G(X)$ and $\mathcal{E}_G(Y)$, we just compute the number of solutions of $y \cdot T \in \{0,2\}^h$. It is a simple task to preform it (e.g. using GAP), and to conclude that the number of solutions for the matrix of G is 32, while it is 16 for the matrix of L, i.e.

$$|\mathcal{E}_G(X)| = 32 < 2^{\kappa(X)} = 64$$

$$|\mathcal{E}_G(Y)| = 16 = 2^{\kappa(X)} = 16.$$

Thus even for 2-groups the bound of Corollary 4.2 cannot be improved, in general.

EXAMPLE 6.2. Using the same ideas of the previous example, it is possible to compute $\kappa(X)$ and $E_G(X)$ for the unit spheres in the regular real representation of the small groups (that is, the units of the Burnside groups). Because of Corollary 5.4, the degree homomorphism d_G is mono, and therefore we have directly the order of $\mathcal{E}_G(X)$. We state the results of this computation in the tables at the end of the paper. We preformed the computation for all the 213 non-abelian groups of order less than 63 (the time needed to perform the computations can be long: for example, an Intel Pentium II with 386M of RAM, took 97 hours to compute the order of $\mathcal{E}_G(X)$ for the regular G-sphere of the group `[54, 14]`).

In the tables, the first column is the Id of the groups, as stored in the library of small groups in GAP. The second column is the 2-logarithm of the order of $E_G(X)$, while the third column is the number of subgroups of G with Weyl group of order 1 or 2 (which is equal to $\kappa(X)$, in this case).

A small recursive function which computes the number of solutions of the equation $yT \in \{0,\}$ might be the following (it does not print out the percentage of the computation in progress, so that it might take long time without warning.

```
rsols:=function(M,K)
# M is a nxn matrix
# K is a n x 2 matrix
local nsols,n,j,k,l,ll,RedMat,RedK,RedK1,RedK2;
n:=Length(M);
l:=Filtered(K[n], i -> IsInt(i / M[n][n]  ));
ll:=Length(l);
if (n=1 ) or (ll=0) then
   nsols:=ll;
elif ( ll = 1 ) then
   RedMat:=List([1..n-1],j ->List([1..n-1],k->M[j][k]));
   RedK:=List([1..n-1],j -> K[j]-(l[1])* M[n][j]/M[n][n]);
   nsols:= rsols(RedMat,RedK);
elif ( ll = 2 ) then
   RedMat:= List([1..n-1],j -> List([1..n-1],k->M[j][k]));
   RedK1:= List([1..n-1],j -> K[j]-(l[1])*M[n][j]/M[n][n]);
   RedK2:= List([1..n-1],j -> K[j]-(l[2])*M[n][j]/M[n][n]);
   nsols:= rsols(RedMat , RedK1) + rsols(RedMat ,RedK2);
```

```
fi;
return nsols;
end;

NumberOf02Solutions:=function(M)
return rsols(M,List([1..Length(M)], i -> [0,2]));
end;
```

Unfortunately the time needed to compute the solutions on large matrixes is too much for ordinary PC's: the needed time as an exponential growth. In case the representation is the regular representation, we can optimize the algorithm by doing a more selective search, as in the following function. But this improves considerably the needed time only if there are enough normal subgroups of index 2.

```
NumberOf02SolutionsRegular:=function(M)
local l,n,RM;
n:=Length(M);
l:=Filtered([1..(n-1)], i ->
    not(
        (M[i][i] =  2)
and
        (M[i][1] = 2)
and
        ( Length( Filtered([(i+1)..(n-1)], j -> M[j][i]<> 0))=0 )
    )
);
RM:= List(l, i -> List(l, j -> M[i][j]));
return 2^(n-Length(l))*rsols(RM,List(l, i -> [0,2]));
end;
```

EXAMPLE 6.3. Let G be the dihedral group of order 6, acting on the plane in the canonical way, and X the corresponding unit sphere. The isotropy types of G are (H_1) and (1), where H_1 denotes a subgroup of order 2 of G. The Table of Marks is

```
1:  6
2:  3 1
3:  2 . 2
4:  1 1 1 1
```

and therefore the Table of Isotropy Marks T is

```
1:  6
2:  3 1
```

The number of solutions is therefore 2, and thus $\operatorname{Im} d_G = \mathbb{Z}_2 = \{+1, -1\}$.

Now let us compute the Kernel $\ker d_G$: $\gamma_1 = 1$ and $\gamma_{H_1} = 2$. The dimension $\dim X_{H_1} = 0$, so that the number of generators of $\ker d_G$ is 0, and thus

$$E_G(X) = \mathbb{Z}_2.$$

EXAMPLE 6.4. As in the previous example, consider the dihedral group G of order 6, but this time consider the representation given by the sum of the canonical one and one copy of the trivial one. That is, G acts on the 3-dimensional Euclidean space, and it is generated by reflections of three planes crossing at angles of $\pi/3$ along the fixed line. The G-sphere X now has three isotropy types: (1), (H_1) as before, and (G). Thus the Table of Isotropy Marks is

```
1:  6
2:  3 1
4:  1 1 1
```

The number of solutions of $y \cdot T \in \{0,2\}^3$ is 4, and thus $d_G(\mathcal{E}_G(X)) \cong \mathbb{Z}_2^2$.

About the Kernel of d_G: $\gamma_1 = 1$, $\gamma_{H_1} = 2$ and $\gamma_G = 2$, with $\dim X_{H_1} = 1$ and $\dim X_G = 0$. Thus $\ker d_G$ has 1 generator, and this means that it is isomorphic to $\mathbb{Z}$. Therefore there is an exact sequence

$$0 \to \mathbb{Z} \to \mathcal{E}_G(X) \to \mathbb{Z}_2^2 \to 0$$

of groups and homomorphisms.

EXAMPLE 6.5. Consider now the alternating group A_4, and its non-trivial real irreducible representation on the 3-dimensional space (that is, orientation preserving symmetries of a standard Euclidean 3-simplex). This is the Table of Marks of G.

```
1:  12
2:   6 2
3:   4 . 1
4:   3 3 . 3
5:   1 1 1 1 1
```

By computing the isotropy groups, it is possible therefore to see that the isotropy types correspond to the subgroups H_1, H_2 and H_3 of the previous list (implicit in the Table of Marks): the Table of Isotropy Marks is the following.

```
1:  12
2:   6 2
3:   4 . 1
```

As a consequence, $\operatorname{Im} d_G \cong \mathbb{Z}_2^1$ (and here we can see another example in which $2^{\kappa(X)} > |\operatorname{Im} d_G|$). About the Kernel of d_G, first note that $\dim X^{H_1} = 2$, $\dim X^{H_2} = 0$ and $\dim X^{H_3} = 0$. Thus, the codimension of X_s^H in X^H is at least 2, for every isotropy type H. This implies that the Kernel of d_G is trivial, and therefore

$$\mathcal{E}_G(X) \cong \mathbb{Z}_2$$

EXAMPLE 6.6. Let G be the alternating group A_5, acting on the 4-dimensional Euclidean space as permutations of the standard 4-simplex in the canonical way (the representation is irreducible). We already computed its Table of Marks: now we need to compute the Table of Isotropy Marks. The dimension fixed by the subgroups are

```
[ 4, 2, 2, 1, 0, 1, 0, 1, 0 ]
```

and therefore the isotropy groups are

$$[H_1, H_2, H_3, H_6, H_8].$$

Thus the Table of Isotropy Marks is

```
1:  60
2:  30 2
3:  20 . 2
6:  10 2 1 1
8:   5 1 2 . 1
```

and thus

$$d_G \mathcal{E}_G(X) \cong \mathbb{Z}_2^2.$$

Now we consider the Kernel of d_G: because $\dim X^{H_1} = 3$, $\dim X^{H_2} = 1$, $\dim X^{H_3} = 1$, $\dim X^{H_6} = 0$ and $\dim X^{H_8} = 0$, it is $\gamma_{H_1} = 1$. It is now needed to compute γ_{H_2} and γ_{H_3}. From the Table of Isotropy Marks it is possible to deduce that X^{H_2} is a 1-sphere, with the $\mathbb{Z}_2$-action of the Weyl group $W_G H_2$. The Weyl group acts on the subgroups H_6 and H_8 by conjugacy, and this action corresponds to the action of fixed subspaces X^{H_6} and X^{H_8}. The singular set $X_s^{H_2}$ is the union of three disjoint 0-spheres: 2 copies (by the $W_G H_2$-action) of X^{H_6}, and 1 copy of X^{H_8}. The action of $W_G H_2$ must send X^{H_8} into itself, and must swap the two copies of X^{H_6}. Therefore it acts as a reflection along X^{H_8}. It is now a simple task to check that X_{H_2} has 6 components, and therefore $X_{H_2}/W_G H_2$ has 3 components. Thus $\gamma_{H_2} = 3$. Note that we even didn't properly need to know the action of the Weyl group $W_G H_2$, because we only need the number of components of $X_{H_2}/W_G H_2$. That is, we have used the following formula.

$$\gamma_{H_2} = \frac{\#\pi_0(X_{H_2})}{|W_G H_2|} = \frac{6}{2} = 3.$$

The same arguments hold for H_3, and therefore

$$\gamma_{H_3} = \frac{\#\pi_0(X_{H_3})}{|W_G H_3|} = \frac{6}{2} = 3.$$

This now concludes the computations: in the exact sequence of the degree homomorphism

$$0 \to K \to \mathcal{E}_G(X) \to \mathbb{Z}_2^2 \to 0$$

the kernel is a torsion-free, solvable group, with 4 generators, and derived length at most 2.

7. Concluding remarks

One can expect, by looking at the examples, that it is reasonable to search for a solution of the following problems:

1. To find an algorithm that computes the exact group structure of the kernel K, once given the usual data of the representation sphere (Table of Marks, character values, group structure). I don't know whether $E_G(X)$ need to be commutative or not, for G-spheres.
2. To ask something easier than in the previous point: to find an algorithm to compute just the number γ_H, for each H. I have started working on it, using some easy techniques about arrangements, but it is still in development stage.
3. To find a way to avoid computing the solutions of the equation $y \cdot T \in \{0,2\}^h$, to know their number. For example, the similar equation

$$y \cdot T \in \{0,1\}^h$$

gives rise to the Dress' famous characterization theorem [**Dre69**], and the number of its solutions can be computed simply by noting that if H and K are two subgroups with $[H : K] = p > 1$ prime, then the corresponding H-column and K-column in the Table of Marks are congruent mod p (and this allows to compute exactly the number of solutions, in term of perfect subgroups of G). This relation in the set of conjugacy classes of subgroups of G generates an equivalence relation, whose classes are named Dress classes. In our case, we can use the same argument, but for all primes $p \neq 2$. Unfortunately, in our settings the number of Dress classes doesn't give the number of solutions of the equation, but just an upper bound. Anyhow, it can be computed by a GAP function like the following:

```
DressClasses:= function(M)
local i,j,k,ll,n,p, Mp, Mpt,labeli,labelj ;
n:=Length(M);
ll:=[1..n];
for p in Filtered(Set(FactorsInt(M[1][1])),p ->p<>2)
do
Mp:=M*One(GF(p));
Mpt:=TransposedMat(Mp);
for i in [1..n] do
for j in [1..i] do
if ( Mpt[j] = Mpt[i]   ) then
   labeli:=ll[i]; labelj:=ll[j];
   for k in [1..n] do
      if ( ll[k] = labelj )  then ll[k]:=labeli; fi;
   od;
fi;
od;od;od;
return ll;
end;
```

4. To understand, at least for some class of groups, like nilpotent finite groups, an easy way to compute $\mathcal{E}_G(X)$ from the Table of Isotropy Marks (as in Matsuda's Theorem), or from the group structure of G, without solving the equation $y \cdot T \in \{0, 2\}^h$. There are induction and restriction theorems on Table Of Marks, that might give rise to corresponding induction and restriction properties of groups of self homotopy equivalences (i.e. a Mackey functors prespective).
5. We have seen that there are upper bounds of the order of $\mathcal{E}_G(X)$: $\kappa(X)$ and the number of Dress classes. It is reasonable to investigate under which (algebraic) conditions these bounds become equalities.

References

[CCN+85] J. H. Conway, R. T. Curtis, S. P. Norton, R. A. Parker, and R. A. Wilson, *Atlas of finite groups*, Oxford University Press, Oxford, 1985, Maximal subgroups and ordinary characters for simple groups, With computational assistance from J. G. Thackray.

[Dol65] Albrecht Dold, *Fixed point index and fixed point theorem for Euclidean neighborhood retracts*, Topology **4** (1965), 1–8.

[Dol80] Albrecht Dold, *Lectures on algebraic topology*, second ed., Springer-Verlag, Berlin, 1980.

[Dre69] Andreas Dress, *A characterisation of solvable groups*, Math. Z. **110** (1969), 213–217.

[Fer00] Davide L. Ferrario, *Equivariant deformations of maps and real representations*, Pacific Journal of Mathematics **to appear** (2000).

[IM95] Marek Izydorek and Wacław Marzantowicz, *Equivariant maps between cohomology spheres*, Topol. Methods Nonlinear Anal. **5** (1995), no. 2, 279–289.

[JS78] I. M. James and G. B. Segal, *On equivariant homotopy type*, Topology **17** (1978), no. 3, 267–272.

[Kom87] Katsuhiro Komiya, *The Lefschetz number for equivariant maps*, Osaka J. Math. **24** (1987), no. 2, 299–305.

[Kom89] Katsuhiro Komiya, *Congruences for the Burnside ring*, Transformation groups (Osaka, 1987), Springer, Berlin, 1989, pp. 191–197.

[Kom95] Katsuhiro Komiya, *Equivariant K-theory and maps between representation spheres*, Publ. Res. Inst. Math. Sci. **31** (1995), no. 4, 725–730.

[Kom98] Katsuhiro Komiya, *Equivariant maps between representation spheres of a torus*, Publ. Res. Inst. Math. Sci. **34** (1998), no. 3, 271–276.

[Mat78] Toshimitsu Matsuda, *On the S_n-equivariant self-homotopy equivalences of spheres*, J. Fac. Sci. Shinshu Univ. **13** (1978), no. 1, 43–78.

[Mat79] Toshimitsu Matsuda, *On the equivariant self-homotopy equivalences of spheres*, J. Math. Soc. Japan **31** (1979), no. 1, 69–83.

[Mat82] Toshimitsu Matsuda, *On the unit groups of Burnside rings*, Japan. J. Math. (N.S.) **8** (1982), no. 1, 71–93.

[Mat86] Toshimitsu Matsuda, *A note on the unit groups of Burnside rings as Burnside ring modules*, J. Fac. Sci. Shinshu Univ. **21** (1986), no. 1, 1–10.

[MM83] Toshimitsu Matsuda and Takehiko Miyata, *On the unit groups of Burnside rings of finite groups*, J. Math. Soc. Japan **35** (1983), no. 2, 345–354.

[Pfe97] Götz Pfeiffer, *The subgroups of M_{24}, or how to compute the table of marks of a finite group*, Experiment. Math. **6** (1997), no. 3, 247–270.

[Rub76] Ryszard L. Rubinsztein, *On the equivariant homotopy of spheres*, Dissertationes Math. (1976), no. 134, 48 (English).

[S+95] Martin Schönert et al., *GAP – Groups, Algorithms, and Programming*, Lehrstuhl D für Mathematik, Rheinisch Westfälische Technische Hochschule, Aachen, Germany, fifth ed., 1995.

[tD75] Tammo tom Dieck, *The Burnside ring of a compact Lie group. I*, Math. Ann. **215** (1975), 235–250.

[tD79] Tammo tom Dieck, *Transformation groups and representation theory*, Springer, Berlin, 1979.

[tD87] Tammo tom Dieck, *Transformation groups*, Walter de Gruyter & Co., Berlin, 1987.

[tD78] Tammo tom Dieck, *Idempotent elements in the Burnside ring*, J. Pure Appl. Algebra **10** (1977/78), no. 3, 239–247.

[Tor82] J. Tornehave, *Equivariant maps of spheres with conjugate orthogonal actions*, Current trends in algebraic topology, Part 2 (London, Ont., 1981), Amer. Math. Soc., Providence, R.I., 1982, pp. 275–301.

[Wan87] Stefan Waner, *A note on the existence of G-maps between spheres*, Proc. Amer. Math. Soc. **99** (1987), no. 1, 179–181.

[Wil84] Dariusz Wilczyński, *Fixed point free equivariant homotopy classes*, Fund. Math. **123** (1984), no. 1, 47–60.

[Yos83] Tomoyuki Yoshida, *Idempotents of Burnside rings and Dress induction theorem*, J. Algebra **80** (1983), no. 1, 90–105.

[Yos90] Tomoyuki Yoshida, *The generalized Burnside ring of a finite group*, Hokkaido Math. J. **19** (1990), no. 3, 509–574.

Dipartimento di Matematica, Politecnico di Milano, Piazza Leonardo da Vinci, 32, 20133 Milano (Italy)

E-mail address: `ferrario@mate.polimi.it`

TABLE 1. Self Homotopy Equivalences of regular G-spheres for small non-abelian groups (orders from 6 to 30)

IdGroup	$\log_2 \lvert E_G(X)\rvert$	$\kappa(X)$
[6, 1]	3	3
[8, 3]	5	6
[8, 4]	4	4
[10, 1]	3	3
[12, 1]	3	3
[12, 3]	2	3
[12, 4]	6	7
[14, 1]	3	3
[16, 3]	6	8
[16, 4]	5	6
[16, 6]	4	4
[16, 7]	6	8
[16, 8]	5	7
[16, 9]	5	6
[16, 11]	10	12
[16, 12]	8	8
[16, 13]	8	8
[18, 1]	4	4
[18, 3]	3	3
[18, 4]	7	7
[20, 1]	3	3
[20, 3]	3	4
[20, 4]	6	7
[21, 1]	1	2
[22, 1]	3	3
[24, 1]	3	3
[24, 3]	2	4
[24, 4]	6	7
[24, 5]	6	7
[24, 6]	8	11
[24, 7]	6	7
[24, 8]	7	10
[24, 10]	5	6
[24, 11]	4	4
[24, 12]	6	8
[24, 13]	4	6
[24, 14]	12	15
[26, 1]	3	3
[27, 3]	1	1
[27, 4]	1	1
[28, 1]	3	3
[28, 3]	6	7
[30, 1]	3	3
[30, 2]	3	3
[30, 3]	5	5

TABLE 2. Self Homotopy Equivalences of regular G-spheres for small non-abelian groups (order 32)

IdGroup	$\log_2 \lvert E_G(X) \rvert$	$\kappa(X)$
[32, 2]	7	10
[32, 4]	4	4
[32, 5]	6	8
[32, 6]	7	12
[32, 7]	7	10
[32, 8]	6	8
[32, 9]	7	11
[32, 10]	6	9
[32, 11]	6	8
[32, 12]	5	6
[32, 13]	5	8
[32, 14]	6	8
[32, 15]	5	6
[32, 17]	4	4
[32, 18]	7	10
[32, 19]	6	9
[32, 20]	6	8
[32, 22]	12	16
[32, 23]	10	12
[32, 24]	8	8
[32, 25]	10	12
[32, 26]	8	8
[32, 27]	14	20
[32, 28]	12	16
[32, 29]	10	12
[32, 30]	10	12
[32, 31]	10	12
[32, 32]	8	8
[32, 33]	8	8
[32, 34]	14	20
[32, 35]	10	12
[32, 37]	8	8
[32, 38]	8	8
[32, 39]	12	16
[32, 40]	10	14
[32, 41]	10	12
[32, 42]	10	12
[32, 43]	11	14
[32, 44]	10	12
[32, 46]	20	24
[32, 47]	16	16
[32, 48]	16	16
[32, 49]	17	22
[32, 50]	16	16

TABLE 3. Self Homotopy Equivalences of regular G-spheres for small non-abelian groups (orders from 34 to 46)

IdGroup	$\log_2 \lvert E_G(X) \rvert$	$\kappa(X)$
[34, 1]	3	3
[36, 1]	4	4
[36, 3]	2	3
[36, 4]	8	10
[36, 6]	3	3
[36, 7]	7	7
[36, 9]	5	6
[36, 10]	10	13
[36, 11]	2	3
[36, 12]	6	7
[36, 13]	14	19
[38, 1]	3	3
[39, 1]	1	2
[40, 1]	3	3
[40, 3]	3	4
[40, 4]	6	7
[40, 5]	6	7
[40, 6]	8	11
[40, 7]	6	7
[40, 8]	7	10
[40, 10]	5	6
[40, 11]	4	4
[40, 12]	6	8
[40, 13]	12	15
[42, 1]	3	4
[42, 2]	2	4
[42, 3]	3	3
[42, 4]	3	3
[42, 5]	5	5
[44, 1]	3	3
[44, 3]	6	7
[46, 1]	3	3

TABLE 4. Self Homotopy Equivalences of regular G-spheres for small non-abelian groups (order 48)

IdGroup	$\log_2 \lvert E_G(X) \rvert$	$\kappa(X)$
[48, 1]	3	3
[48, 3]	2	4
[48, 4]	6	7
[48, 5]	6	7
[48, 6]	8	13
[48, 7]	10	15
[48, 8]	8	11
[48, 9]	6	7
[48, 10]	6	7
[48, 11]	6	7
[48, 12]	7	10
[48, 13]	8	11
[48, 14]	9	14
[48, 15]	8	13
[48, 16]	7	11
[48, 17]	7	12
[48, 18]	7	10
[48, 19]	8	13
[48, 21]	6	8
[48, 22]	5	6
[48, 24]	4	4
[48, 25]	6	8
[48, 26]	5	7
[48, 27]	5	6
[48, 28]	6	8
[48, 29]	7	11
[48, 30]	6	9
[48, 31]	4	6
[48, 32]	4	8
[48, 33]	4	6
[48, 34]	12	15
[48, 35]	12	15
[48, 36]	16	23
[48, 37]	12	15
[48, 38]	15	23
[48, 39]	12	15
[48, 40]	12	15
[48, 41]	12	15
[48, 42]	12	15
[48, 43]	14	21
[48, 45]	10	12
[48, 46]	8	8
[48, 47]	8	8
[48, 48]	12	17
[48, 49]	8	12
[48, 50]	6	12
[48, 51]	24	31

TABLE 5. Self Homotopy Equivalences of regular G-spheres for small non-abelian groups (order from 50 to 54)

IdGroup	$\log_2 \lvert E_G(X) \rvert$	$\kappa(X)$
[50, 1]	4	4
[50, 3]	3	3
[50, 4]	9	9
[52, 1]	3	3
[52, 3]	3	4
[52, 4]	6	7
[54, 1]	5	5
[54, 3]	4	4
[54, 4]	3	3
[54, 5]	5	5
[54, 6]	4	4
[54, 7]	11	11
[54, 8]	7	7
[54, 10]	2	2
[54, 11]	2	2
[54, 12]	3	3
[54, 13]	7	7
[54, 14]	29	29
[55, 1]	1	2
[56, 1]	3	3
[56, 3]	6	7
[56, 4]	6	7
[56, 5]	8	11
[56, 6]	6	7
[56, 7]	7	10
[56, 9]	5	6
[56, 10]	4	4
[56, 11]	2	3
[56, 12]	12	15
[57, 1]	1	2
[58, 1]	3	3
[60, 1]	3	3
[60, 2]	3	3
[60, 3]	5	5
[60, 5]	5	7
[60, 6]	3	4
[60, 7]	5	6
[60, 8]	9	12
[60, 9]	2	3
[60, 10]	6	7
[60, 11]	6	7
[60, 12]	10	13
[62, 1]	3	3
[63, 1]	1	2
[63, 3]	1	2

Contemporary Mathematics
Volume **274**, 2001

Two examples to illustrate properties of the group of self-equivalences of a finite CW complex X

Y. Felix

ABSTRACT. We construct an example that shows that the nilpotency of the group of homotopy classes of homotopy self-equivalences of a simply-connected finite CW complex X can be larger than the Toomer invariant of the space. The second example is related to the Arkowitz-Maruyama conjecture: there is an integer N such that the map $\mathrm{Aut}_{\#}(X) \to \mathrm{Aut}_{\#}^N(X)$ is a bijection. We show that the integer N can be arbitrary larger than the dimension of X.

Let X be a 1-connected finite type CW complex. We denote by $AutX$ the group of homotopy classes of self-homotopy equivalences of X and by $Aut_{\#}^N X$ the subgroup of $AutX$ consisting of homotopy classes of self-homotopy equivalences φ such that $\pi_q(\varphi)$ is the identity map for $q \leq N$. The intersection $Aut_{\#}X = \cap_N Aut_{\#}^N X$ consists of the homotopy classes of self-homotopy equivalences that induce the identity in homotopy.

By a result of Dror and Zabrodsky ([**7**]), for $N \geq \dim X$, the group $Aut_{\#}^N X$ is a nilpotent group. If we denote by X_0 the rationalization of X, a result of Maruyama ([**19**]) asserts that the natural map $Aut_{\#}^N(X) \to Aut_{\#}^N(X_0)$ is a rationalization, i.e., $Aut_{\#}^N(X_0) \cong Aut_{\#}^N(X)_0$.

We then have :

THEOREM 1. [**10**] *Let X be a simply connected finite CW complex and $N \geq$ dim X, then Nil $Aut_{\#}^N(X_0) \leq$ cat$(X_0) - 1$.*

Here $\mathrm{cat}(X)$ is the Lusternik-Schnirelmann category of X, i.e., the minimum n such that X can be covered by $n+1$ open sets each contractible in X; the nilpotency of a nilpotent group G, Nil G, is the least integer n such that $G_n = 0$ in the sequence $G_0 = G \supset G_1 = [G, G] \supset G_2 = [G_1, G], \cdots, G_m = [G_{m-1}, G]$.

In [**5**], M. Arkowitz and G. Lupton give a better upper bound for Nil $Aut_{\#}^N(X_0)$ for some classes of spaces. They prove that, under some hypothesis, $Nil\, Aut_{\#}(X_0) < e_0(X)$, where e_0 is the Toomer invariant of X [**24**].

In the first section of this note we exhibit a simply connected finite CW complex X for which $Nil\, Aut_{\#}(X_0) = 2$, $cat(X_0) = 3$ and $e_0(X) = 2$. This shows that we can not replace $\mathrm{cat}(X_0) - 1$ with $e_0(X) - 1$ in general.

The Lusternik-Schnirelmann category and the Toomer invariant can be easily deduced from the Ganea fibrations $p_n : G_n(X) \to X$ [**14**]. Denote by $p_0 : PX \to X$

2000 *Mathematics Subject Classification.* Primary 55P10.

the path space fibration. The fibrations p_n with fiber F_n are then defined inductively as follows. Assume p_{n-1} has been defined, then p_{n-1} extends to a map $\tilde{p}_n : G_n(X) \cup CF_n \to X$ by sending CF_n, the cone on F_n, to the base point. The fibration p_{n+1} is the homotopy fibration associated to $\tilde{p}_n$. The integer $\mathrm{cat}(X)$ is equal to the least integer n such that p_n admits a homotopy section, and $e_0(X)$ is the least integer n such that $H_*(p_n;\mathbb{Q})$ is surjective ([**17**]).

The second part of this note concerns the Arkowitz-Maruyama conjecture:

CONJECTURE [**6**]. Let X be a simply connected finite CW complex. Then there exists an integer N such that the natural homomorphism $\rho_N : Aut_\# X \to Aut_\#^N X$ is an isomorphism.

This conjecture is unsolved in general. The rational version of the conjecture has been proved independently by Felix and Thomas [**12**] and by Maruyama [**20**]. In [**12**] the authors assert to have proved the conjecture in general, but their proof is not correct [**13**]. The conjecture has been proved for some families of spaces, for instance for product of spheres ([**6**]) and for H_0-spaces ([**21**]).

We prove here that the integer N can be arbitrary larger than the dimension.

THEOREM 2. *For each integer $N \geq 4$, there is a simply connected 6-dimensional CW complex X such that $Aut_\# X_0 \neq Aut_\#^N X_0$.*

In the proofs we will use the theory of Sullivan and Quillen minimal models. The reader will find in ([**15**], [**9**], [**23**]) the basic material on this theory.

The invariant $\mathrm{cat}(X_0)$ is a very interesting invariant. First by a result of Toomer ([**25**]), $\mathrm{cat}(X_0) \leq \mathrm{cat}(X)$. The category of X_0 is called the rational category of X and is denoted by $\mathrm{cat}_0(X)$. This integer can be computed by means of the Sullivan minimal model of X. Denote by $(\wedge V, d)$ the Sullivan minimal model of X and let j_n be the relative minimal model for the canonical projection q_n.

$$\begin{array}{ccc} (\wedge V, d) & \xrightarrow{q_n} & (\wedge V / \wedge^{>n} V, \bar{d}) \\ & {\scriptstyle j_n}\searrow & \simeq\uparrow\psi_n \\ & & (\wedge V \otimes \wedge W, D) \end{array}$$

The integer $\mathrm{cat}_0(X)$ is the least integer n such that j_n admits a retraction and the integer $e_0(X)$ is the least integer n such that $H(j_n)$ is injective ([**8**]).

In this paper we will make an intensive use of the theory of inert elements in a graded Lie algebra. For that purpose, we recall in the first section some basic facts about inert elements and inert ideals.

1. Inert elements

Let $(x_1, x_2, \dots)$ be a sequence of homogeneous elements in a graded Lie algebra L. We construct a differential graded Lie algebra $(L_1, d) = (L \amalg \mathbb{L}(y_1, y_2, \dots), d)$ by putting $d(y_i) = x_i$. Here $\coprod$ denotes coproduct of graded Lie algebras and $\mathbb{L}(V)$ denotes the free graded Lie algebra generated by V.

The sequence (x_i) is *inert* if the following equivalent conditions are satisfied ([**16**]):

1. The natural map $L \to H_*(L_1, d)$ is surjective;
2. $H_*(L_1, d) \cong L/(x_i)$;
3. The ideal I generated by the x_i is free and, for the induced action of L/I on the indecomposable elements $QI = I/[I, I]$, QI is a free $U(L/I)$-module generated by the x_i.

The determination of inert sequences is important. For instance:

- A subsequence of an inert sequence is inert ([**2**]).
- The sequence (x_i) is inert in L if and only if the corresponding sequence (x_i) is strongly free (cf [**1**]) in its enveloping algebra UL.
- Suppose $L = \mathbb{L}(V)$. We fix an ordering on V and we denote by $\hat{x}_i$ the high term of x_i for this ordering. Then (x_i) is inert if $\hat{x}_i$ is ([**1**], Theorem 3.2).

Finally an ideal I is called inert if it is generated by inert elements.

Let now (L, d) be a differential graded Lie algebra, and let $(x_1, x_2, \dots)$ be a sequence of homogeneous cycles in L. We construct the differential graded Lie algebras $(L_1, d) = (L \amalg \mathbb{L}(y_1, y_2, \dots), d)$ by putting $d(y_i) = x_i$.

PROPOSITION 1. [**16**] *The sequence* $([x_i])$ *is inert in* $H_*(L, d)$ *if and only if one of the following conditions is satisfied:*

1. *The natural map* $H_*(L, d) \to H_*(L_1, d)$ *is surjective;*
2. $H_*(L_1, d) \cong H_*(L, d)/([x_i])$.

To every simply connected space S, we can associate its Quillen model, that is a differential graded Lie algebra $(\mathbb{L}(V), d)$ with very interesting properties. First of all,

$$V_r \cong H_{r+1}(S; \mathbb{Q}), \qquad H_*(\mathbb{L}(V), d) \cong \pi_*(\Omega S) \otimes \mathbb{Q}.$$

The Quillen model of a sphere is thus a free Lie algebra on one generator with differential equal to 0. On the other hand, let

$$T = S \cup_{\vee_{j \in J} \varphi_j} (\vee_{j \in J} e^{n_j}),$$

denote by $(\mathbb{L}(V), d)$ a Quillen model for S, denote by $\tilde{\varphi}_j : S^{n_j - 2} \to \Omega S$ the adjoint maps, and finally denote by α_j a cycle in $\mathbb{L}(V)$ such that $[\alpha_j] = [\tilde{\varphi}_j] \in H_{n_j-2}(\mathbb{L}(V), d) \cong \pi_{n_j-2}(\Omega S) \otimes \mathbb{Q}$. Then the differential graded Lie algebra

$$(\mathbb{L}(V) \amalg \mathbb{L}(\{z_j\}_{j \in J}), d), \qquad d(z_j) = \alpha_j,$$

is a Quillen model for the space T.

The sequence (φ_j) is called inert if the sequence $(\tilde{\varphi}_j)$ is. Therefore we have:

PROPOSITION 2. *The following conditions are equivalent:*

1. *The induced map* $\pi_*(S) \otimes \mathbb{Q} \to \pi_*(T) \otimes \mathbb{Q}$ *is surjective;*
2. $\pi_*(\Omega T) \otimes \mathbb{Q} \cong (\pi_*(\Omega S) \otimes \mathbb{Q})/(\tilde{\varphi}_j)$;
3. *The elements* $[\tilde{\varphi}_j]$ *are inert in* $H_*(\mathbb{L}(V), d)$.

Consider now the case of the attachment of only one cell. Suppose φ belongs to $\pi_{n-1}(S)$ and $T = S \cup_\varphi e^n$. Denote by

$$P_R(t) = \sum_{n\geq 0} \dim H_n(R;\mathbb{Q})t^n$$

the Poincaré series of a space R, and by

$$P_{UL}(t) = \sum_{n\geq 0} dim(UL)_n t^n$$

the Hilbert series of the enveloping algebra of a graded connected finite type Lie algebra L. Then:

PROPOSITION 3. ([**1**]; Theorem 2.6) *The element φ is inert if and only if*

$$P_{UK}(t)^{-1} = P_{\Omega S}(t)^{-1} + t^{n-2}\,,$$

where K denotes the quotient Lie algebra $K = (\pi_*(\Omega S)\otimes \mathbb{Q})/(\tilde{\varphi})$.

We now have

LEMMA 1. *Let L be a graded Lie algebra and let $\gamma \in L_{even}$, then the element $[\beta,\gamma]$ is inert in $(L \amalg \mathbb{L}(\beta), 0)$.*

PROOF. Since the element β is inert in $L \amalg \mathbb{L}(\beta)$, the ideal generated by β is a free Lie algebra and the indecomposable elements form a free UL-module on one generator β.

We consider therefore the morphism of graded Lie algebras

$$\psi : \mathbb{L}(UL * \beta) \to L \amalg \mathbb{L}(\beta)\,,$$

where $UL * \beta$ denotes a free UL-module generated by β and

$$\psi(\alpha_1\alpha_2\ldots\alpha_n * \beta) = [\alpha_1,[\alpha_2,\ldots[\alpha_n,\beta]\ldots]]\,, \qquad \alpha_i \in L.$$

The image of ψ is the ideal generated by β. It follows that ψ is injective and that there is a short exact sequence of graded Lie algebras:

$$0 \to \mathbb{L}(UL * \beta) \xrightarrow{\psi} L \amalg \mathbb{L}(\beta) \to L \to 0\,.$$

Since $\gamma \in L_{even}$, the element γ is not a zero divisor in UL. Therefore the UL-module generated by γ is free and the ideal generated by $[\beta,\gamma]$ in $L \amalg \mathbb{L}(\beta)$ is isomorphic to the ideal generated by $(UL\cdot\gamma) * \beta$ in $\mathbb{L}(UL * \beta)$. In particular we have an isomorphism of graded vector spaces

$$U\left(L \amalg \mathbb{L}(\beta)\right)/[\beta,\gamma] = UL \otimes T(\,(UL/UL\cdot\gamma) * \beta)\,.$$

Recall that for finite type graded Lie algebras L and L', we have the equalities

$$P_{U(L\amalg L')}(t)^{-1} = P_{UL}(t)^{-1} + P_{UL'}(t)^{-1} - 1 \qquad ([\mathbf{18}], \text{Lemme } 5.1.10)\,,$$

$$P_{TV}(t) = [1 - P_V(t)]^{-1}\,.$$

Denote $K = (L \amalg \mathbb{L}(\beta))/[\beta,\gamma]$, we have

$$P_{UK}(t)^{-1} = P_{UL}(t)^{-1} - t^{|\beta|} + t^{|\beta|+|\gamma|} = P_{U(L\amalg\mathbb{L}(\beta))}(t)^{-1} + t^{|\beta|+|\gamma|}\,.$$

Using Proposition 3, this shows that $[\beta,\gamma]$ is inert. □

As a particular case, we deduce directly the following useful corollary.

COROLLARY 1. *Let S be a space with rational homotopy Lie algebra $(\mathbb{Q}\alpha \times L) \amalg \mathbb{L}(x)$, with $|\alpha|$ even, and let $T = S \cup_\varphi e^n$ with $\tilde{\varphi} = [\alpha, x]$. Then the cell attachment is inert and consequently the rational homotopy Lie algebra of T is $\mathbb{Q}\alpha \times (L \amalg \mathbb{L}(x))$.*

2. A space X with Nil $Aut_\# X > e(X) - 1$

Denote by $T_2\left(S^3_a, S^3, S^3\right)$ the fat wedge of three spheres S^3, i.e., the subset of $S^3_a \times S^3 \times S^3$ consisting of elements (x, y, z) such that at least one of x, y and z is the basis point. It is well known that the product $S^3_a \times S^3 \times S^3$ is obtained by adding an 9-dimensional cell to $T_2\left(S^3_a, S^3, S^3\right)$ along a class ω called the triple (higher order) Whitehead product of the 3-dimensional spheres ([**22**]).

We consider the space

$$X = T_2\left(S^3_a, S^3, S^3\right) \cup_{[S^3_a, \omega]} e^{11} \vee S^6 \vee S^9 \cup_{[S^3_a, S^6]} e^9$$

LEMMA 2. *A (non-minimal) Sullivan model for X is given by the commutative differential graded algebra*

$$(A, d) = \left(\wedge(a, b, c, z, t, x, u)/(A^{>12}, tb, tc, bz, cz, xb, xc, z^2), d\right),$$

with $\mid a \mid = \mid b \mid = \mid c \mid = 3, \quad \mid t \mid = 8, \mid z \mid = 6, \mid x \mid = 9, \mid u \mid = 11, dt = abc, du = xa.$

PROOF. The commutative differential graded algebra

$$(\wedge(a, b, c), 0), \mid a \mid = \mid b \mid = \mid c \mid = 3,$$

is the Sullivan minimal model for $S^3_a \times S^3 \times S^3$. The quasi-isomorphism

$$\rho : \left(\frac{\wedge(a, b, c, t, t_1)}{(tb, tc) + \wedge(a, b, c, t, t_1)^{>11}}, dt = abc, dt_1 = ta\right) \xrightarrow{\simeq} \left(\frac{\wedge(a, b, c)}{(abc)}, 0\right),$$

$$|t| = 8, |t_1| = 10, \rho(t) = \rho(t_1) = 0,$$

shows that the differential graded algebras

$$\left(\frac{\wedge(a, b, c, t, t_1)}{(tb, tc) + \wedge(a, b, c, t, t_1)^{>11}}, dt = abc, dt_1 = ta\right) \text{ and } \left(\frac{\wedge(a, b, c)}{(abc)}, 0\right)$$

are models for the fat wedge $T_2(S^3_a, S^3, S^3)$.

In the same way, a Sullivan (non minimal) model for $T_2(S^3_a, S^3, S^3) \vee S^6 \vee S^9$ is given by

$$\left(\frac{\wedge(a, b, c, t, t_1, z, x, v)}{(\wedge(a, b, c, t, t_1, x, z, v)^{>11} + (tb, tc, bz, cz, va, vb, vc)}, dt = abc, dt_1 = ta, dv = za\right),$$

with $\mid z \mid = 6, \mid x \mid = 9, \mid v \mid = \mid t \mid = 8, \mid t_1 \mid = 10$.

Since the vector space $\pi_*(S) \otimes \mathbb{Q}$ is isomorphic to the dual of the graded vector space formed by the generators of the minimal model of S ([**9**]; Theorem 15.11), and since the Whitehead bracket is dual to the quadratic part of the differential ([**9**]; Proposition 13.6), a Sullivan model for the map

$$[S^3_a, \omega] \vee [S^3_a, S^6] : S^{10} \vee S^8 \;\rightarrow\; T_2(S^3_a, S^3, S^3) \vee S^6 \vee S^9$$

is given by the morphism

$$\psi \,:\, \left(\frac{\wedge(a, b, c, t, t_1, z, x, v)}{(\wedge(a, b, c, t, t_1, x, z, v)^{>11} + (tb, tc, bz, cz, va, vb, vc)}, d\right) \rightarrow \left(\frac{\wedge(s, w)}{(sw, s^2, w^2)}, 0\right),$$

with $\mid s \mid = 8, \mid w \mid = 10, \psi(v) = s, \psi(t_1) = w, \psi(t) = 0$. By ([**15**], Proposition 15.18), a model for the space X is therefore given by $\mathbb{Q}\oplus$ Kerψ.

$$\mathbb{Q} \oplus \mathrm{Ker}\psi \cong \left(\frac{\wedge(a,b,c,t,z,x,)}{\wedge(a,b,c,t,z,x)^{>11} + (tb,tc,bz,cz)}, d \right) .$$

Now the projection $p : (A,d) \to \mathbb{Q}\oplus$ Kerψ defined by $p(u) = 0$ is a surjective map with acyclic kernel $(\mathbb{Q}u \oplus \mathbb{Q}xa, d)$, $d(u) = xa$. Therefore (A,d) is a Sullivan model for X.

□

Denote by $\rho : (\wedge V, d) \xrightarrow{\cong} (A,d)$ the Sullivan minimal model of (A,d). Clearly,

$$(\wedge V, d) \cong (\wedge(a,b,c,z,t,x,u,\cdots),d) \ ,$$

with other generators in degrees ≥ 8. To be more precise, $(\wedge V, d) =$

$$(\wedge(a,b,c,z,t,x_1,x_2,x,u_1,u_2,y_1,y_2,y_3,u,r_2,r_3,m,w_1,w_2,w_3,z_1,z_2,z_3,z_4,..),d) \, ,$$

with other generators in degree ≥ 13. Here

$$|a| = |b| = |c| = 3, |z| = 6, |t| = |x_1| = |x_2| = 8, d(t) = abc, d(x_1) = zb, d(x_2) = zc \, ,$$

$$|u_1| = |u_2| = |y_1| = |y_2| = |y_3| = 10 \, ,$$

$$d(u_1) = tb, d(u_2) = tc, d(y_1) = x_1 b, d(y_2) = x_2 c, d(y_3) = x_1 c + x_2 b \, ,$$

$$|u| = |r_2| = |r_3| = |m| = 11 \, , \ d(u) = xa, d(r_2) = xb, d(r_3) = xc, d(m) = z^2 \, ,$$

$$|w_1| = |w_2| = |w_3| = |z_1| = |z_2| = |z_3| = |z_4| = 12 \, ,$$

$$d(w_1) = u_1 b, d(w_2) = u_2 c, d(w_3) = u_1 c + u_2 b \, ,$$

$$d(z_1) = y_1 b, d(z_2) = y_2 c, d(z_3) = y_1 c + y_3 b, d(z_4) = y_2 b + y_3 c \, .$$

We now recall the homotopy Lie algebra structure of $\pi_* \left(\Omega T_2(S^3_a, S^3, S^3)\right) \otimes \mathbb{Q}$.

LEMMA 3. *There is an isomorphism of graded Lie algebras*

$$\pi_* \left(\Omega T_2(S^3_a, S^3, S^3)\right) \otimes \mathbb{Q} \cong Ab(\alpha,\beta,\gamma) \amalg \mathbb{L}(\omega) \, , \quad |\alpha| = |\beta| = |\gamma| = 2 \, , |\omega| = 7 \, ,$$

where $Ab(\alpha,\beta,\gamma)$ denotes the abelian Lie algebra $\mathbb{Q}\,\alpha \oplus \mathbb{Q}\,\beta \oplus \mathbb{Q}\,\gamma$.

PROOF. Since $\pi_q(\Omega(S^3_a \times S^3 \times S^3)) \otimes \mathbb{Q} = 0$ for $q \neq 2$, the map

$$\pi_*(\Omega T_2(S^3_a, S^3, S^3)) \otimes \mathbb{Q} \to \pi_*(\Omega(S^3_a \times S^3 \times S^3)) \otimes \mathbb{Q}$$

is surjective. By Proposition 2, this implies that the element ω is inert, and that the homotopy Lie algebra $\pi_*(\Omega T_2(S^3_a, S^3, S^3)) \otimes \mathbb{Q}$ is generated by the elements a, b, c and ω. We deduce a surjective map

$$\mathbb{L}(\omega) \amalg Ab(a,b,c) \to \pi_*(\Omega T_2(S^3_a, S^3, S^3)) \otimes \mathbb{Q} \, .$$

A standard computation using Proposition 3 shows that the enveloping algebras on the graded Lie algebras have same Poincaré series

$$P_{U(\mathbb{L}(\omega)\amalg Ab(a,b,c))} = P_{\Omega T_2(S^3_a,S^3,S^3)} = \left((1-t^2)^3 - t^{|\omega|} \right)^{-1} .$$

This implies the result.

□

From Lemma 3, we deduce that

$$\pi_*(\Omega X)\otimes\mathbb{Q}\cong\left(\mathbb{Q}\alpha\times\left[Ab(\beta,\gamma)\amalg\mathbb{L}(\omega,u)\right]\right)\amalg\mathbb{L}(v)\,,$$

with $\mid u\mid=5$ and $\mid v\mid=8$. More precisely, from Lemma 3 and Corollary 1, we deduce

$$\begin{aligned}\pi_*(\Omega(T_2\left(S^3_a,S^3,S^3\right)\,\vee\,S^6\,\vee\,S^9))\otimes\mathbb{Q}&\cong\left[\left[\mathbb{Q}\alpha\times Ab(\beta,\gamma)\right]\amalg\mathbb{L}(u)\right]\amalg\mathbb{L}(\omega,v)\\ \pi_*(\Omega(Y)\otimes\mathbb{Q}&\cong(\mathbb{Q}\alpha\times[Ab(\beta,\gamma)\amalg\mathbb{L}(u)])\;\amalg\;\mathbb{L}(\omega,v)\\ &\cong((\mathbb{Q}\alpha\times[Ab(\beta,\gamma)\amalg\mathbb{L}(u)])\amalg\mathbb{L}(\omega))\amalg\mathbb{L}(v)\\ \pi_*(\Omega X)\otimes\mathbb{Q}&\cong(\mathbb{Q}\alpha\times(Ab(\beta,\gamma)\amalg\mathbb{L}(u,\omega)))\;\amalg\;\mathbb{L}(v)\,,\end{aligned}$$

with $Y=T_2\left(S^3_a,S^3,S^3\right)\,\vee\,S^6\,\vee\,S^9\cup_{[S^3_a,S^6]}e^9)$. In particular $\pi_*(\Omega X)\otimes\mathbb{Q}$ is generated by ω and by elements in degree ≤ 8.

LEMMA 4. *Nil* $Aut_\#(X_0)=2$

PROOF. We define two maps $\varphi,\psi:(A,d)\to(A,d)$, $\varphi(x)=x+az$, $\psi(z)=z+bc$; the maps φ and ψ are the identity on the other generators of A.

Denote $\theta=\psi^{-1}\varphi^{-1}\psi\,\varphi$, we have $\theta(x)=x+abc$. The map θ is not homotopic to the identity map, because otherwise by ([**15**]) there should exist a morphism of differential graded algebras

$$H:(\wedge V\otimes\wedge\bar{V}\otimes\wedge\hat{V},D)\longrightarrow(\wedge V,d)$$

such that $H(v)=v$ and $H\left(e^{sD+Ds}v\right)=\theta(v)$ for all v in V. For degree reasons, we have $H(\bar{a})=0$. Since $x+abc=e^{sD+Ds}x=x+D\,H(\bar{x})$, we have $H(\bar{x})=t$. Now

$$e^{sD+Ds}u=u+D\bar{u}+\bar{x}a-x\bar{a}+\frac{1}{2}(-\hat{x}\bar{a}+\bar{x}\hat{a}),$$

so that $u=\theta(u)=H(e^{sD+Ds}\,u)=u+d\,H\bar{u}+ta$, which is impossible because $[ta]\neq 0$.

A short computation shows that φ and ψ are the identity on the indecomposable elements in degrees ≤ 9 and on the element t, the homotopy classes $[\varphi]$ and $[\psi]$ define elements in $Aut_\#(X_0)$. As shown above $[\varphi,\psi]\neq 0$, so that Nil $Aut_\#(X_0)\geq 2$.

The space X has a natural cone decomposition in 3 steps which gives $\text{cat}(X_0)\leq\text{cat}(X)\leq\text{cl}(X)\leq 3$. By ([**10**]), this yields Nil $Aut_\#(X_0)\leq 2$. Therefore, we have Nil $Aut_\#(X_0)=2$. □

LEMMA 5. $e_0(X)=2$

PROOF. We consider the continuous map

$$\theta:Y=T_2(S^3_a,S^3,S^3)\vee S^9\vee(S^3_x\times(S^8\vee S^6))\to X$$

that is the identity on $T_2(S^3_a,S^3,S^3)\vee S^9\vee S^6$ and maps S^3_x and S^8 respectively to S^3_a and along the higher order Whitehead bracket ω. The space Y is a two-cone and thus $\text{cat}(Y)\leq 2$. Therefore the Ganea fibration $p_2(Y):G_2(Y)\to Y$ has a section σ. The naturality up to homotopy of G_2 gives the homotopy commutative diagram diagram

$$\begin{array}{ccc} G_2(Y) & \xrightarrow{G_2(\theta)} & G_2(X) \\ {\scriptstyle p_2(Y)}\downarrow & & \downarrow{\scriptstyle p_2(X)} \\ Y & \xrightarrow{\theta} & X \end{array}$$

This shows that $\theta \sim p_2(X)G_2(\theta)\sigma$. Now since $H_*(\theta;\mathbb{Q})$ is surjective, the same is true for $H_*(p_2(X);\mathbb{Q})$. Therefore $e_0(X) \leq 2$. In fact $e_0(X) = 2$ because there are non trivial cup products in $H^*(X;\mathbb{Q})$. □

3. Relations between $Aut^N_\# X$ and $Aut_\# X$

The example.

Let us fix an integer n. We construct a CW complex X with cells in dimensions 3 and 6 such that $\pi_q(\Omega X) \otimes \mathbb{Q}$ is generated via Whitehead brackets by the classes in degree 2 for $q \leq 4+2n$, but not for $q = 5+2n$.

Denote then by $S = X \vee S^6$, and consider the automorphism φ of S that is the identity on the cells of dimension 3 and on all the cells of dimension 6 except on one cell in X, e, whose image is obtained by pinching : $e \to e + S^6$. Then $\pi_{\leq 4+2n}(\Omega\varphi) \otimes \mathbb{Q}$, being the identity on classes in degree 2 and 5, is the identity map, but $\pi_{5+2n}(\Omega\varphi)$ is not the identity map by our particular choice of cell e.

The space X and the automorphism φ.

$$\begin{aligned} X = \quad & S^3_\alpha \vee S^3_{\gamma_1} \vee \ldots \vee S^3_{\gamma_n} \vee S^3_{\beta_1} \vee \ldots \vee S^3_{\beta_n} \cup_{[\alpha,\gamma_1]} e^6_1 \cup_{[\alpha,\gamma_2]} e^6_2 \ldots \cup_{[\alpha,\gamma_n]} e^6_n \\ & \cup_{[\alpha,\beta_1]} e^6 \cup_{[\alpha,\beta_2]-[\beta_1,\gamma_1]} e'^6_1 \ldots \cup_{[\alpha,\beta_n]-[\beta_{n-1},\gamma_{n-1}]} e'^6_{n-1} \cup_{[\beta_n,\gamma_n]} e'^6_n \,. \end{aligned}$$

The automorphism φ of $X \vee S^6$ is the identity on all the cells except on e^6,

$$\varphi(e^6) = e^6 + S^6 \,.$$

Write $X = Y \cup_{[\alpha,\beta_1]} e^6$. In order to prove our assertion we will compute $\pi_*(\Omega X) \otimes \mathbb{Q}$ from the Quillen model of X:

$$\begin{aligned} & (L_X, d) \cong (\mathbb{L}(\alpha, \gamma_1, \ldots, \gamma_n, \beta_1, \ldots, \beta_n, x_1, \ldots, x_n, y_1, \ldots, y_n, x), d)\,, \\ & d(\alpha) = d(\gamma_i) = d(\beta_i) = 0\,, d(x_i) = [\alpha, \gamma_i]\,, \\ & d(y_j) = -[\alpha, \beta_{j+1}] + [\beta_j, \gamma_j]\,, \text{ for } j < n\,, d(y_n) = [\beta_n, \gamma_n]\,, d(x) = [\alpha, \beta_1]\,. \end{aligned}$$

LEMMA 6. *The ideal $I \subset \mathbb{L}(\alpha, \beta_1, \ldots, \beta_n, \gamma_1, \ldots, \gamma_n)$ generated by the elements*

$$[\alpha, \gamma_i]_{1 \leq i \leq n}, [\beta_n, \gamma_n], [\alpha, \beta_j] - [\beta_{j-1}, \gamma_{j-1}]_{2 \leq j \leq n}$$

is inert.

PROOF. Let us fix an ordering on the variables $\alpha, \beta_i, \gamma_i$ by putting the γ_i before the α and the β_i. The high terms are the $\gamma_i \cdot \alpha$ and the $\gamma_i \cdot \beta_i$. They form a combinatorially free set ([**1**]), and a result of Anick ([**1**], Theorem 3.2) yields the result. □

Therefore we have a quasi-isomorphism

$$(\mathbb{L}(\alpha,\gamma_1,\dots,\gamma_n,\beta_1,\dots,\beta_n,x_1,\dots,x_n,y_1,\dots,y_n),d)\xrightarrow{\simeq}(\mathbb{L}(\alpha,(\gamma_i,\beta_i)_{1\le i\le n})/I,0)\,.$$

In particular, a (non minimal) Quillen model for X is given by

$$\mathcal{L}=((\mathbb{L}(\alpha,\gamma_1,\dots,\gamma_n,\beta_1,\dots,\beta_n)/I)\amalg\mathbb{L}(x),d)\,,$$
$$|\alpha|=|\gamma_i|=|\beta_i|=2,|x|=5,d(\alpha)=d(\gamma_i)=d(\beta_i)=0,d(x)=[\alpha,\beta_1]\,.$$

The map $\pi_{5+2n}(\Omega\varphi)\otimes\mathbb{Q}$ is not the identity map.

LEMMA 7. *The element*

$$[\gamma_n,[\gamma_{n-1},\cdots,[\gamma_3,[\gamma_2,[\gamma_1,x]]]\cdots]]$$

is a cycle in $\mathcal{L}_{5+2n}$ whose homology class in nonzero in $H(\mathcal{L})$.

PROOF.

$$\begin{aligned}
&d([\gamma_n,[\gamma_{n-1},\cdots,[\gamma_3,[\gamma_2,[\gamma_1,x]]\cdots]])\\
&=[\gamma_n,[\gamma_{n-1},\cdots,[\gamma_3,[\gamma_2,[\gamma_1,[\alpha,\beta_1]]]\cdots]]\\
&=\pm[\gamma_n,[\gamma_{n-1},\cdots,[\gamma_3,[\gamma_2,[\alpha,[\gamma_1,\beta_1]]\cdots]]\\
&=\pm[\gamma_n,[\gamma_{n-1},\cdots,[\gamma_3,[\gamma_2,[\alpha,[\alpha,\beta_2]]\cdots]]\\
&=\pm[\gamma_n,[\gamma_{n-1},\cdots,[\gamma_3,[\alpha,[\alpha,[\beta_2,\gamma_2]]\cdots]]\\
&\dots\\
&=\pm[\gamma_n,[\gamma_{n-1},\cdots,[\gamma_i,[\alpha,\dots,[\alpha,[\beta_{i-1},\gamma_{i-1}]]\cdots]]\\
&\dots\\
&=\pm[\alpha,[\alpha,\cdots,[\alpha,[\alpha,[\beta_n,\gamma_n]]\cdots]]=0
\end{aligned}$$

On the other hand $[\gamma_n,[\gamma_{n-1},\cdots,[\gamma_3,[\gamma_2,[\gamma_1,x]]\cdots]]$ is not a boundary because the boundaries belong to the ideal generated by α and the β_i. □

The map φ is represented by the automorphism $\bar\varphi$ of $(\mathcal{L}\amalg\mathbb{L}(z),|z|=5,d(z)=0)$, defined by $\bar\varphi(\alpha)=\alpha$, $\bar\varphi(\gamma_i)=\gamma_i$, $\bar\varphi(\beta_j)=\beta_j$, $\bar\varphi(x)=x+z$, $\bar\varphi(z)=z$. In homology we have

$$\bar\varphi[\gamma_n,[\gamma_{n-1},\cdots,[\gamma_1,x]\cdots]]=[\gamma_n,[\gamma_{n-1},\cdots,[\gamma_1,x]\cdots]]+[\gamma_n,[\gamma_{n-1},\cdots,[\gamma_1,z]\cdots]]\,.$$

Since the class of $[\gamma_n,[\gamma_{n-1},\cdots,[\gamma_1,z]\cdots]]$ is nonzero, the map $\pi_{5+2n}(\Omega\varphi)\otimes\mathbb{Q}$ is not the identity.

The map $\pi_{\le 4+2n}(\Omega\varphi)\otimes\mathbb{Q}$ is the identity map

Denote by J the sub Lie algebra of $\pi_*(\Omega X)\otimes\mathbb{Q}$ generated by the elements of degree 2. Clearly $\pi_*(\Omega\varphi)$ is the identity on J. It is therefore enough to prove Lemma 8.

LEMMA 8. $J_r=\pi_r(\Omega X)\otimes\mathbb{Q}$ *for* $r<5+2n$.

PROOF. Denote by Z the homotopy fiber of the map $q : X \to S^3_\alpha$ that is the identity on S^3_α and is trivial on the other cells. Following ([**23**], Proposition VI.I. (3)) A Quillen model for Z is given by the kernel of the model of q:

$$\tilde{q} : L_X = (\mathbb{L}(\alpha, \gamma_1, \dots, \gamma_n, \beta_1, \dots, \beta_n, x_1, \dots, x_n, y_1, \dots, y_n, x), d) \longrightarrow (\mathbb{L}(\alpha), 0)\,,$$
$$\tilde{q}(\alpha) = \alpha, \tilde{q}(\gamma_i)\tilde{q}(\beta_i) = \tilde{q}(x_i) = \tilde{q}(y_i) = \tilde{q}(x) = 0\,.$$

The kernel is a free graded Lie algebra of the form

$$L' = (\mathbb{L}(\mathbb{Q}[\alpha] \otimes (\gamma_1, \dots, \gamma_n, \beta_1, \dots, \beta_n, x_1, \dots, x_n, y_1, \dots, y_n, x)), D)\,,$$

where $\alpha^r \otimes u$ is sent into L_X onto the rth iterated bracket of u by α, $[\alpha, \dots, [\alpha, u] \dots]$. Therefore we have $D(\alpha^r \otimes x_i) = \alpha^{r+1} \otimes \gamma_i$. This means that the ideal of L' generated by the $\alpha^r x_i$ and the $\alpha^{r+1}\gamma_i$, for $r \geq 0$ is acyclic. We denote by L_2 the quotient differential graded Lie algebra.

$$L_1 \xrightarrow{\simeq} L_2\,.$$

Now in the quotient we have

$$D(\alpha^r \otimes x) = \alpha^{r+1} \otimes \beta_1$$
$$D(\alpha^r \otimes y_i) = -\alpha^{r+1} \otimes \beta_{i+1} + [\alpha^r \otimes \beta_i, \gamma_i] \qquad \text{for } i \leq n-1\,.$$

In particular the ideal generated by the elements $\alpha^r \otimes x$, $\alpha^r \otimes y_i$ and their differentials is acyclic. A quillen model for the fiber Z is thus given by

$$(\mathbb{L}(\gamma_1, \dots, \gamma_n, \beta_1, \dots, \beta_n, y_n, \alpha \otimes y_n, \alpha^2 \otimes y_n \dots)), D)\,,$$

with

$$D(y_n) = [\beta_n, \gamma_n]$$
$$D(\alpha \otimes y_n) = [\alpha, [\beta_n, \gamma_n]] = [[\alpha, \beta_n], \gamma_n] = [[\beta_{n-1}, \gamma_{n-1}], \gamma_n]$$
$$\dots$$
$$D(\alpha^r \otimes y_n) = [[[\beta_{n-r}, \gamma_{n-r}], \gamma_{n-r+1}], \dots, \gamma_{n-1}], \gamma_n] \qquad \text{for } r < n$$

This shows that the $(5+2n)$-skeleton of Z is

$$S^3_{\beta_1} \vee \dots \vee S^3_{\beta_n} \vee S^3_{\gamma_1} \vee \dots \vee S^3_{\gamma_n} \cup_{[\beta_n, \gamma_n]} e^6 \cup_{[\gamma_n, [\gamma_{n-1}, \beta_{n-1}]]} e^8 \cup$$
$$\dots \cup_{[\gamma_n, [\gamma_{n-1}, [\dots, [\gamma_2, [\gamma_1, \beta_1]], \dots]]} e^{2n+3}\,.$$

SUBLEMMA. *The elements*

$$[\beta_n, \gamma_n], [\gamma_n, [\gamma_{n-1}, \beta_{n-1}]], \cdots, [\gamma_n, [\gamma_{n-1}, [\cdots, [\gamma_2, [\gamma_1, \beta_1]] \cdots]]]$$

form an inert sequence.

PROOF OF THE SUBLEMMA. Recall ([**16**],Theorem 4.1) that in an evenly graded Lie algebra, if an element α is inert and β belongs to the ideal generated by α, then the element β is also inert. We prove now by induction on $i \geq 0$ that $\alpha_{n-i} = [\gamma_n, [\cdots \gamma_{n-i+1}, [\gamma_{n-i}, \beta_{n-i}] \dots]$ is inert in $\mathbb{L}(\beta_1, \dots, \beta_n, \gamma_1, \dots, \gamma_n)/(\alpha_0, \dots, \alpha_{n-i-1})$. Suppose this is true for $i-1$, then by Lemma 1, $[\gamma_{n-i}, \beta_{n-i}]$ is an inert element in

$$\mathbb{L}(\beta_{n-i+1}, \dots, \beta_n, \gamma_1, \dots, \gamma_n)/(\alpha_0, \dots, \alpha_{n-i-1}) \amalg \mathbb{L}(\beta_1, \dots, \beta_{n-i})\,.$$

Since α_{n-i} belongs to the ideal generated by $[\gamma_{n-i}, \beta_{n-i}]$, the element α_{n-i} is also inert.

□

END OF THE PROOF OF LEMMA 8. So $\pi_{\leq 3+2n}(\Omega Z) \otimes \mathbb{Q}$ is generated by its elements of degree two. Therefore,

$$\pi_{\leq 3+2n}(\Omega X) \otimes \mathbb{Q} = J_{\leq 3+2n} .$$

On the other hand,

$$\pi_*(\Omega X) \otimes \mathbb{Q} = H(\mathcal{L}) = \oplus_{r \geq 0} H_{(r)}(\mathcal{L})$$

with the new gradation (r) defined by putting x in gradation 1 and the other generators in gradation 0. By ([**11**]), $H_{(0)} = J$ and $H_{(+)} = \mathbb{L}(H_{(1)})$. We observe that $H_{(1)}$ is concentrated in odd degrees because $|x| = 5$. We thus have

$$\pi_{\leq 4+2n}(\Omega X) \otimes \mathbb{Q} = J_{\leq 4+2n} .$$

□

References

[1] D. J. ANICK, Non-Commutative Graded Algebras and Their Hilbert Series, *Journal of Algebra* 78 (1982), 120-140.

[2] D. J. ANICK, A Rational Homotopy Analogue of Whitehead's Problem, in *Algebra, Algebraic Topology and their Interactions.* Springer-Verlag, Lectures Notes in Mathematics 1183 (1986), 28-31.

[3] M. ARKOWITZ, *The Group of Self-Homotopy Equivalences – A Survey.* Springer-Verlag, Lecture Notes in Mathematics 1425 (1990), 170-203.

[4] M. ARKOWITZ and G. LUPTON, *On Finiteness of Subgroups of Self-Homotopy Equivalences.* Proceedings of the Cech Memorial Conference. Contemporary Mathematics 181 (1995), 1-25.

[5] M. ARKOWITZ and G. LUPTON, On the Nilpotency of Subgroups of Self-Homotopy Equivalences, In *Algebraic topology : New trends in localization and periodicity*, Progr. Math. 136, (1995), 1-22.

[6] M. ARKOWITZ and K.-I. MARUYAMA, Self Homotopy Equivalences which Induce the Identity on Homology, Cohomology or Homotopy Groups, *Topology and its Applications* 27 (1998), 133-154.

[7] E. DROR and A. ZABRODSKY, Unipotency and Nilpotency in Homotopy Equivalences, *Topology* 18 (1979), 187-197.

[8] Y. FELIX and S. HALPERIN, Rational L.S. Category and its Applications, *Trans. Amer. Math. Soc.* 372 (1982), 1-37.

[9] Y. FELIX, S. HALPERIN and J.-C. THOMAS, Rational Homotopy Theory, Graduate Texts in Mathematics 205 (2000).

[10] Y. FELIX and A. MURILLO, A Note on the Nilpotency of Subgroups of Self-Homotopy Equivalences, *Bull. London Math. Soc.* 29 (1997), 486-488.

[11] Y. FELIX and J.-C. THOMAS, Sur la Structure des Espaces be L.S. Catégorie Deux, *Illinois Journal of Mathematics*, 30 (1986), 574-593.

[12] Y. FELIX and J.-C. THOMAS, On spaces of the same strong n-type, *Homology, Homotopy and Applications* 1 (1999), 205-217.

[13] Y. FELIX and J.-C. THOMAS, On the Arkowitz-Maruyama conjecture – correction to the paper : spaces of the same strong n-type. To appear.

[14] T. GANEA, Lusternik-Schnirelmann category and strong category, *Illinois J. of Math.* 11 (1967), 417-427.

[15] S. HALPERIN. *Lectures on Minimal Models.* Mémoire be la Société Mathématique be France 9/10 (1983).

[16] S. HALPERIN and J.M. LEMAIRE, Suites Inertes dans les Algèbres be Lie Graduées ("Autopsie d'un Meurtre II"), *Math. Scand.* 61 (1987), 39-67.

[17] I. M. JAMES, On Category in the Sense of Lusternik-Schnirelmann, *Topology* 17 (1978), 341-378.

[18] J.-M. LEMAIRE, *Algèbres Connexes et Homologie des Espaces de Lacets.* Springer-Verlag, Lecture Notes in Mathematics 422 (1974).

[19] K.-I. MARUYAMA, Localization of a Certain Subgroup of Self-Homotopy Equivalences, *Pacific Journal of Mathematics* 136 (1989), 293-301.

[20] K.-I. MARUYAMA, A subgroup of self-homotopy equivalences which satisfies the M-L condition, *Bulletin of the Faculty of Education, Chiba University* 48 - 02/29/2000.
[21] K.-I. MARUYAMA, Stability properties of maps between Hopf spaces, Preprint.
[22] G. PORTER, Higher order Whitehead products, *Topology* 3 (1965), 123-135.
[23] D. TANRÉ, *Homotopie Rationnelle : Modèles de Chen, Quillen, Sullivan*. Springer-Verlag, Lecture Notes in Mathematics 1025 (1983).
[24] G. TOOMER, Lusternik-Schnirelmann Category and the Moore Spectral Sequence, *Math. Z.* 138 (1974), 123-143.
[25] G. TOOMER. Topological Localization, Category and Cocategory, *Canadian J. of Math.* 27 (1975), 319-322.

INSTITUT MATHÉMATIQUE, UNIVERSITÉ CATHOLIQUE DE LOUVAIN, 2, CHEMIN DU CYCLOTRON, 1348 LOUVAIN-LA-NEUVE, BELGIUM

Contemporary Mathematics
Volume **274**, 2001

Nilpotency and localization of groups of fibre homotopy equivalences

A. Garvín, A. Murillo, P. Pavešić and A. Viruel

ABSTRACT. Let $F \hookrightarrow E \to B$ be a fibration which satisfies suitable finiteness conditions. We prove that the group $\mathcal{E}_\sharp(E,B)$ of self fibre homotopy equivalences which induce the identity on the homotopy groups of the fibre is solvable. Provided it is also nilpotent, we prove that for any set of primes P, $\mathcal{E}_\sharp(E,B)_{(P)} \cong \mathcal{E}_\sharp(E_{(P)},B_{(P)})$. Finally we apply those results to the context of self homotoy equivalences of diagrams.

1. Introduction

The departure point of this paper is the study of the group of self equivalences in the homotopy category of pointed spaces. In what follows we shall always consider connected pointed spaces which have the homotopy type of CW–complexes. For such a space X, denote by $\operatorname{aut} X$ the monoid of self homotopy equivalences, and by $\mathcal{E}(X)$, the group of homotopy classes of self homotopy equivalences of X, i.e., $\mathcal{E}(X) = \pi_0(\operatorname{aut} X)$. This group, together with some distinguished subgroups which preserve additional geometrical structure, has been deeply studied. For a compendium of known results on the subject see [**1**] or [**18**].

This paper is focused on self-equivalences, which fix homotopy groups. Let $\mathcal{E}_\sharp(X)$ be the subgroup of $\mathcal{E}(X)$ formed by all classes $[f]$ such that $\pi_*(f) = 1_{\pi_*(X)}$. In other words $\mathcal{E}_\sharp(X) = \operatorname{Ker}\big(\mathcal{E}(X) \to \Pi_{i\geq 1}\operatorname{aut}\pi_i(X)\big)$. More generally, for any $0 \leq m \leq \infty$ define $\mathcal{E}_\sharp^m(X) = \{[f] \in \mathcal{E}(X) \mid \pi_i(f) = 1_{\pi_i(X)},\ i \leq m\}$.

A celebrated result of Dror and Zabrodsky [**6**] states the following: let X be a space which is either finite dimensional CW-complex or a Postnikov stage, and let $\dim(X)$ denote respectively its topological or homotopical dimension. Then, for any $m \geq \dim(X)$, the group $\mathcal{E}_\sharp^m(X)$, and hence $\mathcal{E}_\sharp(X)$, is nilpotent. Therefore, this group can be localized in the classical sense and Maruyama proved [**13**] that, for any set of primes P (possibly empty), the natural map

$$\mathcal{E}_\sharp^m(X) \longrightarrow \mathcal{E}_\sharp^m(X_{(P)})$$

is the P-localization morphism and thus $\mathcal{E}_\sharp^m(X)_{(P)} \cong \mathcal{E}_\sharp^m(X_{(P)})$.

1991 *Mathematics Subject Classification.* 55P10.

The first, second and fourth author are partially supported by the DGES grant PB97-1095 and by a Junta de Andalucía Grant FQM-0213. The third author is partially supported by Ministry of Science and Technology of the Republic of Slovenia research grant No. J1-0885-0101-98.

The purpose of our paper is to generalize these results to the fibrewise setting. For a given fibration $F \hookrightarrow E \xrightarrow{p} B$ we denote by $\operatorname{aut}_B E$ the monoid of maps $f\colon E \to E$ over B (that is to say $pf = p$) which are fibre homotopy equivalences. By analogy with $\mathcal{E}(X)$ we set $\mathcal{E}(E,B) = \pi_0(\operatorname{aut}_B E)$ the group of fibre homotopy classes of fibre homotopy equivalences.

Now, given a fibration $F \hookrightarrow E \xrightarrow{p} B$ and any map $B' \xrightarrow{f} B$ we obtain a new fibration $F \hookrightarrow E_f \xrightarrow{p_f} B'$ by means of the pullback diagram

$$
\begin{array}{ccccc}
F & \hookrightarrow & E_f & \xrightarrow{p_f} & B' \\
\| & & \downarrow & \lrcorner & \downarrow f \\
F & \hookrightarrow & E & \xrightarrow{p} & B
\end{array}
\qquad (1.1)
$$

which induces a continuous map $\phi : \operatorname{aut}_B E \to \operatorname{aut}_{B'} E_f$, and therefore a group morphism $\pi_0(\phi) : \mathcal{E}(E,B) \to \mathcal{E}(E_f, B')$. For a given G subgroup of $\mathcal{E}(E_f, B')$, define $\mathcal{E}_G(E,B) = \pi_0(\phi)^{-1}(G)$. Then, using the results of James [**10**] and Meiwes [**14**], we prove:

THEOREM 1.1. *Assume that:*

(i) *B has finite Lusternik-Schnirelmann category;*
(ii) *The group G is solvable of solvability class bounded by k.*

Then, $\mathcal{E}_G(E,B)$ is solvable and

$$\operatorname{sol} \mathcal{E}_G(E,B) \leq k + \operatorname{cat} B.$$

PROOF. See Section 2. □

In particular, choosing $B' = *$, we get that $E_f = F$ and $\mathcal{E}(F,*) = \mathcal{E}(F)$. Thus choosing $G = \mathcal{E}_\sharp^m(F)$, we define $\mathcal{E}_\sharp^m(E,B) := \mathcal{E}_{\mathcal{E}_\sharp^m(F)}(E,B)$, i.e., the group of fibre homotopy equivalences of E whose restriction to the fiber are elements of $\mathcal{E}_\sharp^m(F)$. This is the analog to $\mathcal{E}_\sharp^m(X)$ in the fiber setting. As an immediate consequence of Theorem 1.1, we get:

COROLLARY 1.2. *Assume that F is either a finite dimensional complex or a Postnikov stage and that B has finite Lusternik-Schnirelmann category. Then the group $\mathcal{E}_\sharp^m(E,B)$ is solvable for any $m \geq \dim F$ and*

$$\operatorname{sol} \mathcal{E}_\sharp^m(E,B) \leq \operatorname{sol} \mathcal{E}_\sharp^m(F) + \operatorname{cat} B.$$

In the rational category we can be even more accurate:

COROLLARY 1.3. *Let $F \hookrightarrow E \to B$ be a fibration of rational spaces of finite category in which F is 1-connected and either its rational cohomology or homotopy is of finite dimension (call it $\dim(F)$). Then, for any $m \geq \dim F$, $\mathcal{E}_\sharp^m(E,B)$ is nilpotent and*

$$\operatorname{sol} \mathcal{E}_\sharp^m(E,B) \leq \operatorname{cat} F + \operatorname{cat} B - 1.$$

PROOF. By [**8**], $\operatorname{nil} \mathcal{E}_\sharp(F) \leq \operatorname{cat} F - 1$. □

Consider now the group $\mathcal{E}_\Omega(X)$ introduced in [**7**] (see also [**17**]) as the kernel of the natural map $\mathcal{E}(X) \to \mathcal{E}(\Omega X)$. In the fibred setting, for the given fibration $F \hookrightarrow E \xrightarrow{p} B$, we may apply again Theorem 1.1 to the case $B' = *$ and define $\mathcal{E}_\Omega(E,B) = \mathcal{E}_{\mathcal{E}_\Omega(F)}(E,B)$. Then, we get

COROLLARY 1.4. *If F and B have finite Lusternik-Schnirelmann category, then $\mathcal{E}_\Omega(E,B)$ is solvable and*

$$\operatorname{sol}\mathcal{E}_\Omega(E,B) \leq \operatorname{cat} F + \operatorname{cat} B - 1.$$

PROOF. By [**7**], $\mathcal{E}_\Omega(F)$ is nilpotent of nilpotency class bounded by $\operatorname{cat} F - 1$. □

We now turn our interest to the localization of these groups. For a nilpotent fibration of nilpotent spaces $F \hookrightarrow E \to B$ and for any set of primes (possibly empty) P, there is a commutative diagram

$$\begin{array}{ccccc} F & \hookrightarrow & E & \longrightarrow & B \\ \downarrow & & \downarrow & & \downarrow \\ F_{(P)} & \hookrightarrow & E_{(P)} & \longrightarrow & B_{(P)} \end{array}$$

in which the vertical arrows are localization maps and the bottom arrow is also a fibration. Thus there is a natural morphism $\mathcal{E}(E,B) \to \mathcal{E}(E_{(P)}, B_{(P)})$ which restricts to $\mathcal{E}^m_\sharp(E,B) \to \mathcal{E}^m_\sharp(E_{(P)}, B_{(P)})$. Then we prove:

THEOREM 1.5. *Let $F \hookrightarrow E \to B$ be a nilpotent fibration of nilpotent spaces, such that either E is finite dimensional or F is a Postnikov stage, in both case of dimension at most k. If $\mathcal{E}^k_\sharp(E,B)$ is nilpotent then for any $m \geq k$ and for any set of primes P the natural map,*

$$\mathcal{E}^m_\sharp(E,B) \longrightarrow \mathcal{E}^m_\sharp(E_{(P)}, B_{(P)})$$

is the localization morphism, i.e., $\mathcal{E}^m_\sharp(E,B)_{(P)} \cong \mathcal{E}^m_\sharp(E_{(P)}, B_{(P)})$.

PROOF. See Section 3. □

REMARK 1.6. By Theorem 1.1 the group $\mathcal{E}^m_\sharp(E,B)$ is automatically nilpotent if we deal, with fibrations of finite complexes, or when the base has finite category and the fibre is a Postnikov stage. However there are examples of fibrations which satisfy the hypothesis of Theorem 1.5 but not those in Theorem 1.1.

COROLLARY 1.7. *With the same assumptions as in Theorem 1.5, any commutator in $\mathcal{E}^m_\sharp(E,B)$ of length greater or equal than $\operatorname{cat} F + \operatorname{cat} B$ has finite order.*

PROOF. Indeed, by Theorem 1.5 and Corollary 1.3, any such commutator dies under rationalization and thus it has finite order. □

Some of the results above can be applied to a more general context of diagrams. To make exposition as clear as possible, the diagram approach is left for the last section in the note.

The paper is organized as follows: In the Section 2 we prove Theorem 1.1 and give an example which shows that the above inequalities are the best possible. After that we prove Theorem 1.5 in Section 3. Finally, Section 4 is devoted to applications of the previous results to the study of a special case of diagrams.

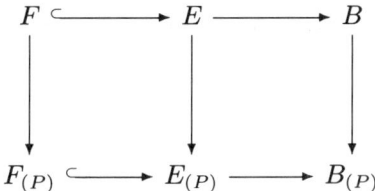

2. Nilpotency

In this section we shall prove Theorem 1.1. Recall the situation: given a fibration $F \hookrightarrow E \xrightarrow{p} B$ and $B' \xrightarrow{f} B$, we obtain a new fibration $F \hookrightarrow E_f \xrightarrow{p_f} B'$ by means of the pullback diagram (1.1), which also induces a group morphism $\pi_0(\phi) : \mathcal{E}(E,B) \to \mathcal{E}(E_f,B')$. For a given G subgroup of $\mathcal{E}(E_f,B')$, define $\mathcal{E}_G(E,B) = \pi_0(\phi)^{-1}(G)$. First we identify the kernel of $\pi_0(\phi)$

LEMMA 2.1. *Define $\mathcal{E}_1(E,B)$ as the subgroup of $\mathcal{E}(E,B)$ formed by classes $\langle g\rangle$ whose restriction to the fibres are homotopic to the identity. Then $Ker\,\pi_0(\phi) \subset \mathcal{E}_1(E,B)$.*

PROOF. Given $g \in \operatorname{aut}_B(E)$, $\hat{g} = \phi(g)$ makes commutative the following cube

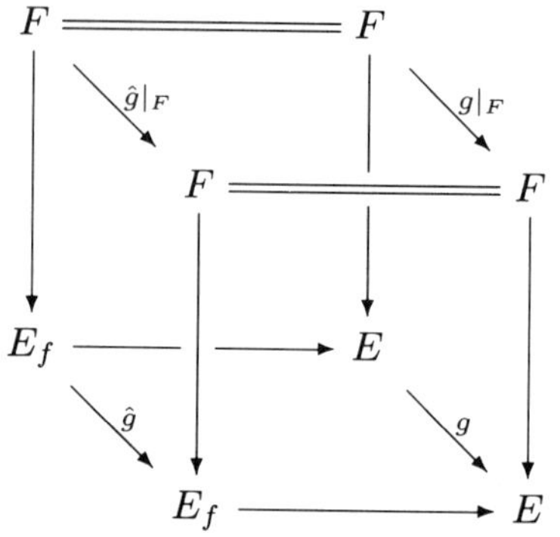

which implies $g|_F = \hat{g}|_F$. Therefore if $\pi_0(\phi)(\langle g\rangle) = \langle\hat{g}\rangle = 1$, then $[\hat{g}|_F] = 1 \in \mathcal{E}(F)$, that is $[g|_F] = 1 \in \mathcal{E}(F)$ and $\langle g\rangle \in \mathcal{E}_1(E,B)$. □

James [**10**] and Meiwes [**14**] proved that $\mathcal{E}_1(E,B)$ is nilpotent and that

$$\operatorname{nil}\mathcal{E}_1(E,B) \leq \operatorname{cat} B.$$

Since $\operatorname{Ker}\pi_0(\phi)|_{\mathcal{E}_G(E,B)} \subset \mathcal{E}_1(E,B)$, it is also nilpotent and

$$\operatorname{nil}_{\operatorname{Ker}\pi_0(\phi)|_{\mathcal{E}_G(E,B)}} \mathcal{E}_G(E,B) \leq \operatorname{cat} B.$$

Finally consider the exact sequence

$$1 \longrightarrow \operatorname{Ker}\pi_0(\phi)|_{\mathcal{E}_G(E,B)} \longrightarrow \mathcal{E}_G(E,B) \longrightarrow G \longrightarrow 1$$

in which G is of solvability class $\leq k$. Then, obviously $\mathcal{E}_G(E,B)$ is solvable and

$$\operatorname{sol}\mathcal{E}_G(E,B) \leq \operatorname{sol} G + \operatorname{sol}\operatorname{Ker}\pi_0(\phi)|_{\mathcal{E}_G(E,B)} \leq k + \operatorname{cat} B.$$

REMARK 2.2. If, instead of having $\mathcal{E}_1(E,B)$ nilpotent, we had $\mathcal{E}(E,B)$ acting nilpotently on $\mathcal{E}_1(E,B)$, then our results could be strengthened. Indeed we could change solvability by nilpotency in theorem 1.1 and corollaries 1.2, 1.3 and 1.4. Assume that $\mathcal{E}(E,B)$ acts nilpotently on $\mathcal{E}_1(E,B)$ with nilpotency order bounded by the category of B. Then $\mathcal{E}_G(E,B)$ also acts nilpotently on $\operatorname{Ker}\pi_0(\phi)|_{\mathcal{E}_G(E,B)} \subset \mathcal{E}_1(E,B)$ and

$$\operatorname{nil}_{\operatorname{Ker}\pi_0(\phi)|_{\mathcal{E}_G(E,B)}} \mathcal{E}_G(E,B) \leq \operatorname{cat} B.$$

Again, as before, consider the exact sequence

$$1 \longrightarrow \operatorname{Ker}\pi_0(\phi)|_{\mathcal{E}_G(E,B)} \longrightarrow \mathcal{E}_G(E,B) \longrightarrow G \longrightarrow 1$$

and, provided G of nilpotency class $\leq k$, by Proposition 4.1, [**11**, Chapter I], $\mathcal{E}_G(E,B)$ is nilpotent and

$$\operatorname{nil}\mathcal{E}_G(E,B) \leq \operatorname{nil} G + \operatorname{nil}_{\operatorname{Ker}\pi_0(\phi)|_{\mathcal{E}_G(E,B)}} \mathcal{E}_G(E,B) \leq k + \operatorname{cat} B.$$

EXAMPLE 2.3. The following easy example shows that the bound attained in Corollary 1.2, thus in Theorem 1.1, is the best possible. Consider the trivial fibration $p\colon S^1 \times S^1 \to S^1$, where p is the projection on the second factor and define a fibre preserving homeomorphism $f\colon S^1 \times S^1 \to S^1 \times S^1$ by $f(a,b) = (ab,b)$. By considering the action of f on the fundamental group we see that the restriction of f induces the identity on $\pi_1(F)$ but it is not globally homotopic to the identity. Hence $\langle f \rangle$ represents a non trivial element in $\mathcal{E}_\sharp(S^1 \times S^1, S^1)$. Thus, since $\mathcal{E}_\sharp(S^1)$ is trivial, we have the equality

$$\mathcal{E}_\sharp(S^1 \times S^1, S^1) = \operatorname{nil}\mathcal{E}_\sharp(S^1) + \operatorname{cat} S^1 = 1.$$

3. Localization

Along this section we prove Theorem 1.5. First we recall the technical machinery used in the proof: a non abelian truncated spectral sequence arising from a Cartan-Eilenberg system. Introduced originally in [**3**], the notion of a Cartan-Eilenberg system has different approaches (see [**16, 19**]). We will use the one in [**12**] which we now briefly describe.

In the sequel we shall consider groups (non necessarily abelian) and pointed sets. Then, it makes sense to talk about exact sequences whose objects are either groups or pointed sets.

A Cartan-Eilenberg system is defined by the following data:

(i) For any couple of integers p, q, with $0 \leq p \leq q \leq \infty$, we are given groups $H^n(p,q)$, $n \leq 0$, and pointed sets $H^1(p,q)$.

(ii) Whenever $p \geq p'$, $q \geq q'$, there are morphisms for any $n \leq 1$

$$\eta\colon H^n(p,q) \longrightarrow H^n(p',q'),$$

such that η is the identity if $p = p'$, $q = q'$, and, if $p \geq p' \geq p''$, $q \geq q' \geq q''$, the following diagram commutes:

$$\begin{array}{ccc} H^n(p,q) & \xrightarrow{\quad\eta\quad} & H^n(p'',q'') \\ & {\scriptstyle\eta}\searrow \quad \nearrow{\scriptstyle\eta} & \\ & H^n(p',q') & \end{array}$$

(iii) Whenever $p \leq q \leq r$ there are morphisms for any $n \leq 0$,

$$\delta\colon H^n(p,q) \to H^{n+1}(q,r)$$

(in particular $H^0(p,q)$ acts on $H^1(q,r)$) satisfying:

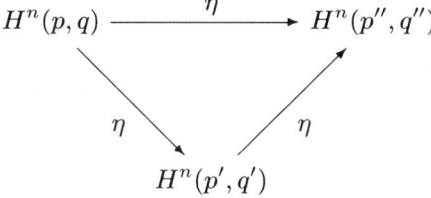

(iii.1) For $p \geq p'$, $q \geq q'$, $r \geq r'$, with $p \leq q \leq r$, $p' \leq q' \leq r'$, the following diagram commutes

$$\begin{array}{ccc} H^n(p,q) & \xrightarrow{\delta} & H^{n+1}(q,r) \\ \Big\downarrow \eta & & \Big\downarrow \eta \\ H^n(p',q') & \xrightarrow{\delta} & H^{n+1}(q',r'). \end{array}$$

(iii.2) For $p \leq q \leq r$, the sequence

$$\cdots \longrightarrow H^n(q,r) \xrightarrow{\eta} H^n(p,r) \xrightarrow{\eta} H^n(p,q) \xrightarrow{\delta} H^{n+1}(q,r) \longrightarrow \cdots$$

$$\cdots \longrightarrow H^0(p,q) \xrightarrow{\delta} H^1(q,r) \xrightarrow{\eta} H^1(p,r) \xrightarrow{\eta} H^1(p,q)$$

is exact.

Given a Cartan-Eilenberg system, the following holds

THEOREM 3.1. *There is a spectral sequence (E_r, d_r) such that:*

(1) *For $1 \leq r \leq \infty$,*

$$E_r^{p,q} = Im\left(H^n(p,p+r) \to H^n(p-r+1,p+1)\right), \quad p+q = n \leq 0,$$
$$E_r^{p,-p-1} = Im\left(H^1(p,p+1) \to H^n(p-r+1,p+1)\right).$$

In particular,

$$E_\infty^{p,q} = Im\left(H^n(p,\infty) \to H^n(0,p+1)\right), \quad p+q = n \leq 0,$$
$$E_\infty^{p,-p+1} = Im\left(H^1(p,p+1) \to H^n(0,p+1)\right).$$

(2) *$d_r \colon E_r^{p,q} \to E_r^{p+r,q-r+1}$ is defined as follows (recall that, for $n = 0$, δ and δ' are actions): Given the diagram*

$$\begin{array}{ccc} H^n(p,p+r) & \xrightarrow{\eta} & H^n(p-r+1,p+1) \\ \Big\downarrow \delta & & \Big\downarrow \delta' \\ H^{n+1}(p+r,p+2r) & \xrightarrow[\eta']{} & H^{n+1}(p+1,p+r+1) \end{array}$$

d_r is by definition $\delta'|_{Im\,\eta} \colon Im\,\eta \to Im\,\eta'$.

(3) *If we consider in $H^n = H^n(0,\infty)$, $n \leq 0$, the decreasing filtration*

$$F^pH^n = Im\left(H^n(p,\infty) \to H^n(0,\infty)\right) = Ker\left(H^n(0,\infty) \to H^n(0,p)\right),$$

then there is an isomorphism $F^pH^n/F^{p+1}H^n \stackrel{\cong}{\to} E_\infty^{p,q}$ $(p+q = n \leq 0)$ induced by $H^n \to H^n(0,p+1)$. In other words this spectral sequence converges from $E_1^{p,q} = H^{p+q}(p,p+1)$ to $H^{p+q}(0,\infty)$, for $p+q = n \leq 0$.

The proof of this result is straightforward from the definitions and only requires some calculations (see **[12]** for details).

REMARK 3.2. Sometimes, and this will be our case, it is possible to shorten the initial data and still being able to construct a spectral sequence as in the theorem above. Namely, assume that in the Cartan-Eilenberg system only the pointed sets $H^1(p, p+1)$, $p \geq 0$, are given. Then, replace (iii.2) by

(iii.2)' For $p \leq q \leq r$, the following sequences are exact:

$$\cdots \longrightarrow H^n(q,r) \longrightarrow H^n(p,r) \longrightarrow H^n(p,q) \longrightarrow H^{n+1}(q,r) \longrightarrow \cdots$$

$$\cdots \longrightarrow H^0(q,r) \longrightarrow H^0(p,r) \longrightarrow H^0(p,q),$$

$$H^0(p,q+1) \longrightarrow H^0(p,q) \longrightarrow H^1(q,q+1).$$

Then, with the convenient modification, only of the term $E_r^{p,-p-1}$, there is still a spectral sequence as before with the same convergence. Indeed, the changes only affect the case $p+q=1$ for which no convergence property has been given.

Proof of theorem 1.5: We will apply this algebraic machinery to our topological setting, similarly as in [**4**]. Let $F \hookrightarrow E \xrightarrow{p} B$ be the given fibration and let

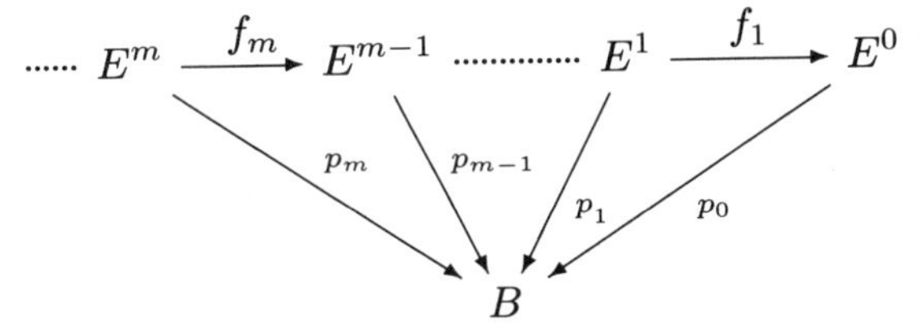

be its Moore-Postnikov decomposition. In particular each $f_m \colon E^m \to E^{m-1}$ is a fibration with fibre $K(\pi_m(F), m)$.

Assume that E is a finite complex of dimension at most k and for $m \geq k$ consider the fibration $F^m \to E^m \overset{p_m}{\to} B$ corresponding to the m-th stage of the Moore-Postnikov decomposition. E^m is obtained from E by attaching cells in dimensions greater or equal to $m+2$, so by obstruction theory $[E,E]_B \cong [E^m, E^m]_B$ (where $[\,,\,]_B$ denotes fibre homotopy classes of maps over B). Since the fibre F^m of the fibration $E^m \to B$ is precisely the m-th stage of the Postnikov decomposition of F, this bijection restricts to the isomorphism $\mathcal{E}_\sharp^m(E,B) = \mathcal{E}_\sharp(E^m, B)$. It follows that it suffices to consider only the case when F is a Postnikov stage and m is bigger or equal to its homotopical dimension.

Since the fibration is nilpotent, its Moore-Postnikov decomposition admits a principal refinement

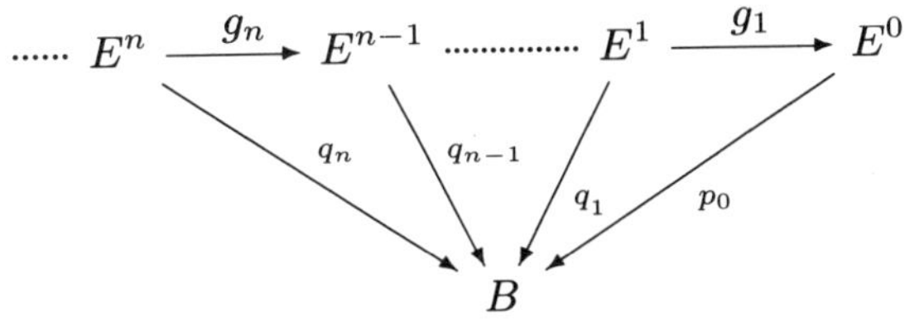

that is to say, for each $n \geq 1$, $g_n \colon E^n \to E^{n-1}$ is a principal fibration of type (G_n, s_n), classified by the Postnikov invariant $\xi_n \colon E^{n-1} \to K(G_n, s_n + 1)$, in which G_n is an abelian group and $1 \leq s_1 \leq \cdots \leq s_n \leq s_{n+1} \leq \cdots$.

We will need the following notation. Let $\mathrm{aut}_{\sharp B} E$ be the monoid of self-fibre homotopy equivalences of E over B that induce the identity on the homotopy groups of the fibre, with the identity map of E as the base point. Observe that $\pi_0(\mathrm{aut}_{\sharp B}E) = \mathcal{E}_\sharp(E, B)$. For each $p \geq 0$ the fibration

$$\mathrm{map}\,(E, E) \to \mathrm{map}\,(E^p, E^p)$$

restricts to fibrations $\mathrm{aut}_{\sharp B}E \to \mathrm{aut}_{\sharp B}E_p$ whose fibre will be denoted by X_p^∞. More generally, for $0 \leq p \leq q \leq \infty$ we consider fibrations

$$X_p^q \longrightarrow \mathrm{aut}_{\sharp B}E^q \longrightarrow \mathrm{aut}_{\sharp B}E^p,$$

and observe that for $p \leq q \leq r$ the induced map $X_p^r \to X_p^q$ is a fibration with fibre X_q^r.

To begin building a Cartan-Eilenberg system define

$$H^n(p, q) = \pi_{-n}(X_p^q), \quad 0 \leq p \leq q \leq \infty, \quad n \leq 0.$$

Note that whenever $p \geq p'$, $q \geq q'$, with $p \leq q$, $p' \leq q'$, there is a natural map $X_p^q \to X_{p'}^{q'}$ that induces morphisms

$$\eta\colon H^n(p, q) \to H^n(p', q').$$

On the other hand, if $p \leq q \leq r$ let

$$\delta\colon H^n(p, q) = \pi_{-n}(X_p^q) \longrightarrow \pi_{-n-1}(X_q^r) = H^{n+1}(q, r)$$

be the connecting homomorphism in the homotopy exact sequence of $X_q^r \hookrightarrow X_p^r \to X_p^q$. To complete the data we define

$$H^1(p, p+1) = [E^p, K(G_{p+1}, s_{p+1} + 1]$$

with the $(p+1)$-Postnikov invariant ξ_{p+1} as base point.

Clearly, an element in $\langle f\rangle \in \mathcal{E}_\sharp(E^q, B)$ can be extended to $\mathcal{E}_\sharp(E^{q+1}, B)$ if and only if $\xi_q \circ f$ is fibre homotopic to ξ_q, which shows that the sequence

$$H^0(p, q+1) \to H^0(p, q) \to H^1(q, q+1)$$

is exact for all $p \leq q$.

We conclude that the above data form a Cartan-Eilenberg system so, by theorem 3.1, we have a spectral sequence $I_r^{p,q}$ whose initial term is:

$$I_1^{p,q} = H^n(p, p+1) = \pi_{-n}(X_p^{p+1}), \quad p + q = n \leq 0.$$
$$I_1^{p,-p+1} = [E^p, K(G_{p+1}, s_{p+1} + 1] = H^{s_{p+1}+1}(E^p; G_{p+1}).$$

To effectively compute $H^{-1}(p, p+1)$ and $H^0(p, p+1)$ we use [**4**, Lemma 3.2]:

$$X_p^{p+1} = \mathrm{aut}_{\sharp E^{p+1}}E^p \cong \mathrm{map}\,\big(E^p, K(G_{p+1}, s_{p+1})\big).$$

Therefore,

$$I_1^{p,-p} = \pi_0(X_p^{p+1}) \cong [E^p, K(G_{p+1}, s_{p+1})] \cong H^{s_{p+1}}(E^p; G_{p+1}),$$
$$I_1^{p,-p-1} = \pi_1(X_p^{p+1}) \cong \pi_1\mathrm{map}\,(E^p, K(G_{p+1}, s_{p+1})) \cong H^{s_{p+1}-1}(E^p; G_{p+1}),$$

so we have the general formula

$$I_1^{p,p+j} = H^{s_{p+1}+j}(E^p; G_{p+1}), \qquad j = -1, 0, 1. \tag{$*$}$$

Choose now a set of prime numbers P and denote by $J_r^{p,q}$ the corresponding spectral sequence induced by the localized fibration $F_{(P)} \hookrightarrow E_{(P)} \to B_{(P)}$. In view of $(*)$, the induced morphism of spectral sequences at the first level $I_1^{p,q} \to J_1^{p,q}$ for $p + q = -1, 0$ is the P-localization. On the other hand, since F is a Postnikov stage, the differential d_r is trivial for r big. Thus, by comparison, $I_\infty^{p,q} \to J_\infty^{p,q}$ is also the P-localization for $p+q=-1,0$.

Finally, consider the filtration A^p of $H^0(0,\infty) = \pi_0(\operatorname{aut}_{\sharp B}) = \mathcal{E}_\sharp(E,B)$ given in theorem 3.1, whose associated graded group is $I_\infty^{p,q}$ for $p+q=0$. Since

$$A^p = \operatorname{Im}\big(\mathcal{E}_\sharp(E,E^p) \to \mathcal{E}_\sharp(E,B)\big),$$

and since $E = E^p$ for p big, $\mathcal{E}_\sharp(E,E^p)$ and thus A^p eventually vanishes. Therefore, $\mathcal{E}_\sharp(E,B) \to \mathcal{E}_\sharp(E_{(P)},B_{(P)})$ is the P-localization morphism, which concludes the proof of Theorem 1.5.

REMARK 3.3. Similarly as in the non fibred case, we do not know whether Theorem 1.5 holds for $m=\infty$, i.e., if for a nilpotent fibration $F \hookrightarrow E \to B$ of finite complexes, $\mathcal{E}_\sharp(E,B) \to \mathcal{E}_\sharp(E_{(P)},B_{(P)})$ is the localization morphism in general. Observe however, that the theorem is true, whenever $\mathcal{E}_\sharp^m(E,B)$ is finite for some m, which is often the case in the applications.

4. An aplication to self homotopy equivalences of diagrams

In this section we apply the machinery developed in the previous sections to a more general context of diagrams. Given a diagram $\mathcal{D}$ of spaces and continuous maps $\{X_i;\, X_i \xrightarrow{f_{(i,j)}} X_j\}_{(i,j)\in I\times I}$ over a partially orderes set $(I,\leq)$, the monoid of self homotopy equivalences is defined as

$$\operatorname{aut}\mathcal{D} = \Big\{(g_i)_{i\in I} \in \prod_{i\in I} \operatorname{aut} X_i \,:\, f_{(i,j)}g_i = g_j f_{(i,j)}\Big\},$$

thus the group of homotopy classes of self homotopy equivalences of $\mathcal{D}$ is $\mathcal{E}(\mathcal{D}) = \pi_0(\operatorname{aut}\mathcal{D})$. Also given a space X_{i_0} in $\mathcal{D}$, we can define

$$\operatorname{aut}_{X_{i_0}}\mathcal{D} = \Big\{(g_i)_{i\in I} \in \operatorname{aut}\mathcal{D} : g_{i_0} = 1_{X_{i_0}}\Big\},$$

and therefore $\mathcal{E}(\mathcal{D},X_{i_0}) = \pi_0(\operatorname{aut}_{X_{i_0}}\mathcal{D})$. Then, the inverse limit in the category of pointed spaces gives rise to a continuous map $\operatorname{aut}\mathcal{D} \xrightarrow{\phi} \operatorname{aut}\lim\mathcal{D}$, respectively $\operatorname{aut}_{X_{i_0}}\mathcal{D} \xrightarrow{\phi} \operatorname{aut}\lim\mathcal{D}$, which induces a group morphism $\mathcal{E}(\mathcal{D}) \xrightarrow{\pi_0(\phi)} \mathcal{E}(\lim\mathcal{D})$, respectively $\mathcal{E}(\mathcal{D},X_{i_0}) \xrightarrow{\pi_0(\phi)} \mathcal{E}(\lim\mathcal{D})$. Then, given $G \subset \mathcal{E}(\lim\mathcal{D})$, we can define $\mathcal{E}_G(\mathcal{D}) = \pi_0(\phi)^{-1}(G)$ and $\mathcal{E}_G(\mathcal{D},X_{i_0}) = \pi_0(\phi)^{-1}(G)$. In particular, for $G = \mathcal{E}_\sharp^m(\lim\mathcal{D})$ (resp. $G = \mathcal{E}_\Omega(\lim\mathcal{D})$), define $\mathcal{E}_\sharp^m(\mathcal{D},X_{i_0}) := \mathcal{E}_G(\mathcal{D},X_{i_0})$ (resp. $.\mathcal{E}_\Omega(\mathcal{D},X_{i_0}) := \mathcal{E}_G(\mathcal{D},X_{i_0})$)

Consider now $\mathcal{D}$ a star-shaped diagram with final object B, that is, $\mathcal{D}$ is a diagram with objects $\{E_i\}_{i\in I}\cup\{B\}$ such that there is only morphism $E_i \xrightarrow{f_i} B$ for all $i\in I$. Then $\lim\mathcal{D}$ appears as the pullback

(4.1)
$$\begin{array}{ccc} \lim\mathcal{D} & \longrightarrow & \prod_{i\in I} E_i \\ \big\downarrow & \lrcorner & \big\downarrow \\ B & \xrightarrow{\ \Delta\ } & \prod_{i\in I} B \end{array}$$

where Δ is the diagonal map, and we can easily identify $\mathcal{E}_\sharp^m(\mathcal{D}, B)$ as well as $\mathcal{E}_\Omega(\mathcal{D}, B)$

PROPOSITION 4.1. *Let $\mathcal{D}$ be a star-shaped diagram with final object B such that any arrow $E_i \xrightarrow{f_i} B$ in $\mathcal{D}$ is a fibration (with fibre F_i). Then:*

(i) $\mathcal{E}_\sharp^m(\mathcal{D}, B) = \prod_{i\in I} \mathcal{E}_\sharp^m(E_i, B)$;

(ii) $\mathcal{E}_\Omega(\mathcal{D}, B) = \prod_{i\in I} \mathcal{E}_\Omega(E_i, B)$.

PROOF. Notice that as all $F_i \hookrightarrow E_i \xrightarrow{f_i} B$ are fibrations,

$$\prod_{i\in I} F_i \hookrightarrow \prod_{i\in I} E_i \xrightarrow{\prod f_i} \prod_{i\in I} B$$

is fibration too, and diagram (4.1) gives rise to a fibration $\prod_{i\in I} F_i \hookrightarrow \lim \mathcal{D} \to B$. Moreover,

a) by definition $\operatorname{aut}_B \mathcal{D} = \prod_{i\in I} \operatorname{aut}_B E_i$;

b) by diagram (4.1), $\operatorname{aut}_B \mathcal{D} \xrightarrow{\phi} \operatorname{aut} \lim \mathcal{D}$ factors through $\operatorname{aut}_B \lim \mathcal{D}$.

Therefore, given $(g_i)_{i\in I} \in \operatorname{aut}_B \mathcal{D}$, we obtain $\hat{g} = \phi((g_i)_{i\in I}) \in \operatorname{aut}_B \lim \mathcal{D}$ such that the diagram

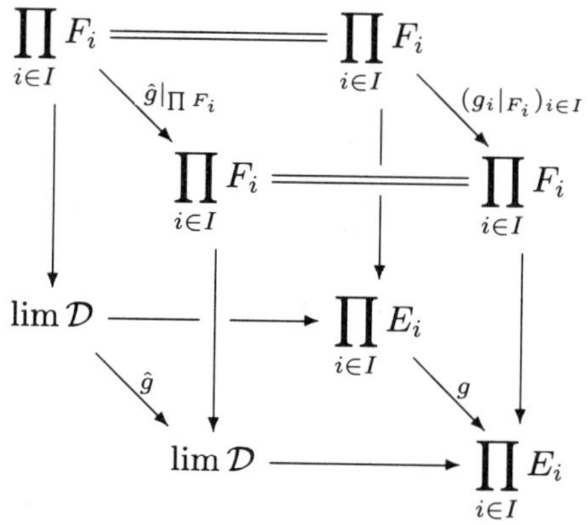

commutes, which implies $\hat{g}|_{\prod F_i} = (g_i|_{F_i})_{i\in I}$. Finally, if $[\hat{g}] \in \mathcal{E}_\sharp^m(\lim \mathcal{D})$ (resp. $[\hat{g}] \in \mathcal{E}_\Omega(\lim \mathcal{D})$), then $[\hat{g}|_{\prod F_i}] \in \mathcal{E}_\sharp^m(\prod F_i)$ (resp. $[\hat{g}|_{\prod F_i}] \in \mathcal{E}_\Omega(\lim \mathcal{D})$) and therefore $< g_i > \in \mathcal{E}_\sharp^m(E_i, B)$ (resp. $< g_i > \in \mathcal{E}_\Omega(E_i, B)$). □

A combination of Proposition 4.1 and Corollary 1.2 gives rise to

COROLLARY 4.2. *Let $\mathcal{D}$ be a star-shaped diagram with final object B such that any arrow $E_i \xrightarrow{f_i} B$ in $\mathcal{D}$ is a fibration with fibre F_i being either a finite dimensional complex or a Postnikov stage. If B has finite Lusternik-Schnirelmann category, then the group $\mathcal{E}_\sharp^m(\mathcal{D}, B)$ is nilpotent for any $m \geq \sup_{i\in I}\{\dim F_i\}$ and*

$$\operatorname{nil} \mathcal{E}_\sharp^m(\mathcal{D}, B) \leq \sup_{i\in I}\{\operatorname{nil} \mathcal{E}_\sharp^m(F_i)\} + \operatorname{cat} B.$$

PROOF. By Proposition 4.1 we obtain

$$\operatorname{nil} \mathcal{E}_\sharp^m(\mathcal{D}, B) = \sup_{i\in I}\{\operatorname{nil} \mathcal{E}_\sharp^m(E_i, B)\}$$

and by Corollary 1.2

$$\operatorname{nil} \mathcal{E}_\sharp^m(E_i, B) \leq \operatorname{nil} \mathcal{E}_\sharp^m(F_i) + \operatorname{cat} B,$$

which produces the desired result. □

A proof similar to that above leads to an extension of Corollary 1.3.

COROLLARY 4.3. *Let $\mathcal{D}$ be a star-shaped diagram of rational spaces of finite category with final object B such that any arrow $E_i \xrightarrow{f_i} B$ in $\mathcal{D}$ is a fibration with fibre F_i being 1-connected and either its rational cohomology or homotopy is of finite dimension (call it* $\dim(F)$*). Then, for any $m \geq \sup_{i\in I}\{\dim F_i\}$, the group $\mathcal{E}^m_\sharp(\mathcal{D}, B)$ is nilpotent and*

$$\operatorname{nil} \mathcal{E}^m_\sharp(\mathcal{D}, B) \leq \sup_{i\in I}\{\operatorname{cat} F_i\} + \operatorname{cat} B - 1.$$

Proposition 4.1 gives rise to a natural group morphism

$$\mathcal{E}^m_\sharp(\mathcal{D}, B) \xrightarrow{f_\sharp} \mathcal{E}^m_\sharp(\mathcal{D}_{(P)}, B_{(P)}),$$

that in view of Proposition 4.1 and Theorem 1.5 proves

COROLLARY 4.4. *Let $\mathcal{D}$ be a star-shaped diagram of nilpotent spaces with final object B such that any arrow $E_i \xrightarrow{f_i} B$ in $\mathcal{D}$ is a nilpotent fibration with fibre F_i. Assume that for each $i \in I$ either E_i is finite dimensional or F_i is a Postnikov stage, in both case of dimension at most k. If $\mathcal{E}^k_\sharp(\mathcal{D}, B)$ is nilpotent then for any $m \geq k$ and for any set of primes P the natural map,*

$$\mathcal{E}^m_\sharp(\mathcal{D}, B) \xrightarrow{f_\sharp} \mathcal{E}^m_\sharp(\mathcal{D}_{(P)}, B_{(P)})$$

is the localization morphism, i.e., $\mathcal{E}^m_\sharp(\mathcal{D}, B)_{(P)} \cong \mathcal{E}^m_\sharp(\mathcal{D}_{(P)}, B_{(P)})$.

Notice that along the proof of Proposition 4.1 we observed that the morphism $\operatorname{aut}_B \mathcal{D} \xrightarrow{\phi} \operatorname{aut} \lim \mathcal{D}$ factors through $\operatorname{aut}_B \lim \mathcal{D}$, hence it is also interesting to study the group $\mathcal{E}(\lim \mathcal{D}, B)$. First we prove that in our star-shaped diagrams situation, localizing commutes with taking inverse limits

PROPOSITION 4.5. *Let $\mathcal{D}$ be a star-shaped diagram of nilpotent spaces with final object B such that any arrow $E_i \xrightarrow{f_i} B$ in $\mathcal{D}$ is a nilpotent fibration. Then for any set of primes (possibly empty) P, the localization maps η induce a commutative diagram*

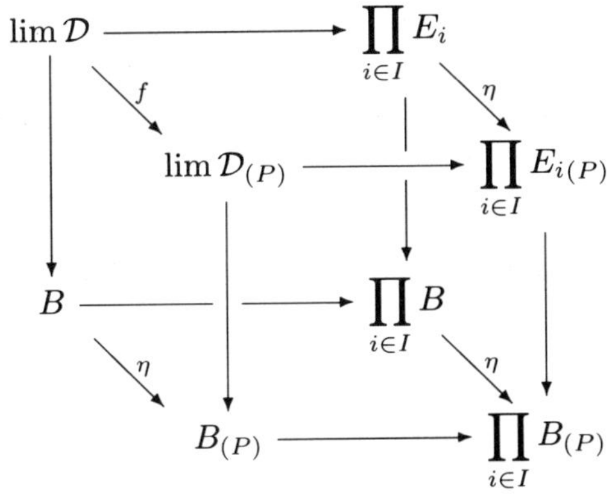

in which f is a localization map.

PROOF. As all $E_i \xrightarrow{f_i} B$ are nilpotent fibrations, $\prod_{i\in I} E_i \to \prod_{i\in I} B$ is so, and its localization $\prod_{i\in I} E_{i(P)} \to \prod_{i\in I} B_{(P)}$ is fibration again. Therefore $\lim \mathcal{D}$ and $\lim \mathcal{D}_{(P)}$ appear as pullbacks of diagrams in which one of the arrows is a fibration.

According to XI.4.1.(iv) in [**2**], $\lim \mathcal{D}$ and $\lim \mathcal{D}_{(P)}$ are indeed the homotopy pullbacks of those diagrams. Call $\underline{D}$ and $\underline{D}_{(P)}$ the pullback diagram involving $\lim \mathcal{D}$ and $\lim \mathcal{D}_{(P)}$ respectively

Now, $\lim \mathcal{D}_{(P)}$ is P-local as it is an inverse limit of P-local spaces. To finish the proof we show that f induces iso in $\pi_* \otimes \mathbb{Z}_{(P)}$. In order to do so we compare the Bousfield-Kan spectral sequences (see XI.§7 [**2**]) associated to $\underline{D}$ and $\underline{D}_{(P)}$. The second stage of the Bousfield-Kan spectral sequence associated to diagram $\underline{D}$ is given by $E_2^{s,t}(\underline{D}) = \lim^s(\pi_t(\underline{D}))$, where $\lim^s$ is the s-th derived inverse limit, and it converges to $\pi_* \text{holim}\, \underline{D} = \pi_* \lim \mathcal{D}$. By Lemma 6.1 in [**15**] and since $\_ \otimes \mathbb{Z}_{(P)}$ is an exact the spectral sequence with second stage $E_2^{s,t} \otimes \mathbb{Z}_{(P)} = \lim^s(\pi_t(\underline{D}) \otimes \mathbb{Z}_{(P)})$ converges to $\pi_* \text{holim}\, \underline{D} \otimes \mathbb{Z}_{(P)}$.

Now, the P-localization map gives an isomorphism

$$\eta_\sharp : \pi_*(\text{holim}\, \underline{D}) \otimes \mathbb{Z}_{(P)} \longrightarrow \pi_*(\text{holim}\, \underline{D}_{(P)})$$

which also induces an isomorphism between the second stages of the spectral sequences

$$E_2^{*,*}(\underline{D}) \otimes \mathbb{Z}_{(P)} \xrightarrow{\eta_\sharp} E_2^{*,*}(\underline{D}_{(P)}).$$

Hence the E_∞ terms are isomorphic, that is,

$$\pi_*(\lim \mathcal{D}) \otimes \mathbb{Z}_{(P)} \xrightarrow{f_\sharp} \pi_*(\lim \mathcal{D}_{(P)}).$$

□

Applying the proposition above and Theorem 1.5 we can obtain the following corollary

COROLLARY 4.6. *Let $\mathcal{D}$ be a star-shaped diagram of nilpotent spaces with final object B such that any arrow $E_i \xrightarrow{f_i} B$ in $\mathcal{D}$ is a nilpotent fibration with fibre F_i being a Postnikov stage of dimension at most k. If $\mathcal{E}_\sharp^k(\lim \mathcal{D}, B)$ is nilpotent then for any $m \geq k$ and for any set of primes P the natural map,*

$$\mathcal{E}_\sharp^m(\lim \mathcal{D}, B) \longrightarrow \mathcal{E}_\sharp^m(\lim \mathcal{D}_{(P)}, B_{(P)})$$

is the localization morphism, i.e., $\mathcal{E}_\sharp^m(\lim \mathcal{D}, B)_{(P)} \cong E_\sharp^m(\lim \mathcal{D}_{(P)}, B_{(P)})$.

PROOF. As all the fibrations $E_i \xrightarrow{f_i} B$ are nilpotent fibrations, the fibration $\prod_{i\in I} E_i \to \prod_{i \in I} B$ is so. Hence by diagram (4.1) we obtain that $\lim \mathcal{D} \longrightarrow B$ is a nilpotent fibration with fibre $\prod_{i\in I} F_i$ which is again a Postnikov stage of dimension at most k. Applying Theorem 3 we obtain that

$$\mathcal{E}_\sharp^m(\lim \mathcal{D}, B) \longrightarrow \mathcal{E}_\sharp^m\big((\lim \mathcal{D})_{(P)}, B_{(P)}\big)$$

is the localization morphism, but in view of Proposition 4.5, $\mathcal{E}_\sharp^m\big((\lim \mathcal{D})_{(P)}, B_{(P)}\big) = \mathcal{E}_\sharp^m(\lim \mathcal{D}_{(P)}, B_{(P)})$. □

References

[1] ARKOWITZ, M., *The Group of Self-Homotopy Equivalences-A Survey.* Springer-Verlag **1425** (1990), 170-203.

[2] BOUSFIELD, A.K. AND KAN D.M., *Homotopy limits, completions and localizations*, Lectures Notes in Mathematics **304**, 1972.

[3] CARTAN, H. AND EILENBERG, S., *Classes d'applications d'un espace dans un groupe topologique.* Seminaire H. Cartan exp. 6. (1962/63).

[4] DIDIERJEAN G., *Homotopie de l'espace des équivalences de'homotopie fibrées.* Ann. Inst. Fourier, **35** (1985), 33-47.

[5] DOLD, A., *Partitions of unity in the theory of fibrations.* Annals of Math. **78** (1963), 223-255.
[6] DROR, E. AND ZABRODSKY, A., *Unipotency and Nilpotency in Homotopy Equivalences.* Topology **18** (1979), 187-197.
[7] FÉLIX, Y. AND MURILLO, A., *A note on the nilpotency of subgroups of self-homotopy equivalences.* Bull. London Math. Soc. **29** (1997), 486-488.
[8] FÉLIX, Y. AND MURILLO, A., *A bound for the nilpotency of a group of self homotopy equivalences.* Proc. Amer. Math. Soc. **126** (1998), 625-627.
[9] FÉLIX, Y. AND THOMAS, J.C., *Nilpotent subgroups of the group of fibre homotopy equivalences.* Pub. Mat UAB. **39** (1995), 95-106.
[10] JAMES, I.M., *On Fibre Spaces and Nilpotency II.* Math. Proc. Camb. Phil. Soc. **86** (1979), 215-217.
[11] HILTON, P., MISLIN, G. AND ROITBERG, J., *Localization of Nilpotent Groups and Spaces.* Mathematics Studies 15, North-Holland (1975).
[12] LEGRAND, A., *Homotopie des Espaces de Sections.* Lecture notes in Math. **941**, Springer-Verlag (1988).
[13] MARUYAMA, K.I., *Localization of a certain subgroup of self-homotopy equivalences.* Pacific J. Math. **136** (1989), 293-301.
[14] MEIWES, H., *On Fibrations and Nilpotency-some remarks upon two articles by I.M. James.* Manuscripta Math. **39** (1982), 263-270.
[15] MURILLO, A. AND VIRUEL, A., *Lusternik-Schnirelmann Cocategory: A Whitehead dual approach,* To appear at Progress in Mathematics, Birkhäuser Verlag.
[16] PAVEŠIĆ, P., *A spectral sequence for the group of self-maps which induce identity automorphisms of homology groups.* Preprint.
[17] PAVEŠIĆ, P., *On the group* $Aut_{\Omega}(X)$. Preprint.
[18] RUTTER, J.W., *Spaces of Self-Homotopy Equivalences. A survey.* Lecture notes in Math. **1662**, Springer-Verlag (1997).
[19] SHIH, W., *On the group* $\epsilon[X]$ *of homotopy equivalence maps.* Bull. Amer. Math. Soc. **70** (1964), 293-296.

DEPARTAMENTO DE ÁLGEBRA, GEOMETRÍA Y TOPOLOGÍA,
UNIVERSIDAD DE MÁLAGA,
AP. 59,
29080 MÁLAGA,
SPAIN.

FAKULTETA ZA MATEMATIKO IN FIZIKO,
UNIVERZA V LJUBLJANI,
JADRANSKA 19,
1111 LJUBLJANA,
SLOVENIJA.

Contemporary Mathematics
Volume **274**, 2001

The Homotopy Groups of the Homotopy Fibre of an Induced Map of Function Spaces

K.A. Hardie and K.H. Kamps

ABSTRACT. Elements of the n'th u-based track group $\pi_n^X(Y;u)$ in the sense of M.G. Barratt are interpreted as 2-morphisms of the homotopy 2-groupoid of the $(n-2)$-fold loop space of Y^X based at u. This enables a reinterpretation of the exact homotopy sequence of the map of function spaces $Y^X \to B^X$ induced by a fixed map $g : Y \to B$.

Introduction

We work for the most part in a convenient category $\mathcal{C}$ of pointed Hausdorff spaces. Let $g : Y \to B$ be a map in $\mathcal{C}$. Then there is a long exact *homotopy sequence*

$$\to \pi_2(B) \to \pi_1(F_g) \to \pi_1(Y) \to \pi_1(B) \to \pi_0(F_g) \to \pi_0(Y) \to \pi_0(B) \tag{0.1}$$

of groups and pointed sets [DKP] where F_g denotes the homotopy fibre of g. Given a space X in $\mathcal{C}$, if we consider the map of function spaces

$$g_. : Y^X \to B^X$$

induced by composition with g then the homotopy sequence for $g_.$

$$\to \pi_2(B^X) \to \pi_1(F_{g.}) \to \pi_1(Y^X) \to \pi_1(B^X) \to \pi_0(F_{g.}) \to \pi_0(Y^X) \to \pi_0(B^X)$$

is equivalent to the dual Barratt-Puppe sequence studied by Nomura [N]

$$\to \pi(\Sigma X, Y) \to \pi(\Sigma X, B) \to \pi(X, F_g) \to \pi(X, Y) \to \pi(X, B)\ . \tag{0.2}$$

Note that so far the base-point $* \in Y^X$ (respectively $* \in B^X$) has been understood to be the constant map to $* \in Y$ (respectively to $* \in B$).

1991 *Mathematics Subject Classification.* Primary 55Q05; Secondary 55Q35, 55P10, 55R05.
Key words and phrases. Homotopy 2-groupoid, n-track, homotopy fibre.
Research support to the first author from the University of Cape Town and from the University of the Western Cape is acknowleged.

However, we are also interested in the homotopy groups of the underlying unpointed space of Y^X based at some other map $u : X \to Y$ in $\mathcal{C}$. (For example X^X based at $1 : X \to X$.) In this case there is an associated sequence

$$(0.2.1) \qquad \to \pi_n(Y^X, u) \to \pi_n(B^X, gu) \to \pi_{n-1}(F_{g.}, \tilde{u}) \to \pi_{n-1}(Y^X, u) \to$$

but, unfortunately, no conversion of the sequence 0.2.1 analogous to 0.2 seems to be known.

Homotopy groups such as $\pi_n(Y^X, u)$ have been studied by Barcus and Barratt [BB], Federer [F], Rutter [R1] and others. In particular, Barratt [B; 6.4] points out that $\pi_n(Y^X, u)$ is equivalent to the n-th u-based *track group* $\pi_n^X(Y; u)$ comprising the relative homotopy classes of maps $F : X \times I^n \to Y$ such that

$$(0.3) \qquad F(x,t) = u(x) \ (x \in X,\ t \in \dot{I}^n)\ , F(*,t) = * \ (t \in I^n)\ ,$$

where $\dot{I}^n$ refers to the boundary of I^n. (The notation $\pi_n^X(Y; u)$ can be found in [R1].)

A corresponding conversion (interpretation) of the groups (and final homotopy set) associated with the homotopy fibre of $g.$ would certainly be desirable.

An indication that there might be an alternative homotopy sequence based on the methods of coherent homotopy theory is given by the five term exact sequence (of groups and pointed sets)

$$(0.4) \quad \pi_1^X(Y; u) \xrightarrow{g.} \pi_1^X(B; gu) \xrightarrow{m_u} \pi(X, Y/B)_{\tilde{u}} \xrightarrow{d} \pi(X, Y)_u \xrightarrow{g.} \pi(X, B)_{gu}$$

associated with a homotopy commutative triangle

$$(0.5) \qquad \begin{array}{ccc} X & \xrightarrow{\quad u \quad} & Y \\ & \Uparrow u_t & \\ {\scriptstyle f} \searrow & & \swarrow {\scriptstyle g} \\ & B & \end{array}$$

described in [HK1, Theorem 4.3]. (The suffices $\tilde{u}$, u and gu decorating the last three pointed sets specify the base-points.)

When the sequence 0.4 was discovered there seemed to be no obvious prolongation to the left but now with the conceptual machinery of the homotopy 2-groupoid $\mathbf{G}_2E$ of a Hausdorff space E [HKK], we are able to present a long sequence

$$(0.6) \qquad \longrightarrow \pi_n^X(Y; u) \xrightarrow{g.} \pi_n^X(B; gu) \xrightarrow{m} \pi_{n-1}(X, Y/B) \xrightarrow{d} \pi_{n-1}^X(Y; u) \longrightarrow$$

terminating with the special case of 0.4 in which $f = gu$ and $\tilde{u}$ is the morphism of the *track homotopy set over* B, $\pi(X, Y/B)$ [HK3], determined by the constant homotopy at gu. The novelty lies in the description of the groups $\pi_{n-1}(X, Y/B)$ and the associated homomorphisms. Since we describe also a ladder diagram

$$\begin{array}{ccccccccc} \longrightarrow & \pi_n^X(Y; u) & \xrightarrow{g.} & \pi_n^X(B; gu) & \xrightarrow{m} & \pi_{n-1}(X, Y/B) & \xrightarrow{d} & \pi_{n-1}^X(Y; u) & \longrightarrow \\ & \approx \downarrow & & \approx \downarrow & & \phi \downarrow & & \approx \downarrow & \\ \longrightarrow & \pi_n(Y^X, u) & \longrightarrow & \pi_n(B^X, gu) & \longrightarrow & \pi_{n-1}(F_{g.}, \tilde{u}) & \longrightarrow & \pi_{n-1}(Y^X, u) & \longrightarrow \end{array}$$

it follows from the 5-lemma that our sequence is the desired conversion.

The prolongation requires a new application of $\mathbf{G}_2E$ to interpret the notion of n-track ($n \geq 2$) in the sense of 0.3 as a 2-morphism of

$$\mathbf{G}_2\Omega_u^{n-2}Y^X,$$

where Y^X refers to the (unpointed) function space of all pointed maps $X \to Y$ and $\Omega_u^n Y^X$ is essentially its n-fold loopspace based at the point u. The details are given, together with a description of 0.6 and a discussion of the exactness, in section 2, preceded (section 1) by a brief reminder of the definition and basic results concerning $\mathbf{G}_2E$.

Regarding applications, we see the sequence 0.6 as being well adapted to computation of the groups $\pi_n(X, Y/B)$, given situations in which the relevant track groups (and the homomorphisms $g_.$) are known. To defend this claim, we describe in section 3 a secondary operation designed to settle problems of group extension arising in such computations. Finally in section 4 we discuss a specialisation of 0.6 adapted to computation of homotopy groups of certain spaces of homotopy equivalences over B. We acknowledge that the discussion in section 4 is not essentially new and that the ideas here go back to the early work on fibre-homotopy self-equivalences. For details, see the monograph [R2, Chapter 21]. There is some reason to hope, however, that the simple form of the secondary operation given in section 3 will facilitate computation in the future.

The interpretation of n-tracks as 2-morphisms of the homotopy 2-groupoid of a loop space given in (2.3) is of independent interest and may be expected to lead to a notion of *thin* n-track of use in the theory of the higher-dimensional homotopy groupoids.

§1. The homotopy 2-groupoid

In this section we recall some details of the homotopy 2-groupoid $\mathbf{G}_2E$ of a Hausdorff space E as given in [HKK]. If p and q are points of E, a *path* f in E from p to q is a continuous map $f : I \to E$ from the unit interval I into E such that $f(0) = p$ and $f(1) = q$. If g is another path in E from q to r, we denote their *concatenation* by $g \bullet f$.

Let $f, f' : p \simeq q$ be paths in E. A *relative homotopy* $f_s : f \simeq f' : p \simeq q$ is a homotopy such that the initial and final points of f and f' remain fixed during the homotopy. Let $f_s, f'_s : f \simeq f' : p \simeq q$ be two relative homotopies. We consider f_s and f'_s themselves to be *relatively homotopic*, if they are homotopic via a homotopy $(I \times I) \times I \longrightarrow E$ which is constant on the boundary of $I \times I$. The relative homotopy class $\{f_s\}$ of f_s is called a *2-track*.

Concatenation of the relative homotopies f_s and f'_s (defined if $f_1 = f'_0$) induces a *vertical pasting* operation on 2-tracks, denoted $+$, yielding a groupoid structure (with identities denoted 0 or 0_f) on the set $\Pi_2E(p, q)$ of 2-tracks between paths in E from p to q. Similarly if $f_s : f \simeq f' : p \simeq q$ and $g_s : g \simeq g' : q \simeq r$, the *horizontal pasting* of homotopies

$$g_s \bullet f_s : g \bullet f \simeq g' \bullet f' : p \simeq r \ ,$$

obtained by concatenation of the respective paths at each stage of the homotopy, induces a corresponding operation on 2-tracks:

$$(\{f_s\},\{g_s\}) \longrightarrow \{g_s\} \bullet \{f_s\} \ ,$$

satisfying the *interchange* property

$$(\{g_s\}+\{g'_s\}) \bullet (\{f_s\}+\{f'_s\}) = (\{g_s\}\bullet\{f_s\}) + (\{g'_s\}\bullet\{f'_s\}) \ .$$

A relative homotopy $\psi_s : f \simeq f' : p \simeq q$ is *thin* if it can be factored

$$\psi_s : I \times I \xrightarrow{\phi_s} J \xrightarrow{p} E \ ,$$

where J is a tree, $\phi_s : \phi \simeq \phi'$ is a relative homotopy, ϕ and ϕ' are paths in J which (i) have the same initial and the same final points, (ii) are finitely piecewise linear and (iii) satisfy $p\phi = f$, $p\phi' = f'$.

The *underlying groupoid* $\mathbf{G}E$ of $\mathbf{G}_2E$ is the groupoid of $\sim$ classes of paths in E, where $f \sim f'$ if there exists a thin relative homotopy from f to f' and where the operation $\bullet$ is induced by concatenation of paths. Let $\mathbf{N}E(p,q)$ denote the subgroupoid of $\Pi_2E(p,q)$ whose morphisms are the relative homotopy classes of thin relative homotopies. Then $\mathbf{N}E(p,q)$ is a normal subgroupoid of $\Pi_2E(p,q)$ and we define $\mathbf{G}_2E(p,q)$ to be the quotient groupoid $\Pi_2E(p,q)/\mathbf{N}E(p,q)$. We use $\langle f\rangle$ to denote the *semitrack* (i.e. $\sim$ class) of a path f and $0_{\langle f\rangle}$ to denote the identity 2-track in $\mathbf{G}_2(E,\langle f\rangle) = \mathbf{G}_2E(\langle f\rangle,\langle f\rangle)$. The main result of [HKK] may be stated as follows.

PROPOSITION 1.1. *The sets $\mathbf{N}E(p,q)(f,f')$ are singletons or empty. $\mathbf{G}_2E$ is a 2-groupoid with underlying groupoid $\mathbf{G}E$, 2-morphism sets $\mathbf{G}_2E(p,q)(\langle f\rangle,\langle f'\rangle)$ and horizontal composition $\bullet$. $\mathbf{G}_2E$ is functorial in E. For each path f in E, there is a natural isomorphism (of abelian groups)*

$$\sigma_{\langle f\rangle} : \mathbf{G}_2(E,\langle f\rangle) \xrightarrow{\approx} \pi_2(E,f(0)) \ ,$$

where $\pi_2(E,f(0))$ refers to the second homotopy group of E based at the point $f(0)$.

§2. n-tracks as 2-morphisms

So far the homotopy 2-groupoid of a Hausdorf space $\mathbf{G}_2E$, given in [HKK], has been used in the case $E = Y^X$ of the (unpointed) function space of all pointed maps $X \to Y$ to study 2-tracks inhabiting cubical diagrams. We now indicate that an element α of $\pi_n^X(Y;u)$ $(n \geq 2)$ can be regarded as a 2-morphism of the homotopy 2-groupoid of a suitable function space. Let

$$\Omega_u^n Y^X \subseteq Y^{X\times I^n} \tag{2.1}$$

be the space of all maps F satisfying the conditions 0.3 and recall that α is the relative homotopy class of such a map. Then, if we set

$$F''(s)(x,s') = F(x,(s,s')) = u(x) \ , \tag{2.2}$$

we obtain a map $F'' : I^2 \to \Omega_u^{n-2}Y^X$. Thus α can be considered as a 2-morphism of the homotopy 2-groupoid $\mathbf{G}_2\Omega_u^{n-2}Y^X$ with domain and codomain the semitrack

$\langle c_{up_n}\rangle$, where c_{up_n} is the constant path at the point up_n (which is the composite of $u : X \to Y$ with the projection $p_n : X \times I^{n-2} \to X$). Clearly in this way we have

(2.3) $$\pi_n^X(Y;u) \approx \mathbf{G}_2\Omega_u^{n-2}Y^X(\langle c_{up_n}\rangle, \langle c_{up_n}\rangle) .$$

Further, we define

$$\pi_{n-1}(X, Y/B) = M_n(X, Y/B)/\sim ,$$

where $M_n(X, Y/B)$ is the set of ordered pairs $(\lambda, \langle k\rangle)$ with $\langle k\rangle$ the semitrack of a path k from up_n to up_n and λ a 2-morphism inhabiting the diagram

(2.3.1) $$gup_n \;\underset{\langle c_{gup_n}\rangle}{\overset{g.\langle k\rangle}{\Downarrow\lambda}}\; gup_n ,$$

where $\langle c_{gup_n}\rangle$ refers to the semitrack of the constant homotopy at gup_n. We define

(2.4) $$(\lambda, \langle k\rangle) \sim (\lambda', \langle k'\rangle)$$

if there exists a 2-track κ, where

(2.5) $$up_n \;\underset{\langle k\rangle}{\overset{\langle k'\rangle}{\Downarrow\kappa}}\; up_n \qquad \text{and} \qquad \lambda' = g.\kappa + \lambda .$$

In this way we obtain an equivalence relation $\sim$ and we denote the equivalence class of $(\lambda, \langle k\rangle)$ by $\{\lambda, \langle k\rangle\}$. The binary operation on $\pi_{n-1}(X, Y/B)$ is defined by the rule

$$\{\lambda, \langle k\rangle\}\{\lambda'', \langle k''\rangle\} = \{\lambda'' \bullet \lambda, \langle k''\rangle \bullet \langle k\rangle\} ,$$

where $\lambda'' \bullet \lambda$ is the composite 2-track in the diagram

$$gup_n \;\underset{\langle c_{gup_n}\rangle}{\overset{g.\langle k\rangle}{\Downarrow\lambda}}\; gup_n \;\underset{\langle c_{gup_n}\rangle}{\overset{g.\langle k''\rangle}{\Downarrow\lambda''}}\; gup_n .$$

Since $\lambda''\bullet\lambda$ is the horizontal composition in a 2-groupoid, we have the two equivalent expressions

(2.5.1) $$\lambda'' \bullet \lambda = (g.\langle k\rangle)^*\lambda'' + \lambda = (g.\langle k''\rangle)_*\lambda + \lambda''$$

in terms of the 'star' operations, where

$$\langle f\rangle_*\alpha = 0_{\langle f\rangle} \bullet \alpha \quad \text{and} \quad \langle g\rangle^*\alpha = \alpha \bullet 0_{\langle g\rangle} .$$

The homomorphism $m : \pi_n^X(B; gu) \to \pi_{n-1}(X, Y/B)$ is defined by setting

$$m(\mu) = \{\mu, \langle c_{up_n}\rangle\} .$$

To define the boundary homomorphism $d : \pi_{n-1}(X, Y/B) \to \pi_{n-1}^X(Y; u)$, recall that $k : I \to \Omega_u^{n-2}Y^X \subseteq Y^{X\times I^{n-2}}$ is a path with initial and final point up_n. Hence

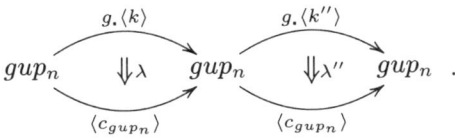

k defines a map $K : X \times I^{n-1} \to Y$ satisfying the conditions 0.3 (with n replaced by $n-1$). We set

$$d\{\lambda, \langle k \rangle\} = \{\langle k \rangle\} = \{K\} \in \pi_{n-1}^X(Y; u) \ .$$

Then we have

THEOREM 2.6. *The groups* $\pi_{n-1}(X, Y/B)$ *are abelian for* $n > 2$ *and the sequence* 0.6 *is exact.*

One may use properties of the horizontal composition in a 2-category to check that the given definitions give rise to a group structure. In particular, we note that the inverse of $\{\lambda, \langle k \rangle\}$ is given by

$$\{\lambda, \langle k \rangle\}^{-1} = \{-(g_. \langle k \rangle^{-1})_* \lambda, \langle k \rangle^{-1}\} \ ,$$

where $\langle k \rangle^{-1}$ refers to the groupoid inverse. That the operation is abelian for $n > 2$, will follow later from Proposition 2.7. To check the exactness, it will be convenient to use multiplicative notation.

(i) Suppose $\nu \in \pi_n^X(Y; u)$. Then $m(g_. \nu) = \{g_. \nu, \langle c_{up_n} \rangle\}$, which is equal to 1 in view of the relation $\sim$. Suppose that $m\mu = 1$. Then again via $\sim$, there exists ν such that $g_. \nu = \mu$.

(ii) Since $\{\langle c_u \rangle\} = 1$ in $\pi_{n-1}^X(Y; u)$, we have $dm = 1$. Suppose that $d\{\lambda, \langle k \rangle\} = 1$. Then there exists a 2-morphism $\theta : \langle c_{up_n} \rangle \Rightarrow \langle k \rangle$. It follows that $m(g_. \theta + \lambda) = \{\lambda, \langle k \rangle\}$.

(iii) It is easy to check that $g_. d = 1$. Suppose that $g_. \{\langle k \rangle\} = 1$. Then there exists $\lambda' : g_. \langle k \rangle \Rightarrow \langle c_{gup_n} \rangle$ and hence $d\{\lambda', \langle k \rangle\} = \{\langle k \rangle\}$, as required.

To define an isomorphism

$$\phi : \pi_{n-1}(F_{g.}) \to \pi_{n-1}(X, Y/B) \ ,$$

suppose that the map $h : I^{n-1}, \dot{I}^{n-1} \to F_{g.}, (u, c_{gu})$ is a representative of an element η of $\pi_{n-1}(F_{g.})$ and consider the diagram

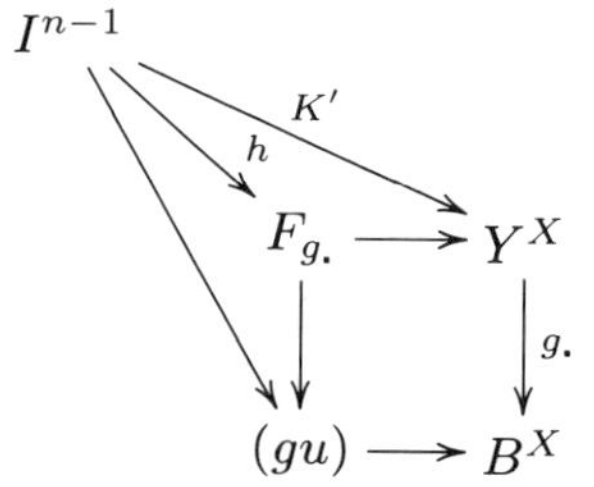

in which the rectangle is the standard (pointed) homotopy pullback. Then $K'(\dot{I}^{n-1}) = u$ defines a path $k : I \to \Omega_u^{n-2} Y^X$ and the standard homotopy yields a relative homotopy of $g_. k$ to c_{gup_n}, hence an element $(\lambda, \langle k \rangle)$ of form 2.3.1, whose $\sim$ class $\phi(\eta)$ is clearly independent of the choice of representatives. It is routine to check that ϕ is compatible with the homomorphisms of the homotopy sequences and so we have proved the following.

PROPOSITION 2.7. *ϕ is an isomorphism.*

§3. A secondary operation

Suppose that the groups and homomorphisms $g_{.} : \pi_n^X(Y;u) \to \pi_n^X(B;gu)$, $(n \geq 1)$ are known. The exactness of the sequence 0.6 implies that there exist short exact sequences

$$Cok_n \to \pi_{n-1}(X,Y/B) \to Ker_{n-1} \quad (n \geq 2)\ , \tag{3.1}$$

where Cok_n (respectively Ker_{n-1}) refer to the cokernel of $g_{.}$ in dimension n (respectively the kernel of $g_{.}$ in dimension $n-1$). These enable the groups $\pi_{n-1}(X,Y/B)$ to be computed modulo problems of group extension. Moreover, in the case $n = 2$ the extension may be non-abelian.

To assist in resolving these problems the following secondary operation may be useful. Suppose that $\alpha,\ \alpha' \in Ker_{n-1}$ are elements satisfying

$$\alpha\alpha' = 1 \in Ker_{n-1} \tag{3.2}$$

and suppose we have

$$\beta,\ \beta' \in \pi_{n-1}(X,Y/B) \ \text{ with } \ d\beta = \alpha \ \text{ and } \ d\beta' = \alpha'.$$

Then it becomes important to be able to decide whether or not $\beta\beta' = 1$.

Let $\beta = \{\lambda, \langle k\rangle\}$, $\beta' = \{\lambda', \langle k'\rangle\}$ and consider the element $\gamma(\kappa) = g_{.}\kappa + \lambda' \bullet \lambda$ of $\pi_n^X(B;gu)$, i.e. $\gamma(\kappa)$ is the total 2-track inhabiting the diagram

(3.3)

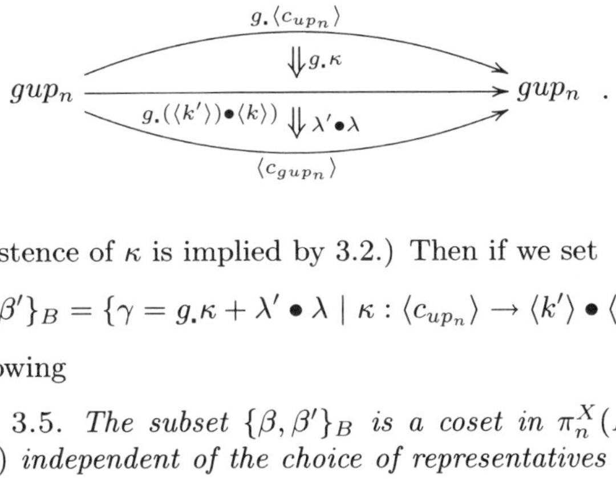

(Note that the existence of κ is implied by 3.2.) Then if we set

$$\{\beta,\beta'\}_B = \{\gamma = g_{.}\kappa + \lambda' \bullet \lambda \mid \kappa : \langle c_{up_n}\rangle \to \langle k'\rangle \bullet \langle k\rangle\} \tag{3.4}$$

we obtain the following

PROPOSITION 3.5. *The subset $\{\beta,\beta'\}_B$ is a coset in $\pi_n^X(B;gu)$ of the subgroup $g_{.}(\pi_n^X(Y;u))$ independent of the choice of representatives $(\lambda,\langle k\rangle)$, $(\lambda',\langle k'\rangle)$. Moreover*

$$m\{\beta,\beta'\}_B = \beta\beta'\ .$$

PROOF. If $\gamma(\kappa)$ and $\gamma(\kappa')$ are two elements of the subset then clearly $\gamma(\kappa)(\gamma(\kappa'))^{-1} = g_{.}(\kappa - \kappa')$. The independence of $\{\beta,\beta'\}_B$ on the choice of representatives is a consequence of the $\sim$ relation and the remaining equality is a direct consequence of the definition of m.

Note that the effect of the secondary operation is to transform the problem of deciding whether $\beta\beta' = 1$ in $\pi_{n-1}(X,Y/B)$, which is a group concerning which we have no a priori knowledge, into a similar problem in $\pi_n^X(B;gu)$, a group which we have assumed to be known.

§4. The space of homotopy equivalences over B

In this final section we specialise the preceding theory by selecting $Y = X$ and choosing $u = 1$, the identity map $X \to X$. Since every pointed self-map of X belonging to the component of 1 is a homotopy equivalence, we may expect to recover an analog of the sequence 0.6 of form

$$(4.1) \qquad \to \pi_n(EX;1) \xrightarrow{g.} \pi_n^X(B;g) \xrightarrow{m} \pi_{n-1}(EX/B) \xrightarrow{d} \pi_{n-1}(EX;1) \to \quad ,$$

where EX refers to the identity component of the space of self-homotopy equivalences of X. To interpret the groups $\pi_{n-1}(EX/B)$, let

$$(4.2) \quad EX_n = \{F \in X^{X \times I^n} \mid F \text{ satisfies 0.3 with } u = 1,\ F(-,t) \in EX,\ t \in I^n\}$$

and note that an element of $\pi_n(EX;1)$ can be regarded as a 2-morphism of $\mathbf{G}_2(EX_{n-2})$ with domain and codomain $\langle c_{up_n} \rangle$ and an element of $\pi_{n-1}(EX/B)$ as an equivalence class $\{\lambda, \langle k \rangle\}$ under the relation $\sim$, as in section 3 above, except that this time the path k lies in EX_{n-2}. With this modification we recover exactness of the sequence 4.1 and an isomorphism

$$\phi : \pi_{n-1}(F_{g.}) \approx \pi_{n-1}(EX/B) ,$$

where this time $g.$ is the map $EX \to B^X$ induced by composition with g.

A definition of an associated secondary operation can be given as before, so that when the relevant information is available computation of $\pi_{n-1}(EX/B)$ is possible. The motivation for doing this is that a point of $F_{g.}$ defines a diagram

$$\begin{array}{ccc} X & \xrightarrow{\ v\ } & X \\ & \Uparrow v_t & \\ {\scriptstyle g} \searrow & & \swarrow {\scriptstyle g} \\ & B & \end{array} \quad ,$$

where $v \in EX$ $(t \in I)$. The totality of such diagrams may be regarded as a 'space of homotopy self-equivalences of X over the fixed space B'.

References

[BB] W.D. Barcus and M.G. Barratt, *On the homotopy classification of the extensions of a fixed map*, Trans. Amer. Math. Soc. **88** (1958), 57–74.

[B] M.G Barratt, *Track groups I*, Proc. Lond. Math. Soc. (3) **5** (1955), 71-106.

[BHK] H.J. Baues, K.A. Hardie and K.H. Kamps, *The self-equivalence groups in certain coherent homotopy categories*, Tsukuba J. Math. **21** (1997), 213-228.

[DKP] T. tom Dieck, K.H. Kamps and D. Puppe, *Homotopietheorie*, Lecture Notes in Math., vol. 157, Springer-Verlag, 1970.

[F] H. Federer, *A study of function spaces by spectral sequences*, Trans. Amer. Math. Soc. **82** (1956), 340-361.

[HK1] K.A. Hardie and K.H. Kamps, *Exact sequence interlocking and free homotopy theory*, Cahiers de Top. et Géom. Diff. Cat. **26** (1985), 3-31.

[HK2] K.A. Hardie and K.H. Kamps, *Homotopy over B and under A*, Cahiers de Top. et Géom. Diff. Cat. **28** (1987), 183-196.

[HK3] K.A. Hardie and K.H. Kamps, *Track homotopy over a fixed space*, Glasnik Mat. **24** (1989), 161-179.

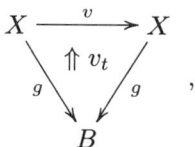

[HK4] K.A. Hardie and K.H. Kamps, *Coherent homotopy over a fixed space* (1955), Elsevier Science B.V., 195-211.
[HKK] K.A. Hardie, K.H. Kamps and R.W. Kieboom, *A homotopy 2-groupoid of a Hausdorff space*, Appl. Cat. Struc. **8** (2000), 209-234.
[N] Y. Nomura, *On mapping sequences*, Nagoya Math. J. **17** (1960), 111-145.
[R1] J.W. Rutter, *A homotopy classification of maps into an induced fibre space*, Topology **6** (1967), 379-403.
[R2] J.W. Rutter, *Spaces of Homotopy Self-Equivalences*, Lecture Notes in Math., vol. 1662, Springer-Verlag, 1997.

DEPARTMENT OF MATHEMATICS, UNIVERSITY OF CAPE TOWN, 7700 RONDEBOSCH, SOUTH AFRICA

UNIVERSITY OF THE WESTERN CAPE, PRIVATE BAG X17, 7535 BELLVILLE, SOUTH AFRICA
E-mail address: `hardieka@iafrica.com`

FACHBEREICH MATHEMATIK, FERNUNIVERSITÄT, POSTFACH 940, D-58084 HAGEN, GERMANY
E-mail address: `Heiner.Kamps@fernuni-hagen.de`

Contemporary Mathematics
Volume **274**, 2001

Fibrations, self homotopy equivalences and negative derivations

Volker Hauschild

ABSTRACT. An elliptic space X with $H^{od}(X;\mathbb{Q}) = 0$ is called of type F_0. In this note we consider the connected component $E(\xi, 1)$ of the identity of the monoid $E(\xi)$ of fiber homotopy equivalences of oriented fibrations $\xi\colon X \to E \to B$ where X is F_0 showing that under some conditions $E(\xi, 1)$ is a finite $\mathbb{Q}$–Hopf space. In particular it is shown that this is true for the fibration $G/H \to B_H \to B_G$ where G is a compact connected Lie group and $H \subset G$ a closed connected subgroup of maximal rank. Other results concern Borel fibrations $\beta_G[X] : X \to X_G \to B_G$ where a torus G acts smoothly on a closed oriented manifold X of F_0–type such that the fixed set X^G consists of isolated points. It is shown that $E(\beta_G[X], 1)$ is rationally homotopy equivalent to a finite product of odd spheres.

1. Introduction

Suppose X is a topological space with the homotopy type of a not necessarily finite CW–complex. In the following (co)homology is taken with coefficients in a fixed field $\mathbb{Q}$ of vanishing characteristic.

DEFINITION 1.1. A 1-connected space X is called of type F_0 if the following conditions are satisfied.

(1) $dim\, H^*(X) < \infty$
(2) $dim\, \pi_*(X) \otimes_{\mathbb{Z}} \mathbb{Q} < \infty$
(3) $H^{od}(X) = 0.$

Examples of such spaces are among the most popular and natural in algebraic topology. For example, all homogeneous spaces G/U, where G is a compact connected Lie group and $U \subset G$ is a closed connected subgroup of maximal rank, are of this type. It is a result of S. Halperin [**3**] that the rational cohomology ring $H^*(X)$ of such a space is a quasihomogeneous complete intersection algebra over $\mathbb{Q}$. This means that the $\mathbb{Q}$–algebra $A_0 = H^*(X)$ can be written as $A_0 = P/I_0$, where P is a graded polynomial ring in generators of even degree and I_0 is an ideal

---

1991 *Mathematics Subject Classification.* Primary 55P62; Secondary 14BO7, 14M10, 13E10.
Key words and phrases. Weighted complete intersections, negatively graded derivations, spectral sequences.

generated by a prime series $\{f_1, \ldots, f_n\}$ of elements f_j homogeneous with respect to the gradation given on P.
Let $E(X,1)$ be the connected component of the identity in the monoid of $E(X)$ of self homotopy equivalences of a space X. It is well known that oriented Hurewicz fibrations $X \to E \to B$ are classified up to fiber homotopy equivalence by the homotopy classes of maps $f\colon B \to BE(X,1)$. In the case X is a space of type F_0, W. Meier found formulas for the rational homotopy groups of the classifying space $BE(X,1)$ purely in terms of the cohomology ring structure of X [**11, 12**]. These formulas have a deformation theoretic interpretation as was explained by the author in [**5**]. It is shown that the oddly graded part $\pi_{od}(X) \otimes_{\mathbb{Z}} \mathbb{Q}$ is canonically isomorphic to the subspace $Der(A_0)_-$ of the derivations of negative degree in $Der(A_0)$, see [**5**], Th. B., whereas the evenly graded part $\pi_{ev}(X) \otimes_{\mathbb{Z}} \mathbb{Q}$ can be identified with the subspace $T^1(A_0)_-$ of negative degree in the finite $\mathbb{Q}-$vector space $T^1(A_0)$ of infinitesimal deformations of the $\mathbb{Q}-$algebra A_0. It is a longstanding conjecture that $Der(A_0)_-$ is zero for quasihomogeneous complete intersections A_0 of dimension zero. The conjecture would imply that the universal classifying space $BE(X,1)$ has no odd rational homotopy, i.e., $E(X,1)$ is a finite $\mathbb{Q}-$Hopf space, and then is rationally homotopy equivalent to a finite product of odd spheres. It would also follow that its cohomology ring over the rationals is a polynomial algebra in as many generators as there are linearly independent infinitesimal deformations of negative grade. This in turn would imply that every $\mathbb{Q}-$orientable Hurewicz fibration with a fiber X, $H^*(X) = A_0$, has a collapsing Serre spectral sequence. This was indeed the form Halperin gave to his conjecture (see also [**1**], conjecture 14):

CONJECTURE 1.2. *Let X be a space of elliptic type with vanishing odd rational cohomology. Then the Serre spectral sequence of any $\mathbb{Q}-$oriented fibration $X \to E \to B$ satisfies $E_2 \cong E_\infty$.*

The conjecture has been proved for a series of important elliptic spaces as the mentioned homogeneous spaces of maximal rang G/H by Shiga and Tezuka in [**14**]. An older result concerning Kähler manifolds can be found in [**12**].

In this note we do not consider the conjecture of Halperin directly but we consider $\mathbb{Q}-$oriented fibrations $\xi = \{X \to E \to B\}$ where X is a space of type F_0 and we shall study the rational homotopy of the monoids of the fiber homotopy equivalences exploiting some algebraic results. The following two theorems are implicit in the note of M. Markl [**9**], but I shall present here a more streamlined proof using the language of Kähler modules. In the following theorems we consider only commutative algebras with grading, i.e., we do not consider the category of graded commutative algebras as is usual in rational homotopy theory.

THEOREM 1.3. *Let k be a field, let R be a positively graded noetherian $k-$algebra and let A be a positively graded flat $R-$algebra such that $\bar{A} = A/\underline{m}_R A = A \otimes_R k$ is a graded $k-$algebra without negatively graded derivations. Then $(Der_R A)_- = 0$.*

As a corollary we obtain

THEOREM 1.4. *Let R be a noetherian graded $k-$algebra, let $A|R$ be a flat graded $R-$algebra and let $\bar{A} = A/\underline{m}_R A$. If $Der_k(\bar{A})_- = 0$ and $Der_k(R)_- = 0$, then $Der_k(A)_- = 0$.*

The reader should observe that these results do not afford that $\bar{A}$ is a complete intersection. These are simply two quite general results on the behaviour of negative derivations under deformations. If we specialize to the case when $\bar{A}$ is a complete intersection we obtain the following wellknown result of M. Markl [9]

THEOREM 1.5 (**M. Markl** [9]). *Let $X \to E \to B$ be an oriented fibration where X, E and B are elliptic spaces. Let $\pi_{ev}(E(X,1)) \otimes_{\mathbb{Z}} \mathbb{Q} = 0$ and $\pi_{ev}(E(X,1)) \otimes_{\mathbb{Z}} \mathbb{Q} = 0$. Then $\pi_{ev}(E(X,1)) \otimes_{\mathbb{Z}} \mathbb{Q} = 0$.*

But in the light of the main result (Thm. A) in [5] Thms. 1.3 and 1.4 have more general applications.

Let ξ be a fibration, $\xi = \{X \to E \to B\}$ where X is a space of type F_0. Let $E(\xi)$ be the topological monoid of the fiber homotopy equivalences of ξ. Let $E(\xi, 1) \subset E(\xi)$ be the submonoid of those fiber homotopy equivalences which are homotopic to the identity by a family of fiber homotopy equivalences, i.e., we consider the "connected component of the identity" of $E(\xi)$. Then we prove:

THEOREM 1.6. *Let ξ be a fibration as above. Then the topological monoid $E(\xi, 1)$ is rationally homotopy equivalent to a finite product of odd spheres if $E(X, 1)$ is rationally homotopy equivalent to a finite product of odd spheres.*

We see that Thm. 1 gives sufficient criteria for the vanishing of $(Der_R A)_-$. Another sufficient condition for the vanishing of $(Der_R A)_-$ is that the Jacobi determinant of the defining relations of A as an $R-$algebra, should be a nonzero divisor of A, see Lemma 6.1. The application of this simple observation to fibrations needs a description of their Euler classes by this determinant. As a direct application we consider Borel fibrations. Suppose X is a closed oriented smooth manifold of F_0-type. Let a torus G act smoothly on X and consider the Borel fibration

$$\beta_G[X] \,:\, X \to E_G \times_G X \to B_G.$$

THEOREM 1.7. *Let G act smoothly on X with only isolated fixed points. Then the topological monoid $E(\beta_G[X], 1)$ is rationally homotopy equivalent to a finite product of odd spheres.*

2. Universal Fibrations and Self Homotopy Equivalences

Let X be a topological space, then we denote by

$$E(X) = \{f : X \to X \mid f \text{ is a homotopy equivalence}\}$$

the monoid of self-homotopy equivalences of X. We take a base point $\{*\} \subset X$ and consider the submonoid $E^*(X) \subset E(X)$ of the self homotopy equivalences which preserve the base point. Let $ev : E(X) \to X$ be the evaluation map $\alpha \to \alpha(*)$, then there is the fibration

$$E^*(X) \to E(X) \to X.$$

Applying the Dold-Lashof-May classifying space functor from the category of topological monoids into the category of topological spaces gives a fibration

$$X \to BE^*(X) \to BE(X).$$

If X is connected, we can also consider the connected component containing an identity $E(X,1) \subset E(X)$ which gives us the fibration

$$X \to BE^*(X,1) \to BE(X,1).$$

This is the universal fibration which classifies oriented Hurewicz fibrations with fiber X up to fiber homotopy equivalence. We state now W. Meiers theorem.

THEOREM 2.1. *(see* [**11**]*) Suppose X to be a finite simply connected CW-complex with evenly graded rational cohomology and finite dimensional rational homotopy, which is formal in the sense of rational homotopy theory. Then the following statements are equivalent:*

(1) *All derivations of negative degree of $H^*(X)$ vanish.*
(2) *The Serre spectral sequence of every orientable fibration with fiber X collapses at the 2-term.*
(3) $\pi_{2i-1}(BE(X,1)) \otimes_{\mathbb{Z}} \mathbb{Q} = 0$ *for all $i \in \mathbb{N}$.*

REMARK 2.2. If k is a field, let $A|k$ be a k-algebra. Then a k-derivation $d : A \to A$ is a k-linear map with $d(a \cdot b) = ad(b) + d(b)b$. If A has a $\mathbb{Z}$-graduation, i. e.,

$$A = \sum_{i \in \mathbb{Z}} A^i, \quad A^i \cdot A^j \subset A^{i+j},$$

then a derivation d is called of degree n if $d(A^i) \subset A^{i+n}, n \in \mathbb{Z}$. If n is negative, d is called a derivation of negative degree or simply a negative derivation.

Example: Let k be a field and let $A = k[x]$, $|x| = 2$, then $d = \frac{\partial}{\partial x}$ is a derivation of degree -2 on A.

3. On the rational homotopy of $E(X_1 \times X_2, 1)$

In this paragraph I will show that the formulas I gave for the rational homotopy of an elliptic space in [**5**] provide a very effective way to calculate the rational homotopy groups of the monoid $E(X_1 \times X_2, 1)$ if the corresponding groups for $E(X_1, 1)$ and $E(X_2, 1)$ are known. We recapitulate a little bit of the commutative algebra used in [**5**]. Let k be a field of any characteristic, let R be a (positively graded) noetherian k-algebra and let A be a finitely generated graded R-algebra, which is free over R as a R-module.

We consider a presentation $A = P/I$, where $P = R[x_1, \dots, x_n]$ is a graded polynomial algebra and $I \subset P$ is an ideal generated by homogeneous elements. We then have the exact sequence of A-modules [**10**], p. 188,

$$I/I^2 \longrightarrow \Omega_{P|R} \otimes_P A \longrightarrow \Omega_{A|R} \to 0.$$

Dualizing, gives the standard exact sequence:

$$0 \to Der_R(A) \longrightarrow Hom_P(\Omega_{P|R}, A) \xrightarrow{Jac} Hom_A(I/I^2, A) \longrightarrow T^1_R(A) \to 0.$$

Suppose now that A is a R-relative complete intersection of relative dimension zero, i.e., the ideal I is generated by a regular series $F_1, \dots, F_n$. Then it follows that both middle terms in the above sequence are free A-modules. In the following sums indicate direct sums of modules. We have therefore

$$\Omega_{P|R} \cong \sum_{i=1}^{n} P dx_i$$

and

$$I/I^2 \cong \sum_{j=1}^{n} A\, dF_j.$$

The dual modules are then generated by the corresponding partial derivatives, i.e.,

$$Hom_P(\Omega_{P|R}, A) \cong \sum_{i=1}^{n} A\frac{\partial}{\partial x_i}$$

and

$$Hom_A(I/I^2, A) \cong \sum_{j=1}^{n} A\frac{\partial}{\partial F_j}.$$

The map Jac is given on the generators by

$$Jac\left(\frac{\partial}{\partial x_i}\right) = \sum_{j=1}^{n}\left(\frac{\partial F_j}{\partial x_i} + I\right)\frac{\partial}{\partial F_j}$$

PROPOSITION 3.1.

$$Der_k(A_1 \otimes_k A_2) \cong Der_k(A_1) \otimes_k A_2 \oplus A_1 \otimes_k Der_k(A_2)$$

and

$$T_k^1(A_1 \otimes_k A_2) \cong T_k^1(A_1) \otimes_k A_2 \oplus A_1 \otimes_k T_k^1(A_2).$$

PROOF. The first formula is a consequence of the product law for derivatives. The second formula follows from the standard exact sequence for the case $R = k$. Let

$$A = A_1 \otimes A_2 = P_1/I_1 \otimes P_2/I_2 = P_1 \otimes P_2/(I_1 \otimes P_2 + P_1 \otimes I_2),$$

then

$$\Omega_{P|k} \cong \Omega_{P_1|k} \otimes P_2 \oplus P_1 \otimes \Omega_{P_2|k}.$$

Moreover, if $I = I_1 \otimes P_2 \oplus P_1 \otimes I_2$, then

$$I/I^2 \cong I_1/I_1^2 \otimes A_2 \oplus A_1 \otimes I_2/I_2^2.$$

The Jacobian homomorphism is splitting into a direct sum

$$Jac = Jac_1 \otimes Id_{A_2} \oplus Id_{A_1} \otimes Jac_2.$$

So, it follows that the standard exact sequence for the algebra $A = A_1 \otimes A_2$ splits into a sum of two exact sequences: The standard sequence for A_1 tensorized over k with A_2 and the standard sequence for A_2 tensorized with A_1. This implies the splittings claimed in the proposition. □

One of the main results in [**5**] is a topological interpretation of the negatively graded part of the above exact sequence. In particular the negatively graded parts of the first and the last term of the sequence can be interpreted as rational homotopy groups of certain topological monoids. Let $X \to E \to B$ be a fibration, then denote by $E(\xi, 1)$ the connected component of the identity in the monoid of fiber homotopy equivalences of ξ. If M is a $\mathbb{Z}-$graded module, let $M_- \subset M$ be its negatively graded part. Moreover, denote the rationalization of a $\mathbb{Z}-$module M by a superscript $M^{\mathbb{Q}}$. The following is Thm. A in [**5**].

THEOREM 3.2. *Let $\xi : X \to E \to B$ be a $\mathbb{Q}-$oriented Hurewicz fibration where X is an elliptic space and B is formal with $H^{od}(B) = 0$. Then there are canonical isomorphisms*

$$\pi_{ev}^{\mathbb{Q}}(E(\xi, 1)) \cong Der_R H^*(E)_-$$

and

$$\pi_{od}^{\mathbb{Q}}(E(\xi, 1)) \cong T_R^1 H^*(E)_-.$$

As a corollary of this formulas together with Prop. 1. we obtain formulas for the rational homotopy of $E(X_1 \times X_2, 1)$ when X_1 and X_2 are F_0.

THEOREM 3.3 (**W. Meier**). *Let X_1 and X_2 be F_0-spaces, then*

$$\pi_\nu^{\mathbb{Q}}(E(X_1 \times X_2, 1) = \sum_{i=0}^{\nu} \pi_{\nu-i}^{\mathbb{Q}}(E(X_1, 1)) \otimes H^i(X_2) + H^i(X_1) \otimes \pi_{\nu-i}^{\mathbb{Q}}(E(X_2, 1)).$$

PROOF. The above formula is shorthand for the two formulas for even and odd rational homotopy. Clearly, these formulas correspond to the two formulas in Prop. 3.1. Taking negatively graded parts gives the result. □

As a corollary of Thm. 3.3 we obtain

THEOREM 3.4. *Let X_1 and X_2 be elliptic spaces such that $\pi_{ev}^{\mathbb{Q}}(E(X_1, 1)) = 0$ and $\pi_{ev}^{\mathbb{Q}}(E(X_2, 1)) = 0$. Then $\pi_{ev}^{\mathbb{Q}}(E(X_1 \times X_2, 1)) = 0$.*

Historical remark: The first time I saw these formulas was in 1982 at Oberwolfach during a talk of W. Meier. So I attribute the formula of Thm. 3.3 to him.

4. Some results on modules of derivations

Let k be a field, let R be positively graded noetherian $k-$algebra and let A be a graded $R-$algebra which is free as an $R-$module. Write $A = P/I$, where $P = R[x_1, \dots, x_n]$. Let $Q = P \otimes_R k = k[x_1, \dots, x_n]$, let $\bar{A} = A \otimes_R k = A/\underline{m}_R A$ be the residue algebra of A on the closed point corresponding to the maximal ideal $\underline{m}_R \subset R$ of the positively graded elements. (Recall that in the following we do not necessarily assume that $\bar{A}$ is a complete intersection.) Let $D \in Der_R(A)$ and write

$$D = \sum_{i=1}^{n} a_i \frac{\partial}{\partial x_i}.$$

Denote the remainder mod $\underline{m}_R P$ with a bar and write

$$\bar{D} = \sum_{i=1}^{n} \bar{a}_i \frac{\partial}{\partial x_i}$$

as an element of $Hom_Q(\Omega_Q, \bar{A})$. We begin the proof of Thm. 1.3 with a lemma.

LEMMA 4.1. *The expression $\bar{D}$ is a derivation of $\bar{A}$.*

PROOF. Let the ideal I be generated by the homogeneous polynomials $F_1, \dots, F_m$, $m \geq n$. Let $f_j \in Q$ be a remainder $mod\, \underline{m}_R P$ of F_j. If we write $P = R \otimes_k Q$ and use the standard section of the projection $P \to Q$ we can consider f_j an element of P.

Then we can write $F_j = f_j + r_j$, where $r_j \in \underline{m}_R P$. Now we have

$$\frac{\partial F_j}{\partial x_i} \equiv \frac{\partial f_j}{\partial x_i}\, mod\, \underline{m}_R P.$$

Let $J = (f_1, \dots, f_m) \subset Q$, then $\bar{A} = Q/J$. The fact that D is a derivation of A is equivalent to a system of m equations (in following simply called an equation):

$$\sum_{i=1}^{n} a_i \left(\frac{\partial F_j}{\partial x_i} + I \right) = 0$$

in A. It follows that

$$\sum_{i=1}^{n} \bar{a}_i \left(\frac{\partial f_j}{\partial x_i} + J \right) = 0$$

as an equation in $\bar{A}$ which means that $\bar{D}$ is a derivation of $\bar{A}$. □

Now we present the proof of Thm. 1.3.

PROOF. We prove the result by induction on the embedding dimension $edim\, R$ of R. Let $edim\, R = 1$, i.e., $R = k[t]$ or $R = k[t]/(t^m)$ for some homogeneous generator t of positive degree. We observe that $Der_R(A)$ is torsion free on R since it is a submodule of the free A–module $Hom_A(\Omega_{P|R}, A)$ and A is free on R. Let $D \in Der_R(A)$ be homogeneous of negative degree such that $deg\, D$ is minimal, i.e., for every D' with $deg\, D' < deg\, D$ it follows that $D' = 0$. Write

$$D = \sum_{i=1}^{n} a_i \frac{\partial}{\partial x_i}$$

with $a_i \in A$. Then a_i is homogeneous of degree $deg\, a_i - deg\, x_i = deg\, D$. The fact that D is a derivation is equivalent to the equation in A :

$$\sum_{i=1}^{n} a_i \left(\frac{\partial F_j}{\partial x_i} + I \right) = 0.$$

So, $\bar{D}$ is a derivation of negative degree of $\bar{A}$ and so by hypothesis must vanish. Thus it follows $a_i \in \underline{m}_R A = (t)A$. Therefore we can write $a_i = tb_i$ for some $b_i \in A$. Let

$$D' = \sum_{i=1}^{n} b_i \frac{\partial}{\partial x_i}.$$

Since A is torsion free on R we have

$$\sum_{i=1}^{n} b_i \left(\frac{\partial f_j}{\partial x_i} + I \right) = 0,$$

and so D' must be a derivation again. But $deg\, D' < deg\, D$ and by hypothesis it must vanish. Thus it follows $D = tD' = 0$.

Let now R be of embedding dimension n, i.e., it can be written as a ring extension $R = S[t_n]$, with a homogeneous generator t_n where S is a graded ring of embedding dimension $n - 1$. It follows $S = R/(t_n)$. Let $D \in Der_R(A)$ be a derivation of minimal negative degree. Write

$$D = \sum_{i=1}^{n} a_i \frac{\partial}{\partial x_i}$$

with $a_i \in A$ homogeneous. Denote the remainder mod t_n by a bar. Let $\bar{P} = P/t_n P$ and $A' = A \otimes_R S = A/t_n A$. By an argument analogous to that used in the proof of Lemma 4.1 we see that

$$\bar{D} = \sum_{i=1}^{n} \bar{a}_i \frac{\partial}{\partial x_i}$$

is an element of $Der_S(A')$. It follows by induction $\bar{D} = 0$ and therefore $D = t_n D'$ for some expression

$$D' = \sum_{i=1}^{n} b_i \frac{\partial}{\partial x_i}$$

as an element of $Hom_P(\Omega_{P|R}, A)$ where $b_i \in A, a_i = t_n b_i$. But since D is a derivation we have the usual equation

$$\sum_{i=1}^{n} a_i \left(\frac{\partial f_j}{\partial x_i} + I \right) = 0$$

or

$$\sum_{i=1}^{n} t_n b_i \left(\frac{\partial f_j}{\partial x_i} + I \right) = 0.$$

Now t_n is a nonzero divisor of A and therefore one has

$$\sum_{i=1}^{n} b_i \left(\frac{\partial f_j}{\partial x_i} + I \right) = 0,$$

i.e., D' is also a derivation of A. But then $deg\, D' < deg\, D$ and so by the minimality of the degree of D one has $D' = 0$. It follows $D = t_n D' = 0$. □

Now we use the former result to prove Thm. 1.4.

PROOF. Let T be an $A-$module. Corresponding to the succession of ring homomorphisms $k \to R \to A$ one has an exact sequence [**10**], p. 187

$$Der_R(A,T) \to Der_k(A,T) \to Der_k(R,T).$$

In the case $T = A$ we get the exact sequence

$$Der_R(A) \to Der_k(A) \to Der_k(R,A).$$

We show $Der_k(R,A)_- = 0$. By definition one has $Der_k(R,A) = Hom_R(\Omega_{R|k}, A)$. Since A is a flat graded $R-$module it is also free on R. Therefore one has

$$Hom_R(\Omega_{R|k}, A) = Hom_R(\Omega_{R|k}, R) \otimes_R A = Der_k(R) \otimes_R A.$$

This shows $Der_k(R,A)_- = 0$ if $Der_k(R)_- = 0$. It follows by the above exact sequence that

$$Der_R(A)_- \to Der_k(A)_-$$

is surjective and therefore by Thm. 1.3 and the hypothes is $Der_k(A)_- = 0$. □

5. Iterated fibrations

In the following we use the results of the previous paragraph to show that certain classes of artinian $k-$algebras do not have negative derivations. We define what in the sequel is called a tower algebra.

DEFINITION 5.1. We say that a graded $k-$algebra A_0 is an iterated deformation or tower algebra if there is a chain $B_0 \to B_1 \to \cdots \to B_n$ of graded (local) $k-$algebras and graded homomorphisms such that

(1) $B_0 \cong k, B_n \cong A_0$
(2) B_i is flat on B_{i-1} and $\bar{B}_i \cong B_i/\underline{m}_{i-1}B_i = B_i \otimes_{B_{i-1}} k$ is artinian with generators all of the same degree. Here $\underline{m}_i \subset B_i$ are the respective augmentation ideals.

By an iterative application of Thm. 1.4 we get the following result.

THEOREM 5.2. *Let k be a field and let A_0 be a tower algebra such that the step quotients $\bar{B}_i$ have no negative derivations. Then A_0 has no negative derivations.*

The following result attributed to the author of [**15**] shows that the vanishing of negative derivations is merely a consequence of the artinian property and not of the fact that the algebra is a complete intersection.

LEMMA 5.3. *Let k be a field of characteristic zero, let B be a graded artinian strict commutative $k-$algebra B generated by homogeneous elements $y_1, \ldots, y_n \in B$, let $d\colon B \to B$ be a $k-$derivation with $d(y_i) = a_i \in k$. Then $d = 0$.*

In particular it follows that artinian $k-$algebras which are multiplicatively generated by elements which are all of the same degree do not have negative derivations.

PROOF. Since B is artinian there exists for every i a minimal $m_i \in \mathbb{N}$ such that $y_i^{m_i} = 0$. It follows $D(y_i^{m_i}) = m_i y_i^{m_i-1} D(y_i) = 0$. If $D(y_i) = a_i \neq 0$, then by hypothesis it follows $y_i^{m_i-1} = 0$, which is a contradiction. It follows $d = 0$. □

As a consequence we obtain the following theorem.

THEOREM 5.4. *Let A_0 be a graded $k-$algebra with a tower structure $B_0 \to \cdots \to B_n$ such that the successive quotients $B_i/\underline{m}_{i-1}B_i$ are artinian $k-$algebras of the type*

$$\bar{B}_i = \frac{k[y_1, \ldots, y_n]}{(p_1, \ldots, p_n)}$$

with $deg\, y_1 = \cdots = deg\, y_n = d_i$. Then A_0 has no negative derivations.

We can observe that the class of artinian $k-$algebras without negative derivations seems not to be confined to complete intersections. This raises the question about a general characterization of graded artininian algebras without negative derivations in terms of their weighting data.

In the following we consider a simple application of the previous results.

THEOREM 5.5. *Let $E_n \to \cdots \to E_1 \to E_0$ be a sequence of oriented fibrations $X_i \to E_i \to E_{i-1}$ where $X_i = G_i/U_i$ is a 1-connected homogeneous space of non-vanishing Euler number, i.e., $rk\, G_i = rk\, U_i$ for all i and $E_0 = \{*\}$ a point. Then with $E = E_n$ the monoid $E(E,1)$ does not have even rational homotopy groups and $H^*(BE(E,1))$ is a polynomial algebra in even generators.*

PROOF. By hypothesis $H^*(E_i; \mathbb{Q})$ is a free $H^*(E_{i-1}; \mathbb{Q})-$module and

$$H^*(X_i) \cong H^*(E_i)/H^*_+(E_{i-1}) \cdot H^*(E_i)$$

does not have negative derivations by [**14**]. Then apply Theorem 5.2. □

6. Monoids of fiber homotopy equivalences

Let us now consider monoids of fiber homotopy equivalences. As mentioned before we are considering oriented Hurewicz fibrations $\xi\colon X \to E \to B$ such that the following conditions are satisfied.

(1) X is F_0
(2) B is a 1-connected formal space with $H^{od}(B) = 0$.

It follows that the main result of [**5**] can be applied We therefore consider again the exact sequence

$$0 \to Der_R(A) \longrightarrow Hom_P(\Omega_{P|R}, A) \xrightarrow{Jac} Hom_A(I/I^2, A) \longrightarrow T^1_R(A) \to 0$$

with $R = H^*(B), A = H^*(E)$. Let $\Delta = |Jac|$ be the determinant of the Jacobian homomorphism. Then by Cramers rule Jac is injective if and only if Δ is a nonzero divisor of A.

LEMMA 6.1. *Let Δ be a nonzero divisor of A, then $Der_R(A)_- = 0$.*

We have therefore two sufficient criterions which guarantee that $Der_R(A)_-$ vanishes. The first is given by Thm. 1.3 whereas the second is a simple consequence of the fact that often Δ is a nonzero divisor of A.

In the following let $E(\xi)$ be the topological monoid of the fiber homotopy equivalences of ξ. Let $E(\xi,1) \subset E(\xi)$ be the connected component containing the identity. In the note [**5**] the this monoid is called $G_0(\xi)$.

THEOREM 6.2. *Let ξ be a fibration as above and suppose one of the following conditions is satisfied:*

A) *The element Δ is a nonzero divisor of $H^*(E)$.*
B) *The ring $H^*(X)$ does not have negative derivations.*

Then $\pi_{ev}^{\mathbb{Q}}(E(\xi,1)) = 0$ and the topological monoid $E(\xi,1)$ is a finite $\mathbb{Q}-$Hopf space, i.e., is rationally homotopy equivalent to a finite product of odd spheres.

As an example we consider the fibration

$$\xi_{G/H} \,:\, G/H \to B_H \to B_G.$$

where G is a compact connected Lie group and $H \subset G$ a closed connected subgroup of maximal rank. Then $R_H = H^*(B_H)$ is a domain and therefore the corresponding Δ is a nonzero divisor. As an immediate corollary of the previous theorem we get the following result.

THEOREM 6.3. *The topological monoid $E(\xi_{G/H},1)$ is a finite $\mathbb{Q}-$Hopf space, i.e., $E(\xi_{G/H},1)$ is rationally homotopy equivalent to a finite product of odd spheres.*

The consideration of the cokernel, i.e., the T^1-term of the exact sequence and its good behaviour under base change (tensor product) gives the following result.

THEOREM 6.4. *Let ξ be an oriented fibration where X is an F_0-space. If $b \in B$ and if $e_b :\ E(\xi,1) \to E(X_b,1)$ is the corresponding evaluation map, then the induced map*

$$(e_b)_*\colon \pi_{od}^{\mathbb{Q}}(E(\xi,1)) \to \pi_{od}^{\mathbb{Q}}(E(X_b,1))$$

is surjective.

PROOF. By the usual exact sequence defining $T^1_R(A)$ as the cokernel of the Jacobian homomorphism and by the invariance of the Jacobian homomorphism under base change we conclude $T^1_k(\bar{A}) \cong T^1_R(A) \otimes_R k$, where $k \cong R/\underline{m}_R$ and $\bar{A} = A/\underline{m}_R A$. By reasons of grading the projection map modulo the submodule $\underline{m}_R T^1_R(A)$ induces a surjective linear map $T^1_R(A)_- \to T^1_k(\bar{A})_-$ which is easily seen to be the map $(e_b)_*$ by the topological interpretation of the negatively graded part of the exact sequence in [**5**]. □

Recall however that $Der_R(A) = 0$ does not necessarily imply $Der_k(\bar{A}) = 0$. This is precisely the problem behind a possible proof of the Halperin conjecture.

The previous result gives also information on the Gottlieb group. Traditionally, the Gottlieb group is the image of $\pi_*(E(X))$ in $\pi_*(X)$ under the evaluation homomorphism $ev\,:\, E(X) \to X$. Consider the fibration

$$E(X,1) \to E^*(X,1) \to X$$

where $E^*(X,1) \subset E(X,1)$ is the monoid of pointed self homotopy equivalences. When $H^*(X)$ does not have negative derivations, then both monoids $E(X,1)$ and $E^*(X,1)$ are rationally homotopy equivalent to a finite product of odd spheres, see [**5**]. The long exact homotopy sequence of the above fibration is of the form

$$0 \to \pi_{od}^{\mathbb{Q}}(X) \to \pi_{od}^{\mathbb{Q}}(E(X,1)) \to \pi_{od}^{\mathbb{Q}}(E^*(X,1)) \to \pi_{od}^{\mathbb{Q}}(X) \to 0$$

As a corollary of the last theorem we obtain

THEOREM 6.5. *The image of the map induced in rational homotopy by the composition* $ev \circ e_b : E(\xi,1) \to E(X_b,1) \to X_b$ *is identical to* $\pi_{od}^{\mathbb{Q}}(X)$.

7. On Borel fibrations and equivariant Euler classes

In this paragraph we want to apply Lemma 6.1 and Thm. 6.2 (A). First we give an explicit formula for the equivariant Euler class of a smooth action of a compact connected Lie group G on a closed oriented manifold X of F_0-type. As usual write $H^*(X) = P/I_0$ with $P = \mathbb{Q}[x_1,\dots,x_n]$, $deg\, x_i = d_i \equiv 0(2)$ and I_0 generated by a regular series $\{f_1,\dots,f_n\}$ of quasihomogeneous polynomials. Then let the equivariant cohomology of the $G-$action be given by $H_G^*(X) = R_G[X_1,\dots,X_n]/I$ with $I = (F_1,\dots,F_n)$ such that $F_j - 1 \otimes f_j \in R_G^+ R_G[(X)]$, see e.g. [**4**], Thm. 1.1.

THEOREM 7.1. *If G acts smoothly on X, then the equivariant Euler class $e_G(X)$, i.e., the $n-$th Chern class of the complexified equivariant tangent bundle*

$$(E_G \times_G TX) \otimes_{\mathbb{R}} \mathbb{C} \to E_G \times_G X$$

is given by the class Δ of $det\left(\frac{\partial F_j}{\partial x_i}\right)$.

For the proof we need the following classical result.

LEMMA 7.2. *Let X be a closed connected oriented smooth manifold of F_0-type with $A_0 = H^*(X) = \mathbb{Q}[(x)]/I_0$ where $I_0 = (f_1,\dots,f_n)$ as above. Then the Euler class $e(X)$ of the complexified tangent bundle $TX \otimes_{\mathbb{R}} \mathbb{C}$ is given by Δ.*

PROOF. The Euler class $e(X)$ (or orientation class) is uniquely determined by the condition $e(X)[X] = \chi(X)$. Let $Tr : A_0 \to \mathbb{Q}$ be the trace of A_0 given by $Tr(u) = u[X]$. Then by [**8**], 2.8 Prop. and 2.9 Prop. one has $Tr\,\Delta = lg_{\mathbb{Q}} A_0 = \chi(X)$. The result follows. □

We recall that the equivariant Euler class is defined as the top Chern class of the complexification of the bundle $E_G \times_G TX \to X_G$.

Let B be a closed oriented smooth manifold and let $f\colon B \to B_G$ be a continuous map into the classifying space B_G of G. In the following let $T_G X = E_G \times_G TX, f^*T_G X = \tilde{f}^* T_G X$ and $E_f = f^* X_G$ where $\tilde{f}\colon E_f \to X_G$ is the fiber map induced by f. The pullback of the above commutative fiber diagram with respect to f gives a commutative diagram. [1]

---

[1]Typeset by P. Taylors diagram-package - the author (V. H.) takes full responsibility for the imperfect realization.

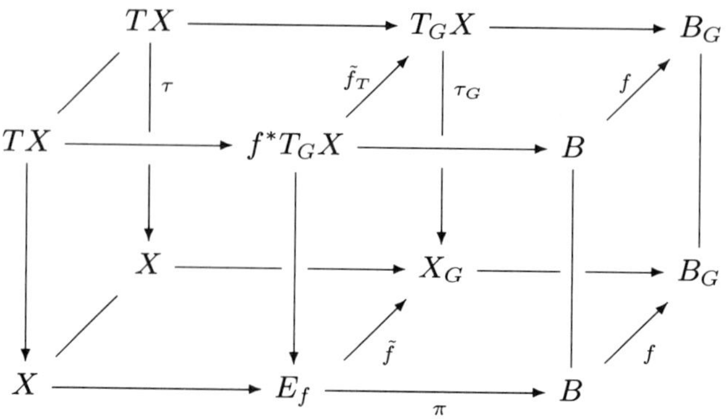

THEOREM 7.3. *Suppose there is a class $u_G(X) \in H^*_G(X)$ with the following property: For all oriented manifolds B and all orientation preserving maps $f : B \to B_G$ one has $\pi^* e(B) \cdot \tilde{f}^* u_G = e(E_f)$ where $e(B)$ and $e(E_f)$ are the corresponding orientation classes and $\pi : E_f \to B$ is the projection. Then $u_G(X) = \Delta$.*

PROOF. It is easy to see that it suffices to prove the result when $G = T^s$ is a torus. Then $B_G = \mathbb{C}P^\infty \times \cdots \times \mathbb{C}P^\infty$ and let

$$B_G^{N_1,\ldots,N_s} = \mathbb{C}P^{N_1} \times \mathbb{C}P^{N_2} \times \cdots \times \mathbb{C}P^{N_s}.$$

Then with $N \geq \sum 2N_i$ the space $B_G^{N_1,\ldots,N_s}$ can be considered a subspace of the $N-$skeleton B_G^N of B_G. Let therefore $B = B_G^{N_1,\ldots,N_s}$ and put $f = i$ where i is the inclusion of B into B_G. Then one has $H^*(B) = H^*(B_G)/\underline{a}_f$, where $\underline{a}_f \subset H^*(B_G) = \mathbb{Q}[t_1, \ldots, t_s]$ is the ideal $\underline{a}_f = (t_1^{n_1+1}, \ldots, t_s^{N_s+1})$. Since $H^*_G(X) = H^*(X_G)$ is flat over $R_G = H^*(B_G)$, the Eilenberg-Moore spectral sequence gives $H^*(E_f) = H^*_G(X) \otimes_{R_G} H^*(B)$. Write as above $H^*_G(X) = P_G/I$, where $P_G = R_G[X_1, \ldots, X_n]$, then it follows

$$H^*(E_f) = P_G/(I + \underline{a}_f P_G) = P_G/(F_1, \ldots, F_n, t_1^{N_1+1}, \ldots, t_s^{N_s+1}).$$

In particular we see that $H^*(E_f)$ is a complete intersection. By the formula of Lemma 4 we obtain

$$e(E_f) = det \begin{pmatrix} \tilde{f}^* \partial F/\partial x & 0 & 0 & 0 \\ 0 & (N_1+1)t_1^{N_1} & 0 & 0 \\ 0 & & \ddots & \\ 0 & 0 & 0 & (N_s+1)t_s^{N_s} \end{pmatrix}.$$

Thus it follows that

$$e(E_f) = \prod_{i=1}^{s} (N_i + 1) t_i^{N_i} \cdot \tilde{f}^* \Delta$$

and therefore

$$e(E_f) = \pi^* e\left(P^{N_1}\mathbb{C} \times \cdots \times P^{N_s}\mathbb{C}\right) \cdot \tilde{f}^* \Delta.$$

So we can write

$$e(E_f) = \pi^* e(B) \cdot \tilde{f}^* \Delta.$$

Therefore, if $u_G(X) \in H^*_G(X)$ is a class satisfying the hypothesis of Thm. 7.3 then the following equation holds:

$$\pi^* e(B) \cdot \tilde{f}^* u_G(X) = \pi^* e(B) \cdot \tilde{f}^* \Delta.$$

Now by the flatness of $H^*(E_f)$ on $H^*(B)$ one has

$$\tilde{f}^* u_G(X) = \tilde{f}^* \Delta.$$

in the quotient $H^*(E_f) = H_G^*(X)/(t_1^{N_1+1}, \dots, t_s^{N_s+1})H_G^*(X)$ for every $N_1, \dots, N_s \in \mathbb{N}$ greater than a fixed number N. It follows

$$u_G(X) - \Delta \in (t_1, \dots, t_s)^i H_G^*(X)$$

for all $i \geq N+1$ with $N = max\, N_i$. By the Krull intersection theorem one has

$$\bigcap_i (t_1, \dots, t_s)^i = (0)$$

and consequently

$$u_G(X) = \Delta.$$

□

Now we are able to prove Thm. 7.1.

PROOF. It suffices to show that $e_G(X)$ satisfies the property in Thm. 7.3. Let $E_f = f^*(E_G \times_G X) = (f^* E_G) \times_G X$. Now $\tilde{f}^* T_G X$ is the tangent bundle of E_f along the fibers. Consequently the tangent bundle TE_f splits as $TE_f = \tilde{f}^* T_G X \oplus \pi^* TB$. So we get $e(E_f) = e(\tilde{f}^* T_G X) \cdot e(\pi^* TB) = \tilde{f}^* e_G(X) \cdot \pi^* e(B)$. □

Suppose X is a closed oriented smooth manifold of F_0−type. Let a torus G act smoothly on X. Consider the Borel fibration

$$\beta_G[X] \, : \, X \to E_G \times_G X \to B_G.$$

THEOREM 7.4. *Let G act on X with only isolated fixed points. Then the topological monoid $E(\beta_G[X], 1)$ is rationally homotopy equivalent to a finite product of odd spheres.*

PROOF. By the above results it suffices to show that $e_G(X)$ is a nonzero divisor of $H_G^*(X)$. Let DTX be the disc bundle of TX and let STX be the corresponding sphere bundle. Since $H_G^*(X) \cong H_G^*(DTX)$ we get as a part of the exact sequence of the pair $(DTX, STX)_G = E_G \times_G (DTX, STX)$

$$H_G^*(DTX, STX) \to H_G^*(X) \to H_G^{ev}(STX) \to 0.$$

Now by the Thom isomorphism in equivariant cohomology $H_G^*(DTX, STX)$ can be identified with $H_G^*(X)$ shifted by $d = dim\, X$ whereas its image in $H_G^*(X)$ can be identified with the ideal in $H_G^*(X)$ generated by the equivariant Euler class. We get therefore the exact sequence

$$0 \to e_G(X) H_G^*(X) \to H_G^*(X) \to H_G^{ev}(STX) \to 0.$$

Now $H_G^*(X)$ is a flat graded R_G−module. This means that it is free over R_G. If G acts on X with isolated fixed points, then the fixed set $(STX)^G$ is empty. By the localization theorem in equivariant cohomology $H_G^*(STX)$ must be a torsion module on R_G. But this shows that the ideal $e_G(X)H_G^*(X)$ has the same R_G−rank as $H_G^*(X)$. It follows that $e_G(X)H_G^*(X)$ is free on $H_G^*(X)$. This means that $e_G(X)$ is a nonzero divisor of $H_G^*(X)$. The rest of the proof follows from Thms. 3.2, 6.2 (A). □

References

[1] Dupont, N.: Problems and conjectures in rational homotopy theory: a survey, Expo. Math. 12 (1994), 323-352

[2] Friedlander, J. B., Halperin, S.: An arithmetic characterization of the rational homomotpy groups of certain spaces, Inv. math. 53, 117-133, 1979

[3] Halperin, S.: Finiteness in the minimal models of Sullivan, T.A.M.S., Vol. 230, 173-199, 1977

[4] Hauschild, V.: The Euler characteristic as an obstruction to compact Lie group actions, T. A. M. S., Vol. 298, Number 2, Dec. 1986

[5] Hauschild, V.: Deformations and the Rational Homotopy of the Monoid of Fiber Homotopy Equivalences, Illinois Journal of Mathematics, Vol. 37, Number 4, 537 - 560, Winter 1993

[6] Hauschild, V.: Effective Actions and Weyl Degrees of Compact Lie Groups, Mathematische Nachrichten 161 (1993) 171-183

[7] Hauschild, V.: Rational Homotopy of Circle Actions, Pacific Journal of Math., Vol. 191, No. 2, 1999

[8] Kreuzer, M., Kunz, E.: Traces in strict Frobenius algebras and strict complete intersections, Journal für reine und angewandte Mathematik, 381, 181-204, 1987

[9] Markl, M.: Towards one conjecture on collapsing of the Serre spectral sequence, Suppl. ai Rend. Circ. Matem. Palermo, Ser. II, II.22 (1989), 152-159

[10] Matsumura, H.: Commutative Algebra, Second edition, The Benjamin/Cummings Publishing Company, Inc., Reading, Massachusetts, 1980

[11] Meier, W.: Rational universal fibrations and flag manifolds, Math. Ann. 258, 329-340, 1982

[12] Meier, W.: Kähler manifolds and homogeneous spaces, Math. Z. 183, 473-481, 1983

[13] Scheja, G., Storch, U.: Über Spurfunktionen bei vollständigen Durchschnitten, Journal für reine und angewandte Mathematik, 278/279, 174-190, 1975

[14] Shiga, H., Tezuka, M.: Rational fibrations, homogeneous spaces with positive Euler characteristics and Jacobians, Ann. Inst. Fourier, Grenoble, 37, 1, 81-106, 1987

[15] Thomas, J.-C.: Homotopie Rationelle des Fibrés de Serre, Thèse,Université des Sciences et Techniques de Lille I, 1980

Dipartimento di Matematica, UNICAL, I-87036 Rende, Italy,
E-mail address: `hausch@unical.it`

Contemporary Mathematics
Volume **274**, 2001

CLASSIFYING SPACES AND A SUBGROUP OF THE EXCEPTIONAL LIE GROUP G_2

BY
KENSHI ISHIGURO

ABSTRACT. We consider a problem on the conditions of a compact Lie group that its loop space of the p–completed classifying space be a p–compact group, as well as some related problems. A previously obtained necessary condition is shown to be not sufficient. Our counterexample is given by a quotient group of a subgroup of the exceptional Lie group G_2 at $p = 3$. The K-theory of the space is isomorphic to $K(BG_2; \mathbb{Z}_3^\wedge)$, though its loop space is not a 3-compact group.

The notion of a p–compact group X, [4], is a good generalization of a compact Lie group G at the prime p. The structure of the classifying space BX is similar to that of $(BG)_p^\wedge$. Here we say that a space is a *p–compact classifying space* if its loop space is a p–compact group. It is well-known that $(BG)_p^\wedge$ is p–compact if $\pi_0(G)$ is a p–group. In [14] the author has tried to find the conditions on G that $(BG)_p^\wedge$ be a p–compact classifying space, and obtained some results mostly for a special case. Theorem 2 of [14] implies that the loop space $\Omega(BG)_p^\wedge$ is a p–compact toral group if and only if the compact Lie group G is p–nilpotent, [8]. Thus the connected component of G is necessarily a torus. For the general case, necessary conditions are stated in [14, Proposition 3.1]. Our work in this paper has been motivated by showing that the converse is false, even though the rational cohomology of $(BG)_p^\wedge$ is expressed as an invariant ring of pseudoreflections. We will use a subgroup of the exceptional Lie group G_2 to find a counterexample and to see simplicity of the p–completed classifying space of a non-connected compact Lie group.

Suppose that G is simple and simply–connected, and that the order of its Weyl group is divisible by a prime p. According to the results of [11] and [12], a map $f : BG \longrightarrow BX$ is essential if and only if $Ker\ f$ is included in the center of G, except that G is the exceptional Lie group G_2 at $p = 3$. The Lie group G_2 contains $SU(3)$. Assume $H = SU(3) \rtimes \mathbb{Z}/2$ is the

2000 *Mathematics Subject Classification.* Primary 55R35; Secondary 55P15, 55P60.

subgroup of G_2 discussed in [11, p220, proof of Theorem 2]. The center of $SU(3)$, isomorphic to $\mathbb{Z}/3$, is a normal subgroup of H. Let Γ_2 denote the quotient group $H/(\mathbb{Z}/3)$. Then we see $(BH)^\wedge_3 \simeq (BG_2)^\wedge_3$ and the quotient map induces a map $f : (BH)^\wedge_3 \longrightarrow (B\Gamma_2)^\wedge_3$ with $f|_{B\mathbb{Z}/3} = 0$. Note that the center of G_2 is trivial. According to [14, Proposition 3.1], if G is a compact Lie group and $(BG)^\wedge_p$ is p–compact, then $\pi_0 G$ is p–nilpotent and $\pi_1((BG)^\wedge_p)$ is isomorphic to a p–Sylow subgroup of $\pi_0 G$. These two conditions are satisfied if p does not divide the order of $\pi_0(G)$. Consequently the 3–completed classifying space $(B\Gamma_2)^\wedge_3$, rationally equivalent to $(BG_2)^\wedge_3$, satisfies the necessary conditions.

Theorem 1. *Let $G = \Gamma_2$, the quotient group of a subgroup $SU(3) \rtimes \mathbb{Z}/2$ of the exceptional Lie group G_2. For $p = 3$, the following hold:*

1. *$\pi_0 G$ is p–nilpotent and $\pi_1((BG)^\wedge_p)$ is isomorphic to a p–Sylow subgroup of $\pi_0 G$.*
2. *$(BG)^\wedge_p$ is rationally equivalent to $(BG_2)^\wedge_p$.*
3. *$(BG)^\wedge_p$ is not a p-compact classifying space.*

Next we discuss a minimality of this example. As usual, for a compact Lie group G, the connected component with the identity is denoted by G_0, and $\pi_0 G = G/G_0$. When $rank(G_0) = 1$, the following result implies the converse of [14, Proposition 3.1] is true.

Theorem 2. *Suppose G is a compact Lie group with $rank(G_0) = 1$. If the group $\pi_0 G$ is p–nilpotent, then $(BG)^\wedge_p$ is a p–compact classifying space.*

We recall a few more examples. Suppose a compact connected Lie group G is simple, and NT denotes the normalizer of a maximal torus T of G. In [14] it is determined exactly when $(BNT)^\wedge_p$ is p–compact. The Weyl group being p–nilpotent is sufficient except the following cases. At $p = 2$, it is shown that $(BNT)^\wedge_2$ is not 2–compact if the Weyl group is one of the following four cases: $W(A_2)$, $W(B_3) = W(C_3)$, $W(D_3)$, $W(G_2)$, though $\pi_0(NT)$ is 2–nilpotent in each case.

We turn back to the subgroup $H = SU(3) \rtimes \mathbb{Z}/2$ of G_2 and its quotient group $\Gamma_2 = H/(\mathbb{Z}/3)$. Suppose T_1 is a maximal torus of $H_0 = SU(3)$ and T_2 is the image of T_1 under the quotient homomorphism $H \longrightarrow \Gamma_2$. Considering the normalizers, we obtain the following commutative diagram of groups:

$$\begin{array}{ccccc}
\mathbb{Z}/3 & = & \mathbb{Z}/3 & \longrightarrow & * \\
\downarrow & & \downarrow & & \downarrow \\
T_1 & \longrightarrow & N_H(T_1) & \longrightarrow & W_1 \\
\downarrow & & \downarrow & & \| \\
T_2 & \longrightarrow & N_{\Gamma_2}(T_2) & \longrightarrow & W_2
\end{array}$$

Here $W_1 = N_H(T_1)/T_1$ and $W_2 = N_{\Gamma_2}(T_2)/T_2$, and both groups are isomorphic to the Weyl group of G_2. The map $T_1 \longrightarrow T_2$ is the admissible map, [1], for $SU(3) \longrightarrow PU(3)$. We notice that the W_2–action on T_2 is the dual representation of the W_1–action on T_1. The main result of [13] states that, in general, if the dual representation of $W(SU(n))$ is denoted by $W(SU(n))^*$, then $K(BPU(n);\mathbb{Z}_p^\wedge) = K(BT^{n-1};\mathbb{Z}_p^\wedge)^{W(SU(n))^*}$ as λ–rings for any p, and the integral representation of $W(SU(n))$ is not isomorphic to $W(SU(n))^*$.

Since the exceptional Lie group G_2 is 3–torsion free, the mod 3 cohomology $H^*(BG_2,\mathbb{F}_3)$ is isomorphic to the ring of invariants $H^*(BT^2;\mathbb{F}_3)^{W(G_2)}$. The Weyl group $W(G_2)$ is the dihedral group of order 12 presented as $D_{12} =< r, s \mid r^6 = s^2 = 1, srs = r^5 >$. The matrix (integral) representation can be taken as follows:

$$r = \begin{pmatrix} 1 & -1 \\ 1 & 0 \end{pmatrix}, \quad s = \begin{pmatrix} 1 & -1 \\ 0 & -1 \end{pmatrix}$$

Consequently, if we write $H^*(BT^2;\mathbb{F}_3) = \mathbb{F}_3[t_1, t_2]$ with $deg(t_i) = 2$, then the ring of invariants is the following polynomial ring:

$$H^*(BT^2;\mathbb{F}_3)^{W(G_2)} = \mathbb{F}_3[x_4, x_{12}]$$

where $x_4 = (t_1 - t_2)^2$ and $x_{12} = t_1^2 t_2^2 (t_1 + t_2)^2$.

The group G_2 contains $SU(3)$, and $W(SU(3))$ can be a subgroup of $W(G_2)$ such that $W(SU(3)) = \{1, r^2, r^4, s, sr^2, sr^4\}$. Consider the following invariant rings:

$$H^*(BT^2;\mathbb{F}_3)^{W(SU(3))} = \mathbb{F}_3[y_4, y_6]$$

where $y_4 = (t_1 - t_2)^2$ and $y_6 = t_1 t_2 (t_1 + t_2)$, and for the dual representation we have

$$H^*(BT^2;\mathbb{F}_3)^{W(SU(3))^*} = \mathbb{F}_3[z_2, z_{12}]$$

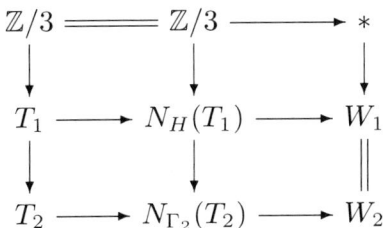

where $z_2 = t_1 + t_2$ and $z_{12} = t_1^2 t_2^2 (t_1 - t_2)^2$. Recall [20, §3] and [21, Ch10] that an unstable algebra is said to be *realizable* if it is isomorphic to the mod p cohomology of a space over the Steenrod algebra. Obviously the unstable algebra $H^*(BT^2;\mathbb{F}_3)^{W(SU(3))}$ is realizable, however, $H^*(BT^2;\mathbb{F}_3)^{W(SU(3))^*}$ is not realizable, [13]. As the following result indicates, the case of G_2 is different. In fact, we see $H^*(BT^2;\mathbb{F}_3)^{W(G_2)} = H^*(BT^2;\mathbb{F}_3)^{W(G_2)^*}$ as unstable algebras.

Theorem 3. *Let Γ_2 be the compact Lie group as in Theorem 1. Then the following hold:*

(1) *The 3-adic K-theory $K(B\Gamma_2;\mathbb{Z}_3^\wedge)$ is isomorphic to $K(BG_2;\mathbb{Z}_3^\wedge)$ as a λ-ring.*
(2) *Let Γ be a compact Lie group such that $\Gamma_0 = PU(3)$ and the order of $\pi_0(\Gamma)$ is not divisible by 3. Then any map from $(B\Gamma)_3^\wedge$ to $(BG_2)_3^\wedge$ is null homotopic. In particular $[(B\Gamma_2)_3^\wedge, (BG_2)_3^\wedge] = 0$.*

We recall that if a connected compact Lie group G is simple, the following results hold:

(1) For any prime p, the space $(BG)_p^\wedge$ has no nontrivial retracts. ([9])
(2) Assume $|W(G)| \equiv 0 \bmod p$. If a self-map $(BG)_p^\wedge \longrightarrow (BG)_p^\wedge$ is not null homotopic, it is a homotopy equivalence. ([18])
(3) Assume $|W(G)| \equiv 0 \bmod p$. For a compact Lie group K, if a map $f : (BG)_p^\wedge \longrightarrow (BK)_p^\wedge$ is trivial in mod p cohomology, then f is null homotopic. ([11])

Replacing G by Γ_2 at $p = 3$, we will see that (3) still holds. On the other hand it is not known if (1) and (2) hold, though on the level of K-theory they do.

The author would like to thank Clarence Wilkerson for his comments.

1. The p–completion of BG and p–compact groups.

We will prove Theorem 1 and Theorem 2 in this section. Our proof of Theorem 1 uses a result of [15]. This result implies that there is no fibration of the type $B\mathbb{Z}/3 \longrightarrow (BG_2)_3^\wedge \longrightarrow Y$, despite the existence of a map $f : (BG_2)_3^\wedge \longrightarrow Y$ with $f|_{B\mathbb{Z}/3} = 0$. The proof of Theorem 2 uses a result of [7], which determines topology and algebra of the non-modular unstable polynomial algebras at any odd prime.

Lemma 1. *Suppose G is a compact Lie group such that a prime p does not divide the order of $\pi_0(G)$. Then $(BG)_p^\wedge$ is 1-connected.*

Proof. Recall that there exists a Postnikov fibration $F \longrightarrow (BG)^\wedge_p \xrightarrow{q} B\pi$ where $\pi = \pi_1((BG)^\wedge_p)$ and $q_* : \pi_1((BG)^\wedge_p) \longrightarrow \pi_1(B\pi)$ is isomorphic. Consider the following diagram:

$$\begin{array}{ccc}
BG_0 & & F \\
\downarrow & & \downarrow{\scriptstyle i} \\
BG & \longrightarrow & (BG)^\wedge_p \\
\downarrow & & \downarrow{\scriptstyle q} \\
B\pi_0(G) & \longrightarrow & B\pi
\end{array}$$

The map $B\pi_0(G) \longrightarrow B\pi$ is induced from $BG \longrightarrow (BG)^\wedge_p$, since the composite $BG_0 \longrightarrow B\pi$ is null homotopic. We note that p does not divide $|\pi_0(G)|$, and that π is a finite p–group, [3]. Consequently the composite $BG \longrightarrow B\pi$ is null homotopic. So there is a map $r : BG \longrightarrow F$ such that the p–completion $(i \cdot r)^\wedge_p$ is homotopy equivalent to the identity map of $(BG)^\wedge_p$, since BG is $\mathbb{F}_p$–good, [3, Ch VII Proposition 5.1]. Thus $\pi_1(F) = 0$ implies $\pi_1((BG)^\wedge_p) = 0$. This completes the proof. $\square$

Lemma 2. *Let Γ_2 be the compact Lie group as in Theorem 1. Then we have $\pi_1((B\Gamma_2)^\wedge_3) = 0$ and $\pi_2((B\Gamma_2)^\wedge_3) = \mathbb{Z}/3$.*

Proof. Without 3–completion, we have $\pi_1(B\Gamma_2) = \mathbb{Z}/2$ and $\pi_2(B\Gamma_2) = \mathbb{Z}/3$ from the exact sequences of homotopy groups for the following commutative diagram of groups:

$$\begin{array}{ccccc}
\mathbb{Z}/3 & = & \mathbb{Z}/3 & \longrightarrow & * \\
\downarrow & & \downarrow & & \downarrow \\
SU(3) & \longrightarrow & H & \longrightarrow & \mathbb{Z}/2 \\
\downarrow & & \downarrow & & \| \\
PU(3) & \longrightarrow & \Gamma_2 & \longrightarrow & \mathbb{Z}/2
\end{array}$$

Thus Lemma 1 shows $\pi_1((B\Gamma_2)^\wedge_3) = 0$.

Next consider the Postnikov system $\{X_n\}$ of the space $B\Gamma_2$ so that

$$\pi_i(X_n) = \begin{cases} 0 & \text{if } i > n \\ \pi_i(B\Gamma_2) & \text{if } i \leq n \end{cases}$$

and we have the fibrations $K(\pi_n(B\Gamma_2), n) \longrightarrow X_n \longrightarrow X_{n-1}$. We note that $X_1 = K(\mathbb{Z}/2, 1)$ and $X_2 = K(\mathbb{Z}/3, 2) \times X_1$. For $n \geq 2$, we see $\pi_2((X_n)^\wedge_3) = \mathbb{Z}/3$. Tower lemma [3] implies, therefore, that $\pi_2((B\Gamma_2)^\wedge_3) = \varprojlim_n \pi_2((X_n)^\wedge_3) = \mathbb{Z}/3$. $\square$

Proof of Theorem 1. From the definition of Γ_2 together with Lemma 2, it is easy to see that the first two conditions are satisfied. Consequently it suffices to show that the space $(B\Gamma_2)^\wedge_3$ is not 3-compact.

If $X = \Omega(B\Gamma_2)^\wedge_3$ is a 3-compact group, Lemma 2 implies that there is a Postnikov fibration $B\mathbb{Z}/3 \longrightarrow B\widetilde{X} \longrightarrow BX$ where $\widetilde{X}$ is a 1–connected 3-compact group, [19]. Notice that the projective group $PSU(3)$ is a subgroup of Γ_2, and that the map $(BH)^\wedge_3 \longrightarrow (BG_2)^\wedge_3$ is homotopy equivalent.

Consider the following diagram:

$$\begin{array}{ccccc}
B\mathbb{Z}/3 & \longrightarrow & B\widetilde{X} & \longrightarrow & BX \\
 & & \uparrow & & \| \\
(BG_2)^\wedge_3 & \xleftarrow{\ \sim\ } & (BH)^\wedge_3 & \longrightarrow & (B\Gamma_2)^\wedge_3 \\
 & & \uparrow & & \uparrow \\
 & & BSU(3) & \longrightarrow & BPSU(3)
\end{array}$$

where the map $(BH)^\wedge_3 \longrightarrow B\widetilde{X}$ is the lift of $(BH)^\wedge_3 \longrightarrow BX$, since $(BH)^\wedge_3$ is 2–connected.

First, suppose that the 3–completion of the composite $BSU(3) \longrightarrow B\widetilde{X}$ is a monomorphism of 3-compact groups. Since the composition of monomorphisms is also a monomorphism, the map $(BG_2)^\wedge_3 \longrightarrow B\widetilde{X}$ is a monomorphism. According to a result of Møller, [18, Lemma 2.5], this map is an isomorphism. This means that there is a fibration $B\mathbb{Z}/3 \longrightarrow (BG_2)^\wedge_3 \longrightarrow BX$. The result [15, Theorem 4] shows that this is a contradiction, since the center of the exceptional Lie group G_2 is trivial.

Next, suppose that the 3–completion of the composite $BSU(3) \longrightarrow B\widetilde{X}$ is not a monomorphism of 3–compact groups. If we consider the upper right of the previous diagram on the level of the maximal tori of 3–compact groups, we obtain the following diagram:

$$\begin{array}{ccccc}
B\mathbb{Z}/3 & \longrightarrow & (BT^2)^\wedge_3 & \xrightarrow{B\phi} & (BT^2)^\wedge_3 \\
\uparrow & & \uparrow{\scriptstyle B\psi} & & \| \\
B\gamma & \longrightarrow & (BT^2)^\wedge_3 & \longrightarrow & (BT^2)^\wedge_3 \\
\uparrow & & \uparrow & & \uparrow \\
B\alpha & = & B\alpha & \longrightarrow & *
\end{array}$$

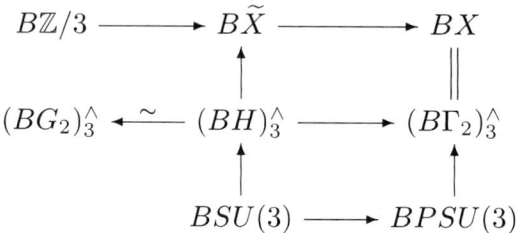

where ϕ and ψ are the admissible map for $B\widetilde{X} \longrightarrow BX$ and $(BH)^\wedge_3 \longrightarrow B\widetilde{X}$ respectively, and α and γ are suitable finite 3–groups. Since $B\psi$ is not a monomorphism, the group α is non–trivial, [4, Proposition 9.11]. Consequently the kernel of the map $BSU(3) \longrightarrow BX$ contains $\mathbb{Z}/3$ as a proper subgroup. This is a contradiction, since a map $f : BSU(3) \longrightarrow BX$ is essential if and only if $Ker\ f$ is included in the center of $SU(3)$. Consequently $\Omega(B\Gamma_2)^\wedge_3$ can not be a 3–compact group. □

Proof of Theorem 2. First assume $p \not| \ |\pi_0 G|$. Then $H^*(BG;\mathbb{F}_p)$ is isomorphic to $H^*(BG_0;\mathbb{F}_p)^{\pi_0 G}$. Since $rank(G_0) = 1$, the connected compact Lie group is one of S^1, S^3, $SO(3)$. Recall that, if $H^*(BS^1;\mathbb{F}_p) = \mathbb{F}_p[t]$ with $deg(t) = 2$ and $H^*(B(\mathbb{Z}/2)^2;\mathbb{F}_2) = \mathbb{F}_2[x,y]$ with $deg(x) = deg(y) = 1$, then for odd p, we see $H^*(BS^3;\mathbb{F}_p) = H^*(BSO(3);\mathbb{F}_p) = \mathbb{F}_p[t]^{\mathbb{Z}/2} = \mathbb{F}_p[t^2]$, and for $p = 2$, $H^*(BS^3;\mathbb{F}_2) = \mathbb{F}_2[t^2]$ and $H^*(BSO(3);\mathbb{F}_2) = \mathbb{F}_2[x,y]^{GL(2,\mathbb{F}_2)} = \mathbb{F}_2[c_2,c_3]$. The mod p cohomology of each of classifying spaces of these Lie groups is a polynomial algebra, and $H^*(BG;\mathbb{F}_p)$ is also a polynomial algebra. If p is odd, a result of Dwyer–Miller–Wilkerson [7, Theorem 1.2] shows $(BG)^\wedge_p$ is a p–compact classifying space. For the case $p = 2$, the unstable algebra $H^*(BG;\mathbb{F}_2)$ must be one of the following:

$$H^*(BS^1;\mathbb{F}_2), H^*(BS^3;\mathbb{F}_2), H^*(BSO(3);\mathbb{F}_2)$$

It is known [6] that, in these cases, the mod 2 cohomology determines the homotopy type of each of the classifying spaces. Thus $(BG)^\wedge_2$ is homotopy equivalent to one of $(BS^1)^\wedge_2$, $(BS^3)^\wedge_2$, $(BSO(3))^\wedge_2$.

Next consider the general case. Since the group $\pi_0 G$ is p–nilpotent, if $\gamma = \pi_0 G$, then $\gamma = \nu \rtimes \gamma_p$ where ν is the subgroup generated by all elements of order prime to p, and where γ_p is the p–Sylow subgroup. We obtain the following commutative diagram of groups:

$$\begin{array}{ccccc}
G_0 & \longrightarrow & H & \longrightarrow & \nu \\
\| & & \downarrow & & \downarrow \\
G_0 & \longrightarrow & G & \longrightarrow & \pi_0 G \\
\downarrow & & \downarrow & & \downarrow \\
* & \longrightarrow & \gamma_p & = & \gamma_p
\end{array}$$

where H is the subgroup of G induced by the inclusion $\nu \longrightarrow \pi_0 G$. Since p does not divides $|\nu|$, the previous argument implies $(BH)^\wedge_p$ is a p–compact classifying space. Notice that the fibration $BH \longrightarrow BG \longrightarrow B\gamma_p$ is preserved by the p–completion, since γ_p is a finite p–group. Consequently $(BG)^\wedge_p$ is a p–compact classifying space. □

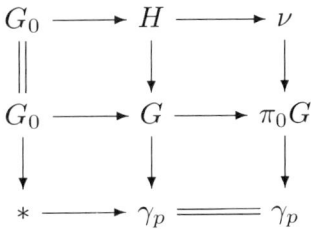

2. Invariant rings and some properties of $B\Gamma_2$ and BG_2 at $p = 3$.

We will prove Theorem 3 in this section. Suppose G is a compact connected Lie group. The Weyl group $W(G)$ acts on the maximal torus T^n, and the integral representation $W(G) \longrightarrow GL(n, \mathbb{Z})$ is obtained. If such Lie groups are locally isomorphic, the representations of the Weyl groups are equivalent over $\mathbb{Q}$. They need not be, however, equivalent as $\mathbb{Z}$–representations. For instance, as mentioned before [13], the integral representation of $W(PU(n))$ is not equivalent to $W(SU(n))$, since the $\mathbb{Q}$–equivalence is induced by admissible map as follows:

$$\phi = \begin{pmatrix} 2 & 1 & \dots & 1 \\ 1 & \ddots & \ddots & \vdots \\ \vdots & \ddots & \ddots & 1 \\ 1 & \dots & 1 & 2 \end{pmatrix}$$

with $\det \phi \neq \pm 1$. Turning to the rings of invariant, we note that an isomorphism of cohomology rings need not imply the equivalence of the two representations. Such an example will be given by $W(G_2)$ and $W(G_2)^*$.

Lemma 3. *There is $\psi \in GL(2, \mathbb{Z})$ such that $\psi W(G_2)\psi^{-1} = W(G_2)^*$.*

Proof. As mentioned before, the representation of the Weyl group $W(G_2)$ is generated by $r = \begin{pmatrix} 1 & -1 \\ 1 & 0 \end{pmatrix}$ and $s = \begin{pmatrix} 1 & -1 \\ 0 & -1 \end{pmatrix}$. Let $\psi = \begin{pmatrix} -1 & 1 \\ 1 & 0 \end{pmatrix}$. A calculation shows

$$\psi r \psi^{-1} = {}^t r \quad \text{and} \quad \psi s \psi^{-1} = {}^t(sr)$$

where ${}^t r = \begin{pmatrix} 1 & 1 \\ -1 & 0 \end{pmatrix}$ and ${}^t(sr) = \begin{pmatrix} 0 & -1 \\ -1 & 0 \end{pmatrix}$. This completes the proof. □

Remark. From Lemma 3 we see that, over the Steenrod algebra,

$$H^*(BT^2; \mathbb{F}_3)^{W(G_2)} \cong H^*(BT^2; \mathbb{F}_3)^{W(G_2)^*}$$

We notice, however, that the mod 3 reductions of the integral representations of $W(G_2)$ and $W(G_2)^*$ are not equivalent. In fact, the composition series for the $W(G_2)$–module V is

$$0 \longrightarrow \mathbb{F}_3 < t_1 - t_2 > \longrightarrow V \longrightarrow \mathbb{F}_3 < [t_1 + t_2] > \longrightarrow 0\ ,$$

while the composition series for the $W(G_2)^*$–module U is

$$0 \longrightarrow \mathbb{F}_3 < t_1 + t_2 > \longrightarrow U \longrightarrow \mathbb{F}_3 < [t_1 - t_2] > \longrightarrow 0\ .$$

The actions on $\mathbb{F}_3 < t_1 + t_2 >$ and $\mathbb{F}_3 < [t_1 + t_2] >$ are trivial, and those on $\mathbb{F}_3 < t_1 - t_2 >$ and $\mathbb{F}_3 < [t_1 - t_2] >$ are non-trivial.

As in [13], the $\mathbb{Q}$–representations of $W(G_2)$ and $W(G_2)^*$ are equivalent. The equivalence is given by the admissible map $\begin{pmatrix} 2 & 1 \\ 1 & 2 \end{pmatrix}$. Consequently, under the projection $\mathbb{Z} \longrightarrow \mathbb{F}_p$ for $p \neq 3$, the mod p reductions of the integral representations are equivalent. For $p = 2$, the kernel of each of the reduction is $\mathbb{Z}/2$, and the image is $GL(2, \mathbb{F}_2)$. Otherwise the kernel is trivial.

Proof of Theorem 3. (1) Recall [2] that if G is a compact connected Lie group and T is its maximal torus, we have $K(BG; \mathbb{Z}_p^\wedge) \cong K(BT; \mathbb{Z}_p^\wedge)^{W(G)}$ for any prime p. Hence $K(BG_2; \mathbb{Z}_3^\wedge) \cong K(BT^2; \mathbb{Z}_3^\wedge)^{W(G_2)}$. Next, for the group extension $PU(3) \longrightarrow \Gamma_2 \longrightarrow \mathbb{Z}/2$, applying the spectral sequence $H^*(B\mathbb{Z}/2; K(BPU(3); \mathbb{Z}_3^\wedge)) \Rightarrow K(B\Gamma_2; \mathbb{Z}_3^\wedge)$ we see the following:

$$K(B\Gamma_2; \mathbb{Z}_3^\wedge) \cong K(BPU(3); \mathbb{Z}_3^\wedge)^{\mathbb{Z}/2} \cong K(BT^2; \mathbb{Z}_3^\wedge)^{W(G_2)^*}$$

Lemma 3 shows $K(BT^2; \mathbb{Z}_3^\wedge)^{W(G_2)} \cong K(BT^2; \mathbb{Z}_3^\wedge)^{W(G_2)^*}$. In fact, the map ψ is the admissible map between the two λ–rings, [23].

(2) A map $\alpha : (B\Gamma)_3^\wedge \longrightarrow (BG_2)_3^\wedge$ is null homotopic if and only if so is the restriction of α on $BPU(3)$, since $PU(3)$ contains all 3–toral subgroups of Γ. Let $f = \alpha|_{BPU(3)}$. It suffices to show that the induced homomorphism of mod 3 cohomology

$$f^* : H^*(BG_2; \mathbb{F}_3) \longrightarrow H^*(BPU(3); \mathbb{F}_3)$$

is trivial. We note that

$$H^*(BG_2; \mathbb{F}_3) = \mathbb{F}_3[x_4, x_{12}]$$

and that

$$H^*(BPU(3); \mathbb{F}_3) = \mathbb{F}_3[y_2, y_8, y_{12}] \otimes \Lambda(y_3, y_7)/J$$

where J is the ideal generated by $y_2 y_3$, $y_2 y_7$, and $y_3 y_7 + y_2 y_8$, [16, Theorem 3.10] . For the map f, there is $\phi : BT^2 \longrightarrow (BT^2)_3^\wedge$ which makes the following diagram commutative:

$$\begin{array}{ccc} BPU(3) & \xrightarrow{\ f\ } & (BG_2)_3^\wedge \\ \uparrow & & \uparrow{\scriptstyle Bi} \\ BT^2 & \xrightarrow{\ \phi\ } & (BT^2)_3^\wedge \end{array}$$

If f was not null homotopic, then $Ker\ f$ should be trivial and hence the map ϕ would be mod 3 homotopy equivalence. Consequently we obtain the following commutative diagram:

$$\begin{array}{ccc} H^*(BG_2;\mathbb{F}_3) & \xrightarrow{f^*} & H^*(BPU(3);\mathbb{F}_3) \\ {\scriptstyle (Bi)^*}\downarrow & & \downarrow \\ H^*(BT^2;\mathbb{F}_3) & \xrightarrow{\cong} & H^*(BT^2;\mathbb{F}_3) \end{array}$$

Since $(Bi)^*$ is a monomorphism, so is f^*. Thus we have $f^*(x_4) = \pm y_2^2$ and $f^*(x_{12}) = \pm y_{12} + a y_2^6$ for some $a \in \mathbb{F}_3$. Thus the image of f^* would be included in the algebra generated by y_2 and y_{12}. According to [16, Theorem 3.20] , however, we see that $\mathcal{P}^1(y_{12}) = y_8^2 + y_{12}y_2^2$. This means that $Im\ f^*$ would not be closed under the action of a Steenrod operation. This contradiction implies the desired result. □

References

1. J.F. ADAMS and Z.MAHMUD, *Maps between classifying spaces*, Inventiones Math. **35** (1976), 1–41.
2. M. F. ATIYAH and F. HIRZEBRUCH, *Vector bundles and homogeneous spaces*, Proc. Sympos. Pure Math., AMS **3** (1961), 7–38.
3. A. BOUSFIELD and D. KAN, *Homotopy limits, completions and localisations*, LNM **304** (1972).
4. W.G. DWYER and C.W. WILKERSON, *Homotopy fixed-point methods for Lie groups and finite loop spaces*, Ann. of Math. **139 (2)** (1994), 395–442.
5. W.G. DWYER and C.W. WILKERSON, *Centers and Coxeter elements*, Preprint.
6. W.G. DWYER, H. R. MILLER and C.W. WILKERSON, *The homotopic uniqueness of BS^3*, Proc. of 1986 Barcelona conference, LNM **1298** (1987), 90–105.
7. W.G. DWYER, H. R. MILLER and C.W. WILKERSON, *Homotopical uniqueness cf classifying spaces*, Topology **31 (1)** (1992), 29–45.
8. H.W. HENN, *Cohomological p–nilpotence criteria for compact Lie groups*, Théorie de l'homotopie, Astérisque 191, Soc. Math. France (1990), 211–220.
9. K. ISHIGURO, *Classifying spaces and p-local irreducibility*, Journal of Pure and Applied Algebra **49** (1987), 253–258.
10. K. ISHIGURO, *Rigidity of p-completed classifying spaces of alternating groups and classical groups over a finite field*, Transaction of AMS **329 (2)** (1992), 697–713.
11. K. ISHIGURO, *Classifying spaces of compact simple Lie groups and p–tori*, Proc. of 1990 Barcelona conference, LNM **1509** (1992), 210–226.
12. K. ISHIGURO, *Retracts of classifying spaces*, Adams memorial symposium on algebraic topology volume 1, London Math. Soc. Lecture Notes 175, Cambridge Univ. Press (1992), 271–280.
13. K. ISHIGURO, *Projective unitary groups and K-theory of classifying spaces*, Fukuoka Univ. Sci. Rep. **28 (1)** (1998), 1–6.
14. K. ISHIGURO, *Toral groups and classifying spaces of p–compact groups*, Preprint.

15. K. ISHIGURO and D. NOTBOHM, *Fibrations of classifying spaces*, Transaction of AMS **343 (1)** (1994), 391–415.
16. A. KONO and N. YAGITA, *Brown-Peterson and ordinary cohomology theories of classifying spaces for compact Lie groups*, Transaction of AMS **339 (2)** (1993), 781–798.
17. J. MØLLER, *Homotopy Lie groups*, Bull. of AMS **32 (4)** (1995), 413–428.
18. J. MØLLER, *Rational isomorphisms of p–compact groups*, Topology **35 (1)** (1996), 201–225.
19. J. MØLLER and D. NOTBOHM, *Centers and finite coverings of finite loop spaces*, J. reine angew. Math. **456** (1994), 99–133.
20. D. NOTBOHM, *Classifying spaces of compact Lie groups and finite loop spaces, Handbook of Algebraic Topology*, North-Holland (1995), 1049–1094.
21. L. SMITH, *Polynomial invariants of finite groups*, A K Peters, Ltd., Wellesley, MA (1995).
22. C.W. WILKERSON, *Self-maps of classifying spaces*, LNM **418** (1974), 150–157.
23. C.W. WILKERSON, *Lambda-rings, binomial domains, and vector bundles over* $\mathbb{CP}^\infty$, Comm. Algebra **10(3)** (1982), 311–328.

Fukuoka University, Fukuoka 814-0180, Japan
e-mail: kenshi@ cis.fukuoka-u.ac.jp

Contemporary Mathematics
Volume **274**, 2001

THE STRUCTURE OF THE HUREWICZ HOMOMORPHISM

DONALD KAHN AND CHRISTOPHER SCHWARTZ

1. Introduction

The classical Hurewicz homomorphism is a natural transformation from homotopy groups to homology groups. It was originally an effective tool for studying the first nontrivial homotopy groups, but it also is related to questions in the group of homotopy-self-equivalences. Early results on the group of homotopy-self-equivalences depended on the representation of this group in the automorphisms of homology or homotopy [7] and more generally [13]. But the kernel and image of the Hurewicz homomorphism - for example - are invariant under the maps induced by a homotopy-self-equivalence. In particular, the Hurewicz homomorphism may play a role in the study of the general question of André Legrand [9] on the relationship between the subgroups of the group of homotopy-self-equivalences which fix homotopy groups or which fix homology groups.

In this paper, we propose to study the Hurewicz homomorphism from the viewpoint of the spectral sequence of a Postnikov tower. We then give a complete analysis of the five cases where the Hurewicz homomorphisms are a monomorphism, a split monomorphism, an isomorphism, an epimorphism or a split epimorphism in <u>all</u> dimensions. In the second section, we look at examples which distinguish these five cases. We offer a particularly rich collection of spaces where the Hurewicz map is mono (but not split mono), in terms of stable spaces with finitely - many non-vanishing homotopy groups. In the third section we look at examples of other "Hurewicz" homomorphisms as

1991 *Mathematics Subject Classification.* Primary 55Q05, ; Secondary 55S45.

well as the important (but very difficult) non-simply-connected case. We close with some problems that we believe merit further research.

2. The Spectral Sequence

We shall work in the category of pointed spaces which have the homotopy-type of a simply-connected (or simple) based CW-complex with locally - finite homology. The most general approach to the Hurewicz homomorphisms is the spectral sequence of a Postnikov Tower (see [7]). This first quadrant spectral sequence was actually first discovered by J.F. Adams [2] in his work on BU. The extraordinary richness of this marvelous mathematician has led to a situation where we cannot - for fear of confusion - add his name to this spectral sequence; hence the rather prosaic "spectral sequence of a Postnikov Tower."

For a space X in our category. The homology spectral sequence $\{E^r_{p,q}, d^r\}$ has the following properties:

i) If X is $(m-1)$ - connected, $E^2_{p,q} = 0$, if $p < 0$ or $q < m$
ii) $E^2_{p,q} = H_{p+q}(K(\pi_q(X), q)), 0 \leq p \leq q$
iii) $E^r_{p,q}$ converges to $H_*(X)$.
iv) The edge homomorphism (inclusion of y-axis) identifies with Hurewicz homomorphism $h_n : \pi_n(X) \to H_n(X)$.

For clarity we display E^2 as follows:

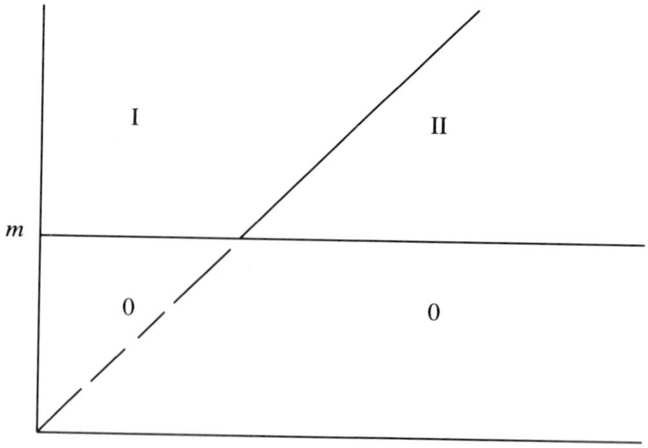

Figure 1. *The E^2-term.*

Region I is the stable region, where the homology of Eilenberg-MacLane spaces is displayed on horizontal intervals. Region II is largely unknown. There are many easy applications. For example, let $x \epsilon \ker(h_q), q < 2m-1$ be a p - torsion element. Then the order of

x is bounded by p raised to a power which is the number of dimensions for which $\pi_j(X)$ has elements of infinite order or p-torsion.

Although much is unknown, we can prove some general results.

Theorem 2.1. *X has the homotopy - type of a product of Eilenberg-MacLane spaces if and only if the cohomology spectral sequence of a Postnikov tower for X collapses with all possible coefficients.*

Proof. The proof consists of examining the differentials in the exact couple. If X is a product, the inclusions of the filtration which give the exact couple, $\cdots \subseteq F_m \subseteq \cdots \subseteq X$, are all split. This quickly leads to the vanishing of the differentials, in either homology or cohomology.

Conversely, if all the differentials vanish in cohomology with G coefficients then the image of $H^n(K(\pi_m(X), m), G)$ in $H^n(F_m; G)$ pulls back to $H^n(X, G)$. This is illustrated by the diagram:

$$H^n(X;G) \to \cdots \to H^n(F_m;G) \leftarrow H^n(K(\pi_m(X), m); G)$$

By the Serre exact sequence of the fibration

$$F_m \subseteq X \to X_{m-1},$$

the transgression annihilates the fundamental class of F_m, as well as the fundamental class of $K(\pi_m(X), m) \subseteq X_m \to X_{m-1}$. Hence, all k-invariants vanish and X is the desired product (see [12]). $\square$

In the abstract, the spectral sequence contains the answers to all questions about the Hurewicz homomorphism $h_n : \pi_n(X) \to H_n(X)$. We set out to answer, as best possible, the cases for which h_n is, in all dimensions, either a monomorphism, a split monomorphism, an isomorphism, a split epimorphism, or just an epimorphism. We shall see by example that in the first and last cases, one cannot expect a much more specific result, beyond the appropriate statement for the edge homomorphism. However, the split mono case was analyzed, many years ago, by Jean-Pierre Meyer [10]. In fact, it was Meyer who first gave a clear geometric version of a Postnikov tower, the first to understand Whitehead products within a Postnikov framework, and was the first to prove the following:

Proposition 2.1. *h_n is a split monomorphism in all dimensions, if and only if X is product of Eilenberg-MacLane spaces.*

Proof. Consider a single fibration in a Postnikov tower for X:

$$K(\pi_n(X), n) \xrightarrow{i} X_n \to X_{n-1}$$

It is easy to see that the map on homology $(i)_*$ identifies with the Hurewicz homomorphism h_n. Note that the functor $Hom(-, G)$ takes a split monomorphism to a split epimorphism. Using the Universal

Coefficient Theorem, we see that i^* is onto. Using the Serre exact sequence, we see that the transgression τ vanishes and hence the k-invariant is zero.

The converse is trivial because the inclusion of a factor in a product has a left inverse. □

We may now analyze the middle case.

Proposition 2.2. *The only simple, non-contractible space, for which h_n is an isomorphism in all dimensions, is the circle S^1.*

Proof. Generalizing the previous proposition to the simple case, we see that X is a product of $K(\pi, n)$-spaces. If there is one factor $K(\pi_n(X), n), n > 1$ and $\pi_n(X)$ non-trivial the usual results on the homology of Eilenberg-MacLane spaces [6] show that h_n is not onto in all dimensions. Thus, $X = K(G, 1)$ with G Abelian.

Suppose G has a finite summand. Then the standard results on homology of Abelian groups show that h_n cannot be isomorphic in all dimensions.

Finally, suppose G is a direct sum of copies of Z. If there is more than one summand, we conclude - by the Kunneth Theorem - that h_n is not isomorphic in all dimensions. Therefore, $X = K(Z, 1)$, as desired. □

The following result is due to H. J. Baues [4] and [5].

Proposition 2.3. *$h_n : \pi_n(X) \to H_n(X)$ is a split epimorphism for all n, if and only if X has the homotopy-type of a wedge of Moore spaces (spaces with a single non-vanishing homology group).*

Proof. The proof uses the following lemma:

Lemma 2.1. *For any X satisfying our basic assumptions, put $M_n = M(\pi_n(X), n)$ to be the Moore space with a single non-vanishing homology group $\pi_n(X)$. Then there is a map $M_n \to X$ which induces the map h_n on homology.*

Proof. M_n may be thought of as the $(n+1)$skeleton of $K(\pi_n(X), n)$. The inclusion $K(\pi_n(X), n) \to X_n$, in the nth Postnikov term, induces h_n on homology. □

Now if h_n is split epi in every dimension, the map

$$\vee_n h_n : \vee_n M_n \to X$$

may be rewritten as wedge of Moore spaces which are mapped isomorphically on homology, and a wedge summand which are mapped by the trivial maps. The result follows from Whitehead's theorem on maps which induce isomorphisms on homology. □

Finally, one might ask for similar results in a single dimension, or in an interval of dimensions. The following two propositions are tedious but easy applications of the Serre exact sequences of the various fibrations in a Postnikov tower:

Proposition 2.4. *a)* $h_m : \pi_m(X) \to H_m(X)$ *is surjective, if and only if* $H_m(X_{m-1}) = 0$
b) h_m *is injective if and only if* $H_{m+1}(X_m) \to H_{m+1}(X_{m-1})$ *is an isomorphism.*

Proposition 2.5. h_j *is isomorphic for* $0 < j \leq m$ *and epimorphic for* $j = m+1$, *if and only if* $H_{j+1}(X_{j-1}) = 0, 0 < j \leq m$.

This last result generalizes the Hurewicz theorem, but it may be of more interest when homotopy and homology are taken with coefficients.

3. The basic examples

We now address the question of whether all our five cases are district. Cellular examples were found by Baues [5] and examples involving Postnikov sections were found independently by Schwartz [12]. We are able to give much more general results which imply that there are entire families of spaces with finitely-many non-vanishing homotopy groups, for which the Hurewicz homomorphism is a non-split monomorphism.

Theorem 3.1. *Let* X *be an* $(n-1)$ *connected space,* $n > 2$, *with finitely-many non-vanishing, finitely-generated homotopy groups* $\pi_j(X), n \leq j < 2n-1$. *Suppose every* $\pi_j(X), n < j$, *is a free Abelian group.*

Then the Hurewicz homomorphism is a monomorphism in each dimension. It is a split monomorphism in each dimension, if and only if X *is a product of Eilenberg-MacLane spaces.*

Proof. The edge homomorphism in the spectral sequence of a Postnikov system for X identifies with the Hurewicz homomorphism. The classic results of H. Cartan and J.P. Serre (see [6]) imply that $H_{m+i}(K(\pi, m))$ is finite when $0 < i < m$. Referring to the basic properties of the spectral sequence, in section 1, we see that the differentials which end at $E^2_{0,q}$ must vanish because their ranges are free Abelian groups while their domains are finite. Hence, the edge homomorphism is mono. □

Remark 3.1. *1) Let p be an odd prime number, $2(p-1) < n$ with n odd. Let X be a space with two non-vanishing homotopy groups $\pi_n(X)$ and $\pi_{n+2(p-1)}(X)$, both isomorphic to the integers. Suppose that the k-invariant is a non-zero integer class, see H. Cartan [6].*

Then $h_{n+2(p-1)}$ is a non-split monomorphism (see Proposition 2.1 and Theorem 2.1).

2) Let p be an odd prime and X have two non-vanishing homotopy groups $\pi_n(X)$ and $\pi_m(X), n < m < 2n-1$, with $\pi_m(X) \approx Z/p^j, j > 1$. Then $h_n \otimes 1_{Z/p}$ is a monomorphism. In fact, by H. Cartan [6] , the domain of the only non-zero differential ending with $E^2_{0,m}$ contains only summands of prime order. Hence, if we tensor with Z/p , such differentials vanish.

For examples of non-split epimorphisms, let us consider a map $\phi : K(Z,2) \to K(Z,4)$ corresponding to twice (more generally p times) a generator of $H^4(K(Z,2)) = Z$. We form the pull-back of the path-loop fibration ϕ^* over $K(Z,4)$ to get a fibre space

$$K(Z,3) \to E \xrightarrow{p} K(Z,2).$$

By naturality, the homology transgression

$$\tau : H_4(K(Z,2)) \to H_3(K(Z,3))$$

is multiplication by 2. It follows from the Serre exact sequence that the Hurewicz map h_3 is the reduction map $Z \to Z/2$ (more generally reduction mod p).

Because $H_4(E) = 0$, one can find a CW-structure E with a 4-dimensional subcomplex whose homology is isomorphic to that of E through dimension 4. See [12] for details. For this subcomplex, h_1 and h_2 are isomorphisms, h_3 is a non-split epimorphism, and $h_i = 0$ for $i > 3$.

4. Extensions of the work

Lastly, we consider possible extensions of this work.

a) Localizations of the Hurewicz map. We look at the rationalization of the Hurewicz map. The following proposition is a re-writing in modern terms of a result of Arkowitz and Curjel [1].

Proposition 4.1. *With our general assumptions (X is a simply-connected CW-complex), $h_n \otimes 1_Q$ is an isomorphism in all dimension, if and only if there is a fibration*

$$F \to X \to K(Z, 2n+1)$$

where all the homotopy groups of F are finite.

Proof. Apply a general Proportion 2.1 to the rationalization X_Q of X. X_Q is a Cartesian product of $K(V_j, j)$ spaces, where V_j is a rational vector space. If there is more than one factor, the Kunneth Theorem shows at once that some h_m is not onto. Also, if $V_j \neq 0, j$ even, because the rational cohomology is then a polynomial algebra.

We quickly conclude that there is a single $V_j \neq 0$ and $\dim V_j = 1$. To complete the proof, take a rational homotopy equivalence $P : X \to K(Z, 2n+1)$ and convert it to a fibration. □

b) The non-simply connected case. This is truly terra incognito. Of course, h_1 is always onto. Even when $H_i(X) = 0$ for $i > 1$, a wedge of several circles is a case where h_1 is a non-split epimorphism, because of free group of rank > 1 does not contain a free abelian subgroup of the same rank.

More striking are the spaces X for which $H_n(X) = 0, n > 0$, often called the acyclic spaces. In a trivial sense, the Hurewicz map is always a split epimorphism. They deserves further study.

In our opinion, the most interesting, largely un-explored family of such spaces consists of the acyclic $K(G,1)$ spaces, that is spaces X for which $\pi_1(X) \neq \{e\}, \pi_i(X) = 0$ when $i > 1$, yet $H_i(X) = 0$ for $i > 0$. It is known that G must be infinite, if it is non-trivial, and L. Evens and others have a purely algebraic construction. However, one may also get examples from geometric topology. Let M be an open, non-simply-connected component of the complement of an Alexander horned sphere [3] in S^3. The sphere theorem of C. Papakyriakopoulos [11] implies that $\pi_2(M) = 0$. The universal cover of M say $\widetilde{M}$, is an open 3-mainfold with π_1 $(\widetilde{M})=$ π_2 $(\widetilde{M})=$ 0. But it is an open 3-mainfold, so H_3 $(\widetilde{M})=$ 0. Therefore, π_3 $(\widetilde{M})=$ 0 and $\widetilde{M}$ is contractible. Hence, M is a $K(G,1)$ space. It also follows from the Alexander duality theorem that $H_*(M)$ is the same as the homology of a point.

The class of acyclic $K(G,1)$ spaces is closed under both wedge and product. It measures the non-uniqueness of the maps $K(G,1) \to X$ inducing isomorphisms on homology. All such spaces have h_n (trivially) split epimorphisms. They deserve further study.

We end with two conjectures:

1. The edge homomorphism in the spectral sequence for a Postnikov system, in an extraordinary homology theory, will play the role of the Hurewicz homomorphism, and much of the present paper will be true in more general setting.

2. Much of this work can be extended to nilpotent spaces, for which the Postnikov techniques still work.

REFERENCES

[1] Arkowitz, M. and Curjel, C., *The Hurewicz Homomorphism and Finite Homotopy Invariants*, Trans. A.M.S., **110**, (1964), pp. 538–551.

[2] Adams, J.F, *On Chern Characters and the Structure of the Unitary Group*, Proc. Camb. Plilo. Soc., **57**, 2, (1961).

[3] Alexander, J., *An example of a simply-connected surface bounding a region which is not simply-connected*, Proc. Nat. Acd. of Sciences, **10**, (1924), pp. 8–10.

[4] Baues, H. J., *Homotopy-types*, Handbook of Algebraic Topology, Ed. I. James, Oxford, (1995).

[5] Baues, H. J., *Homotopy-types and Homology*, Oxford V. Press, (1996).

[6] Cartan, H., *Sur les Groupes d' Eilenberg-MacLane*, Proc. Nat Acad. Sciences, **40**, (1954), pp. 467–471 and **40**, (1954), pp. 704–707.

[7] Kahn, D.W, *The Group of Stable Self-Equivalences*, Topology 11, (1972 A), pp. 133–140.

[8] Kahn, D.W, *The Spectral Sequence of a Postnikov System*, Comm. Math. Helv, **40**, fasc 3, (1966).

[9] Legrand, A., Problems in Proceeding of Conference on *Groups of Self-Equivalences and Related Topics*, Montreal (1988), Springer Lecture Notes in Math. 1425, (1990 A).

[10] Meyer, J. P., *Whitehead Products and Postnikov systems*, Amer. Journal of Math., **82**, (1960), pp. 271–280.

[11] Papakyriakopoulos, C., *On Dehn's Lemma and the Asphericity of Knots*, Annals of Math., **66**, (1957), pp. 1–16.

[12] Schwartz, C., *The Hurewicz Homomorphism*, Thesis University of Minnesota, July (1997).

[13] Wilkerson, C.W., *Applications of Minimal Simplicial Groups*, Topology 15, (1976 A), pp. 111–130.

DONALD KAHN, SCHOOL OF MATHEMATICS, UNIVERSITY OF MINNESOTA, MINNEAPOLIS, MN 55455

E-mail address: dkahn@isystems.net

CHRISTOPHER SCHWARTZ, DEPARTMENT OF MATHEMATICS, AUGSBURG COLLEGE, MINNEAPOLIS, MN 55454

Contemporary Mathematics
Volume **274**, 2001

Joins, Diagonals and Hopf Invariants

HOWARD J. MARCUM

0. Introduction and motivation

In the category of well-pointed based topological spaces and based maps we consider a fixed diagram of spaces and maps

$$
(0.1) \qquad
\begin{array}{ccccc}
A & \xleftarrow{f} & C & \xrightarrow{g} & B \\
{\scriptstyle\alpha}\downarrow & \overset{F}{\Leftarrow} & \downarrow & \overset{G}{\Rightarrow} & \downarrow{\scriptstyle\beta} \\
X & \longleftarrow & * & \longrightarrow & Y
\end{array}
$$

with specified null homotopies F and G. Taking double mapping cylinders of horizontal pairs yields a map $c\colon \mathcal{M}(f,g) \to X \vee Y$ and so in the the manner of [4, Definition 1.4] a Hopf invariant $c\mathrm{HI}\colon \Sigma\Omega\mathcal{M}(f,g) \to \Omega X * \Omega Y$ is obtained. It will be convenient to call (0.1) a *Ganea situation* and to refer to the map c as the *associated Ganea map*. In certain situations computation of $c\mathrm{HI}(u)$ for a given class $u\colon \Sigma U \to \mathcal{M}(f,g)$ is of interest. Yet as restriction to those cases classically studied already shows this can be quite difficult indeed (as evidenced for example by the Hopf invariant one problem for spheres).

An example more relevant to our present purposes concerns the following special case of (0.1) with D denoting the defining homotopy for the suspension of C in the left square and D_g the defining homotopy for the mapping cone of g (taken with vertex at parameter $t = 0$) in the right square.

$$
\begin{array}{ccccc}
* & \longleftarrow & C & \xrightarrow{g} & B \\
\downarrow & \overset{D}{\Leftarrow} & \downarrow & \overset{D_g}{\Rightarrow} & \downarrow{\scriptstyle i_1} \\
\Sigma C & \longleftarrow & * & \longrightarrow & C_g
\end{array}
$$

The Hopf invariant obtained in this case is denoted $\mathrm{GHI}\colon \Sigma\Omega C_g \to \Omega\Sigma C * \Omega C_g$ and called the Ganea Hopf invariant for the map $g\colon C \to B$. Thus given an arbitrary

2000 *Mathematics Subject Classification.* Primary 55Q15, 55Q25, 55Q35.

Key words and phrases. Hopf invariant, dual product, Whitehead product map, exterior join, extended join product.

map $u: \Sigma U \to C_g$ we are here in the situation of a 2-stage mapping cone

$$\begin{array}{ccccc} C & & \Sigma U & & \\ {\scriptstyle g}\downarrow & & \downarrow{\scriptstyle u} & & \\ B & \xrightarrow[i_1]{} & C_g & \xrightarrow[i_1]{} & C_u \end{array}$$

and the problem arises of computing $\mathrm{GHI}(u): \Sigma U \to \Omega\Sigma C * \Omega C_g$. Computations of this kind are relevant in recent studies of cone length and Lusternik-Schnirelmann category (cf. [3], [6], [12]). While such specific applications motivate our work we shall only be concerned with the general theory in this paper. Actually the Ganea Hopf invariant GHI is defined for the more general case of an arbitrary cotraid pair, rather than just for a single map (see [4, §4] and also §4 below). Also we point out that the Hopf invariant $\mathrm{GHI}: \Sigma\Omega C_g \to \Omega\Sigma C * \Omega C_g$ above differs slightly from that [5, §4] considered by Ganea (see [4, Remark 4.2]); the latter would use $-D$ in place of D.

Returning to the general case of (0.1) one might raise instead the question of computing the composite

$$\Sigma\Omega\mathcal{M}(f,g) \xrightarrow{c\mathrm{HI}} \Omega X * \Omega Y \xrightarrow{\partial_\alpha * \partial_\beta} F_\alpha * F_\beta \tag{0.2}$$

or even the composite

$$\Sigma\Omega\mathcal{M}(f,g) \xrightarrow{c\mathrm{HI}} \Omega X * \Omega Y \xrightarrow{\Omega i_1 * \Omega i_1} \Omega C_\alpha * \Omega C_\beta \tag{0.3}$$

where ∂_α and ∂_β are the boundary maps in the respective long homotopy sequences. Of course (0.2) is a factor of (0.3); this follows by virtue of the fact that for any map, say $\alpha: A \to X$, the map $\Omega i_1: \Omega X \to \Omega C_\alpha$ can be factored up to homotopy as $\Omega X \xrightarrow{\partial_\alpha} F_\alpha \xrightarrow{\mu_\alpha'} \Omega C_\alpha$ where $\mu_\alpha: \Sigma F_\alpha \to C_\alpha$ is the homotopy Thom class of α and μ_α' its adjoint. In fact, recalling the more general theory of Hopf invariants developed in [10], we may use naturality to recognize (0.2) as the Hopf invariant $\{j_\bigstar \circ c\}$HIand (0.3) as the Hopf invariant $\{w \circ j_\bigstar \circ c\}$HIwhere $j_\bigstar: X \vee Y \to E(\alpha \bigstar \beta)$ is the canonical inclusion in the total space of the exterior join fibration $\alpha \bigstar \beta$ of α and β, and where $w: E(\alpha \bigstar \beta) \to C_\alpha \vee C_\beta$ is the generalized Whitehead product map defined in [8, Theorem 7.3].

In this paper we approach the problem of computing (0.2) by considering a related but somewhat different Hopf invariant which may also be obtained from data (0.1). This Hopf invariant, denoted $\lambda\mathrm{HI}: \Sigma\Omega\mathcal{M}(f,g) \to F_\alpha * F_\beta$, is constructed by means of a class $\lambda: \mathcal{M}(f,g) \to E(\alpha \bigstar \beta)$ which admits a factorization through the "diagonal" map $\delta: \mathcal{M}(f,g) \to \mathcal{M}(f \times f, g \times g)$. Consequently one may attempt to deduce computational information about λHI from properties of the map δ. We implement this approach to study λHI for the important case where the domain cotriad in (0.1) is taken to be the projection cotriad $A \xleftarrow{p_A} A \times B \xrightarrow{p_B} B$. In that case the corresponding diagonal map δ may be identified up to homeomorphism as the join of diagonal maps $\Delta * \Delta: A * B \to (A \times A) * (B \times B)$. We proceed to develop an appropriate decomposition of $\Delta * \Delta$ (which seems of independent interest). For that purpose we utilize an operation called the extended join product of maps (a version of this operation was considered previously by Baues [1]). As a consequence we obtain in Theorem 7.9 a five term expression for the λHI Hopf invariant associated with the projection cotriad.

The Hopf invariants λHI and $\{j_\bigstar \circ c\}$HI are related but we do not determine their full and precise relationship; it seems somewhat subtle. What we do show however (Theorem 5.5 below) is that at the level of (0.3) each of $(\Omega i_1 * \Omega i_1)\circ c$HIand $\{w \circ \lambda\}$HI may be related to the Ganea Hopf invariant GHI of the domain cotraid. In the expression for $\{w \circ \lambda\}$HI there is also involved the dual product operation [4, Definition 2.5] of Arkowitz (or, as it is sometimes called, the co-Whitehead product). These expressions are obtained by a careful study of naturality and universal examples.

1. The diagonal map $\delta\colon \mathcal{M}(f,g) \to \mathcal{M}(f \times f, g \times g)$

We work in the category of well-pointed based topological spaces. All constructions involving parameters are reduced. In particular $\mathcal{M}$ denotes the *reduced* double mapping cylinder.

Primarily to fix notation we review in this section some basic constructions. We use the convention that mapping cones have vertices at parameter $t = 0$. Hence the mapping cone C_α of a map $\alpha\colon A \to X$ is defined by the following homotopy pushout square.

$$\begin{array}{ccc} A & \xrightarrow{\alpha} & X \\ \downarrow & \overset{D_\alpha}{\Longrightarrow} & \downarrow{\scriptstyle i_1} \\ * & \longrightarrow & C_\alpha \end{array} \qquad D_\alpha(a,t) = [a,t] \in C_\alpha \text{ for } a \in A,\ 0 \le t \le 1$$

Similarly the homotopy pullback square

$$\begin{array}{ccc} F_\alpha & \xrightarrow{\ell_\alpha} & A \\ \downarrow & \overset{\Psi}{\Longrightarrow} & \downarrow{\scriptstyle \alpha} \\ * & \longrightarrow & X \end{array} \qquad \begin{aligned} &F_\alpha = \{(\tau,a) \in X^{[0,1]} \times A \mid \tau(0) = *, \tau(1) = \alpha(a)\} \\ &\Psi((\tau,a),t) = \tau(t),\ \ell_\alpha(\tau,a) = a \end{aligned}$$

defines the homotopy fiber F_α of the map α.

For induced maps on fibers and cofibers we use a systemic notation. Specifically, suppose that the homotopy commutative square on the left below with homotopy F is given. Then two induced homotopy commutative squares as given on the right below may be constructed.

$$(1.1) \qquad \begin{array}{ccc} A & \xrightarrow{h} & B \\ {\scriptstyle\alpha}\downarrow & \overset{F}{\Longrightarrow} & \downarrow{\scriptstyle\beta} \\ X & \xrightarrow[s]{} & Y \end{array} \qquad \begin{array}{ccc} F_\alpha & \xrightarrow{\sigma_F} & F_\beta \\ {\scriptstyle\ell_\alpha}\downarrow & \Longrightarrow & \downarrow{\scriptstyle\ell_\beta} \\ A & \xrightarrow[h]{} & B \end{array} \qquad \begin{array}{ccc} X & \xrightarrow{s} & Y \\ {\scriptstyle i_1^\alpha}\downarrow & \Longrightarrow & \downarrow{\scriptstyle i_1^\beta} \\ C_\alpha & \xrightarrow[\zeta_F]{} & C_\beta \end{array}$$

By definition σ_F is the map given by $\sigma_F(\tau,a) = (F(a,\bullet) + s \circ \tau, h(a))$ while ζ_F is induced by the double mapping cylinder functor $\mathcal{M}$ as indicated below.

$$\begin{array}{ccccc} * & \longleftarrow & A & \xrightarrow{\alpha} & X \\ \downarrow & & {\scriptstyle h}\downarrow & \overset{-F}{\Longrightarrow} & \downarrow{\scriptstyle s} \\ * & \longleftarrow & B & \xrightarrow[\beta]{} & Y \end{array} \quad \overset{\mathcal{M}}{\rightsquigarrow} \quad \begin{array}{c} C_\alpha \\ \downarrow{\scriptstyle \zeta_F} \\ C_\beta \end{array}$$

Explicitly,

$$\zeta_F[a,t] = \begin{cases} [h(a), 2t], & 0 \le t \le \frac{1}{2} \\ i_1 F(a, 2-2t), & \frac{1}{2} \le t \le 1 \end{cases}$$

where $a \in A$ and $i_1: Y \to C_\beta$ is the inclusion at parameter $t = 1$. Note that if F is a static homotopy (i.e., $s \circ \alpha = \beta \circ h$) then the definition of ζ_F can be taken simply to be $\zeta_F[a,t] = [h(a), t]$. Also note that when $X = *$ then $\zeta_F: \Sigma A \to C_\beta$ is usually called a *coextension* of h relative to β.

We recall that the homotopy pushout square

$$(1.2) \qquad \begin{array}{ccc} C & \xrightarrow{g} & B \\ {\scriptstyle f}\downarrow & \overset{D}{\Longrightarrow} & \downarrow{\scriptstyle i_1} \\ A & \xrightarrow[i_0]{} & \mathcal{M}(f,g) \end{array} \qquad D(c,t) = [c,t] \in \mathcal{M}(f,g) \text{ for } c \in C,\ 0 \le t \le 1$$

defines the double mapping cylinder. The notation $\lambda_g: \mathcal{M}(f,g) \to C_g$ will denote the canonical map (or operator) which collapses the subset A at parameter $t = 0$ of $\mathcal{M}(f,g)$ to a point. Let $r: \mathcal{M}(f,g) \to \mathcal{M}(g,f)$, $r[c,t] = [c, 1-t]$, denote *parameter reversal.* We observe, with reference to (1.1), that the square

$$(1.3) \qquad \begin{array}{ccc} \mathcal{M}(h,\alpha) & \xrightarrow{\mu_{-F}} & Y \\ {\scriptstyle \lambda_\alpha}\downarrow & & \downarrow{\scriptstyle i_1^\beta} \\ C_\alpha & \xrightarrow{\zeta_F} & C_\beta \end{array}$$

is homotopy commutative and is in fact a homotopy pushout. This follows by an application of [9, Lemma 3.3].

Also, for arbitrary spaces X and Y, recall from [10, Example 1.7] the *canonical splitting map*

$$\chi := i_X \circ \varepsilon_X \circ \Sigma\Omega p_X + i_Y \circ \varepsilon_Y \circ \Sigma\Omega p_Y : \Sigma\Omega(X \times Y) \to X \vee Y$$

where p_X, p_Y are projections, i_X, i_Y are inclusions and $\varepsilon_X, \varepsilon_Y$ are evaluation maps. The composite $\Sigma\Omega(X \times Y) \xrightarrow{\chi} X \vee Y \xrightarrow{\text{inc}} X \times Y$ is homotopic to $\varepsilon_{X \times Y}$.

Now using (1.2) a product square

$$\begin{array}{ccc} C\times C & \xrightarrow{g\times g} & B\times B \\ {\scriptstyle f\times f}\downarrow & \overset{D\times D}{\Longrightarrow} & \downarrow{\scriptstyle i_1\times i_1} \\ A\times A & \xrightarrow[i_0\times i_0]{} & \mathcal{M}(f,g)\times\mathcal{M}(f,g) \end{array}$$

may be formed. This yields an induced map

$$P := \mu_{(D\times D)}: \mathcal{M}(f \times f, g \times g) \to \mathcal{M}(f,g) \times \mathcal{M}(f,g)$$

having the property that $P = P_1 \triangle P_2$ where the maps $P_i: \mathcal{M}(f\times f, g\times g) \to \mathcal{M}(f,g)$ for $i = 1, 2$, are the "projections" defined componentwise. Also there is a "diagonal" map $\delta: \mathcal{M}(f,g) \to \mathcal{M}(f \times f, g \times g)$ whose components are the respective diagonal maps. Moreover there are obvious inclusion maps $i_1, i_2: \mathcal{M}(f,g) \to \mathcal{M}(f \times f, g \times g)$ onto the "first" and "second" factors. Taken together these yield an inclusion map

$$i := i_1 \triangledown i_2: \mathcal{M}(f,g) \vee \mathcal{M}(f,g) \to \mathcal{M}(f \times f, g \times g)$$

with the property that $P \circ i$ is the usual inclusion of the wedge into the product. Moreover P is a coretractile map ([10], Definition 1.5); a splitting map for P may be taken to be the composite

$$\Sigma\Omega(\mathcal{M}(f,g) \times \mathcal{M}(f,g)) \xrightarrow{\chi} \mathcal{M}(f,g) \vee \mathcal{M}(f,g) \xrightarrow{i} \mathcal{M}(f \times f, g \times g).$$

Let F_P denote the homotopy fiber of P and $\ell\colon F_P \to \mathcal{M}(f\times f, g\times g)$ the fiber projection. Unfortunately it doesn't seem possible to present F_P in a more recognizable form.

We assemble the above data in the following diagram.

$$\text{(1.4)} \qquad \begin{array}{ccccc} & & F_P & & \\ & & \downarrow{\scriptstyle \ell} & & \\ \mathcal{M}(f,g) & \xrightarrow{\delta} & \mathcal{M}(f\times f, g\times g) & \xleftarrow{i\circ\chi} & \Sigma\Omega(\mathcal{M}(f,g)\times\mathcal{M}(f,g)) \\ & {\scriptstyle \Delta}\searrow & \downarrow{\scriptstyle P} & \swarrow{\scriptstyle \varepsilon_{\mathcal{M}(f,g)\times\mathcal{M}(f,g)}} & \\ & & \mathcal{M}(f,g)\times\mathcal{M}(f,g) & & \end{array}$$

From such data, in accordance with [10, Definition 1.17], there is obtained a Hopf invariant $\delta\mathrm{HI}_{(\ell,P;i\circ\chi)}\colon \Sigma\Omega\mathcal{M}(f,g) \to F_P$. In view of the specific form of the splitting map $i \circ \chi$ it follows that the characterizing equation for $\delta\mathrm{HI}_{(\ell,P;i\circ\chi)}$ takes the form

$$\text{(1.5)} \qquad \ell \circ \delta\mathrm{HI}_{(\ell,P;i\circ\chi)}(u) = \delta \circ u - i_2 \circ u - i_1 \circ u$$

for each $u\colon \Sigma U \to \mathcal{M}(f,g)$. However because a more specific description of F_P has not been achieved, explicit computation of $\delta\mathrm{HI}_{(\ell,P;i\circ\chi)}$ is lacking.

2. The Hopf invariant λHI

Throughout this section we assume that the data in diagram (0.1) is given. This information gives rise to a map $\lambda\colon \mathcal{M}(f,g) \to E(\alpha\bigstar\beta)$ induced by application of the double mapping cylinder functor $\mathcal{M}$ as indicated below.

$$\text{(2.1)} \qquad \begin{array}{ccccc} A & \xleftarrow{f} & C & \xrightarrow{g} & B \\ {\scriptstyle 1\triangle o}\downarrow & \overset{f\triangle - G}{\Leftarrow} & \downarrow{\scriptstyle f\triangle g} & \overset{-F\triangle g}{\Rightarrow} & \downarrow{\scriptstyle o\triangle 1} \\ A\times Y & \xleftarrow[1\times\beta]{} & A\times B & \xrightarrow[\alpha\times 1]{} & X\times B \end{array} \quad \overset{\mathcal{M}}{\rightsquigarrow} \quad \begin{array}{c} \mathcal{M}(f,g) \\ \downarrow{\scriptstyle\lambda} \\ E(\alpha\bigstar\beta) \end{array}$$

Recall that the total space $E(\alpha\bigstar\beta)$ of the exterior join is just by definition the indicated double mapping cylinder $\mathcal{M}(1 \times \beta, \alpha \times 1)$. By properties of the double mapping cylinder functor the homotopy class of λ depends only on the track classes of the homotopies F and G.

Also we induce a map ξ as follows.

$$\text{(2.2)} \qquad \begin{array}{ccccc} A\times A & \xleftarrow{f\times f} & C\times C & \xrightarrow{g\times g} & B\times B \\ {\scriptstyle 1\times o}\downarrow & \overset{H}{\Leftarrow} & \downarrow{\scriptstyle f\times g} & \overset{K}{\Rightarrow} & \downarrow{\scriptstyle o\times 1} \\ A\times Y & \xleftarrow[1\times\beta]{} & A\times B & \xrightarrow[\alpha\times 1]{} & X\times B \end{array} \quad \overset{\mathcal{M}}{\rightsquigarrow} \quad \begin{array}{c} \mathcal{M}(f\times f, g\times g) \\ \downarrow{\scriptstyle\xi} \\ E(\alpha\bigstar\beta) \end{array}$$

The homotopies used here are defined in terms of the original homotopies F and G by $H(c, c', t) = (f(c), G(c', 1-t))$ and $K(c, c', t) = (F(c, 1-t), g(c'))$ for $c, c' \in C$ and $0 \leq t \leq 1$. The next proposition is immediate.

PROPOSITION 2.3. *The composite* $\mathcal{M}(f,g) \xrightarrow{\delta} \mathcal{M}(f \times f, g \times g) \xrightarrow{\xi} E(\alpha \bigstar \beta)$ *equals* λ.

Now let $\mu_F \colon C_f \to X$ and $\mu_G \colon C_g \to Y$ be the maps induced by the homotopies F and G. Then a direct verification using the definitions shows that the square

$$\begin{array}{ccc} \mathcal{M}(f \times f, g \times g) & \xrightarrow{\xi} & E(\alpha \bigstar \beta) \\ {\scriptstyle P} \downarrow & & \downarrow {\scriptstyle p_\star} \\ \mathcal{M}(f,g) \times \mathcal{M}(f,g) & \xrightarrow[(\mu_F \circ \lambda_f \circ r) \times (\mu_G \circ \lambda_g)]{} & X \times Y \end{array}$$

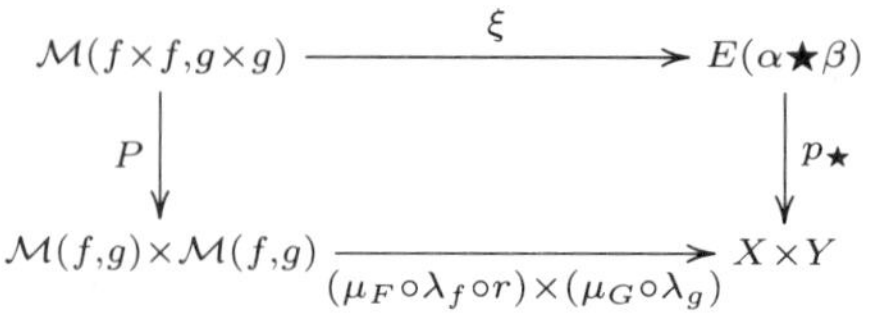

is homotopy commutative. As a consequence we have:

PROPOSITION 2.4. *The map λ has projection-type $(\mu_F \circ \lambda_f \circ r, \mu_G \circ \lambda_g)$ with respect to $p_{\alpha \bigstar \beta} \colon E(\alpha \bigstar \beta) \to X \times Y$.*

We collect the above data in the following diagram

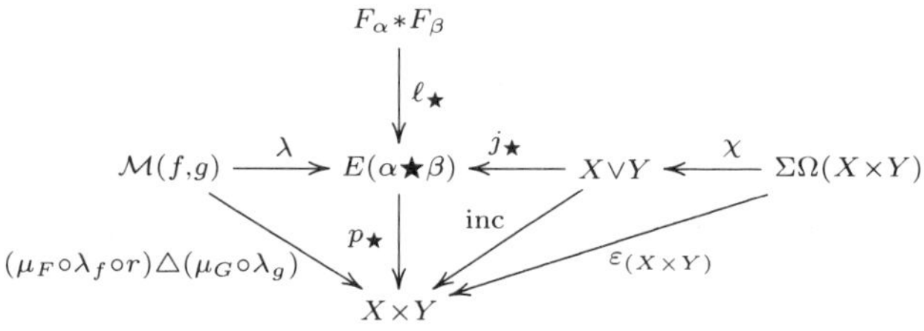

and apply [10, Definition 1.17] to obtain a Hopf invariant

(2.5) $\qquad \lambda \mathrm{HI} := \lambda \mathrm{HI}_{(\ell_\star, p_\star; j_\star \circ \chi)} \colon \Sigma \Omega \mathcal{M}(f,g) \to F_\alpha * F_\beta.$

This Hopf invariant associated with diagram (0.1) is our primary object of study in this paper. If a class $u \colon \Sigma U \to \mathcal{M}(f,g)$, with adjoint $u^{\cdot} \colon U \to \Omega \mathcal{M}(f,g)$, is given then

$$\ell_\star \circ \lambda \mathrm{HI}_{(\ell_\star, p_\star; j_\star \circ \chi)}(u) = \lambda \circ u - j_\star \circ \chi \circ \Sigma \Omega(p_\star) \circ \Sigma \Omega(\lambda) \circ \Sigma(u^{\cdot})$$
$$= \lambda \circ u - j_Y \circ \mu_G \circ \lambda_g \circ u - j_X \circ \mu_F \circ \lambda_f \circ r \circ u$$

specifies the *characterizing equation* for $\lambda \mathrm{HI}$. An alternate form for the characterizing equation is given in the next proposition.

PROPOSITION 2.6. *Let $\phi \colon F_P \to F_\alpha * F_\beta$ denote the map induced on homotopy fibers by the square following Proposition 2.3. Then the equality*

$$\lambda \mathrm{HI}_{(\ell_\star, p_\star; j_\star \circ \chi)} = \phi \circ \delta \mathrm{HI}_{(\ell, P; i \circ \chi)}$$

is valid as homotopy classes $\Sigma \Omega \mathcal{M}(f,g) \to F_\alpha * F_\beta$. In particular we note that

$$\ell_\star \circ \lambda \mathrm{HI}(u) = \lambda \circ u - \xi \circ i_2 \circ u - \xi \circ i_1 \circ u$$

for a given class $u \colon \Sigma U \to \mathcal{M}(f,g)$.

PROOF. By Proposition 2.3 $\lambda = \xi \circ \delta$ so we construct the following diagram.

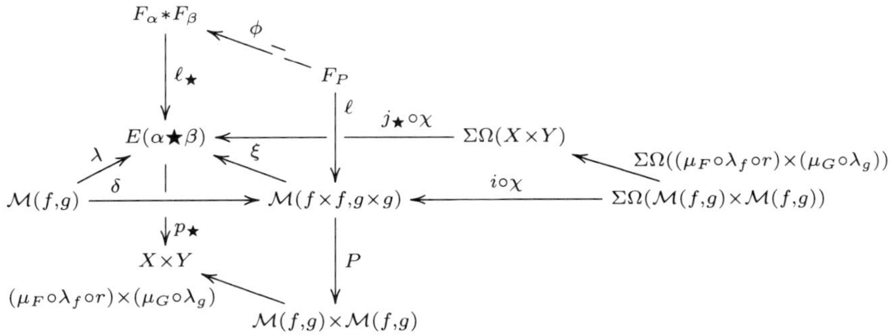

To show that this diagram is homotopy commutative it remains only to check that the square

$$\begin{array}{ccc} \mathcal{M}(f,g) \vee \mathcal{M}(f,g) & \xrightarrow{(\mu_F \circ \lambda_f \circ r) \times (\mu_G \circ \lambda_g)} & X \vee Y \\ \downarrow i & & \downarrow j_\star = j_X \nabla j_Y \\ \mathcal{M}(f \times f, g \times g) & \xrightarrow{\xi} & E(\alpha \bigstar \beta) \end{array}$$

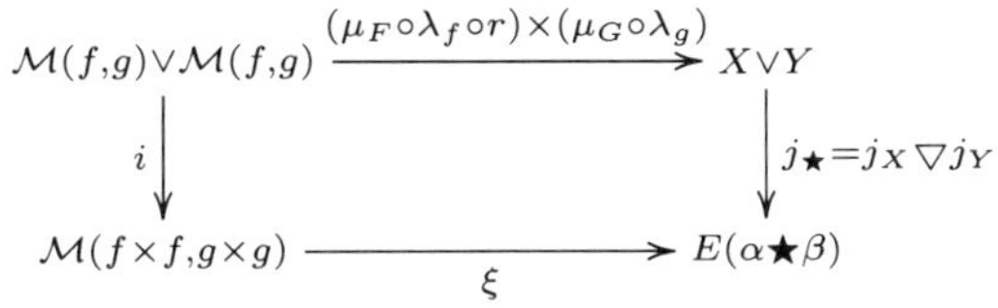

is homotopy commutative. But this fact is readily seen directly from the definitions. Consequently by the general theory

$$\lambda \mathrm{HI}_{(\ell_\star, p_\star; j_\star \circ \chi)} = \{\xi \circ \delta\} \mathrm{HI}_{(\ell_\star, p_\star; j_\star \circ \chi)} = \phi \circ \delta \mathrm{HI}_{(\ell, P; i \circ \chi)}$$

is obtained. $\square$

The following observation will be useful later in expressing $\lambda \mathrm{HI}$ as a sum.

PROPOSITION 2.7. *Let $V_1, \ldots, V_n : \Sigma U \to E(\alpha \bigstar \beta)$ denote classes each of which has projection-type (o, o) with respect to $p_{\alpha \bigstar \beta}$. For $0 \le i \le n$ let $\widetilde{V}_i : \Sigma V \to F_\alpha * F_\beta$ be the unique lifting such that $\ell_\star \circ \widetilde{V}_i = V_i$. Now suppose that $u : \Sigma U \to \mathcal{M}(f, g)$ satisfies*

$$\lambda \circ u = \xi \circ i_1 \circ u + V_1 + \cdots + V_n + \xi \circ i_2 \circ u.$$

Then $\lambda \mathrm{HI}(u) = (\widetilde{V}_1)^{\xi \circ i_1 \circ u} + \cdots + (\widetilde{V}_n)^{\xi \circ i_1 \circ u}$ as classes $\Sigma U \to F_\alpha * F_\beta$. Here the conjugate $(\widetilde{V}_i)^{\xi \circ i_1 \circ u}$ is uniquely characterized by

$$\ell_\star \circ (\widetilde{V}_1)^{\xi \circ i_1 \circ u} = \xi \circ i_1 \circ u + V_i - \xi \circ i_1 \circ u.$$

Also note that if the suspension comultiplication on ΣU is homotopy abelian then conjugation by $\xi \circ i_1 \circ u$ is not needed and we may write $\lambda \mathrm{HI}(u) = \widetilde{V}_1 + \cdots + \widetilde{V}_n$.

PROOF. We have

$$\begin{aligned} \ell_\star \circ &\{(\widetilde{V}_1)^{\xi \circ i_1 \circ u} + \cdots + (\widetilde{V}_n)^{\xi \circ i_1 \circ u}\} \\ &= (\xi \circ i_1 \circ u + V_1 - \xi \circ i_1 \circ u) + \cdots + (\xi \circ i_1 \circ u + V_n - \xi \circ i_1 \circ u) \\ &= \xi \circ i_1 \circ u + V_1 + \cdots + V_n - \xi \circ i_1 \circ u \\ &= \xi \circ i_1 \circ u + V_1 + \cdots + V_n + \xi \circ i_2 \circ u - \xi \circ i_2 \circ u - \xi \circ i_1 \circ u \\ &= \lambda \circ u - \xi \circ i_2 \circ u - \xi \circ i_1 \circ u. \end{aligned}$$

Thus the result follows from the form of the characterizing equation for λHI given in Proposition 2.6. □

3. Naturality for λHI and cHI

In this section we formulate naturality results for the two Hopf invariants λHI and cHI. For this purpose we suppose given a diagram with homotopies as indicated.

$$\begin{array}{ccccccccc} A' & \xleftarrow{a} & A & \xleftarrow{f} & C & \xrightarrow{g} & B & \xrightarrow{b} & B' \\ {\scriptstyle\alpha'}\downarrow & \overset{H}{\Leftarrow} & {\scriptstyle\alpha}\downarrow & \overset{F}{\Leftarrow} & \downarrow & \overset{G}{\Rightarrow} & \downarrow{\scriptstyle\beta} & \overset{K}{\Rightarrow} & \downarrow{\scriptstyle\beta'} \\ X' & \xleftarrow[x]{} & X & \longleftarrow & * & \longrightarrow & Y & \xrightarrow[y]{} & Y' \end{array} \tag{3.1}$$

We set $f' = a \circ f$, $g' = b \circ g$, $F' = Hf + xF$ and $G' = Kg + yG$. Denote by $\mu\colon \mathcal{M}(f,g) \to \mathcal{M}(f',g')$ the evident map induced on double mapping cylinders. Let $\sigma_H\colon F_\alpha \to F_{\alpha'}$ and $\sigma_K\colon F_\beta \to F_{\beta'}$ be the respective maps induced on homotopy fibers.

We first establish naturality for λHI. For clarity of notation let λ denote the defining class in (3.1) determined by the homotopies F, G and λ' the defining class for the homotopies F', G'.

PROPOSITION 3.2. *With reference to diagram* (3.1) *each of the following squares is homotopy commutative, where the map ξ will be constructed in the proof.*

$$\begin{array}{ccc} \mathcal{M}(f,g) & \xrightarrow{\lambda} & E(\alpha\bigstar\beta) \\ {\scriptstyle\mu}\downarrow & & \downarrow{\scriptstyle\xi} \\ \mathcal{M}(f',g') & \xrightarrow{\lambda'} & E(\alpha'\bigstar\beta') \end{array} \qquad \begin{array}{ccc} \Sigma\Omega\mathcal{M}(f,g) & \xrightarrow{\lambda\mathrm{HI}_{(\ell_\bigstar,p_\bigstar;j_\bigstar\circ\chi)}} & F_\alpha * F_\beta \\ {\scriptstyle\Sigma\Omega\mu}\downarrow & & \downarrow{\scriptstyle\sigma_H*\sigma_K} \\ \Sigma\Omega\mathcal{M}(f',g') & \xrightarrow{\lambda'\mathrm{HI}_{(\ell_\bigstar,p_\bigstar;j_\bigstar\circ\chi)}} & F_{\alpha'} * F_{\beta'} \end{array}$$

PROOF. We begin by applying the exterior join functor as follows.

$$\begin{array}{ccc} F_\alpha & \xrightarrow{\sigma_H} & F_{\alpha'} \\ {\scriptstyle\ell_\alpha}\downarrow & \Rightarrow & \downarrow{\scriptstyle\ell_{\alpha'}} \\ A & \xrightarrow{a} & A' \\ {\scriptstyle\alpha}\downarrow & \overset{H}{\Rightarrow} & \downarrow{\scriptstyle\alpha'} \\ X & \xrightarrow[x]{} & X' \end{array} \qquad \begin{array}{ccc} F_\beta & \xrightarrow{\sigma_K} & F_{\beta'} \\ {\scriptstyle\ell_\beta}\downarrow & \Rightarrow & \downarrow{\scriptstyle\ell_{\beta'}} \\ B & \xrightarrow{b} & B' \\ {\scriptstyle\beta}\downarrow & \overset{K}{\Rightarrow} & \downarrow{\scriptstyle\beta'} \\ Y & \xrightarrow[y]{} & Y' \end{array} \qquad \rightsquigarrow \qquad \begin{array}{ccc} F_\alpha * F_\beta & \xrightarrow{\sigma_H*\sigma_K} & F_{\alpha'} * F_{\beta'} \\ {\scriptstyle\ell_{\alpha\bigstar\beta}}\downarrow & \Rightarrow & \downarrow{\scriptstyle\ell_{\alpha'\bigstar\beta'}} \\ E(\alpha\bigstar\beta) & \xrightarrow{\xi} & E(\alpha'\bigstar\beta') \\ {\scriptstyle p_{\alpha\bigstar\beta}}\downarrow & \Rightarrow & \downarrow{\scriptstyle p_{\alpha'\bigstar\beta'}} \\ X\times Y & \xrightarrow[x\times y]{} & X'\times Y' \end{array}$$

This defines the map ξ (but see below for a more explicit construction). Futhermore since as an induced map ξ is compatible with the canonical splitting $j_\bigstar \circ \chi$ of the exterior join we obtain the relation

$$\{\xi \circ \lambda\}\mathrm{HI}_{(\ell_\bigstar,p_\bigstar;j_\bigstar\circ\chi)} = (\sigma_H * \sigma_K) \circ \lambda\mathrm{HI}_{(\ell_\bigstar,p_\bigstar;j_\bigstar\circ\chi)}.$$

Also we have $\{\lambda' \circ \mu\}\mathrm{HI}_{(\ell_\bigstar,p_\bigstar;j_\bigstar\circ\chi)} = \lambda'\mathrm{HI}_{(\ell_\bigstar,p_\bigstar;j_\bigstar\circ\chi)} \circ \Sigma\Omega\mu$ directly from the definitions so it remains to verify that $\xi \circ \lambda = \lambda' \circ \mu$. For this purpose we note that both composite maps $\xi \circ \lambda$ and $\lambda' \circ \mu$ are induced under the double mapping cylinder functor $\mathcal{M}$, as we indicate in the two diagrams below.

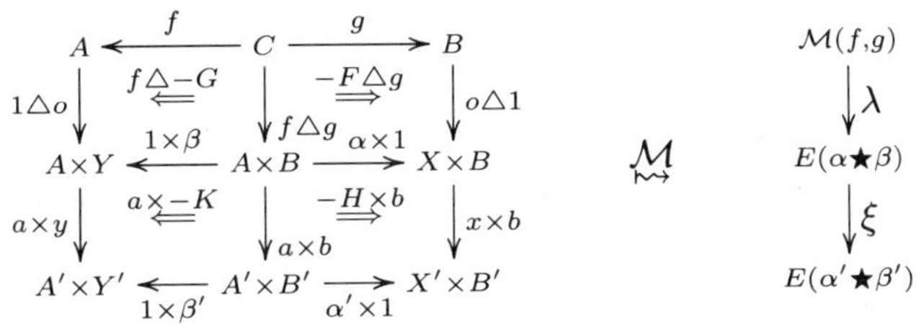

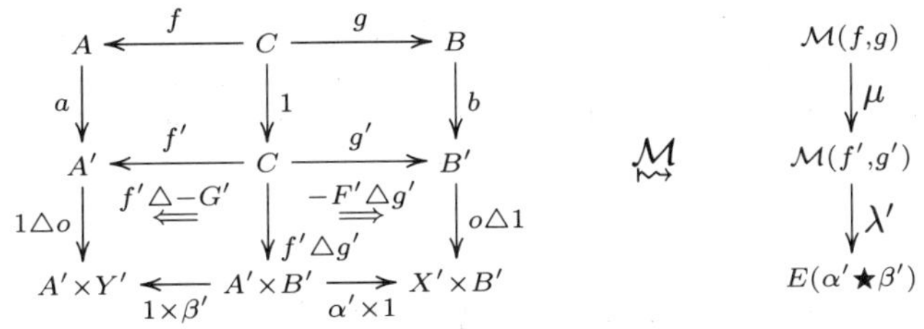

Because $G' = Kg + yG$ we have equalities of homotopies

$$
\begin{aligned}
(a \times y)(f\triangle(-G)) + (a \times -K)(f\triangle g) &= (a \circ f)\triangle(-yG) + (a \circ f)\triangle(-Kg)\\
&= (a \circ f)\triangle(-yG - Kg)\\
&= f'\triangle(-G').
\end{aligned}
$$

Similarly $(x \times b)(-F\triangle g) + (-H \times b)(f\triangle g) = -F\triangle g'$. These considerations imply that the composite morphisms to the left in these diagrams are actually equal before application of the functor $\mathcal{M}$. Hence indeed $\xi \circ \lambda = \lambda' \circ \mu$. $\square$

Next we present naturality for cHI. For clarity we use c and c' for the respective Ganea maps defined by (3.1).

PROPOSITION 3.3. *With reference to diagram* (3.1) *each of the following squares is homotopy commutative.*

$$\begin{array}{ccc} \mathcal{M}(f,g) & \xrightarrow{c} & X \vee Y \\ \mu \downarrow & & \downarrow x \vee y \\ \mathcal{M}(f',g') & \xrightarrow{c'} & X' \vee Y' \end{array} \qquad \begin{array}{ccc} \Sigma\Omega\mathcal{M}(f,g) & \xrightarrow{c\mathrm{HI}_{(\ell,\mathrm{inc};\chi)}} & \Omega X * \Omega Y \\ \Sigma\Omega\mu \downarrow & & \downarrow \Omega x * \Omega y \\ \Sigma\Omega\mathcal{M}(f',g') & \xrightarrow{c'\mathrm{HI}_{(\ell,\mathrm{inc};\chi)}} & \Omega X' * \Omega Y' \end{array}$$

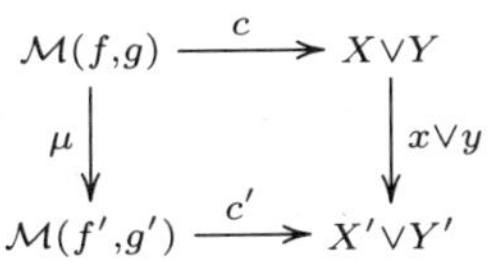

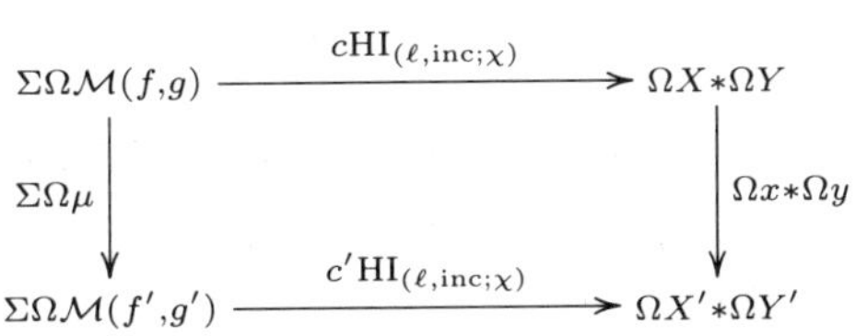

PROOF. By the general theory homotopy commutativity of the right square follows from that of the left so we consider only the case of the left square. Now both composites in the left square are induced under the double mapping cylinder functor $\mathcal{M}$ as indicated below.

$$\begin{array}{ccc} A \xleftarrow{f} C \xrightarrow{g} B & & \mathcal{M}(f,g) \\ \alpha \downarrow \stackrel{F}{\Leftarrow} \downarrow 1 \stackrel{G}{\Rightarrow} \downarrow \beta & & \downarrow c \\ X \xleftarrow{} * \xrightarrow{} Y & \stackrel{\mathcal{M}}{\rightsquigarrow} & X \vee Y \\ x \downarrow \quad \downarrow \quad \downarrow y & & \downarrow x \vee y \\ X' \xleftarrow{} * \xrightarrow{} Y' & & X' \vee Y' \end{array}$$

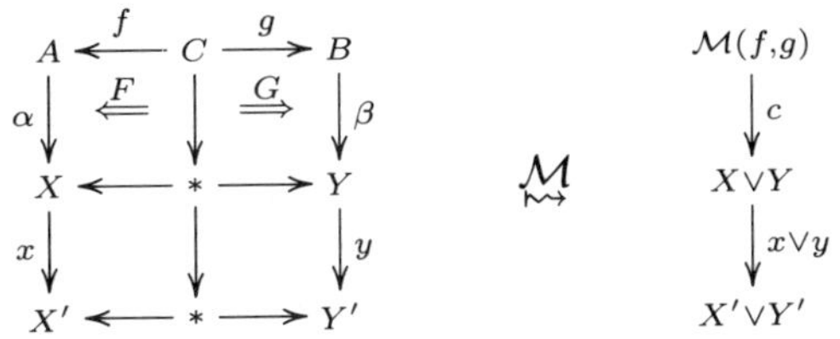

$$\begin{array}{ccc}
A \xleftarrow{f} C \xrightarrow{g} B & & \mathcal{M}(f,g) \\
{\scriptstyle a}\downarrow \quad {\scriptstyle 1}\downarrow \quad {\scriptstyle b}\downarrow & & \downarrow{\scriptstyle \mu} \\
A' \xleftarrow{f'} C \xrightarrow{g'} B' & \overset{\mathcal{M}}{\rightsquigarrow} & \mathcal{M}(f',g') \\
{\scriptstyle \alpha'}\downarrow \; {\scriptstyle F'}\Leftarrow \; \downarrow \; {\scriptstyle G'}\Rightarrow \downarrow {\scriptstyle \beta'} & & \downarrow{\scriptstyle c'} \\
X' \xleftarrow{} * \xrightarrow{} Y' & & X' \vee Y'
\end{array}$$

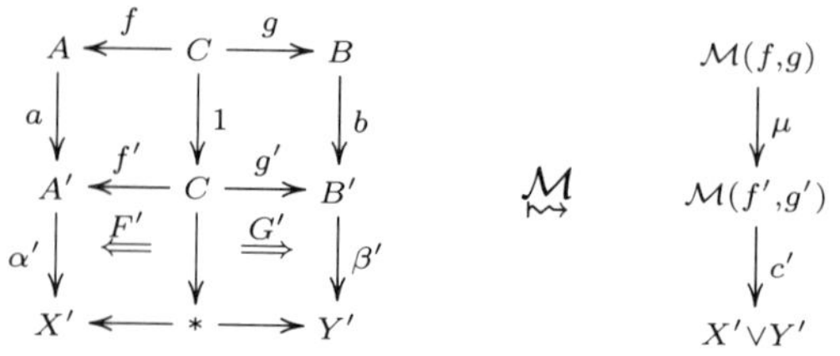

Since $F' = Hf + xF$ and $G' = Kg + yG$ the triple of homotopies $(H, 1_{(C \to *)}, K)$, with $1_{(C \to *)}$ denoting $C \times [0,1] \to *$, constitutes a modification from the morphism inducing $(x \vee y) \circ c$ to the morphism inducing $c' \circ \mu$. It is a property of the double mapping cylinder functor $\mathcal{M}$ that morphisms related in this way induce homotopic maps under $\mathcal{M}$. Therefore $(x \vee y) \circ c$ and $c' \circ \mu$ are homotopic and the left square in the proposition is homotopy commutative. □

4. Universal examples for λHI and cHI

In this section we use the naturality results of §3 to identify appropriate "universal" examples for λHI and cHI. Let diagram (0.1) be given; we may use it to construct the following diagram (3.1) situation.

(4.1)
$$\begin{array}{cccccc}
A \xleftarrow{1} & A \xleftarrow{f} & C \xrightarrow{g} & B \xrightarrow{1} & B \\
{\scriptstyle \alpha}\downarrow & {\scriptstyle i_1^f}\downarrow \; {\scriptstyle D_f}\Leftarrow \downarrow & {\scriptstyle D_g}\Rightarrow \downarrow {\scriptstyle i_1^g} & \downarrow {\scriptstyle \beta} \\
X \xleftarrow[\mu_F]{} & C_f \xleftarrow{} & * \xrightarrow{} & C_g \xrightarrow[\mu_G]{} & Y
\end{array}$$

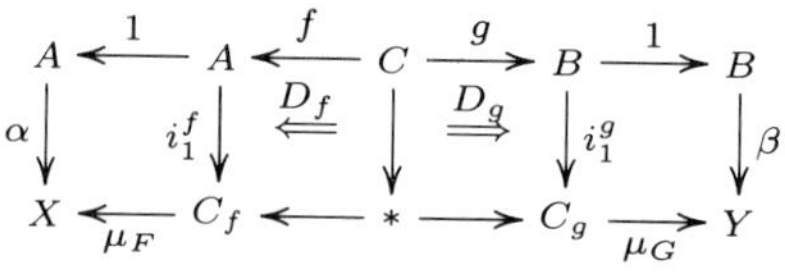

We first consider the case of λHI and apply Proposition 3.2 to diagram (4.1). There results a homotopy commutative square of the form:

(4.2)
$$\begin{array}{ccc}
\mathcal{M}(f,g) & \longrightarrow & E(i_1^f \bigstar i_1^g) \\
{\scriptstyle id}\downarrow & & \downarrow{\scriptstyle \xi} \\
\mathcal{M}(f,g) & \xrightarrow{\lambda} & E(\alpha \bigstar \beta)
\end{array}$$

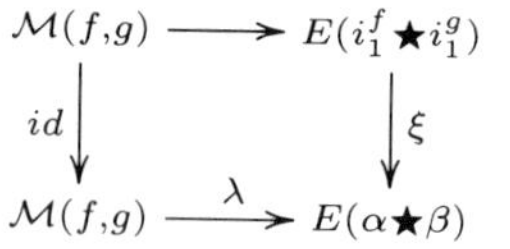

To better interpret the map $\mathcal{M}(f,g) \to E(i_1^f \bigstar i_1^g)$ we recall the following definition.

DEFINITION 4.3. If (X, X_0) and (Y, Y_0) are pairs of spaces then their *partial product* is defined by

$$(X, X_0) \dot\times (Y, Y_0) := \{(x,y) \in X \times Y \mid x \in X_0 \text{ or } y \in Y_0\}.$$

Note that $(X, *) \dot\times (Y, *) = X \vee Y$ and that $(X, X_0) \dot\times (Y, Y_0) = X \times Y$ if either $X_0 = X$ or $Y_0 = Y$.

Observe also that the pushout square

$$\begin{array}{ccc}
X_0 \times Y_0 & \xrightarrow{\text{inc} \times 1} & X \times Y_0 \\
{\scriptstyle 1 \times \text{inc}}\downarrow & & \downarrow{\scriptstyle \text{inc}} \\
X_0 \times Y & \xrightarrow[\text{inc}]{} & (X, X_0) \dot\times (Y, Y_0)
\end{array}$$

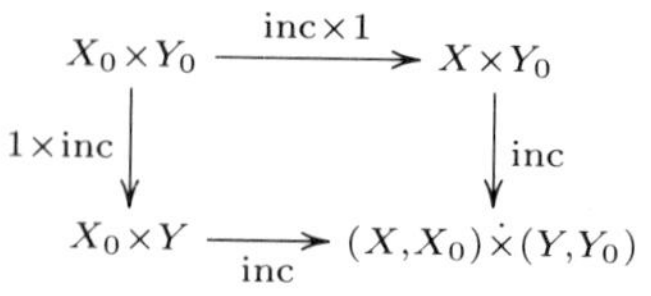

will be a homotopy pushout if the inclusions $X_0 \subset X$ and $Y_0 \subset Y$ are closed cofibrations and thus in this case that the induced map from the exterior join (the homotopy pushout) to the partial product (the topological pushout)

$$\phi: E((X_0 \subset X)\bigstar(Y_0 \subset Y)) \to (X, X_0)\dot{\times}(Y, Y_0)$$

is a homotopy equivalence.

We now return to (4.2). Since we work in the category of well-pointed spaces, the inclusions $i_1^f: A \to C_f$ and $i_1^g: B \to C_g$ are closed fibrations so by the above observation the induced map $\phi: E(i_1^f \bigstar i_1^g) \to (C_f, A)\dot{\times}(C_g, B)$ is a homotopy equivalence. We abbreviate $C_f \dot{\times} C_g := (C_f, A)\dot{\times}(C_g, B)$ and rewrite (4.2) as a square of the following form.

$$\begin{array}{ccc} \mathcal{M}(f,g) & \xrightarrow{\widetilde{\lambda}} & C_f \dot{\times} C_g \\ {\scriptstyle id}\downarrow & & \downarrow{\scriptstyle \widetilde{\xi}} \\ \mathcal{M}(f,g) & \xrightarrow{\lambda} & E(\alpha\bigstar\beta) \end{array}$$

We refer to the Hopf invariant

$$\text{(4.4)} \qquad \dot{\times}\mathrm{HI} := \widetilde{\lambda}\mathrm{HI}_{(\ell,\mathrm{inc};\mathrm{inc}\circ\chi)}: \Sigma\Omega\mathcal{M}(f,g) \to F(i_1^f) * F(i_1^g)$$

as the *partial product Hopf invariant* of the cotriad (f, g). We remark that if $u: \Sigma U \to \mathcal{M}(f, g)$ is given then

$$\ell \circ (\dot{\times}\mathrm{HI})(u) = \lambda \circ u - i_{C_g} \circ \lambda_g \circ u - i_{C_f} \circ \lambda_f \circ r \circ u$$

as classes $\Sigma U \to C_f \dot{\times} C_g$. This is called the *characterizing equation* for $\dot{\times}$HI. It uniquely specifies $\dot{\times}$HI since $\ell: F(i_1^f) * F(i_1^g) \to C_f \dot{\times} C_g$ is a monomorphism on homotopy.

Additionally let $\sigma_0: F_{i_1^f} \to F_\alpha$ and $\sigma_1: F_{i_1^g} \to F_\beta$ be the maps induced on homotopy fibers. Then by application of Proposition 3.2 we have:

PROPOSITION 4.5. $\lambda\mathrm{HI} = (\sigma_0 * \sigma_1) \circ (\dot{\times}\mathrm{HI})$ *as classes* $\Sigma\Omega\mathcal{M}(f, g) \to F_\alpha * F_\beta$.

This proposition shows that $\dot{\times}$HI serves as a universal example for Hopf invariants of the type λHI.

We move next to consideration of a universal example for cHI. We recall [4, Definition 4.1] that the data in the following homotopy commutative diagram

$$\begin{array}{ccccc} & & \Omega C_f * \Omega C_g & & \\ & & \downarrow{\scriptstyle \ell} & & \\ \mathcal{M}(f,g) & \xrightarrow{\psi} & C_f \vee C_g & \xleftarrow{\chi} & \Sigma\Omega(C_f \times C_g) \\ & {\scriptstyle (\lambda_f \circ r)\triangle\lambda_g}\searrow & \downarrow{\scriptstyle \mathrm{inc}} & \swarrow{\scriptstyle \varepsilon(C_f \times C_g)} & \\ & & C_f \times C_g & & \end{array}$$

defines the *Ganea-Hopf invariant*

$$\mathrm{GHI} := \psi\mathrm{HI}_{(\ell,\mathrm{inc};\chi)}: \Sigma\Omega\mathcal{M}(f, g) \to \Omega C_f * \Omega C_g$$

associated to the cotriad $A \xleftarrow{f} C \xrightarrow{g} B$. Observe that ψ is just the Ganea map associated to the Ganea situation:

$$\begin{array}{ccccc}
A & \xleftarrow{f} & C & \xrightarrow{g} & B \\
{\scriptstyle i_1^f}\downarrow & \overset{D_f}{\Leftarrow} & \downarrow & \overset{D_g}{\Rightarrow} & \downarrow{\scriptstyle i_1^g} \\
C_f & \longleftarrow & * & \longrightarrow & C_g
\end{array}$$

We apply Proposition 3.3 to diagram (4.1). This yields the following homotopy commutative square

$$\begin{array}{ccc}
\mathcal{M}(f,g) & \xrightarrow{\psi} & C_f \vee C_g \\
{\scriptstyle id}\downarrow & & \downarrow{\scriptstyle \mu_F \vee \mu_G} \\
\mathcal{M}(f,g) & \xrightarrow{c} & X \vee Y
\end{array}$$

from which we conclude that cHI= $(\Omega\mu_F * \Omega\mu_G) \circ$ GHI. Consequently GHI serves as a universal example for Hopf invariants of the type cHI.

5. Comparison results

In this section we assume diagram (0.1) is given and proceed to embed it in the following diagram (3.1) situation

$$(5.1) \qquad \begin{array}{ccccccccc}
* & \longleftarrow & A & \xleftarrow{f} & C & \xrightarrow{g} & B & \longrightarrow & * \\
\downarrow & \overset{-D_\alpha}{\Leftarrow} & {\scriptstyle \alpha}\downarrow & \overset{F}{\Leftarrow} & \downarrow & \overset{G}{\Rightarrow} & \downarrow{\scriptstyle \beta} & \overset{-D_\beta}{\Rightarrow} & \downarrow \\
C_\alpha & \xleftarrow[i_1]{} & X & \longleftarrow & * & \longrightarrow & Y & \xrightarrow[i_1]{} & C_\beta
\end{array}$$

where D_α and D_β denote defining homotopies in the respective mapping cones. Our interest is in the following pairs of composite maps associated with this data.

$$(5.2) \qquad \begin{cases} \mathcal{M}(f,g) \xrightarrow{\lambda} E(\alpha\bigstar\beta) \xrightarrow{w} C_\alpha \vee C_\beta \\ \mathcal{M}(f,g) \xrightarrow{\psi} C_f \vee C_g \xrightarrow{T} C_g \vee C_f \xrightarrow{\kappa\vee\kappa} \Sigma C \vee \Sigma C \xrightarrow{\zeta_F \vee \zeta_G} C_\alpha \vee C_\beta \end{cases}$$

$$(5.3) \qquad \begin{cases} \mathcal{M}(f,g) \xrightarrow{c} X \vee Y \xrightarrow{i_1 \vee i_1} C_\alpha \vee C_\beta \\ \mathcal{M}(f,g) \xrightarrow{\psi} C_f \vee C_g \xrightarrow{\kappa\vee\kappa} \Sigma C \vee \Sigma C \xrightarrow{r \vee r} \Sigma C \vee \Sigma C \xrightarrow{\zeta_F \vee \zeta_G} C_\alpha \vee C_\beta \end{cases}$$

Here T denotes interchange of wedge factors and w is as explicitly constructed in [8, Theorem 7.3]. The map w is a generalized Whitehead product map. The canonical operator κ denotes a collapsing map. The maps $\zeta_F: \Sigma C \to C_\alpha$ and $\zeta_G: \Sigma C \to C_\beta$, which we refer to as coextensions, use notation as introduced in (1.1).

PROPOSITION 5.4. *The maps in* (5.2) *and in* (5.3) *are homotopic pairs respectively.*

PROOF. We first apply Proposition 3.2 to diagram (5.1). In the corresponding square

$$
\begin{array}{ccc}
\mathcal{M}(f,g) & \xrightarrow{\ \lambda\ } & E(\alpha\bigstar\beta) \\
{\scriptstyle \mu=\kappa}\downarrow & & \downarrow{\scriptstyle \xi} \\
\Sigma C & \xrightarrow{\ \lambda'\ } & C_\beta \vee C_\alpha = E\left(\begin{array}{c} * \\ \downarrow \\ C_\alpha \end{array} \bigstar \begin{array}{c} * \\ \downarrow \\ C_\beta \end{array}\right)
\end{array}
$$

it may be checked (cf [10, Lemma 3.5]) that λ' equals the composite

$$\Sigma C \xrightarrow{\ \omega_{\Sigma C}\ } \Sigma C \vee \Sigma C \xrightarrow{\ \mu_{G'} \vee \mu_{-F'}\ } C_\beta \vee C_\alpha$$

where $F' = -D_\alpha f + i_1 F$ and $G' = -D_\beta g + i_1 G$. The relations $\zeta_F = \mu_{-F'}$ and $\zeta_G = \mu_{-G'}$ are readily established. Then the following diagram is seen to be homotopy commutative.

$$
\begin{array}{ccccccc}
\mathcal{M}(f,g) & \xrightarrow{\ \kappa\ } & \Sigma C & & & & \\
 & & {\scriptstyle \omega}\downarrow & & & & \\
{\scriptstyle \psi}\downarrow & & \Sigma C \vee \Sigma C & & \searrow{\scriptstyle \mu_{G'} \vee \mu_{-F'}} & & \\
 & & {\scriptstyle r\vee 1}\downarrow & & & & \\
C_f \vee C_g & \xrightarrow[\kappa\vee\kappa]{} & \Sigma C \vee \Sigma C & \xrightarrow[r\vee r]{} & \Sigma C \vee \Sigma C & \xrightarrow[\mu_{G'} \vee \mu_{F'}]{} & C_\beta \vee C_\alpha \\
{\scriptstyle T}\downarrow & & {\scriptstyle T}\downarrow & & & & \downarrow{\scriptstyle T} \\
C_g \vee C_f & \xrightarrow{\ \kappa\vee\kappa\ } & \Sigma C \vee \Sigma C & & \xrightarrow{\ \zeta_F \vee \zeta_G\ } & & C_\alpha \vee C_\beta
\end{array}
$$

Sine $E(\alpha\bigstar\beta) \xrightarrow{\xi} C_\beta \vee C_\alpha \xrightarrow{T} C_\alpha \vee C_\beta$ may be identified with the map w, we conclude that $w \circ \lambda = (\zeta_F \vee \zeta_G) \circ (\kappa \vee \kappa) \circ T \circ \psi$, as claimed.

Next we apply Proposition 3.3 to diagram (5.1). This yields the square

$$
\begin{array}{ccc}
\mathcal{M}(f,g) & \xrightarrow{\ c\ } & X \vee Y \\
{\scriptstyle \mu=\kappa}\downarrow & & \downarrow{\scriptstyle i_1 \vee i_1} \\
\Sigma C & \xrightarrow{\ c'\ } & C_\alpha \vee C_\beta
\end{array}
$$

in which c' is, by use of [10, Lemma 3.5], just the composite

$$\Sigma C \xrightarrow{\ \omega_{\Sigma C}\ } \Sigma C \vee \Sigma C \xrightarrow{\ \mu_{-F} \vee \mu_{G'}\ } C_\alpha \vee C_\beta.$$

The following diagram

$$
\begin{array}{ccccccc}
\mathcal{M}(f,g) & \xrightarrow{\ \kappa\ } & \Sigma C & & & & \\
 & & {\scriptstyle \omega}\downarrow & & & & \\
{\scriptstyle \psi}\downarrow & & \Sigma C \vee \Sigma C & & \searrow{\scriptstyle \mu_{-F} \vee \mu_{G'}} & & \\
 & & {\scriptstyle r\vee 1}\downarrow & & & & \\
C_f \vee C_g & \xrightarrow[\kappa\vee\kappa]{} & \Sigma C \vee \Sigma C & \xrightarrow[r\vee r]{} & \Sigma C \vee \Sigma C & \xrightarrow[\zeta_F \vee \zeta_G]{} & C_\alpha \vee C_\beta
\end{array}
$$

is homotopy commutative and gives $(i_1 \vee i_1) \circ c = (\zeta_F \vee \zeta_G) \circ (r \vee r) \circ (\kappa \vee \kappa) \circ \psi$, as claimed. □

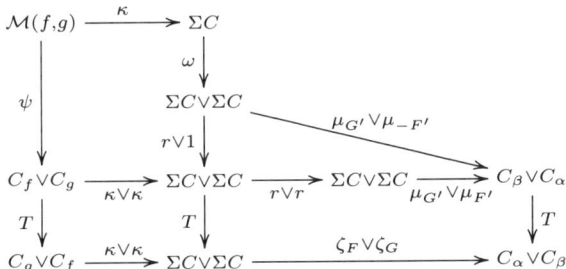

THEOREM 5.5. *With reference to diagram* (0.1) *the equalities*

$$\{w \circ \lambda\}\mathrm{HI} = (\Omega\zeta_F * \Omega\zeta_G) \circ (\Omega\kappa * \Omega\kappa) \circ T \circ \mathrm{GHI} - [\zeta_F \circ \kappa \circ \varepsilon, \zeta_G \circ r \circ \kappa \circ \varepsilon]'$$

and

$$(\Omega i_1^\alpha * \Omega i_1^\beta) \circ c\mathrm{HI} = (\Omega(-\zeta_F) * \Omega(-\zeta_G)) \circ (\Omega\kappa * \Omega\kappa) \circ \mathrm{GHI}$$

are valid as classes $\Sigma\Omega\mathcal{M}(f,g) \to \Omega C_\alpha * \Omega C_\beta$. *In this instance* T *of course denotes the interchange map* $\Omega C_f * \Omega C_g \to \Omega C_g * \Omega C_f$ *(see* [4, *Definition* 1.7]*) while* $\varepsilon = \varepsilon_{\mathcal{M}(f,g)}\colon \Sigma\Omega\mathcal{M}(f,g) \to \mathcal{M}(f,g)$ *is the evaluation map.*

PROOF. We consider the following homotopy commutative diagram of vertical fibrations in which $\mu^{\cdot}_\alpha\colon F_\alpha \to \Omega C_\alpha$ and $\mu^{\cdot}_\beta\colon F_\beta \to \Omega C_\beta$ are adjoints of the respective homotopy Thom classes.

$$\begin{array}{ccccc}
F_\alpha * F_\beta & \xrightarrow{\mu^{\cdot}_\alpha * \mu^{\cdot}_\beta} & \Omega C_\alpha * \Omega C_\beta & \xrightarrow{1*1} & \Omega C_\alpha * \Omega C_\beta \\
\downarrow{\scriptstyle \ell_\star} & & \downarrow{\scriptstyle \ell_\star} & & \downarrow{\scriptstyle \ell} \\
E(\alpha\bigstar\beta) & \xrightarrow{\xi} & C_\beta \vee C_\alpha & \xrightarrow{T} & C_\alpha \vee C_\beta \\
\downarrow{\scriptstyle p_\star} & & \downarrow{\scriptstyle p_\star} & & \downarrow{\scriptstyle \mathrm{inc}} \\
X \times Y & \xrightarrow{i_1 \times i_1} & C_\alpha \times C_\beta & \xrightarrow{1\times 1} & C_\alpha \times C_\beta
\end{array}$$

Compatibility of splittings is determined by the diagram:

$$\begin{array}{ccccc}
\Sigma\Omega(X \times Y) & \xrightarrow{\Sigma\Omega(i_1 \times i_1)} & \Sigma\Omega(C_\alpha \times C_\beta) & \xrightarrow{\Sigma\Omega(1\times 1)} & \Sigma\Omega(C_\alpha \times C_\beta) \\
\downarrow{\scriptstyle j_\star \circ \chi} & & \downarrow{\scriptstyle T \circ \chi} & & \downarrow{\scriptstyle \chi} \\
E(\alpha\bigstar\beta) & \xrightarrow{\xi} & C_\beta \vee C_\alpha & \xrightarrow{T} & C_\alpha \vee C_\beta
\end{array}$$

Hence and by Proposition 5.4 we may write

$$\begin{aligned}
(\mu^{\cdot}_\alpha * \mu^{\cdot}_\beta) \circ \lambda\mathrm{HI}_{(\ell_\star, p_\star; j_\star \circ \chi)} &= \{T \circ \xi \circ \lambda\}\mathrm{HI}_{(\ell, \mathrm{inc}; \chi)} \\
&= \{w \circ \lambda\}\mathrm{HI}_{(\ell, \mathrm{inc}; \chi)} \\
&= \{(\zeta_F \vee \zeta_G) \circ (\kappa \vee \kappa) \circ T \circ \psi\}\mathrm{HI}_{(\ell, \mathrm{inc}; \chi)} \\
&= (\Omega\zeta_F * \Omega\zeta_G) \circ (\Omega\kappa * \Omega\kappa) \circ \{T \circ \psi\}\mathrm{HI}_{(\ell, \mathrm{inc}; \chi)}
\end{aligned}$$

Next by using [4, Theorem 3.3] we have

$$\{T \circ \psi\}\mathrm{HI}_{(\ell, \mathrm{inc}; \chi)} = T \circ \psi\mathrm{HI}_{(\ell, \mathrm{inc}; \chi)} - [\lambda_g \circ \varepsilon, \lambda_f \circ r \circ \varepsilon]'$$

as classes $\Sigma\Omega\mathcal{M}(f,g) \to \Omega C_g * \Omega C_f$. The first equality now readily follows from these observations.

Also by Proposition 5.4,

$$\begin{aligned}
(\Omega i_1 * \Omega i_1) \circ c\mathrm{HI}_{(\ell, \mathrm{inc}; \chi)} &= \{(i_1 \vee i_1) \circ c\}\mathrm{HI}_{(\ell, \mathrm{inc}; \chi)} \\
&= \{(\zeta_F \vee \zeta_G) \circ (r \vee r) \circ (\kappa \vee \kappa) \circ \psi\}\mathrm{HI}_{(\ell, \mathrm{inc}; \chi)} \\
&= (\Omega(-\zeta_F) * \Omega(-\zeta_G)) \circ (\Omega\kappa * \Omega\kappa) \circ \psi\mathrm{HI}_{(\ell, \mathrm{inc}; \chi)}
\end{aligned}$$

which yields the second equality since $\mathrm{GHI} = \psi\mathrm{HI}_{(\ell, \mathrm{inc}; \chi)}$ by definition. □

6. The extended join operation and an identity in $[A*B, (A\times A)*(B\times B)]$

PROPOSITION 6.1. *For any space X the equality*

$$\Sigma\triangle = \Sigma i_1 + \kappa \circ d_X + \Sigma i_2 \colon \Sigma X \to \Sigma(X \times X)$$

is valid. Here $\triangle\colon X \to X \times X$ is the diagonal map, $i_1, i_2\colon X \to X \times X$ the two inclusions, $\kappa\colon X*X \to \Sigma(X\times X)$ the canonical collapsing map and $d_X\colon \Sigma X \to X*X$ the diagonal class (defined in [4, Definition 2.2]).

PROOF. The homotopy class d_X is characterized by the equation $\nu\circ d_X = \Sigma\widetilde{\triangle}$ where ν is the composite

$$X * X \xrightarrow{\kappa} \Sigma(X \times X) \xrightarrow{\Sigma q} \Sigma(X \wedge X)$$

and $\widetilde{\triangle} = q\circ\triangle\colon X \to X\wedge X$ denotes the reduced diagonal. Because ν is a homotopy equivalence there is also a unique class $\widetilde{\kappa}\colon \Sigma(X\wedge X) \to \Sigma(X\times X)$ such that $\widetilde{\kappa}\circ\nu = \kappa$. Moreover, since κ may be regarded as the Hopf construction on the identity map $1\colon X \times X \to X \times X$, the relation

$$\widetilde{\kappa} \circ \Sigma q = -\Sigma(i_1 \circ p_1) + 1_{\Sigma(X\times X)} - \Sigma(i_2 \circ p_2)$$

is valid. This last equation uses a general fact about Hopf constructions (cf. [7,(2.1)]). Here $p_1, p_2\colon X \times X \to X$ are the two projection maps. Hence

$$\begin{aligned} \kappa \circ d_X &= \widetilde{\kappa} \circ \nu \circ d_X = \widetilde{\kappa} \circ \Sigma\widetilde{\triangle} = \widetilde{\kappa} \circ \Sigma q \circ \Sigma\triangle \\ &= \{-\Sigma(i_1 \circ p_1) + 1_{\Sigma(X\times X)} - \Sigma(i_2 \circ p_2)\} \circ \Sigma\triangle \\ &= -\Sigma i_1 + \Sigma\triangle - \Sigma i_2 \end{aligned}$$

and the proposition is established. □

NOTATION 6.2. For spaces A and B belonging to the category in which we work (that is, well-pointed spaces), the canonical operator

$$\begin{aligned} \varphi_{A,B}\colon A * B &\longrightarrow (\Sigma A) \wedge B \\ [a, b, t] &\longmapsto [[a, t], b] \end{aligned}$$

for $a \in A$, $b \in B$, $0 \le t \le 1$, is a homotopy equivalence. In fact the diagram

(6.3)
$$\begin{array}{ccc} A*B & \xrightarrow{\quad\varphi_{A,B}\quad} & (\Sigma A)\wedge B \\ & {\scriptstyle\nu}\searrow \quad \nearrow{\scriptstyle \mathrm{sh}} & \\ & \Sigma(A\wedge B) & \end{array}$$

is commutative where sh is a "shuffle" homeomorphism (cf. [2]). Recall that the cogroup structure on $A*B$ is obtained from the suspension comultiplication $\omega_{\Sigma(A\wedge B)}$ on $\Sigma(A \wedge B)$ by requiring that ν be homotopy linear. In fact a comultiplication ω' on $(\Sigma A) \wedge B$ can be defined by the composite

$$\omega'\colon (\Sigma A) \wedge B \xrightarrow{\omega_{\Sigma A} \wedge 1} (\Sigma A \vee \Sigma A) \wedge B \xrightarrow{\cong} ((\Sigma A) \wedge B) \vee ((\Sigma A) \wedge B)$$

where $\omega_{\Sigma A}\colon \Sigma A \to \Sigma A \vee \Sigma A$ is the suspension comultiplication on ΣA. Then the maps sh and $\varphi_{A,B}$ are readily seen to be homotopy linear with respect to these comultiplications. A canonical operator $\rho_{A,B}: A * B \to A \wedge (\Sigma B)$ may also be defined and allows a similar discussion; details are left to the reader.

DEFINITIONS 6.4. (a) Let maps $\Omega: \Sigma A \to \Sigma Z$ and $k: B \to W$ be given. We define a homotopy class $\Omega \circledast k: A * B \to Z * W$ by requiring that the square

$$\begin{array}{ccc} A*B & \xrightarrow{\Omega \circledast k} & Z*W \\ \Big\downarrow{\scriptstyle \varphi_{A,B}} & & \Big\downarrow{\scriptstyle \varphi_{Z,W}} \\ (\Sigma A)\wedge B & \xrightarrow[\Omega \wedge k]{} & (\Sigma Z)\wedge W \end{array}$$

be homotopy commutative. This construction is well-defined since the operator φ is a homotopy equivalence. Moreover, in view of diagram (6.3), it is clear that if $\Omega = \Sigma h$ with $h: A \to Z$ then $\Omega \circledast k = h * k$, the ordinary join product of maps.

(b) Similarly for given maps $h: A \to Z$ and $\eta: \Sigma B \to \Sigma W$ there is a homotopy class $h \circledast \eta: A * B \to Z * W$ defined by use of the operator ρ mentioned above. Details are omitted. And if $\eta = \Sigma k$ with $k: B \to W$ then $h \circledast \eta = h * k$, the ordinary join product of maps.

REMARK 6.5. The operations $\Omega \circledast k$ and $h \circledast \eta$ are versions of operations discussed in [1, p. 40].

The operation $\circledast$ is linear in the following sense.

PROPOSITION 6.6. *(a) For maps $\Omega_1, \Omega_2: \Sigma A \to \Sigma Z$ and $k: B \to W$ the equation $(\Omega_1 + \Omega_2) \circledast k = \Omega_1 \circledast k + \Omega_2 \circledast k$ holds.*

(b) For maps $h: A \to Z$ and $\eta_1, \eta_2: \Sigma B \to \Sigma W$ the equation $h \circledast (\eta_1 + \eta_2) = h \circledast \eta_1 + h \circledast \eta_2$ holds.

PROOF. We verify only part (a); part (b) is similar. Let $\widehat{+}$ denote the group operation in $[(\Sigma A) \wedge B, (\Sigma Z) \wedge W]$ induced by the above comultiplication ω' on $(\Sigma A) \wedge B$. Then it is immediate from the definition of ω' that

$$(**) \qquad \Omega_1 \wedge k \mathbin{\widehat{+}} \Omega_2 \wedge k = (\Omega_1 + \Omega_2) \wedge k.$$

Hence we have

$$\begin{aligned} &\varphi_{Z,W} \circ ((\Omega_1 + \Omega_2) \circledast k) \\ &\qquad = ((\Omega_1 + \Omega_2) \wedge k) \circ \varphi_{A,B} && \text{(by definition of } \circledast) \\ &\qquad = (\Omega_1 \wedge k \mathbin{\widehat{+}} \Omega_2 \wedge k) \circ \varphi_{A,B} && \text{(by relation } (**)) \\ &\qquad = (\Omega_1 \wedge k) \circ \varphi_{A,B} + (\Omega_2 \wedge k) \circ \varphi_{A,B} && \text{(since } \varphi_{A,B} \text{ is homotopy linear)} \\ &\qquad = \varphi_{Z,W} \circ (\Omega_1 \circledast k) + \varphi_{Z,W} \circ (\Omega_2 \circledast k) && \text{(by definition of } \circledast) \\ &\qquad = \varphi_{Z,W} \circ \{\Omega_1 \circledast k + \Omega_2 \circledast k\} \end{aligned}$$

as homotopy classes $A * B \to (\Sigma Z) \wedge W$. It follows that $(\Omega_1 + \Omega_2) \circledast k = \Omega_1 \circledast k + \Omega_2 \circledast k$ since $\varphi_{Z,W}$, being a homotopy equivalence, is a monomorphism on homotopy. □

THEOREM 6.7. *For any spaces A and B the equations*

$$\begin{aligned} \triangle * \triangle &= i_1 * i_1 + (\kappa \circ d_A) \circledast i_1 + i_2 * i_1 + \triangle \circledast (k \circ d_B) \\ &\qquad + i_1 * i_2 + (\kappa \circ d_A) \circledast i_2 + i_2 * i_2 \\ &= i_1 * i_1 + i_1 \circledast (\kappa \circ d_B) + i_1 * i_2 + (k \circ d_A) \circledast \triangle \\ &\qquad + i_2 * i_1 + i_2 \circledast (\kappa \circ d_B) + i_2 * i_2 \end{aligned}$$

hold in the group $[A*B,(A\times A)*(B\times B)]$. Here the various i_1's and i_2's denote appropriate inclusion maps and the $\triangle$'s the appropriate diagonal maps.

PROOF. Using properties of the operation $\circledast$ together with Proposition 6.1 we compute:

$$\begin{aligned}
\triangle*\triangle &= \triangle \circledast \Sigma\triangle \\
&= \triangle \circledast (\Sigma i_1 + \kappa \circ d_B + \Sigma i_2) \\
&= \triangle \circledast \Sigma i_1 + \triangle \circledast (\kappa \circ d_B) + \triangle \circledast \Sigma i_2 \\
&= \triangle * i_1 + \triangle \circledast (\kappa \circ d_B) + \triangle * i_2 \\
&= \Sigma\triangle \circledast i_1 + \triangle \circledast (\kappa \circ d_B) + \Sigma\triangle \circledast i_2 \\
&= (\Sigma i_1 + \kappa \circ d_A + \Sigma i_2) \circledast i_1 + \triangle \circledast (\kappa \circ d_B) + (\Sigma i_1 + \kappa \circ d_A + \Sigma i_2) \circledast i_2 \\
&= \Sigma i_1 \circledast i_1 + (\kappa \circ d_A) \circledast i_1 + \Sigma i_2 \circledast i_1 + \triangle \circledast (k \circ d_B) \\
&\quad + \Sigma i_1 \circledast i_2 + (\kappa \circ d_A) \circledast i_2 + \Sigma i_2 \circledast i_2 \\
&= i_1 * i_1 + (\kappa \circ d_A) \circledast i_1 + i_2 * i_1 + \triangle \circledast (k \circ d_B) \\
&\quad + i_1 * i_2 + (\kappa \circ d_A) \circledast i_2 + i_2 * i_2
\end{aligned}$$

Verification of the second equation is similar and hence omitted. □

COROLLARY 6.8. (a) If A is a suspension space then

$$\triangle*\triangle = i_1*i_1 + i_2*i_1 + \triangle \circledast (k \circ d_B) + i_1*i_2 + i_2*i_2$$

in the abelian group $[A*B,(A\times A)*(B\times B)]$.

(b) If A and B are both suspension spaces then

$$\triangle*\triangle = i_1*i_1 + i_2*i_1 + i_1*i_2 + i_2*i_2$$

in the abelian group $[A*B,(A\times A)*(B\times B)]$.

PROOF. The equations in (a) and (b) follow at once from Theorem 6.7 using the fact that $d_X = 0$ if X is a suspension space. Also note that $A*B$ is a 2-fold suspension provided at least one of A or B is a suspension space; in this case $[A*B,(A\times A)*(B\times B)]$ must be an abelian group. □

7. The case of a projection cotriad

In this section we study the case of a Ganea situation in which the domain cotriad is taken to be a projection cotriad. That is, we assume the following diagram, with induced Ganea map as indicated.

(7.1)
$$\begin{array}{ccccc}
A & \xleftarrow{p_A} A\times B \xrightarrow{p_B} & B & & A*B \\
\alpha\downarrow & {}_F\!\Downarrow \quad \downarrow \quad \Downarrow_G & \downarrow\beta & \mathcal{M} & \downarrow c \\
X & \xleftarrow{\quad} * \xrightarrow{\quad} & Y & & X\vee Y
\end{array}$$

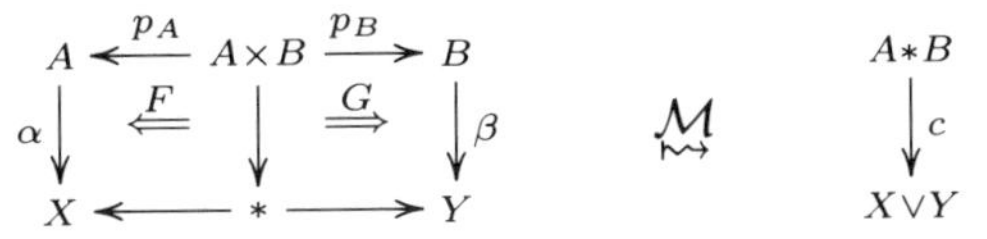

Our approach for studying the Hopf invariant λHI associated to (7.1) will be to examine the class $\lambda: A*B \to E(\alpha\bigstar\beta)$ which defines it in the context of §6.

We construct the following diagram in which $\lambda = \xi \circ \delta$ by Proposition 2.3.

(7.2)
$$\begin{array}{ccc}
A*B & \xrightarrow{\ \delta\ } & \mathcal{M}(p_A \times p_A, p_B \times p_B) \\
{\scriptstyle \Delta*\Delta}\big\downarrow & \overset{\phi}{\swarrow} & \big\downarrow{\scriptstyle \xi} \\
(A \times A)*(B \times B) & \xrightarrow[\ \overline{\xi}\ ]{} & E(\alpha \bigstar \beta)
\end{array}$$

The map ϕ is defined as indicated below; it is a homeomorphism. The map T denotes interchange of factors.

$$\begin{array}{ccccc}
A \times A & \xleftarrow{p_A \times p_A} & (A \times B) \times (A \times B) & \xrightarrow{p_B \times p_B} & B \times B \\
{\scriptstyle 1\times 1}\big\downarrow & & \big\downarrow{\scriptstyle 1\times T\times 1} & & \big\downarrow{\scriptstyle 1\times 1} \\
A \times A & \xleftarrow[p_1]{} & (A \times A) \times (B \times B) & \xrightarrow[p_2]{} & B \times B
\end{array}
\quad \overset{\mathcal{M}}{\rightsquigarrow} \quad
\begin{array}{c}
\mathcal{M}(p_A \times p_A, p_B \times p_B) \\
\big\downarrow{\scriptstyle \phi} \\
(A \times A)*(B \times B)
\end{array}$$

Plainly $\phi \circ \delta = \Delta * \Delta$. The map $\overline{\xi}$ in (7.2) is induced as follows.

$$\begin{array}{ccccc}
A \times A & \xleftarrow{p_1} & (A \times A) \times (B \times B) & \xrightarrow{p_2} & B \times B \\
{\scriptstyle 1\times o}\big\downarrow & \overset{\overline{H}}{\Leftarrow} & \big\downarrow{\scriptstyle p_1 \times p_2} & \overset{\overline{K}}{\Rightarrow} & \big\downarrow{\scriptstyle o\times 1} \\
A \times Y & \xleftarrow[1\times\beta]{} & A \times B & \xrightarrow[\alpha\times 1]{} & X \times B
\end{array}
\quad \overset{\mathcal{M}}{\rightsquigarrow} \quad
\begin{array}{c}
(A \times A)*(B \times B) \\
\big\downarrow{\scriptstyle \overline{\xi}} \\
E(\alpha \bigstar \beta)
\end{array}$$

Here the homotopies used are defined by $\overline{H}(a, a', b, b', t) = (a, G(a', b', 1-t))$ and $\overline{K}(a, a', b, b', t) = (F(a, b, 1-t), b')$ for $a, a' \in A$, $b, b' \in B$ and $0 \leq t \leq 1$. Comparison with the definition of ξ in (2.2) immediately implies that $\overline{\xi} \circ \phi = \xi$.

We also note that the diagram

(7.3)
$$\begin{array}{ccc}
(A*B)\vee(A*B) & \xrightarrow{i = i_1 \triangledown i_2} & \mathcal{M}(p_A \times p_A, p_B \times p_B) \\
& \underset{(i_1 * i_1)\triangledown(i_2 * i_2)}{\searrow} & \big\downarrow{\scriptstyle \phi} \\
& & (A \times A)*(B \times B)
\end{array}$$

is homotopy commutative where i_1 and i_2 denote appropriate inclusion maps.

Now consider the composite $\nu\colon A * B \xrightarrow{\kappa} \Sigma(A \times B) \xrightarrow{\Sigma q} \Sigma(A \wedge B)$ in which κ collapses the ends of the join to a point and $q\colon A \times B \to A \wedge B$ is the canonical quotient map from the cartesian to the smash product. Because we work with well-pointed spaces, ν is a homotopy equivalence; we use it to give the join $A * B$ the structure of an H-cogroup. Thus ν is homotopy linear (i.e., a co-H-map). Let $\overline{\nu}\colon \Sigma(A \wedge B) \to A * B$ be a fixed homotopy inverse for ν. We set:

(7.4)
$$\lambda\mathrm{HI}(1_{A*B}) := \lambda\mathrm{HI}(\overline{\nu}) \circ \nu\colon A * B \to F_\alpha * F_\beta$$

In view of the form of the characterizing equation for λHI given in Proposition 2.6 and by (7.3) we have:

$$\begin{aligned}
\ell_\bigstar \circ \lambda\mathrm{HI}(\overline{\nu}) &= \lambda \circ \overline{\nu} - \xi \circ i_2 \circ \overline{\nu} - \xi \circ i_1 \circ \overline{\nu} \\
&= \lambda \circ \overline{\nu} - \overline{\xi} \circ (i_2 * i_2) \circ \overline{\nu} - \overline{\xi} \circ (i_1 * i_1) \circ \overline{\nu}
\end{aligned}$$

But ν is homotopy linear so

$$\ell_\star \circ \lambda\mathrm{HI}(1_{A*B}) = (\lambda \circ \overline{\nu} - \overline{\xi} \circ (i_2 * i_2) \circ \overline{\nu} - \overline{\xi} \circ (i_1 * i_1) \circ \overline{\nu}) \circ \nu$$
$$= \lambda - \overline{\xi} \circ (i_2 * i_2) - \overline{\xi} \circ (i_1 * i_1)$$
$$= \overline{\xi} \circ (\Delta_A * \Delta_B - i_2 * i_2 - i_1 * i_1).$$

Next by use of Proposition 6.7 we obtain:

(7.5)
$$\begin{aligned}\Delta_A * \Delta_B &- i_2 * i_2 - i_1 * i_1 \\ &= i_1 * i_1 + (\kappa \circ d_A) \circledast i_1 + i_2 * i_1 + \Delta_A \circledast (\kappa \circ d_B) \\ &\quad + i_1 * i_2 + (\kappa \circ d_A) \circledast i_2 - i_1 * i_1\end{aligned}$$

Now in (7.1) it is evident that by necessity both α and β are null homotopic maps. In fact if $i_A: A \to A \times B$ and $i_B: B \to A \times B$ are the inclusions then we have homotopies $Fi_A: o \Rightarrow \alpha$, $Gi_B: o \Rightarrow \beta$, $Fi_B: o \Rightarrow o$ and $Gi_A: o \Rightarrow o$. Using notation introduced in (1.1) we define maps

$$\rho_{AA} := \sigma_{Fi_A}: A \to F_\alpha \qquad \rho_{BB} := \sigma_{Gi_B}: B \to F_\beta$$
$$\rho_{AB} := \sigma_{Gi_A}: A \to F_\beta \qquad \rho_{BA} := \sigma_{Fi_B}: B \to F_\alpha$$

which satisfy $\ell_\alpha \circ \rho_{AA} = 1_A$, $\ell_\beta \circ \rho_{BB} = 1_B$, $\ell_\alpha \circ \rho_{BA} = o: B \to A$ and $\ell_\beta \circ \rho_{AB} = o: A \to B$.

PROPOSITION 7.6. *The equations*
(1) $\overline{\xi} \circ (i_1 * i_2) = \ell_\star \circ (\rho_{AA} * \rho_{BB})$
(2) $\overline{\xi} \circ (i_2 * i_1) = \ell_\star \circ T \circ (\rho_{AB} * \rho_{BA})$
are valid as classes $A * B \to E(\alpha \bigstar \beta)$. *Here* $T: F_\beta * F_\alpha \to F_\alpha * F_\beta$ *is the exchange of factors map.*

PROOF. (1) By direct inspection of definitions the composite $\overline{\xi} \circ (i_1 * i_2)$ is seen to be induced under $\mathcal{M}$ as given below, with homotopies F and G specified by $\widetilde{F}(a,b,t) = (F(a,*,1-t),b)$ and $\widetilde{G}(a,b,t) = (a, G(*,b,1-t))$.

$$\begin{array}{ccccc} A & \xleftarrow{p_A} & A \times B & \xrightarrow{p_B} & B \\ {\scriptstyle i_A}\downarrow & \overset{\widetilde{G}}{\Leftarrow} & \downarrow{\scriptstyle 1\times 1} & \overset{\widetilde{F}}{\Rightarrow} & \downarrow{\scriptstyle i_B} \\ A\times Y & \xleftarrow[1\times\beta]{} & A\times B & \xrightarrow[\alpha\times 1]{} & X\times B \end{array} \qquad \mathcal{M} \rightsquigarrow \qquad \begin{array}{c} A*B \\ \downarrow{\scriptstyle \overline{\xi}\circ(i_1*i_2)} \\ E(\alpha\bigstar\beta) \end{array}$$

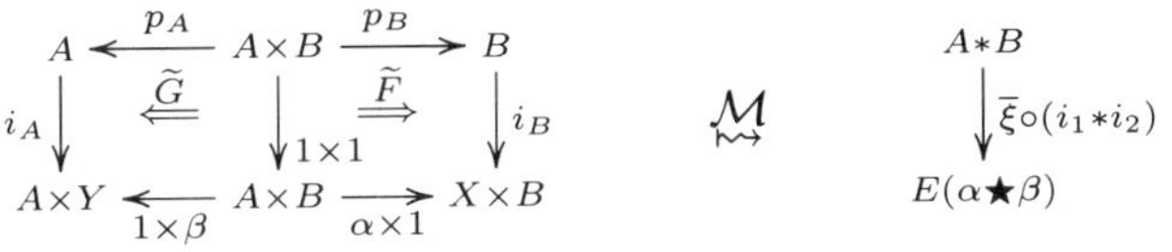

But the diagram of cotriads here on the left is also recognizable as explicitly inducing the map $\widetilde{\xi}$ on total spaces obtained under application of the exterior join functor as indicated below.

$$\begin{array}{ccc} A \xrightarrow{1} A & B \xrightarrow{1} B \\ \downarrow \overset{Fi_A}{\Rightarrow} \downarrow{\scriptstyle \alpha} & \downarrow \overset{Gi_B}{\Rightarrow} \downarrow{\scriptstyle \beta} \\ * \longrightarrow X & * \longrightarrow Y \end{array} \quad \rightsquigarrow \quad \begin{array}{ccc} A*B & \xrightarrow{\widetilde{\xi}} & E(\alpha\bigstar\beta) \\ \downarrow & \Rightarrow & \downarrow{\scriptstyle p_\star} \\ *\times * & \longrightarrow & X\times Y \end{array}$$

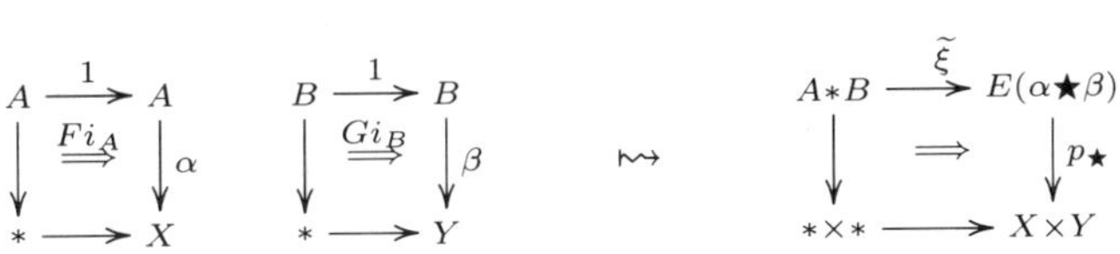

Viewed in this way it is obvious that $\ell_\star \circ (\rho_{AA} * \rho_{BB}) = \widetilde{\xi}$ by the functoriality and so equation (1) is obtained.

(2) By direct inspection of definitions the composite $\bar{\xi} \circ (i_2 * i_1)$ is seen to be induced under $\mathcal{M}$ as follows where the homotopies $\widehat{F}$ and $\widehat{G}$ are given by $\widehat{F}(a, b, t) = (F(*, b, 1-t), *)$ and $\widehat{G}(a, b, t) = (*, G(a, *, 1-t))$.

$$
\begin{array}{ccccc}
A & \xleftarrow{p_A} & A\times B & \xrightarrow{p_B} & B \\
{\scriptstyle o}\downarrow & \overset{\widehat{G}}{\Longleftarrow} & \downarrow{\scriptstyle o\times o} & \overset{\widehat{F}}{\Longrightarrow} & \downarrow{\scriptstyle o} \\
A\times Y & \xleftarrow[1\times\beta]{} & A\times B & \xrightarrow[\alpha\times 1]{} & X\times B
\end{array}
\quad \overset{\mathcal{M}}{\rightsquigarrow} \quad
\begin{array}{c}
A*B \\
\downarrow{\scriptstyle \bar{\xi}\circ(i_2*i_1)} \\
E(\alpha\bigstar\beta)
\end{array}
$$

We also consider the diagram

$$
\begin{array}{ccc}
A & \xrightarrow{o} & B \\
\downarrow & \overset{Gi_A}{\Longrightarrow} & \downarrow{\scriptstyle \beta} \\
* & \longrightarrow & Y
\end{array}
\qquad
\begin{array}{ccc}
B & \xrightarrow{o} & A \\
\downarrow & \overset{Fi_B}{\Longrightarrow} & \downarrow{\scriptstyle \alpha} \\
* & \longrightarrow & X
\end{array}
\quad \rightsquigarrow \quad
\begin{array}{ccccc}
 & \overset{\rho_{AB}*\rho_{BA}}{\nearrow} & F_\beta * F_\alpha & \xrightarrow{T} & F_\alpha * F_\beta \\
 & & \downarrow{\scriptstyle \ell_{\beta\bigstar\alpha}} & & \downarrow{\scriptstyle \ell_{\alpha\bigstar\beta}} \\
A*B & \xrightarrow{\widehat{\xi}} & E(\beta\bigstar\alpha) & \xrightarrow{T} & E(\alpha\bigstar\beta) \\
\downarrow & \Longrightarrow & \downarrow{\scriptstyle p_{\beta\bigstar\alpha}} & & \downarrow{\scriptstyle p_{\alpha\bigstar\beta}} \\
*\times * & \longrightarrow & Y\times X & \xrightarrow{T} & X\times Y
\end{array}
$$

in which the two squares on the left induce the lower left square on the right under the exterior join functor. The three maps T denote interchange of factors maps and yield commutative squares. Furthermore the above diagram of cotriads which induces $\bar{\xi} \circ (i_2 * i_1)$ is readily observed to also induce the composite $T \circ \widehat{\xi}$. Therefore we conclude that $\bar{\xi} \circ (i_2 * i_1) = \ell_{\alpha\bigstar\beta} \circ T \circ (\rho_{AB} * \rho_{BA})$, as claimed. □

PROPOSITION 7.7. *Each of the classes* $(\kappa \circ d_A) \circledast i_1$, $\Delta_A \circledast (\kappa \circ d_B)$ *and* $(\kappa \circ d_A) \circledast i_2$ *has projection-type* (o, o) *with respect to the map*

$$P_1 \triangle P_2 \colon (A \times A) * (B \times B) \to (A * B) \times (A * B)$$

whose components are projections onto the respective factors.

PROOF. The verifications are quite similar; we give the proof only for $(\kappa \circ d_A) \circledast i_1$. For $i = 1, 2$, we may consider the following homotopy commutative diagram

$$
\begin{array}{ccccccc}
A*B & & \xrightarrow{(\kappa\circ d_A)\circledast i_1} & & (A\times A)*(B\times B) & \xrightarrow{P_i} & A*B \\
{\scriptstyle \varphi_{A,B}}\downarrow & & & & \downarrow{\scriptstyle \varphi_{A\times A, B\times B}} & & \downarrow{\scriptstyle \varphi_{A,B}} \\
\Sigma A\wedge B & \xrightarrow[d_A\wedge 1_B]{} & (A*A)\wedge B & \xrightarrow[\kappa\wedge i_1]{} & \Sigma(A\times A)\wedge(B\times B) & \xrightarrow[\Sigma p_i\wedge p_i]{} & \Sigma A\wedge B
\end{array}
$$

in which the vertical maps are homotopy equivalences. Now plainly the composite $A * A \xrightarrow{\kappa} \Sigma(A \times A) \xrightarrow{\Sigma p_i} \Sigma A$ is just the classical Hopf construction on the projection map $p_i \colon A \times A \to A$. But as is well known the Hopf construction on any projection map is null homotopic. Thus the lower composite $\Sigma A \wedge B \to \Sigma A \wedge B$ is homotopic to $o \wedge 1$ for $i = 1$ and to $o \wedge o$ for $i = 2$. In both cases it is null homotopic and consequently the corresponding upper composite $P_i \circ [(\kappa \circ d_A) \circledast i_1] \colon A * B \to A * B$ must be null homotopic as well. □

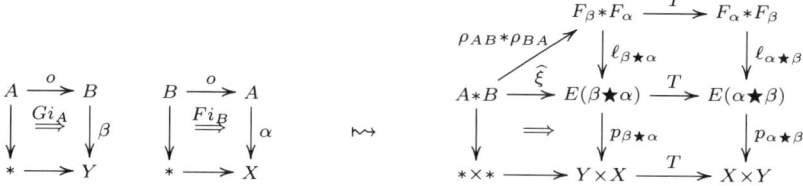

NOTATION 7.8. As a consequence of Proposition 7.7 there exist unique liftings of $\bar{\xi} \circ [(\kappa \circ d_A) \circledast i_1]$, $\bar{\xi} \circ [\Delta_A \circledast (\kappa \circ d_B)]$ and $\bar{\xi} \circ [(\kappa \circ d_A) \circledast i_2]$. We denote these classes by V_1, V and V_2 respectively.

Continuing on from (7.5) we see that

$$\begin{aligned}\ell_\bigstar \circ \lambda\mathrm{HI}(1_{A*B}) &= \bar{\xi} \circ (i_1 * i_1) + \bar{\xi} \circ [(\kappa \circ d_A) \circledast i_1] + \bar{\xi} \circ (i_2 * i_1) \\ &\quad + \bar{\xi} \circ [\Delta_A \circledast (\kappa \circ d_B)] + \bar{\xi} \circ (i_1 * i_2) \\ &\quad + \bar{\xi} \circ [(\kappa \circ d_A) \circledast i_2] - \bar{\xi} \circ (i_1 * i_1) \\ &= \bar{\xi} \circ (i_1 * i_1) + \ell_\bigstar \circ \widetilde{V}_1 + \ell_\bigstar \circ T \circ (\rho_{AB} * \rho_{BA}) + \ell_\bigstar \circ \widetilde{V} \\ &\quad + \ell_\bigstar \circ (\rho_{AA} * \rho_{BB}) + \ell_\bigstar \circ \widetilde{V}_2 - \bar{\xi} \circ (i_1 * i_1)\end{aligned}$$

and thus in the manner of Proposition 2.8 we obtain:

THEOREM 7.9. *Let diagram (7.1) be given. Then, with notation as established above, the equality*

$$\begin{aligned}\lambda\mathrm{HI}(1_{A*B}) &= \widetilde{V}_1^{\bar{\xi} \circ (i_1 * i_1)} + [T \circ (\rho_{AB} * \rho_{BA})]^{\bar{\xi} \circ (i_1 * i_1)} + \widetilde{V}^{\bar{\xi} \circ (i_1 * i_1)} \\ &\quad + (\rho_{AA} * \rho_{BB})^{\bar{\xi} \circ (i_1 * i_1)} + \widetilde{V}_2^{\bar{\xi} \circ (i_1 * i_1)}\end{aligned}$$

is valid as classes $A * B \to F_\alpha * F_\beta$.

REMARKS 7.10. (1) It is to be noted that $i_1 * i_1 \colon A * B \to (A \times A) * (B \times B)$ has projection-type $(1 * 1, o * o)$ with respect to the map $P_1 \triangle P_2$ of Proposition 7.7 and hence $\bar{\xi} \circ (i_1 * i_1)$ will not usually lift to $F_\alpha * F_\beta$.

(2) It would be possible to obtain a similar decomposition using $i_2 * i_2$ in place of $i_1 * i_1$. We leave the pursuit of this to the reader.

Since for an arbitrary space X the diagonal class $d_X \colon \Sigma X \to X * X$ vanishes whenever ΣX is homotopy abelian we obtain the following corollaries. Also recall that $\pi(A * B, Z)$ is abelian if at least one of A or B is a suspension space.

COROLLARY 7.11. *If B is a suspension space then the equality*

$$\lambda\mathrm{HI}(1_{A*B}) = \widetilde{V}_1 + T \circ (\rho_{AB} * \rho_{BA}) + \rho_{AA} * \rho_{BB} + \widetilde{V}_2$$

holds in the abelian group $\pi(A * B, F_\alpha * F_\beta)$.

COROLLARY 7.12. *If A is a suspension space then the equality*

$$\lambda\mathrm{HI}(1_{A*B}) = T \circ (\rho_{AB} * \rho_{BA}) + \widetilde{V} + \rho_{AA} * \rho_{BB}$$

is valid in the abelian group $\pi(A * B, F_\alpha * F_\beta)$.

COROLLARY 7.13. *If A and B are both suspension spaces then the equality*

$$\lambda\mathrm{HI}(1_{A*B}) = T \circ (\rho_{AB} * \rho_{BA}) + \rho_{AA} * \rho_{BB}$$

is valid as classes $A * B \to F_\alpha * F_\beta$.

References

1. H.J. Baues, *Commutator Calculus and Groups of Homotopy Classes*, LMS Lecture Note Series, vol. 50, Cambridge University Press, Campbridge, 1981.
2. M. Boardman and B.J. Steer, *On Hopf invariants*, Comment. Math. Helv. **42** (1967), 180–221.
3. O. Cornea, *Homotopical dynamics* II: *Hopf invariants, smoothings and the Morse complex*, preprint.
4. G. Dula and H.J. Marcum, *Hopf invariants of the Berstein-Hilton-Ganea kind*, Topology Appl. **65** (1995), 179–203.
5. T. Ganea, *A generalization of the homology and homotopy suspension*, Comment. Math. Helv. **39** (1965), 295–322.
6. N. Iwase, *Ganea's conjecture on Lusternik-Schnirelmann category*, Bull. London Math. Soc. **30** (1998), 623–634.
7. H.J. Marcum, *Twisted Whitehead products*, Proc. Amer. Math. Soc. **74** (1979), 358–362.
8. H.J. Marcum, *Fibrations over double mapping cylinders*, Illinois J. Math. **24** (1980), 344–358.
9. H.J. Marcum, *Two results on cofibers*, Pacific J. Math. **95** (1981), 122–142.
10. H.J. Marcum, *Functorial properties of the Hopf invariant.* I, Quaestiones Math. **19** (1996), 537–584.
11. H.J. Marcum, *Cone length of the exterior join*, Glasgow Math. J. **40** (1998), 445–461.
12. D. Stanley, *Spaces with Lusternik-Schnirelmann category n and cone length $n+1$*, Topology **39** (2000), 985–1019.

THE OHIO STATE UNIVERSITY AT NEWARK, 1179 UNIVERSITY DRIVE, NEWARK, OHIO 43055

E-mail address: marcum@math.ohio-state.edu

Contemporary Mathematics
Volume **274**, 2001

A subgroup of self homotopy equivalences which is invariant on genus

Ken-ichi Maruyama

ABSTRACT. In this paper we prove that the subgroup of self homotopy equivalences which consists of maps inducing the identity on homotopy groups is a genus invariant modulo torsion for certain Hopf spaces.

Introduction

For a finitely generated nilpotent group N, the Mislin (localization) genus of N denoted by $G(N)$ is defined to be the isomorphism classes of finitely generated nilpotent groups K with $N_p \cong K_p$ (isomorphic) for each prime number p, where N_p and K_p are the localizations at p (see [Mi], [Wi]). It is known that $G(N)$ is nontrivial in general [BW]. It is also known that there exists a finitely generated torsion free nilpotent group whose Mislin genus has more than one element [Mc],[Og].

For a nilpotent space X of finite type, one can define the Mislin genus $G(X)$ of X analogously, i.e., $G(X)$ is the set of homotopy types X' of finite type with $X_p \cong X'_p$ (homotopy equivalent) at each prime number, where X_p and X'_p are the homotopy theoretic localizations (see [BK], [HMR]). The first example of a nontrivial genus for spaces is found in [HR]; now it is known that there are many spaces with such a property. On the other hand, Wilkerson[Wi] has shown that $G(X)$ is finite for all finite simply connected complexes.

Let $\mathcal{E}(X)$ denote the set of based homotopy classes of self homotopy equivalences of a (based) space X. $\mathcal{E}(X)$ is a group with group operation given by composition of homotopy classes. It is known that $\mathcal{E}(X)$ is not a genus invariant even for finite complexes [Ro]. Let $\mathcal{E}_\#(X)$ be the subgroup of homotopy classes which induce the identity on homotopy groups of X in dimensions $\leq \dim X$. $\mathcal{E}_\#(X)$ is known to be nilpotent by [DZ]. In this paper we study $\mathcal{E}_\#(X)$ where X is a finite H(or H_0)-space. We do not know whether or not $\mathcal{E}_\#(X)$ is a genus invariant. Also we do not know if the genus set $G(\mathcal{E}_\#(X))$ is trivial. However we can show that the group is a genus invariant modulo torsion elements. For a nilpotent group N,

1991 *Mathematics Subject Classification.* Primary 55P10, 55P60; Secondary 55P45.

Key words and phrases. Homotopy equivalences, Genus.

This research was partially supported by the Ministry of Education of Japan, Grant-in-Aid for 10640064, 1999.

we denote by N/τ the quotient group by the subgroup of all the torsion elements. We obtain

THEOREM 0.1. *Let X be a homotopy associative finite H-space and $Y \in G(X)$. Then $\mathcal{E}_\#(Y)/\tau \cong \mathcal{E}_\#(X)/\tau$.*

Let G be a connected compact topological group and $G \to E \to S^{2n+1}$ a principal G-bundle. Let $\theta \in \pi_{2n+1}(BG)$ be the characteristic element of the bundle. We denote by E_f the pullback of the bundle E by a map $f : S^{2n+1} \to S^{2n+1}$. The space E_f is called a sphere extension of E by f. Let $deg(f)$ denote the degree of f.

THEOREM 0.2. *We assume that $\pi_i(G) \otimes \mathbf{Q} = 0$ for $i \geq 2n+1$. If $deg(f)$ is prime to $|\theta|$, then $E_f \in G(E)$ and $\mathcal{E}_\#(E_f)/\tau \cong \mathcal{E}_\#(E)/\tau$, where $|\theta|$ is the order of θ.*

We assume throughout that all spaces are connected and have base points.

§1. Proof of Theorem 0.1

First we recall the work of Zabrodsky[Za] on the genus of an H_0-space which has been generalized by McGibbon [Mc] and prove some lemmas and proposition needed in our proof.

DEFINITION. A finite H_0-space X is a finite nilpotent CW-complex such that X_0 is homotopy equivalent to the product of Eilenberg MacLane spaces, i.e., $X_0 = \prod_{i=1}^r K(\mathbf{Q}, 2n_i - 1)$.

Let X be an n-dimensional finite H_0-space, $X(n)$ the n-th Postnikov stage of X and let $\psi_X : X(n) \to \prod_{i=1}^r K(QH^{2n_i-1}(X,\mathbf{Z})/\tau, 2n_i - 1)$ be a 0-equivalence with fibre F, where $QH^*(X;\mathbf{Z})$ is the module of indecomposables. We define $K = \prod_{i=1}^r K(QH^{2n_i-1}(X,\mathbf{Z})/\tau, 2n_i - 1)$. Let t be an integer divisible by all torsion of $H^m(X;\mathbf{Z})$ for $m \leq n = \dim X$ and by $\prod_{i=1}^n \exp\pi_i(F)$, where exp denotes the exponent of a group, and let $P_t = \{p \in \Pi | p \text{ divides } t\}$, where Π denotes the set of all prime numbers. We write P for P_t. It has been shown ([Za] and [Mc]) that the n-th Postnikov section $Y(n)$ of an element Y of $G(X)$ is given by the following (weak) pullback :

$$\begin{array}{ccc} Y(n) & \longrightarrow & X(n) \\ {\scriptstyle\psi_Y}\downarrow & & \downarrow{\scriptstyle\psi_X} \\ K & \xrightarrow{\Gamma(\mathrm{I}_{d_1,\ldots,d_r})} & K \end{array} \tag{1.1}$$

where $d_1, ..., d_r$ are integers prime to t, and $\Gamma(\mathrm{I}_{d_1,\ldots,d_r})$ is the map induced by the homomorphism $\mathrm{I}_{d_1,\ldots,d_r}$:$QH^*(X;\mathbf{Z})/\tau \to QH^*(X;\mathbf{Z})/\tau$, $\mathrm{I}_{d_1,\ldots,d_r}\ |QH^{2n_i-1}\ (X;\mathbf{Z})/\tau = d_i \cdot 1$.

LEMMA 1.1. *The following is a pullback diagram*

$$\begin{array}{ccccc} Y(n) & & \longrightarrow & & X(n)_{\bar{P}} \\ \downarrow & & & & \downarrow{\scriptstyle\psi_X \circ \ell_0} \\ X(n)_P & \xrightarrow{\psi_X \circ \ell_0} & K_0 & \xrightarrow{\Gamma_0^{-1}} & K_0 \end{array}$$

where $\Gamma_0 = \Gamma(I_{d_1,\ldots,d_r})_0$ and two ℓ_0's are rationalizations of $X(n)_P$ and $X(n)_{\bar{P}}$ respectively.

PROOF. Lemma 1.1 follows from (1.1) and the pullback diagram [HMR]:

$$\begin{array}{ccc} Y & \longrightarrow & Y_{\bar{P}} \\ \downarrow & & \downarrow \\ Y_P & \longrightarrow & Y_0 . \end{array}$$

□

NOTATION. Let $f : X \to Y$ be a homotopy equivalence. We denote by $A_f : \mathcal{E}(X) \to \mathcal{E}(Y)$ the inner automorphism defined by $A_f(h) = fhf^{-1}$.

PROPOSITION 1.2. *Let X be a finite H_0-space, $Y \in G(X)$ and Γ be the map given in Lemma 1.1. Then the following is a pullback diagram*

$$\begin{array}{ccccc} \mathcal{E}_{\#}(Y(n)) & & \longrightarrow & & \mathcal{E}_{\#}(X(n)_{\bar{P}}) \\ \downarrow & & & & \downarrow \\ \mathcal{E}_{\#}(X(n)_P) & \longrightarrow & \mathcal{E}_{\#}(K_0) & \xrightarrow{A_{\Gamma_0^{-1}}} & \mathcal{E}_{\#}(K_0) . \end{array}$$

Here $\mathcal{E}_{\#}(X(n))$ is the subgroup of $\mathcal{E}(X(n))$ which consists of homotopy classes which induce the identity on homotopy group in dimensions $\leq n$ (i.e., all the homotopy groups of $X(n)$).

PROOF. Given $f \in \mathcal{E}_{\#}(Y)$, $(g^{-1}f_P g, h^{-1}f_{\bar{P}}h)$ is an element of the pullback which we are considering since

$$A_{\Gamma_0}((g^{-1}f_P g)_0) = (h^{-1}f_{\bar{P}}h)_0$$

Thus we obtain the assignment from $\mathcal{E}_{\#}(Y)$ to the pullback. Clearly the assignment is an isomorphism. □

Proposition 1.2 yields

LEMMA 1.3. *Let X be a finite H_0-space, $Y \in G(X)$. Then the following is a pullback diagram*

$$\begin{array}{ccccc} \mathcal{E}_{\#}(Y(n))/\tau & & \longrightarrow & & \mathcal{E}_{\#}(X(n)_{\bar{P}})/\tau \\ \downarrow & & & & \downarrow \\ \mathcal{E}_{\#}(X(n)_P)/\tau & \longrightarrow & \mathcal{E}_{\#}(K_0) & \xrightarrow{A_{\Gamma_0^{-1}}} & \mathcal{E}_{\#}(K_0) . \end{array}$$

PROOF. There exists the following pullback for an arbitrary nilpotent group N [HMR]

$$\begin{array}{ccc} N & \longrightarrow & N_{\bar{P}} \\ \downarrow & & \downarrow \\ N_P & \longrightarrow & N_0 . \end{array}$$

The pullback diagram in Proposition 1.2 is equivalent to the above diagram for $N = \mathcal{E}_{\#}(Y(n))$ because $\mathcal{E}_{\#}(X(n)_P)$ is P-local and $\mathcal{E}_{\#}(X(n)_{\bar{P}})$ is $\bar{P}$-local respectively

[Ma],[Mø]. Therefore the diagram in the assertion of Lemma 1.3 is equivalent to the pullback as follows

$$\begin{array}{ccc} N/\tau & \longrightarrow & N_{\bar{P}}/\tau \\ \downarrow & & \downarrow \\ N_P/\tau & \longrightarrow & N_0 \ . \end{array}$$

□

Let $\mathcal{Z}(X)$ consist of all $f \in [X, X]$, the homotopy set of self-maps of X, such that $f_* : \pi_i(X) \to \pi_i(X)$ is the trivial homomorphism for $i \leq dimX$. If X is a homotopy associative H-space, it is known that $[X, X]$ is a nilpotent group. We see easily that $\mathcal{Z}(X)$ is a normal subgroup of $[X, X]$.

LEMMA 1.4. *If X is a homotopy associative finite H-space, $\mathcal{Z}(X_\ell) \cong \mathcal{Z}(X)_\ell$ for a set of prime numbers ℓ.*

PROOF. There exists the following commutative diagram of groups and homomorphisms with exact rows

$$\begin{array}{ccccccc} 0 & \longrightarrow & \mathcal{Z}(X) & \longrightarrow & [X,X] & \xrightarrow{\pi} & \prod_{i=1}^{\dim X} \mathrm{Hom}(\pi_i(X), \pi_i(X)) \\ & & \downarrow & & \downarrow & & \downarrow \\ 0 & \longrightarrow & \mathcal{Z}(X_\ell) & \longrightarrow & [X_\ell, X_\ell] & \xrightarrow{\pi_\ell} & \prod_{i=1}^{\dim X} \mathrm{Hom}(\pi_i(X_\ell), \pi_i(X_\ell)) \ . \end{array}$$

In this diagram, the second and the third vertical homomorphisms are localization maps. In general, for a homomorphism $f : N \to N'$ of nilpotent groups it holds that $(\ker f)_\ell \cong \ker f_\ell$ by Theorem 2.1 of [HMR], hence we obtain the result. □

Now we will prove Theorem 0.1. We will show that A_{Γ_0} induces an isomorphism on $\mathcal{E}_\#(X(n)_P)/\tau$. For this, first we will show that $\mathcal{Z}(X)/\tau$ is invariant under A_{Γ_0} using the fact that $\mathcal{Z}(X)/\tau$ is an abelian group. We then transfer this result to $\mathcal{E}_\#(X(n)_P)/\tau$. By [Mi2], Y has the homotopy type of a finite complex. We should note that $[X, X]$ is bijective with $[X(m), X(m)]$ if $m \geq \dim X = n$. We identify $\mathcal{E}_\#(X)$ with $\mathcal{E}_\#(X(n))$, $\mathcal{Z}(X)$ with $\mathcal{Z}(X(n))$ respectively. Let ι_i be the fundamental class of $H^{2n_i-1}(K(Z, 2n_i - 1))$, then the module of decomposables $DH^*(K)/\tau$ is generated by the elements $\iota_{i_1}\iota_{i_2}\cdots\iota_{i_k}$, $i_1 \leq \cdots \leq i_k$. Note that $[X_0, X_0]$ is isomorphic to $\prod_i H^{2n_i-1}(X)_0$. It is easy to see that $\mathcal{Z}(X)/\tau$ is a finitely-generated abelian group generated by the finitely many elements $\{\sum m_{i_1\cdots i_k} z_{i_1\cdots i_k}\}$, where $m_{i_1\cdots i_k}$ is an integer, $z_{i_1\cdots i_k}$ is the following composition.

$$z_{i_1\cdots i_k} : X_0 \xrightarrow{(a_{i_1\cdots i_k}/b_{i_1\cdots i_k})\iota_{i_1}\iota_{i_2}\cdots\iota_{i_k}} K(\mathbf{Q}, 2n_i - 1) \to X_0$$

Here $a_{i_1\cdots i_k}/b_{i_1\cdots i_k}$ is some rational number. The subgroup L of $[X_0, X_0]$ generated by $\{(1/b_{i_1\cdots i_k})\iota_{i_1}\iota_{i_2}\cdots\iota_{i_k}\}$ is a finitely-generated free abelian group. We have

$$A_{\Gamma_0}(\iota_{i_1}\iota_{i_2}\cdots\iota_{i_k}) = d\iota_{i_1}\iota_{i_2}\cdots\iota_{i_k}$$

for some $d \in \mathbf{Z}_P^*$. Note that the equivalence $\psi_{X0} : X(n)_0 \to K_0$ is an H-map since X is homotopy associative. Therefore, $A_{\Gamma_0}(L_P) \subset L_P$. Now $\mathcal{Z}(X)/\tau$ is a subgroup of L of finite index, so there exists a basis $u_1, ..., u_n$ of L such that $e_1u_1, ..., e_nu_n$ is a basis of $\mathcal{Z}(X)/\tau$, where $e_1, ..., e_n$ are non-zero integers. Consequently, $\mathcal{Z}(X)_P/\tau = \mathcal{Z}(X_P)/\tau$ is invariant under A_{Γ_0}. By the same reason $A_{\Gamma_0}^{-1}(\mathcal{Z}(X_P)/\tau) \subset \mathcal{Z}(X_P/\tau)$.

Hence A_{Γ_0} is an isomorphism of $\mathcal{Z}(X_P)$. Now, by definition of $\mathcal{Z}(X)$, there exists a one to one correspondence between $\mathcal{Z}(X)$ and $\mathcal{E}_\#(X)$, that is

$$I+ : \mathcal{Z}(X) \to \mathcal{E}_\#(X)$$

defined by $I + (x) = I + x$, where I is the identity map of X. It also induces a bijection

$$I+ : \mathcal{Z}(X)/\tau \to \mathcal{E}_\#(X)/\tau$$

Analogously, $\mathcal{E}_\#(X_P)/\tau$ is bijective with $\mathcal{Z}(X_P)$. Let $I + x \in \mathcal{E}_\#(X_P)/\tau$, where $x \in Z(X_P)$. Then we obtain

$$A_{\Gamma_0}(I + x) = I + A_{\Gamma_0}(x)$$

Thus A_{Γ_0} induces an isomorphism on $\mathcal{E}_\#(X_P)/\tau$. Now we have the following commutative diagram

$$\begin{array}{ccccc} \mathcal{E}_\#(X_P)/\tau & \longrightarrow & \mathcal{E}_\#(X_0) & \xrightarrow{A_{\Gamma_0}} & \mathcal{E}_\#(X_0) \\ \downarrow{\scriptstyle A_{\Gamma_0}} & & \downarrow{\scriptstyle A_{\Gamma_0}} & & \downarrow{\scriptstyle I_d} \\ \mathcal{E}_\#(X_P)/\tau & \longrightarrow & \mathcal{E}_\#(X_0) & \xrightarrow{I_d} & \mathcal{E}_\#(X_0) . \end{array}$$

By the fundamental property of pullbacks and Lemma 1.3, we obtain an isomorphism from $\mathcal{E}_\#(Y)/\tau$ to $\mathcal{E}_\#(X)/\tau$.

§2 Proof of Theorem 0.2.

We recall a relation between Postnikov system and homotopy equivalences. For a nilpotent space X, let us denote by $X(n)$ the n-th Postnikov section or its principal refinement.

LEMMA 2.1. ([AC],[No],[Su]) *For each n, the fibering $K(\pi_n(X), n) \to X(n) \xrightarrow{p_n} X(n-1)$ with the k-invariant k^{n+1} induces the following exact sequence*

$$0 \to I(1_{X(n)}) \to p_n^* H^n(X(n-1); \pi_n(X)) \xrightarrow{\kappa} \mathcal{E}_\#(X(n)) \xrightarrow{\sigma_n} \mathcal{E}_\#(X(n-1))$$

Moreover, $\mathrm{Im}\sigma_n = \{h \in \mathcal{E}_\#(X(n-1)), \kappa^{n+1}h = \kappa^{n+1}$ *in* $H^{n+1}(X(n-1); \pi_n(X)\}$.

We obtain

LEMMA 2.2. *Let K be a product of Eilenberg-MacLane spaces. Then there exists the following split exact sequence*

$$0 \to p_n^* H^n(K(n-1); \pi_n(K)) \xrightarrow{\kappa} \mathcal{E}_\#(K(n)) \xrightarrow{\sigma_n} \mathcal{E}_\#(K(n-1)) \to 0.$$

PROOF. Exactness follows from Lemma 2.1. There is an obvious homomorphism (inclusion) $\mathcal{E}(K(n-1)) \to \mathcal{E}(K(n))$ which gives a splitting homomorphism of the exact sequence.

We will prove Theorem 0.2 in the following. By the assumption E_f is in the genus of E. By the facts in the last section, we have the following pullback diagram

$$\begin{array}{ccc} E_f(2n+1) & \longrightarrow & E(2n+1) \\ \downarrow & & \downarrow \\ K & \xrightarrow{\Gamma(I_{1,\dots,1,d})} & K . \end{array}$$

Here $d = deg(f)$. By Lemma 2.1, there exists an exact sequence as follows.

$$p^*H^{2n+1}(E(2n);\pi_{2n+1}(E)) \xrightarrow{\kappa} \mathcal{E}_\#(E(2n+1)) \xrightarrow{\sigma_{2n+1}} \mathcal{E}_\#(E(2n))$$

Let $k_1, ..., k_m$ and $x_{m+1}, ..., x_r$ be generators of $p^*H^{2n+1}(E(2n);\pi_{2n+1}(E))$ and $\mathrm{Im}(\sigma_{2n+1})$ respectively. From these generators, we can define a basis $\{k_i, \bar{x}_i\}$ for $\mathcal{E}_\#(E(2n+1))$ such that $\bar{x}_i^*(e_{2n+1}) = e_{2n+1}$, where $e_{2n+1} \in QH^{2n+1}(E)$ is a generator. This fact can be obtained as follows. For our purpose, it is sufficient to consider generators modulo torsion elements. Assume that we are given $f \in \mathcal{E}_\#(E(2n+1))$ such that $f^*(e_{2n+1}) = e_{2n+1} + \rho$ with a decomposable element ρ (modulo torsion elements). There exists $t \in p^*H^{2n+1}(E(2n);\pi_{2n+1}(E))$ such that

$$(f\kappa(t))^*(e_{2n+1}) = e_{2n+1}$$

modulo torsion elements, since $H^*(E(2n+1))$ is isomorphic to $H^*(G)$ for $* < 2n$ and $H^*(G)/\tau$ is an exterior algebra. Hence we see our claim. Let $P = \{p \in \Pi | (p, |\theta|) = 1\}$. The localization of $\mathcal{E}_\#(E(2n+1))/\tau$ at P is also generated by $\{k_i\}$ and $\{\bar{x}_i\}$ over $\mathbf{Z}_P$. We obtain the equalities.

$$A_{\Gamma_0}^{-1}(k) = dk, \quad A_{\Gamma_0}(k) = -dk$$

for all $k \in p^*H^{2n+1}(E(2n);\pi_{2n+1}(E))/\tau$, and

$$A_{\Gamma_0}(\bar{x}) = \bar{x}$$

for $\bar{x} \in < \bar{x}_i >$, where $d \in \mathbf{Z}_P^*$. For a subgroup $N \subset \mathcal{E}_\#(E(2n+1))$, consider the following exact sequence

$$0 \to N'/N' \cap \ker\kappa \to N \to \sigma_{2n+1}(N) \to 0 \ .$$

Here N' is a subgroup of $p^*H^{2n+1}(E(2n);\pi_{2n+1}(E))$. By an argument analogous to above, N/τ is generated by $\{\tilde{k}_i\}$ and $\{\tilde{x}_i\}$, which satisfy $A_{\Gamma_0}(\tilde{k}_i) = -d\tilde{k}_i$ and $A_{\Gamma_0}(\tilde{x}_i) = \tilde{x}_i$. Therefore, N_p/τ is invariant under A_{Γ_0} and $A_{\Gamma_0}^{-1}$. If $m \geq 2n+1$, $\mathcal{E}_\#(E(m))/\tau$ is the subgroup $\mathcal{E}_\#(E(2n+1))/\tau$ of finite index, and the result follows.

REMARK. We should note that our theorems in this article do not insist that the genus set of the nilpotent group $\mathcal{E}_\#(X)/\tau$, namely $G(\mathcal{E}_\#(X)/\tau)$, is trivial. We do not know if there exists a space such that $G(\mathcal{E}_\#(X)/\tau)$ has more than single element.

PROBLEM. Find a space X such that $\#G(\mathcal{E}_\#(X)/\tau) > 1$.

References

[AC] M. Arkowitz and C. Curjel, *Groups of homotopy classes*, Lecture Notes in Mathematics, vol. 4, Springer-Verlag, 1967.

[BW] V. Belfi and C. Wilkerson, *Some examples in the theory of p-completions*, Indiana Univ. J. of Math. **25** (1976), 565–576.

[BK] A. Bousfield and D. Kan, *Homotopy Limits, Completions and Localizations*, Lecture Notes in Mathematics, vol. 304, Springer, 1972.

[DZ] E. Dror and A. Zabrodsky, *Unipotency and nilpotency in homotopy equivalences*, Topology **18** (1979), 187–197.

[HMR] P. Hilton, G. Mislin and J. Roitberg, *Localization of Nilpotent Groups and Spaces*, Mathematics Studies, vol. 15, North-Holland, 1975.

[HR] P. Hilton and J. Roitberg, *On principal S^3-bundles over spheres*, Ann. of Math. **90** (1969), 91–107.

[Ma] K. Maruyama, *Localization of a certain subgroup of self-homotopy equivalences*, Pacific J. of Math. **136** (1989), 293–301.

[Mc] C. A. McGibbon, *On the localization genus of a space*, Algebraic topology (Progr. Math.), vol. 136, Birkhäuser, 1996, p. 285–306.

[Mi] G. Mislin, *Nilpotent groups with finite commutator subgroups*, Lecture Notes in Math., vol. 418, Springer, 1974, pp. 103–120.

[Mi2] G. Mislin, *Finitely dominated nilpotent spaces*, Ann. Math. **103** (1976), 547–556.

[Mø] J. Møller, *Self-homotopy equivalences of $H_*(-;Z/p)$-local spaces*, Kodai Math.J **12** (1989), 270–281.

[No] Y. Nomura, *Homotopy equivalences in a principal fibre space*, Math. Z. **92** (1966), 380–388.

[Og] F. Oger, *Des groupes nilpotents de classe 2 sans torsion de type fini ayant les mêmes images finies peuvent ne pas être élémentairement équivalents*, Comptes rendus, Acad. Sci. Paris **294** (1982), 1–4.

[Ro] J. Roitberg, *Genus and Symmetry in Homotopy Theory*, Math. Ann **305** (1996), 381–386.

[Su] D. Sullivan, *Infinitesimal computations in topology*, Publ. I.H.E.S **47** (1977), 269–331.

[Wi] C. Wilkerson, *Applications of minimal simplicial groups*, Topology **15** (1976), 115–130.

[Za] A. Zabrodsky, *Hopf Spaces*, Mathematics Studies, vol. 22, North-Holland, 1976.

Department of Mathematics, Faculty of Education, Chiba University, Yayoicho, Chiba, Japan

E-mail address: `maruyama@e.chiba-u.ac.jp`

Contemporary Mathematics
Volume **274**, 2001

Composition structure of the self maps of $SU(3)$ or $Sp(2)$

Kaoru MORISUGI

ABSTRACT. Let G be the special unitary group $SU(3)$ or the symplectic group $Sp(2)$. In [**5**] Mimura and Ōshima determined the group structure of the homotopy set $[G, G]$ of self maps of G. In this note we determine the composition structure of the group $[G, G]$.

1. Introduction and the statement of results

Let G be the special unitary group $SU(3)$ or the symplectic group $Sp(2)$. In [**5**] Mimura and Ōshima determined the group structure of the homotopy set $[G, G]$ of self maps of G. We denote the multiplication of $[G, G]$ by '+', although $[G, G]$ is not abelian. Throughout this note, we consider only the standard multiplication of G and the coefficients of all homology and cohomology groups are always integers, unless otherwise stated.

In this note, we determine the composition structure of the group $[G, G]$.

Recall that G is the total space of S^3-bundle over S^n , where n is 5 or 7 according as $G = SU(3)$ or $Sp(2)$. We denote the bundle projection by $p : G \to S^n$ and the quotient map to the top cell by $q : G \to S^{3+n}$. The group $\pi_{n+3}(G)$ is isomorphic to $\mathbb{Z}/k$ with a generator σ, where $k = 12$ or 120 according as $G = SU(3)$ or $Sp(2)$.

To state our result, we recall Mimura-Ōshima's result[**5**].

Define elements $\alpha, \beta, \gamma \in [G, G]$ by

$$\alpha : G \xrightarrow{p} S^n \xrightarrow{\lambda_2} G, \tag{1.1}$$

$$\beta : G \xrightarrow{id} G \tag{1.2}$$

$$\gamma : G \xrightarrow{q} S^{3+n} \xrightarrow{\sigma} G, \tag{1.3}$$

where $\lambda_2 : S^n \to G$ is the standard generator of $\pi_n(G) \cong \mathbb{Z}$. Let $[\alpha, \beta]$ be the commutator of α and β in the group $[G, G]$. Then the following holds [**5**]:

THEOREM 1.1. *The group $[G, G]$ is generated by α, β and γ, γ is of order k and belongs to the center of $[G, G]$, and the following relation holds: $[\alpha, \beta] = \gamma$ for $G = SU(3)$ and $[\alpha, \beta] = 12\gamma$ for $G = Sp(2)$. Therefore every element of $[G, G]$ can be written as $a\alpha + b\beta + c\gamma$, where a, b, c are integers.*

1991 *Mathematics Subject Classification.* primary55Q10, 55Q52, 55R45.

Since we are working in non-abelian groups, we must be very careful with choices of generators or sign. The choice of generators σ and λ_2 will be explained in §2.

It is well-known that

$$H^*(G) \cong \Lambda_{\mathbb{Z}}(x_1, x_2),$$

where $x_1 \in H^3(G)$, $x_2 \in H^n(G)$ and the both elements are primitive. To state the result, we need some more notation:

For $f \in [G, G]$, define $d_i(f) \in \mathbb{Z}$ by the following equation:

$$f^*(x_i) = d_i(f)x_i,$$

where $f^* : H^*(G) \to H^*(G)$ is the cohomology induced homomorphism by f. Since x_i is primitive, $d_i : [G, G] \to \mathbb{Z}$ is a homomorphism.

In [**5**], the composition structure of $[G, G]$ was partially obtained. We will determine it completely.

Let $m = 1$ or 20 according as $G = SU(3)$ or $Sp(2)$.

Then the main theorem of this note is

THEOREM 1.2. *For any f, $g \in [G, G]$, the following formula holds:*

$$\alpha \circ (f + g) = \alpha \circ f + \alpha \circ g - md_1(f)d_2(g)\gamma.$$

By this theorem, we can determine completely the composition structure of the group $[G, G]$. The algebraic structure of $[G, G]$ looks like the "square ring " of [**2**]. But at present it is not clear whether $[G, G]$ forms a square ring.

2. Recollections

We next explain the notation.

The cell structure of G is given by

$$S^3 \cup_\omega e^n \cup e^{3+n},$$

where the attaching map ω is the standard generator of $\pi_{n-1}(S^3) \cong \mathbb{Z}/l$. Here $l = 2$ or 12 according as $n = 5$ or 7. Let $i : S^3 \to G$ and $p : G \to S^n$ be the fiber inclusion and the bundle projection, respectively. Note that i is a homomorphism of Lie groups.

We fix generators $s_r \in H^r(S^r)$ for $r = n, 3$. Define $x_1 \in H^3(G)$ and $x_2 \in H^n(G)$ by $p^*s_n = x_2$ and $i^*(x_1) = s_3$. Then $H^*(G) = \Lambda(x_1, x_2)$. Let $\beta_1 \in H_3(G)$ and $\beta_2 \in H_n(G)$ be the dual element of x_1 and x_2. Then we can choose $i = \lambda_1 \in \pi_3(G) \cong Z$ and $\lambda_2 \in \pi_n(G) \cong Z$ as $h(\lambda_1) = \beta_1$ and $h(\lambda_2) = l\beta_2$, where h is the Hurewicz homomorphism.

Consider the standard fiber bundles:

$$SU(4) \to SU(5) \to S^9 \tag{2.1}$$

$$Sp(2) \to Sp(3) \to S^{11} \tag{2.2}$$

We fix the generators as $\sigma' = \partial(\iota_9) \in \pi_8(SU(4)) \cong \mathbb{Z}/24$ or $\sigma = \partial(\iota_{11}) \in \pi_{10}(Sp(2)) \cong \mathbb{Z}/120$, where ∂ is the connecting homomorphism in the homotopy exact sequences of the above bundles and $\iota_m \in \pi_m(S^m)$ is the identity element. It is known that $j_* : \pi_8(SU(3)) \to \pi_8(SU(4))$ is monomorphic, where j is the inclusion map. And we choose the generator $\sigma \in \pi_8(SU(3)) \cong \mathbb{Z}/12$ as $j_*(\sigma) = 2\sigma'$.

LEMMA 2.1. *Let $\Delta : G \to G \times G$ be the diagonal map. Then there exists a unique map $\pi : G \to S^n \vee S^{n+3}$ up to homotopy such that the following diagram commutes:*

(2.3)
$$\begin{array}{ccccc} G & \xrightarrow{\Delta} & G \times G & \xrightarrow{p\times q} & S^n \times S^{n+3} \\ & \searrow^{\pi} & & & \uparrow j \\ & & & & S^n \vee S^{n+3}, \end{array}$$

where j is the inclusion. Moreover the following sequence is the homotopy cofiber sequence:

(2.4)
$$S^3 \xrightarrow{i} G \xrightarrow{\pi} S^n \vee S^{n+3}$$

PROOF. The existence of π is clear from the cellular approximation. To show the uniqueness we consider the suspension map $E : X \to \Omega\Sigma X$. Let $f, g : G \to S^n \vee S^{n+3}$ be such that $j \circ f = j \circ g$. Then consider the element $E(f) - E(g) \in [G, \Omega\Sigma(S^n \vee S^{n+3})]$. Clearly $\Omega\Sigma j \circ (E(f) - E(g)) = 0$. Since $(\Omega\Sigma j)_* : [\ \ , \Omega\Sigma(S^n \vee S^{n+3})] \to [\ \ , \Omega\Sigma(S^n \times S^{n+3})]$ is a split monomorphism, we get $E(f) = E(g)$. Since $E : [G, S^n \vee S^{n+3}] \to [G, \Omega\Sigma(S^n \vee S^{n+3})]$ is bijective by the Suspension Theorem, we obtain $f = g$. This proves the uniqueness.

By using a homology calculation, we see that the sequence (2.4) is the homotopy cofiber sequence.

□

Let $f \in [G, G]$. By a similar proof, it is easy to show that there exists a unique function

(2.5)
$$e : [G, G] \to \mathbb{Z}/24$$

which satisfies the following commutative diagram:

(2.6)
$$\begin{array}{ccc} G & \xrightarrow{f} & G \\ \pi \downarrow & & \downarrow p \\ S^n \vee S^{n+3} & \xrightarrow[d_2(f)\iota_n + e(f)\nu_n]{} & S^n, \end{array}$$

where $\nu_n \in \pi_{n+3}(S^n)$ is the $(n-4)$-fold suspension of $\nu_4 \in \pi_7(S^4)$ and ν_4 is defined [**9**] as the classical Hopf construction of the map $\mu' : S^3 \times S^3 \to S^3$ defined by $\mu'(x, y) = x^{-1}y$.

Define the integer $m_G \mod k$ by the following:

(2.7)
$$\lambda_2 \circ \nu_n = m_G \sigma \qquad \text{in } \pi_{n+3}(G) \cong \mathbb{Z}/k.$$

3. Some lemmas

Recall that the product in $[G, G]$ is defined as

$$f + g : G \xrightarrow{\Delta} G \times G \xrightarrow{f \times g} G \times G \xrightarrow{\mu} G,$$

where μ is the standard multiplication of G.

LEMMA 3.1. *Given $f, g, h \in [G, G]$, we have*

$$(f + g) \circ h = f \circ h + g \circ h.$$

Recall the definition of the element α, $\gamma \in [G, G]$. Then we obtain

PROPOSITION 3.2. *For any $h \in [G, G]$,*

$$\gamma \circ h = d_1(h)d_2(h)\gamma \tag{3.1}$$

$$\alpha \circ h = d_2(h)\alpha + m_G e(h)\gamma, \tag{3.2}$$

where m_G is the integer defined in (2.7).

PROOF. $\gamma \circ h = \sigma \circ q \circ h = \sigma \circ (d_1(h)d_2(h)\iota_{n+3}) \circ q = d_1(h)d_2(h)(\sigma \circ q) = d_1(h)d_2(h)\gamma$. This proves (3.1).

(3.2) follows from the following commutative diagram and (2.7).

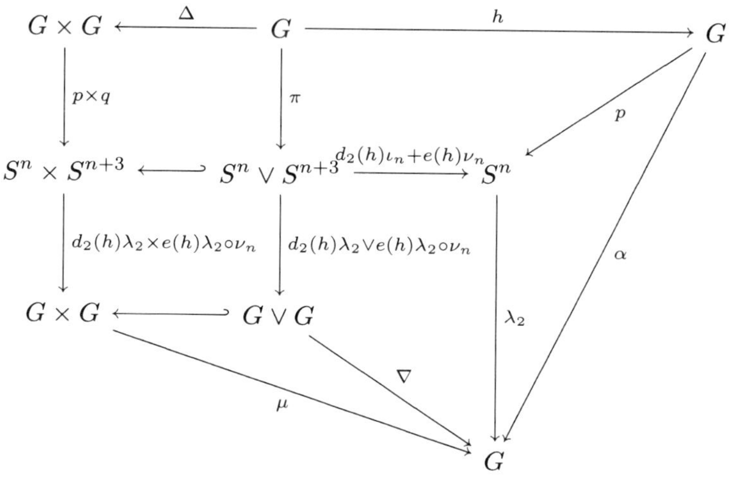

□

LEMMA 3.3. *Let $p \geq q$. Given a map $\varphi : S^p \to S^q$, let K be the mapping cone of φ. Then there exits an element $\theta \in \pi_{p+q+1}(K \wedge K)$ whose Hurewicz image is $a \otimes b + (-1)^{pq+1} b \otimes a$, where $a \in H_q(K)$ and $b \in H_{p+1}(K)$ are generators. In the case that $p = q$ and φ is the map of degree m, the coefficients of homology should be taken as $\mathbb{Z}/m$.*

Applying the above lemma as K being the n-skeleton of G, we get

LEMMA 3.4. *$\pi_{n+3}(G \wedge G) \cong \mathbb{Z} \oplus \mathbb{Z}$ with generators a and b which are characterized by the Hurewicz homomorphism $h : \pi_{n+3}(G \wedge G) \to H_{n+3}(G \wedge G)$:*

$$h(a) = l\beta_2 \otimes \beta_1 \tag{3.3}$$

$$h(b) = \beta_1 \otimes \beta_2 - \beta_2 \otimes \beta_1, \tag{3.4}$$

where l is 2 or 12 according as $G = SU(3)$ or $Sp(2)$. Note that $a = \lambda_2 \wedge i : S^{n+3} = S^n \wedge S^3 \to G \wedge G$.

Given a map $f : X_1 \times X_2 \to Y$, we denote the classical Hopf construction of f by $H(f) : \Sigma X_1 \wedge X_2 \to \Sigma Y$. It is known that $H(f) = \Sigma f \circ H(id)$ for $id : X_1 \times X_2 \to X_1 \times X_2$ and $H(id) : \Sigma X_1 \wedge X_2 \to \Sigma X_1 \times X_2$ is characterized (for example, [**1**] or [**7**]) by

$$H(id) \circ \Sigma q = -\Sigma(i_1 \circ p_1) + \Sigma id - \Sigma(i_2 \circ p_2), \tag{3.5}$$

where $p_k : X_1 \times X_2 \to X_k$ is the k-th projection , $i_k : X_k \to X_1 \times X_2$ is the k-th inclusion and $q : X_1 \times X_2 \to X_1 \wedge X_2$ is the collapsing map.

PROPOSITION 3.5. *Let $H = H(\mu) : \Sigma G \wedge G \to \Sigma G$ be the classical Hopf construction of the multiplication $\mu : G \times G \to G$. Then, in $\pi_{n+4}(S^{n+1})$,*

$$(\Sigma p \circ H)_*(\Sigma a) = 0 \tag{3.6}$$

$$(\Sigma p \circ H)_*(\Sigma b) = -\nu_{n+1} \tag{3.7}$$

PROOF. Since G has a right action of S^3 and the orbit space G/S^3 is equal to S^n, the following diagram commutes:

$$\begin{array}{ccc} G \times S^3 & \xrightarrow{\bar{\mu}} & G \\ {\scriptstyle p\times 1}\downarrow & & \downarrow{\scriptstyle p} \\ S^n \times S^3 & \xrightarrow[p_1]{} & S^n, \end{array} \tag{3.8}$$

where p_1 is the first projection and $\bar{\mu}$ is the restriction of μ. Applying the Hopf construction to the above diagram, we have

$$\begin{aligned}(\Sigma p \circ H)_*(\Sigma a) &= (\Sigma p) \circ H(\mu) \circ (\Sigma \lambda_2 \wedge i) \\ &= (\Sigma p) \circ H(\bar{\mu}) \circ (\Sigma \lambda_2 \wedge 1) \\ &= H(p_1) \circ \Sigma(p \wedge 1) \circ (\lambda_2 \wedge 1) \\ &= 0.\end{aligned}$$

Here we used the fact that $H(p_1) = 0$. This follows from the characterization of the Hopf construction (3.5). This proves (3.6).

For the proof of (3.7), we consider the left actions μ and $\hat{\mu}$ of G on both G and $S^n = G/S^3$. Consider the commutative diagram:

$$\begin{array}{ccc} G \times G & \xrightarrow{\mu} & G \\ {\scriptstyle 1\times p}\downarrow & & \downarrow{\scriptstyle p} \\ G \times S^n & \xrightarrow[\hat{\mu}]{} & S^n. \end{array}$$

Again applying the Hopf construction, we get the commutative diagram:

$$\begin{array}{ccccc} \Sigma S^{n+3} & \xrightarrow{\Sigma b} & \Sigma G \wedge G & \xrightarrow{H(\mu)} & \Sigma G \\ \| & & \downarrow{\scriptstyle \Sigma 1 \wedge p} & & \downarrow{\scriptstyle \Sigma p} \\ \Sigma S^3 \wedge S^n & \xrightarrow[\Sigma i \wedge 1]{} & \Sigma G \wedge S^n & \xrightarrow[H(\hat{\mu})]{} & \Sigma S^n. \end{array}$$

The commutativity of the left square follows from Lemma 3.4. Note that $H(\hat{\mu}) \circ \Sigma i \wedge 1 = J(i)$, where $J : \pi_k(G) \to \pi_{n+k+1}(S^{n+1})$ is the Whitehead J-homomorphism. From Corollary 11.2 in [**8**], we see that $J(i) = J(i_* \iota_3) = \Sigma^{n-3} J(\iota_3) = \Sigma^{n-3}(-\nu_4)$. The last equation follows from the fact that the Whitehead J-homomorphism is just the homotopy induced homomorphism by the Hopf construction of $\mu : S^3 \times S^3 \to S^3$. This completes the proof of Proposition (3.5). □

4. Properties of e

In this section we study the property of the function $e : [G, G] \to \mathbb{Z}/24$. Using the property of e, we have the main theorem in §5.

LEMMA 4.1. *Let $j : S^{n+3} \to S^n \vee S^{n+3}$ be the inclusion. Then, stably, j factors through π, that is, after suitable suspension, there exists a stable map $u : S^{n+3} \to G$ such that $\pi \circ u = j$. Moreover,*

$$\bar{\Delta}_* u = \Sigma^\infty b, \tag{4.1}$$

in $\pi^s_{n+3}(G \wedge G) \cong \pi_{n+3}(G \wedge G)$, where $\bar{\Delta} : G \to G \wedge G$ is the reduced diagonal map, $\pi^s_*()$ denotes the stable homotopy group and b is the element in (3.7).

PROOF. Consider the cofiber sequence:

$$S^3 \xrightarrow{\ i\ } G \xrightarrow{\ \pi\ } S^n \vee S^{n+3} \xrightarrow{\ \partial\ } S^4 \xrightarrow{\ \Sigma i\ } \Sigma G.$$

Note that $\partial \circ j \in \pi^s_{n-1}(S^0)$. If $G{=}SU(3)$ then $n = 5$ and it follows that $\partial \circ j = 0$. When G is $Sp(2)$, the only non-zero possibility of $\partial \circ j$ is ν^2. If $\partial \circ j = \nu^2$, this implies that $\Sigma i_*(\nu^2) = 0$. But, it is well-known that $Sp(2)$ has the following cell structure:

$$Sp(2) = S^3 \cup_{\nu'} e^7 \cup e^{10}$$

and stably $\nu' = 2\nu$. This implies that $\Sigma i_*(\nu^2) \neq 0$ and consequently that $\partial \circ j = 0$. This proves the first assertion. The second one follows from easy homology calculations. □

THEOREM 4.2. *Given $f, g \in [G, G]$, we have*

$$e(f+g) = e(f) + e(g) - d_1(f)d_2(g).$$

PROOF. In order to determine the element $e(f+g)$, we consider the following diagram:

$$\begin{array}{ccccccc}
 & & G & \xrightarrow{\ \Delta\ } G \times G \xrightarrow{\ f\times g\ } & G \times G & \xrightarrow{\ \mu\ } & G \\
 & {\scriptstyle \exists u}\nearrow & \downarrow{\scriptstyle \pi} & & & & \downarrow{\scriptstyle p} \\
S^{n+3} & \xrightarrow{\ j\ } & S^n \vee S^{n+3} & & \xrightarrow{\ d_2(f+g)\iota_n + e(f+g)\nu_n\ } & & S^n.
\end{array}$$

By Lemma (4.1) the map u exists after suitable suspension. Therefore, we study the stable map,

$$\Sigma^\infty(p \circ \mu \circ (f \times g) \circ \Delta) \circ u.$$

There exists a natural splitting,

$$\Sigma G \times G = \Sigma G \vee \Sigma G \wedge G \vee \Sigma G.$$

Under this splitting,

$$\begin{array}{lcr}
\Sigma\Delta & \text{corresponds to} & \Sigma id \vee \Sigma\bar{\Delta} \vee \Sigma id, \\
\Sigma(f \times g) & \text{to} & \Sigma f \vee \Sigma f \wedge g \vee \Sigma g,
\end{array}$$

and

$$\begin{array}{lcr}
\Sigma\mu & \text{to} & \Sigma id + H + \Sigma id,
\end{array}$$

where H is the classical Hopf construction of $\mu : G \times G \to G$.

Therefore

$$\Sigma^\infty(p\circ\mu\circ(f\times g)\circ\Delta)\circ u = \Sigma^\infty(p\circ f)\circ u + \Sigma^\infty(p\circ g)\circ u + \Sigma^\infty(\Sigma p\circ H\circ\Sigma(f\wedge g)\circ\bar{\Delta})\circ u.$$

By the definition of e (2.6) and Lemma 4.1,

$\Sigma^\infty(p \circ f) \circ u = e(f)\Sigma^\infty\nu_n,$
$\Sigma^\infty(p \circ g) \circ u = e(g)\Sigma^\infty\nu_n$

and
$\Sigma^\infty(\Sigma p \circ H \circ \Sigma(f \wedge g) \circ \bar{\Delta}) \circ u = \Sigma^\infty(\Sigma p \circ H \circ \Sigma(f \wedge g) \circ \Sigma b)$.

From (3.4), it follows that $(f \wedge g) \circ b = \dfrac{d_1(f)d_2(g) - d_2(f)d_1(g)}{l} a + d_1(f)d_2(g)b$.

Note that $\dfrac{d_1(f)d_2(g) - d_2(f)d_1(g)}{l}$ is an integer, because that $d_1(h) \equiv d_2(h) \mod l$ for any map $h \in [G, G]$.

By (3.6) and (3.7), it follows that

$$\Sigma^\infty(\Sigma p \circ H \circ \Sigma(f \wedge g) \circ \Sigma b) = -d_1(f)d_2(g)\Sigma^\infty \nu_n.$$

This completes the proof. □

5. Determination of the number m_G

In this section we prove Theorem 1.2. From Proposition 3.2 and Theorem 4.2, it is enough to show

PROPOSITION 5.1.

$$\lambda_2 \circ \nu_n = \begin{cases} \sigma & \textit{if } G = SU(3), \\ 20\sigma & \textit{if } G = Sp(2). \end{cases}$$

PROOF. We first show this in case that $G = Sp(2)$. Then $n = 7$ and $l = 12$. Since $\pi_{n+3}(Sp(3)) = 0$, there exists an integer m such that $\partial(mE) = \lambda_2 \circ \nu_n$:

$$\begin{array}{ccccc} & & & & \Omega S^{n+4} \\ & & \overset{mE}{\nearrow} & & \downarrow \partial \\ S^{n+3} & \xrightarrow[\nu_n]{} & S^n & \xrightarrow[\lambda_2]{} & Sp(2). \end{array}$$

Let $HP^2 = S^4 \cup_{\nu_4} e^8$ be the quaternionic projective plane and Q^k the quaternionic quasi-projective space which is a subcomplex of $Sp(k)$. Recall that there exists [3] a stable splitting map $\theta : Sp(k) \to \Omega^\infty\Sigma^\infty Q^k$. Using the above diagram, there exist a map $f : \Sigma^3 HP^2 \to Sp(3)$ such that the following diagram commutes :

$$\begin{array}{ccccc} S^7 & \xrightarrow{\lambda_2} & Sp(2) & \xrightarrow{\theta} & \Omega^\infty\Sigma^\infty Q^2 \\ \downarrow & & \downarrow & & \downarrow \\ \Sigma^3 HP^2 & \xrightarrow{\exists f} & Sp(3) & \xrightarrow{\theta} & \Omega^\infty\Sigma^\infty Q^3 \\ \downarrow & & \downarrow & & \downarrow \\ S^{11} & \xrightarrow{m\iota_{11}} & S^{11} & \xrightarrow{E^\infty} & \Omega^\infty\Sigma^\infty S^{11}. \end{array}$$

Thus we have a stable map $f : \Sigma^3 HP^2 \to Q^\infty$ such that $f_*(\beta_1) = 12\gamma_2$, where $\beta_i \in H_{4i}(HP^2)$ and $\gamma_k \in H_{4k-1}(Q^\infty)$ are the standard generators. According to [6], from the above data, it follows that $f_*(\beta_2) = 20\gamma_3 \mod 120$. Therefore $m = 20 \mod 120$.

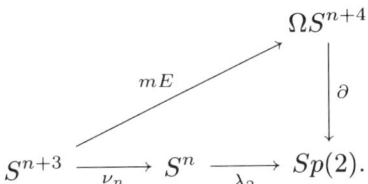

When $G = SU(3)$ and $n = 5$, consider the following diagram:

$$\begin{array}{ccccccc} & & & & & & \Omega S^{n+4} \\ & & & \overset{mE}{\nearrow} & & & \downarrow \partial \\ S^{n+3} & \xrightarrow[\nu_n]{} & S^n & \xrightarrow[\lambda_2]{} & SU(3) & \overset{j}{\hookrightarrow} & SU(4). \end{array}$$

The rest of the proof is almost the same, and so we omit it. □

6. Composition structures in $[G, G]$

From the definition of generators $\alpha, \beta, \gamma \in [G, G]$, it is easy to see

PROPOSITION 6.1. *The following table gives composition structure among generators and values of* $d_i()$*:*

	$\circ\alpha$	$\circ\beta$	$\circ\gamma$	d_1	d_2
α	$l\alpha$	α	0	0	l
β	α	β	γ	1	1
γ	$l\gamma$	γ	0	0	$0,$

where $f \circ g$ *is the element of the* $(f, g)-$*component in the above table.*

Thus using Proposition 3.2, Theorem 1.2 and the above table, we can compute $(a'\alpha + b'\beta + c'\gamma) \circ (a\alpha + b\beta + c\gamma)$, here a, a', b, b', c, c' are integers.

Let $a, b, c \in \mathbb{Z}$. Here we give some corollaries of Theorem 1.2.

COROLLARY 6.2. (1) *Let* $f, g \in [G, G]$. *If* $d_1(f) = 0$ *or* $d_2(g) = 0$, *then* $\alpha \circ (f + g) = \alpha \circ f + \alpha \circ g$,

(2) $\alpha \circ (a\alpha) = a(\alpha \circ \alpha) = al\alpha$,

(3) $\alpha \circ (b\beta) = b\alpha - m\frac{b(b-1)}{2}\gamma$,

(4) $\alpha \circ (c\gamma) = c(\alpha \circ \gamma) = lc\gamma$.

(5) $\alpha \circ (a\alpha + b\beta + c\gamma) = (la + b)\alpha + (lc - m\frac{b(b-1)}{2})\gamma$

References

[1] M. Arkowitz and P. Silberbush. Some properties of Hopf-type constructions. *Math. Proc. Cambridge Philos. Soc.*, 117:287–301, 1995.

[2] H. J. Baues and M. Hartl and T. Pirashvili. Quadratic categories and square rings. *J. Pure and Applied Algebra*, 122:1–40, 1997.

[3] I.M. James. On the homotopy type of Stiefel manifolds. *Proc. Amer. Math. Soc.*, 29:151–158, 1971.

[4] I.M. James and J. H. C. Whitehead. The homotopy theory of sphere bundles over spheres (I). *Proc. London Math. Soc. (3)*, 4:196–218, 1954.

[5] M. Mimura and H. Ōshima. Self homotopy groups of Hopf spaces with at most three cells. *J. Math.Soc. Japan.*

[6] K. Morisugi. Stable maps to projective spaces. *Publ. Res. Inst. Math. Sci.*, 24:301–309, 1988.

[7] K. Morisugi. Hopf construction, Samelson products and suspension maps. *Contemporary Math.*, 239:225–238, 1999.

[8] H. Toda. *Composition Methods in Homotopy Groups of Spheres*, volume 49 of *Ann. Math. Studies.* Princeton Univ. Press, 1962.

[9] G. W. Whitehead. *Elements of Homotopy Theory.* Number 61 in GTM. Springer, 1978.

DEPARTMENT OF MATHEMATICS, WAKAYAMA UNIVERSITY
E-mail address: kaoru-m@math.edu.wakayama-u.ac.jp

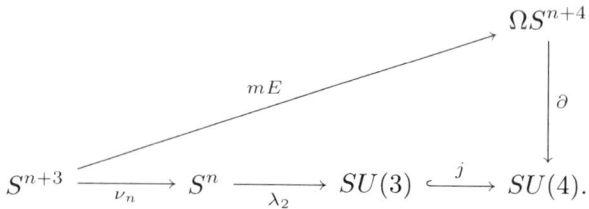

Contemporary Mathematics
Volume **274**, 2001

Self-homotopy of a suspension of the real 4-projective space

Juno MUKAI

ABSTRACT. Let P^n be the real n-dimensional projective space and $\Sigma\mathrm{P}^n$ be the suspension of P^n. Then we determine the group structure of the homotopy set $[\Sigma\mathrm{P}^4, \Sigma\mathrm{P}^n]$ for $n \geq 1$.

1. Introduction

In this note all spaces and homotopies are based. We denote by ΣX the suspension of a space X and by P^n the real n-dimensional projective space. We set $\mathbf{Z}_n = \mathbf{Z}/n\mathbf{Z}$ and the notation $(\mathbf{Z}_m)^r \oplus (\mathbf{Z}_n)^s$ means

$$\underbrace{\mathbf{Z}_m \oplus \cdots \oplus \mathbf{Z}_m}_{r} \oplus \underbrace{\mathbf{Z}_n \oplus \cdots \oplus \mathbf{Z}_n}_{s}.$$

The purpose of this note is to determine the group structure of the homotopy set $[\Sigma\mathrm{P}^4, \Sigma\mathrm{P}^n]$ for $n \geq 1$.

THEOREM 1.1. *The group $[\Sigma\mathrm{P}^4, \Sigma\mathrm{P}^n]$ is abelian for $n \geq 1$ and its structure is given as follows: $[\Sigma\mathrm{P}^4, S^2] \cong \mathbf{Z}_2$, $[\Sigma\mathrm{P}^4, \Sigma\mathrm{P}^2] \cong \mathbf{Z}_4 \oplus \mathbf{Z}_2$, $[\Sigma\mathrm{P}^4, \Sigma\mathrm{P}^3] \cong \mathbf{Z}_4 \oplus (\mathbf{Z}_2)^2$ and $[\Sigma\mathrm{P}^4, \Sigma\mathrm{P}^n] \cong \mathbf{Z}_8 \oplus (\mathbf{Z}_2)^2$ for $n \geq 4$.*

Our method is to use the quasi-fibration $h : \Sigma\mathrm{P}^3 \wedge \mathrm{P}^3 \to \Sigma\mathrm{P}^3$ induced from the Hopf construction for the multiplication of the Hopf space P^3 ([**2**], [**7**]). In fact we use the following direct sum decomposition ([**2**])

$$\pi_n(\Sigma\mathrm{P}^3) = h_*\pi_n(\Sigma\mathrm{P}^3 \wedge \mathrm{P}^3) \oplus \Sigma\pi_{n-1}(\mathrm{P}^3).$$

We denote by $\mathrm{P}^n_k = \mathrm{P}^n/\mathrm{P}^{k-1}$ for $k \leq n$ the stunted real projective space and by $M^n = \Sigma^{n-2}\mathrm{P}^2$ for $n \geq 2$ the Moore space of type $(\mathbf{Z}_2, n-1)$. Let ν' be a generator of the 2-primary component of $\pi_6(S^3) \cong \mathbf{Z}_{12}$ ([**15**]). Then $\Sigma\mathrm{P}^3 \wedge \mathrm{P}^3$ has the following cellular decomposition.

LEMMA 1.2. *$\Sigma\mathrm{P}^3 \wedge \mathrm{P}^3 = ((\Sigma\mathrm{P}^2 \wedge \mathrm{P}^2) \cup_{i\nu'} e^7) \vee M^6 \vee M^6$, where $i : S^3 \hookrightarrow \Sigma\mathrm{P}^2 \wedge \mathrm{P}^2$ is the inclusion.*

To determine the group of self-homotopies of $\Sigma\mathrm{P}^4$, we need the following.

THEOREM 1.3. *$\pi_5(\Sigma\mathrm{P}^4) \cong (\mathbf{Z}_2)^3$.*

2000 *Mathematics Subject Classification.* Primary 55P40, 55Q15, 55Q52.
Key words and phrases. Suspension of real projective space, Moore space, Hopf construction.

The author wishes to thank Morisugi for advising the use of the Hopf fibration $h : \Sigma P^3 \wedge P^3 \to \Sigma P^3$ and for much useful advice. The author also wishes to thank Rees for suggesting that there is no map $S^7 \to \Sigma(P^3 \times P^3)$ of degree 1.

2. Self-homotopy of ΣP^3

For a pair of spaces (X, A), let $i_{A,X} : A \to X$ be the inclusion and $p_{X,A} : X \to X/A$ be the map pinching A to one point. We set $i_{k,n} = i_{P^k,P^n}$ for $k < n$, $i_n = \Sigma^{n-2} i_{1,2}$ and $p_n = \Sigma^{n-2} p_{2,1}$ for $n \geq 2$. We denote by ι_X the identity class of X. We set $\iota_n = \iota_X$ for $X = S^n$ and $\iota'_n = \iota_X$ for $X = M^n$. Let $\eta_n \in \pi_{n+1}(S^n)$ for $n \geq 2$ be the Hopf map. We know $\pi_4(M^3) \cong \mathbf{Z}_4$ and it is generated by an element $\tilde{\eta}_2$ satisfying $p_3\tilde{\eta}_2 = \eta_3$ ([**9**]). We set $\tilde{\eta}_n = \Sigma^{n-2}\tilde{\eta}_2$ for $n \geq 2$.

Let X be a 1-connected finite CW-complex, $\theta : S^{n-1} \to X$ be a mapping and $X^* = X \cup_\theta e^n$ be a complex formed by attaching an n-cell. Let $\omega_n \in \pi_n(X^*, X)$ be the characteristic map of the n-cell e^n of X^* and CY be a cone of a space Y. For an element $\alpha \in \pi_{k-1}(Y)$, we denote by $\hat{\alpha}' \in \pi_k(CY, Y)$ an element satisfying $\partial\hat{\alpha}' = \alpha$, where $\partial : \pi_k(CY, Y) \to \pi_{k-1}(Y)$ is the connecting isomorphism. For $\alpha \in \pi_{k-1}(S^{n-1})$, we set

$$\hat{\alpha} = \omega_n \hat{\alpha}' \in \pi_n(X^*, X).$$

Let $p : (X^*, X) \to (S^n, *)$ be the collapsing map and $\partial : \pi_n(X^*, X) \to \pi_{n-1}(X)$ be the boundary homomorphism. Then we have the following:

$$\partial\hat{\alpha} = \theta \circ \alpha \quad \text{and} \quad p_*\hat{\alpha} = \Sigma\alpha.$$

We know the important relation ([**16**])

$$2\iota'_n = i_n \eta_{n-1} p_n \ (n \geq 3). \tag{2.1}$$

Let $f : M^3 \hookrightarrow M^3 \vee S^3$ and $g : S^3 \hookrightarrow M^3 \vee S^3$ be the canonical inclusions, respectively. Then $P^2 \wedge P^2$ has a cellular decomposition

$$P^2 \wedge P^2 = (M^3 \vee S^3) \cup_{f i_3 \eta_2 + 2g} e^4 = M^3 \cup_{2\iota'_3} CM^3.$$

For the inclusion $i' : M^3 \vee S^3 \hookrightarrow P^2 \wedge P^2$, we set

$$\tilde{\imath}_3 = i' \circ g;\ \tilde{\imath}_n = \Sigma^{n-3}\tilde{\imath}_3 \ (n \geq 3).$$

Since $i' \circ f = \iota'_2 \wedge i_2$ and $(\iota'_2 \wedge i_2) \circ i_3 = i_2 \wedge i_2$, we obtain

$$2\tilde{\imath}_n = -(\Sigma^{n-3}(i_2 \wedge i_2))\eta_{n-1} \ (n \geq 3). \tag{2.2}$$

We denote by $\gamma_n : S^n \to P^n$ the covering map. As is well known, $\Sigma\gamma_2 = 2(i_3\eta_2)$ and $\Sigma^2\gamma_2 = 0$. We denote by $\bar{\eta}_3 \in [M^5, S^3]$ an extension of η_3 and set $\bar{\eta}_n = \Sigma^{n-3}\bar{\eta}_3$ for $n \geq 3$. Now we recall the following ([**9**], [**10**], [**11**]): $\pi_3(M^3) = \mathbf{Z}_4\{i_3\eta_2\}$; $\pi_3(\Sigma P^2 \wedge P^2) = \mathbf{Z}_2\{\Sigma i_2 \wedge i_2\}$; $\pi_4(\Sigma P^2 \wedge P^2) = \mathbf{Z}_4\{\tilde{\imath}_4\}$; $\pi_5(\Sigma P^2 \wedge P^2) = \mathbf{Z}_2\{\tilde{\imath}_4\eta_4\} \oplus \mathbf{Z}_2\{(\iota'_3 \wedge i_2)\bar{\eta}_3\}$; $[M^4, M^3] = \mathbf{Z}_2\{\overline{\Sigma\gamma_2}\} \oplus \mathbf{Z}_2\{\tilde{\eta}_2 p_4\}$, $[\iota'_3, i_3] = \overline{\Sigma\gamma_2} + \tilde{\eta}_2 p_4$; $[M^5, M^4] = \mathbf{Z}_2\{i_4\bar{\eta}_3\} \oplus \mathbf{Z}_2\{\tilde{\eta}_3 p_5\}$. Let $\delta \in \pi_6(M^4)$ be an attaching map of the Stiefel manifold $V_{5,2} = M^4 \cup e^7$. Then we also recall that $\pi_6(M^4) = \mathbf{Z}_4\{\delta\} \oplus \mathbf{Z}_2\{\tilde{\eta}_3\eta_5\}$ and $2\delta = i_4\nu'$ ([**8**]). From the fact that $[\iota'_3, i_3] = \overline{\Sigma\gamma_2} + \tilde{\eta}_2 p_4$ or $\Sigma^2 P^4 = M^4 \cup_{\tilde{\eta}_3 p_5} CM^5$,

$$\Sigma(\overline{\Sigma\gamma_2}) = \tilde{\eta}_3 p_5. \tag{2.3}$$

Let $s : S^5 \hookrightarrow \Sigma^2 P^3 = M^4 \vee S^5$ be the embedding onto the second space. Then, from the cell structure of $\Sigma^2 P^4$ and by using the fact $p_{3,2} \circ \gamma_3 = 2\iota_3$,

(2.4) $$\Sigma^2\gamma_3 = (\Sigma^2 i_{2,3})\tilde{\eta}_3 + 2s.$$

Now we show

LEMMA 2.1. $\pi_3(\Sigma P^3) = \mathbf{Z}_2\{(\Sigma i_{1,3})\eta_2\}$ and $\pi_4(\Sigma P^3) = \mathbf{Z}\{\Sigma\gamma_3\} \oplus \mathbf{Z}_4\{(\Sigma i_{2,3})\tilde{\eta}_2\}$.

PROOF. In the exact sequence ($i = \Sigma i_{2,3}$)

$$\pi_4(\Sigma P^3, M^3) \xrightarrow{\partial} \pi_3(M^3) \xrightarrow{i_*} \pi_3(\Sigma P^3) \longrightarrow 0,$$

we have $\pi_4(\Sigma P^3, M^3) \cong \pi_4(S^4)$ and $\partial\hat{\iota}_3 = \Sigma\gamma_2 = 2(i_3\eta_2)$. This leads to the first half of the lemma.

Next we consider the exact sequence ($i = \Sigma i_{2,3}$)

$$\pi_5(\Sigma P^3, M^3) \xrightarrow{\partial} \pi_4(M^3) \xrightarrow{i_*} \pi_4(\Sigma P^3) \xrightarrow{j_*} \pi_4(\Sigma P^3, M^3).$$

By the argument above, we have $j_*\Sigma\gamma_3 = 2\hat{\iota}_3$. By Theorem 2.1 of [**5**], $\pi_5(\Sigma P^3, M^3) = \mathbf{Z}_2\{\hat{\eta}_3\} \oplus \mathbf{Z}_2\{[\omega_4, i_3]\}$, where $\omega_4 \in \pi_4(\Sigma P^3, M^3)$ is the characteristic map of the 4-cell of ΣP^3. We obtain

$$\partial\hat{\eta}_3 = \Sigma\gamma_2 \circ \eta_3 = i_3\eta_2 \circ 2\iota_3 \circ \eta_3 = 0$$

and

$$\partial[\omega_4, i_3] = -[\Sigma\gamma_2, i_3] = -i_3[2\eta_2, \iota_2] = 0.$$

So i_* is a monomorphism. This leads to the second half, completing the proof. □

Let X be a Hopf space having the homotopy type of a connected CW-complex and $E : X \to \Omega\Sigma X$ be the suspension map. According to [**7**], there exists a fiber sequence

$$\cdots \longrightarrow \Omega\Sigma X \wedge X \xrightarrow{\Omega h} \Omega\Sigma X \xrightarrow{r} X \xrightarrow{i} \Sigma X \wedge X \xrightarrow{h} \Sigma X,$$

which satisfies that $r \circ E = \iota_X$. The connecting map $r : \Omega\Sigma X \to X$ is an H-map if and only if X is a homotopy associative H-space. Thus we have the following.

LEMMA 2.2. For the Hopf fibration $h : \Sigma P^3 \wedge P^3 \to \Sigma P^3$, there exists a direct sum decomposition for a space X:

$$[\Sigma X, \Sigma P^3] = h_*[\Sigma X, \Sigma P^3 \wedge P^3] \oplus \Sigma[X, P^3].$$

Here the direct sum decomposition is given as sets in general and it is given as groups if $X = \Sigma X'$ for a space X'.

The restrictions of h are written as follows: $h' = h\,|_{\Sigma P^2 \wedge P^2}: \Sigma P^2 \wedge P^2 \to \Sigma P^3$, $h'' = h\,|_{M^4}: M^4 \to \Sigma P^3$ and $h''' = h\,|_{S^3}: S^3 \to \Sigma P^3$. We show

LEMMA 2.3. **(i):** $h''' = (\Sigma i_{1,3})\eta_2$ and $h'\tilde{\iota}_4 = \pm(\Sigma i_{2,3})\tilde{\eta}_2$.
(ii): $[M^3, \Sigma P^3] = \mathbf{Z}_4\{\Sigma i_{2,3}\}$ and $2(\Sigma i_{2,3}) = (\Sigma i_{1,3})\eta_2 p_3$.
(iii): $[M^4, \Sigma P^3] = \mathbf{Z}_2\{h''\} \oplus \mathbf{Z}_2\{(\Sigma i_{2,3})\tilde{\eta}_2 p_4\} \oplus \mathbf{Z}_2\{\Sigma\gamma_3 \circ p_4\}$ and $\Sigma\gamma_3 \circ p_4 = (\Sigma i_{2,3})\overline{\Sigma\gamma_2}$.

PROOF. By Lemma 2.2 and the fact that $\pi_2(P^3) = 0$, $\pi_3(\Sigma P^3) = h'_*\pi_3(\Sigma P^2 \wedge P^2)$. Since $\pi_3(\Sigma P^2 \wedge P^2) = \mathbf{Z}_2\{\Sigma i_2 \wedge i_2\}$, we have the first half of (i) by Lemma 2.1. Similarly, by Lemma 2.2, we conclude that $\pi_4(\Sigma P^3) = h_*\pi_4(\Sigma P^3 \wedge P^3) \oplus \Sigma\pi_3(P^3) = h'_*\pi_4(\Sigma P^2 \wedge P^2) \oplus \Sigma\pi_3(P^3) = \mathbf{Z}_4\{h'\tilde{\iota}_4\} \oplus \mathbf{Z}\{\Sigma\gamma_3\}$. This and Lemma 2.1 lead to the second half of (i).

By use of the cofibration starting with $2\iota_3$ and by (2.1), (ii) is obtained.

Now $\Sigma\gamma_3$ is taken as a representative of a Toda bracket $\{\Sigma i_{2,3}, \Sigma\gamma_2, 2\iota_3\}$ because $p_{3,2}\circ\gamma_3 = 2\iota_3$. By the properties of Toda brackets, we have ($i' = \Sigma i_{2,3}$)

$$\begin{aligned} \Sigma\gamma_3\circ p_4 &\in \{i', \Sigma\gamma_2, 2\iota_3\}\circ p_4 \\ &= -(i'\circ\{\Sigma\gamma_2, 2\iota_3, p_3\}) \\ &\ni -(i'\circ\overline{\Sigma\gamma_2}) \\ &\mathrm{mod}\ \ i'_*\pi_4(M^3)\circ p_4 = \{i'\tilde{\eta}_2 p_4\}. \end{aligned}$$

So we can set $\Sigma\gamma_3\circ p_4 = i'(\overline{\Sigma\gamma_2} + x\tilde{\eta}_2 p_4)$ for $x = 0$ or 1. By (2.3) and (2.4),

$$(\Sigma i')\tilde{\eta}_3 p_5 = \Sigma^2\gamma_3\circ p_5 = (1+x)(\Sigma i')\tilde{\eta}_3 p_5.$$

Then $(\Sigma i')\tilde{\eta}_3 p_5$ is a generator of $[M^5, \Sigma^2\mathrm{P}^3] \cong [M^5, M^4]\oplus[M^5, S^5] \cong (\mathbf{Z}_2)^3$. So we obtain $x = 0$, and hence the equality $\Sigma\gamma_3\circ p_4 = (\Sigma i_{2,3})\overline{\Sigma\gamma_2}$ holds. This leads to the second half of (iii).

Obviously we have $[M^3, \mathrm{P}^3] = \mathbf{Z}_2\{\gamma_3 p_3\}$. So, by Lemma 2.2,

$$[M^4, \Sigma\mathrm{P}^3] = h_*[M^4, \Sigma\mathrm{P}^3\wedge\mathrm{P}^3]\oplus\Sigma[M^3, \mathrm{P}^3] = h'_*[M^4, \Sigma\mathrm{P}^2\wedge\mathrm{P}^2]\oplus\mathbf{Z}_2\{\Sigma\gamma_3\circ p_4\}.$$

We note that $\iota'_3\wedge i_2\in[M^4, \Sigma\mathrm{P}^2\wedge\mathrm{P}^2]$ is an extension of $\Sigma i_2\wedge i_2$. So, by use of the cofibration starting with $2\iota_3$,

$$[M^4, \Sigma\mathrm{P}^2\wedge\mathrm{P}^2] = \mathbf{Z}_2\{\iota'_3\wedge i_2\}\oplus\mathbf{Z}_2\{\tilde{\iota}_4 p_4\}.$$

This and (i) lead to the first half of (iii), completing the proof. □

To ensure the commutativity of the group operation of specific elements of $[\Sigma\mathrm{P}^n, Y]$ for a space Y, we use the following criterion which is easily obtained from Theorem X.3.10 of [**17**].

LEMMA 2.4. For elements $\alpha\in[\Sigma\mathrm{P}^n, Y]$ and $\beta\in\pi_{n+1}(Y)$, the equality $\alpha+\beta\Sigma p_{n,n-1} = \beta\Sigma p_{n,n-1}+\alpha$ holds.

Let $q : \mathrm{P}^3 = SO(3)\to SO(3)/SO(2) = S^2$ be the projection which is an extension of $p_2 : \mathrm{P}^2\to S^2$ and satisfies the equation

$$q\gamma_3 = \eta_2.$$

We show

LEMMA 2.5. **(i):** $[\Sigma\mathrm{P}^3, S^3] = \mathbf{Z}_2\{\Sigma q\}\oplus\mathbf{Z}_2\{\eta_3\Sigma p_{3,2}\}$.
(ii): $[\Sigma\mathrm{P}^3, M^3] = \mathbf{Z}_2\{i_3\eta_2\Sigma q\}\oplus\mathbf{Z}_4\{\tilde{\eta}_2\Sigma p_{3,2}\}$.

PROOF. By the Freudenthal suspension theorem, the group $[\Sigma\mathrm{P}^3, S^3]$ is isomorphic to the stable group $\{\mathrm{P}^3, S^2\}\cong\{\mathrm{P}^2, S^2\}\oplus\pi_3^S(S^2)\cong(\mathbf{Z}_2)^2$. This leads to (i). We consider the homotopy exact sequence ($i = \Sigma i_{2,3}$, $p = \Sigma p_{3,2}$):

$$\pi_3(M^3)\xleftarrow{(\Sigma\gamma_2)^*}[M^3, M^3]\xleftarrow{i^*}[\Sigma\mathrm{P}^3, M^3]\xleftarrow{p^*}\pi_4(M^3)\longleftarrow 0.$$

By (2.1) and (i), an extension $i_3\eta_2\Sigma q\in[\Sigma\mathrm{P}^3, M^3]$ of $i_3\eta_2 p_3$ is of order 2. This leads to (ii), completing the proof. □

Now we show

LEMMA 2.6. **(i):** $[\Sigma\mathrm{P}^3, \Sigma\mathrm{P}^3] = \mathbf{Z}\{\iota_{\Sigma\mathrm{P}^3}\}\oplus\mathbf{Z}_4\{(\Sigma i_{2,3})\tilde{\eta}_2\Sigma p_{3,2}\}\oplus\mathbf{Z}_2\{(\Sigma i_{1,3})\eta_2\Sigma q\}$ and

$$2\iota_{\Sigma\mathrm{P}^3} = \Sigma\gamma_3\circ\Sigma p_{3,2}\pm(\Sigma i_{2,3})\tilde{\eta}_2\Sigma p_{3,2}+(\Sigma i_{1,3})\eta_2\Sigma q.$$

(ii): $[\Sigma P^3, \Sigma P^4] = \mathbf{Z}_8\{\Sigma i_{3,4}\} \oplus \mathbf{Z}_2\{(\Sigma i_{1,4})\eta_2\Sigma q\}$ and $2(\Sigma i_{3,4}) = \pm(\Sigma i_{2,4})\tilde{\eta}_2\Sigma p_{3,2} + (\Sigma i_{1,4})\eta_2\Sigma q$.

PROOF. We consider the exact sequence induced from the cofibration starting with $i = \Sigma i_{1,3}$ ($p = \Sigma p_{3,1}$):

$$\pi_2(\Sigma P^3) \xleftarrow{i^*} [\Sigma P^3, \Sigma P^3] \xleftarrow{p^*} [\Sigma P^3_2, \Sigma P^3].$$

We know $\pi_2(\Sigma P^3) = \mathbf{Z}_2\{\Sigma i_{1,3}\}$ and

$$[\Sigma P^3_2, \Sigma P^3] \cong \pi_3(\Sigma P^3) \oplus \pi_4(\Sigma P^3) \cong \mathbf{Z}_2\{(\Sigma i_{1,3})\eta_2\} \oplus \mathbf{Z}\{\Sigma\gamma_3\} \oplus \mathbf{Z}_4\{(\Sigma i_{2,3})\tilde{\eta}_2\}.$$

So we have an equation ($a, b, c \in \mathbf{Z}$)

$$2\iota_{\Sigma P^3} = a\Sigma\gamma_3 \circ \Sigma p_{3,2} + b(\Sigma i_{2,3})\tilde{\eta}_2\Sigma p_{3,2} + c(\Sigma i_{1,3})\eta_2\Sigma q.$$

By composing $\Sigma\gamma_3$ to the equation on the right and by using the fact that $2\tilde{\eta}_2 = i_3\eta_2^2$, we obtain

$$2\Sigma\gamma_3 = 2a\Sigma\gamma_3 + (b+c)(\Sigma i_{1,3})\eta_2^2.$$

This implies $a = 1$ and $b + c \equiv 0 \bmod 2$. On the other hand, by composing $\Sigma i_{2,3}$ to the first equation on the right,

$$2\Sigma i_{2,3} = c(\Sigma i_{1,3})\eta_2 p_3 = 2c\Sigma i_{2,3}.$$

So, by Lemma 2.3.(ii), c is odd. This leads to (i).

Since the stable group $\{P^3, P^3\}$ is isomorphic to $\pi_3^S(S^3) \oplus \pi_3^S(P^2) \oplus \{P^2, P^2\} \cong \mathbf{Z} \oplus \mathbf{Z}_4 \oplus \mathbf{Z}_4$ and $\Sigma^\infty q \equiv \Sigma^\infty p_2 \bmod \eta\Sigma^\infty p_{3,2}$, the homomorphism $\Sigma^\infty : [\Sigma P^3, \Sigma P^3] \to \{P^3, P^3\}$ is a monomorphism. This implies the commutativity of the group $[\Sigma P^3, \Sigma P^3]$, and gives (i).

In the exact sequence ($i = \Sigma i_{3,4}$)

$$[C\Sigma P^3, \Sigma P^3; \Sigma P^4, \Sigma P^3] \xrightarrow{\partial} [\Sigma P^3, \Sigma P^3] \xrightarrow{i_*} [\Sigma P^3, \Sigma P^4] \longrightarrow 0,$$

we have $[C\Sigma P^3, \Sigma P^3; \Sigma P^4, \Sigma P^3] \cong [\Sigma^2 P^3, S^5] = Z\{\Sigma^2 p_{3.2}\}$ and $\partial\hat{\xi} = \Sigma\gamma_3 \circ \xi$ for $\xi = \Sigma p_{3,2}$. So (i) implies (ii). This completes the proof. □

3. A cellular decomposition of $\Sigma P^3 \wedge P^3$

The reduced product $P^3 \wedge P^3$ was studied in [**12**] and [**13**]. First of all, we give a cell structure to $\Sigma P^3 \wedge P^2$.

LEMMA 3.1. $\Sigma P^3 \wedge P^2 = (\Sigma P^2 \wedge P^2) \vee M^6$.

PROOF. We see $\eta_2 \wedge \iota_2' = i_4\bar{\eta}_3 + \tilde{\eta}_3 p_5 \in [M^5, M^4]$, and so its order is 2. Therefore we have

$$\begin{aligned}
\Sigma\gamma_2 \wedge \iota_2' &= (i_3 \circ \eta_2 \circ 2\iota_3) \wedge \iota_2' \\
&= (i_3 \wedge \iota_2') \circ (\eta_2 \wedge \iota_2') \circ (2\iota_3 \wedge \iota_2') \\
&= (i_3 \wedge \iota_2') \circ 2(\eta_2 \wedge \iota_2') \\
&= 0.
\end{aligned}$$

□

Now we prove Lemma 1.2.

LEMMA 3.2. $\Sigma P^3 \wedge P^3 = (\Sigma P^2 \wedge P^2 \cup_{i\nu'} e^7) \vee M^6 \vee M^6$.

PROOF. Our task is to investigate the attaching map $\Sigma\gamma_2 \wedge \iota_{\mathrm{P}^3} : S^3 \wedge \mathrm{P}^3 \to \Sigma\mathrm{P}^2 \wedge \mathrm{P}^3$ in $\Sigma\mathrm{P}^3 \wedge \mathrm{P}^3$. We have

$$\Sigma\gamma_2 \wedge \iota_{\mathrm{P}^3} = (i_3 \wedge \iota_{\mathrm{P}^3}) \circ (\eta_2 \wedge \iota_{\mathrm{P}^3}) \circ (2\iota_3 \wedge \iota_{\mathrm{P}^3}).$$

By Lemma 2.6,

$$\begin{aligned}(\eta_2 \wedge \iota_{\mathrm{P}^3}) \circ (2\iota_3 \wedge \iota_{\mathrm{P}^3}) &= (\eta_2 \wedge \iota_{\mathrm{P}^3}) \circ (2\Sigma^3\iota_{\mathrm{P}^3}) \\ &= (\eta_2 \wedge \iota_{\mathrm{P}^3}) \circ \Sigma^2(\Sigma\gamma_3 \circ \Sigma p_{3,2} \\ &\quad \pm(\Sigma i_{2,3})\tilde{\eta}_2\Sigma p_{3,2} + (\Sigma i_{1,3})\eta_2\Sigma q) \\ &= (\eta_2 \wedge \iota_{\mathrm{P}^3}) \circ \Sigma^3\gamma_3 \circ \Sigma^3 p_{3,2} \\ &\quad \pm(\eta_2 \wedge \iota_{\mathrm{P}^3}) \circ \Sigma^3 i_{2,3} \circ \tilde{\eta}_4\Sigma^3 p_{3,2} \\ &\quad +(\eta_2 \wedge \iota_{\mathrm{P}^3}) \circ \Sigma^3 i_{1,3} \circ \eta_4\Sigma^3 q.\end{aligned}$$

By using (2.4), we obtain

$$\begin{aligned}(\eta_2 \wedge \iota_{\mathrm{P}^3}) \circ \Sigma^3\gamma_3 &= (\eta_2 \wedge \iota_{\mathrm{P}^3}) \circ (\iota_3 \wedge \gamma_3) \\ &= (\iota_2 \circ \eta_2) \wedge (\gamma_3 \circ \iota_3) \\ &= (\Sigma^2\gamma_3) \circ \eta_5 \\ &= (\Sigma^2 i_{2,3})\tilde{\eta}_3\eta_5.\end{aligned}$$

By the fact that $2\Sigma q = 0$,

$$\begin{aligned}(\eta_2 \wedge \iota_{\mathrm{P}^3}) \circ ((\Sigma^3 i_{1,3})\eta_4) \circ \Sigma^3 q &= (\eta_2 \wedge i_{1,3})\eta_4 \circ \Sigma^3 q \\ &= (\Sigma^2 i_{1,3})\eta_3^2 \circ \Sigma^3 q \\ &= 2((\Sigma^2 i_{2,3})\tilde{\eta}_3) \circ \Sigma^3 q \\ &= 0.\end{aligned}$$

Since $\nu' \in \{\eta_3, 2\iota_4, \eta_4\}_1 \bmod 2\nu'$ ([**15**]), we have $\bar{\eta}_3\tilde{\eta}_4 = \pm\nu'$. Then

$$\begin{aligned}(\eta_2 \wedge \iota_{\mathrm{P}^3}) \circ (\Sigma^3 i_{2,3}) \circ \tilde{\eta}_4 &= (\eta_2 \wedge \iota_{\mathrm{P}^3}) \circ (\iota_3 \wedge i_{2,3}) \circ \tilde{\eta}_4 \\ &= (\iota_2 \circ \eta_2) \wedge (i_{2,3} \circ \iota_2') \circ \tilde{\eta}_4 \\ &= (\Sigma^2 i_{2,3}) \circ (\eta_2 \wedge \iota_2') \circ \tilde{\eta}_4 \\ &= (\Sigma^2 i_{2,3}) \circ (i_4\bar{\eta}_3 + \tilde{\eta}_3 p_5) \circ \tilde{\eta}_4 \\ &= (\Sigma^2 i_{2,3}) \circ (i_4\bar{\eta}_3\tilde{\eta}_4) + (\Sigma^2 i_{2,3})\tilde{\eta}_3 p_5 \circ \tilde{\eta}_4 \\ &= (\Sigma^2 i_{1,3})\nu' + (\Sigma^2 i_{2,3})\tilde{\eta}_3\eta_5.\end{aligned}$$

Hence we have $(\eta_2 \wedge \iota_{\mathrm{P}^3}) \circ (\Sigma^3\gamma_3 + (\Sigma^3 i_{2,3})\tilde{\eta}_4) = (\Sigma^2 i_{1,3})\nu'$ and

$$(\eta_2 \wedge \iota_{\mathrm{P}^3}) \circ (2\iota_3 \wedge \iota_{\mathrm{P}^3}) \circ \Sigma^3 p_{3,2} = (\Sigma^2 i_{1,3})\nu'\Sigma^3 p_{3,2}.$$

This yields the equality

$$\Sigma\gamma_2 \wedge \iota_{\mathrm{P}^3} = (i_3 \wedge \iota_{\mathrm{P}^3}) \circ (\Sigma^2 i_{1,3})\nu'\Sigma^3 p_{3,2} = (i_3 \wedge i_{1,3})\nu'\Sigma^3 p_{3,2}.$$

This and Lemma 3.1 lead to the assertion, completing the proof. □

Now we determine $\pi_5(\Sigma\mathrm{P}^3)$. We recall the group $\pi_5(M^3)$. Let F be the homotopy fiber of p_3. Let $j : (M^3, *) \to (M^3, S^2)$ be the inclusion. Then we know the following ([**11**]):

$$\pi_5(M^3) = \mathbf{Z}_2\{\tilde{\eta}_2\eta_4\} \oplus \mathbf{Z}_2\{\overline{\Sigma\gamma_2}\tilde{\eta}_3\} \oplus \mathbf{Z}_2\{\tau\},$$

where τ comes from a generator of $\pi_5(F)$ and $j_*\tau = [[\omega_3, \iota_2], \iota_2]$ with $\omega_3 \in \pi_3(M^3, S^2)$.

We consider the exact sequence ($i' = \Sigma i_{2,3}$):

$$\pi_6(\Sigma P^3, M^3) \xrightarrow{\partial} \pi_5(M^3) \xrightarrow{i'_*} \pi_5(\Sigma P^3) \xrightarrow{j_*} \pi_5(\Sigma P^3, M^3) \xrightarrow{\partial} \pi_4(M^3).$$

In the proof of Lemma 2.1, we know $\pi_5(\Sigma P^3, M^3) = \mathbf{Z}_2\{\hat{\eta}_3\} \oplus \mathbf{Z}_2\{[\omega_4, i_3]\}$ and $\partial\pi_5(\Sigma P^3, M^3) = 0$. So there exists an element $\beta \in \pi_5(\Sigma P^3)$ satisfying $j_*\beta = [\omega_4, i_3]$. By Lemma 1.4 of [**8**], there exists a coextension $\tilde{\eta}''_3 \in \{\Sigma i_{2,3}, \Sigma\gamma_2, \eta_3\} \subset \pi_5(\Sigma P^3)$ of η_3 satisfying $j_*\tilde{\eta}''_3 = \hat{\eta}_3$. We show

LEMMA 3.3. **(i):** $\tilde{\eta}''_3 = h''\tilde{\eta}_3$ and $\Sigma\gamma_3 \circ \eta_4 = (\Sigma i_{2,3})\overline{\Sigma\gamma_2}\tilde{\eta}_3$.
(ii): $\pi_5(\Sigma P^3) = \mathbf{Z}_2\{\Sigma\gamma_3 \circ \eta_4\} \oplus \mathbf{Z}_2\{h''\tilde{\eta}_3\} \oplus \mathbf{Z}_2\{(\Sigma i_{2,3})\tilde{\eta}_2\eta_4\} \oplus \mathbf{Z}_2\{(\Sigma i_{2,3})\tau\} \oplus \mathbf{Z}_2\{\beta\}$.

PROOF. Since h'' is an extension of $(\Sigma i_{1,3})\eta_2$,

$$h''\tilde{\eta}_3 \in \{(\Sigma i_{1,3})\eta_2, 2\iota_3, \eta_3\} \subset \{\Sigma i_{2,3}, 2(i_3\eta_2), \eta_3\}.$$

So $\tilde{\eta}''_3$ is taken as $h''\tilde{\eta}_3$, giving the first half of (i). By Lemma 2.3.(iii), we obtain

$$\Sigma\gamma_3 \circ \eta_4 = (\Sigma\gamma_3) \circ p_4 \circ \tilde{\eta}_3 = (\Sigma i_{2,3})\overline{\Sigma\gamma_2}\tilde{\eta}_3.$$

This leads to the second half of (i).

By Lemma 3.2, we have

$$\pi_5(\Sigma P^3 \wedge P^3) \cong \pi_5(\Sigma P^2 \wedge P^2) \oplus \pi_5(M^6) \oplus \pi_5(M^6).$$

Since $\pi_4(P^3) = \mathbf{Z}_2\{\gamma_3\eta_3\}$ and $\pi_5(\Sigma P^2 \wedge P^2) = \mathbf{Z}_2\{(\iota'_3 \wedge i_2)\tilde{\eta}_3\} \oplus \mathbf{Z}_2\{\tilde{\imath}_4\eta_4\}$, we have $\pi_5(\Sigma P^3) \cong (\mathbf{Z}_2)^5$ by Lemma 2.2. By the definition of h' and h'',

$$h'((\iota'_3 \wedge i_2)\tilde{\eta}_3) = h''\tilde{\eta}_3$$

and by Lemma 2.3.(i),

$$h'(\tilde{\imath}_4\eta_4) = (\Sigma i_{2,3})\tilde{\eta}_2\eta_4.$$

We consider the exact sequence ($i = \Sigma i_{2,3}$)

$$\pi_6(\Sigma P^3, M^3) \xrightarrow{\partial} \pi_5(M^3) \xrightarrow{i_*} \pi_5(\Sigma P^3) \xrightarrow{j_*} \pi_5(\Sigma P^3, M^3) \longrightarrow 0.$$

Since $\pi_5(M^3) \cong (\mathbf{Z}_2)^3$, $\pi_5(\Sigma P^3) \cong (\mathbf{Z}_2)^5$ and $\pi_5(\Sigma P^3, M^3) \cong (\mathbf{Z}_2)^2$, $(\Sigma i_{2,3})_*$ is a monomorphism. So h_* maps the generators of $\pi_5(M^6 \vee M^6) \cong \pi_5(M^6) \oplus \pi_5(M^6) \cong (\mathbf{Z}_2)^2$ to $(\Sigma i_{2,3})\tau$ and β respectively. This leads to (ii), completing the proof. □

4. The group $\pi_5(\Sigma P^\infty)$

Since $P^2 \wedge P^2$ is a mapping cone of $2\iota'_3 = i_3\eta_2 p_3$ and $p_3 \circ (i_3\eta_2 p_3) = 0$, there exists an extension $\bar{p}'_3 \in [P^2 \wedge P^2, S^3]$ of $p_3 : M^3 \to S^3$. By use of the EHP sequence, we see that $[\Sigma P^2 \wedge P^2, \Sigma P^2 \wedge P^2]$ is stable. The following result is a direct consequence of Proposition 7.3 of [**1**] and its proof.

LEMMA 4.1. We set $i' = \iota'_3 \wedge i_2$ and $p' = \iota'_3 \wedge p_2$. Then
(i): $2\iota_{\Sigma P^2 \wedge P^2} = i'(i_4\bar{\eta}_3 + \tilde{\eta}_3 p_5)p'$.
(ii): $[\Sigma P^2 \wedge P^2, \Sigma P^2 \wedge P^2] = \mathbf{Z}_4\{\iota_{\Sigma P^2 \wedge P^2}\} \oplus \mathbf{Z}_4\{\tilde{\imath}_4\Sigma\bar{p}'_3\} \oplus \mathbf{Z}_2\{i'\bar{\eta}_3 p'\}$.

Let $T = T_{X,Y} : X \wedge Y \to Y \wedge X$ be the switching map. We identify the complex $\Sigma P^2 \wedge P^2$ with $P^2 \wedge \Sigma P^2$. Then Theorem 7.4 of [**1**] is stated as follows.

LEMMA 4.2. $T_{\Sigma P^2, P^2} = \pm\iota_{\Sigma P^2 \wedge P^2} \pm \tilde{\imath}_4\Sigma\bar{p}'_3 + i'\bar{\eta}_3 p'$.

Now we show

LEMMA 4.3. $\tilde{\eta}_2 \wedge i_2 + i_2 \wedge \tilde{\eta}_2 = \tilde{\imath}_4\eta_4$.

PROOF. We consider the commutative diagram ($T' = T_{S^4,S^1},\ T = T_{\Sigma \mathrm{P}^2,\mathrm{P}^2}$):

$$\begin{array}{ccc} S^4 \wedge S^1 & \xrightarrow{\tilde{\eta}_2 \wedge i_2} & M^3 \wedge M^2 \\ {\scriptstyle T'}\downarrow & & \downarrow{\scriptstyle T} \\ S^1 \wedge S^4 & \xrightarrow{i_2 \wedge \tilde{\eta}_2} & M^2 \wedge M^3. \end{array}$$

This diagram implies $T(\tilde{\eta}_2 \wedge i_2) = i_2 \wedge \tilde{\eta}_2$ because $T' = \pm\iota_5$. We note the following:

$$\begin{aligned} p'(\tilde{\eta}_2 \wedge i_2) &= (\iota_3' \wedge p_2)(\tilde{\eta}_2 \wedge i_2) \\ &= \tilde{\eta}_2 \wedge (p_2 \circ i_2) \\ &= 0 \end{aligned}$$

and

$$\begin{aligned} (\Sigma \bar{p}_3')(\tilde{\eta}_2 \wedge i_2) &= (\Sigma \bar{p}_3')((\iota_3' \circ \tilde{\eta}_2) \wedge (i_2 \circ \iota_1)) \\ &= (\Sigma \bar{p}_3')((\iota_3' \wedge i_2) \circ (\tilde{\eta}_2 \wedge \iota_1)) \\ &= p_4 \circ \tilde{\eta}_3 \\ &= \eta_4. \end{aligned}$$

It follows from the EHP sequence that

(4.1) $\qquad \Sigma^\infty : \pi_5(\Sigma \mathrm{P}^n \wedge \mathrm{P}^n) \to \pi_4^S(\mathrm{P}^n \wedge \mathrm{P}^n)$ is an isomorphism for $n \geq 2$.

So, by Lemma 4.2, we have

$$T(\tilde{\eta}_2 \wedge i_2) = \tilde{\eta}_2 \wedge i_2 + \tilde{i}_4 \eta_4.$$

This completes the proof. □

Let $h_\infty : \Sigma \mathrm{P}^\infty \wedge \mathrm{P}^\infty \to \Sigma \mathrm{P}^\infty$ be the Hopf construction for the multiplication of the Hopf space P^∞. We show

LEMMA 4.4. $\pi_5(\Sigma \mathrm{P}^\infty) = \mathbf{Z}_2\{(\Sigma i_{2,\infty})\tau\} \oplus \mathbf{Z}_2\{(\Sigma i_{3,\infty})\beta\}$, $(\Sigma i_{3,\infty})h''\tilde{\eta}_3 = 0$ and $(\Sigma i_{2,\infty})\tilde{\eta}_2\eta_4 = 0$.

PROOF. Since $\pi_4(\mathrm{P}^\infty) = 0$, $h_{\infty *} : \pi_5(\Sigma \mathrm{P}^\infty \wedge \mathrm{P}^\infty) \to \pi_5(\Sigma \mathrm{P}^\infty)$ is an isomorophism. By considering the dimension, we have $\pi_5(\Sigma \mathrm{P}^\infty \wedge \mathrm{P}^\infty) \cong \pi_5(\Sigma \mathrm{P}^4 \wedge \mathrm{P}^4)$. So, by the fact that $(\iota_3' \wedge i_2)\tilde{\eta}_3 = \tilde{\eta}_2 \wedge i_2$, by (4.1) and Lemma 4.3,

$$\pi_5(\Sigma \mathrm{P}^2 \wedge \mathrm{P}^2) = \mathbf{Z}_2\{\tilde{\eta}_2 \wedge i_2\} \oplus \mathbf{Z}_2\{i_2 \wedge \tilde{\eta}_2\}$$

and

$$\pi_4^S(\mathrm{P}^2 \wedge \mathrm{P}^2) = \mathbf{Z}_2\{\tilde{\eta} \wedge i_{1,2}\} \oplus \mathbf{Z}_2\{i_{1,2} \wedge \tilde{\eta}\}.$$

Here $i_{k,n}$ stands for $\Sigma^\infty i_{k,n}$ and $\tilde{\eta} = \Sigma^\infty \tilde{\eta}_2$.

We consider the exact sequence ($1_n = \iota_{\mathrm{P}^n},\ i' = i_{3,4},\ p' = p_{4,3}$):

$$\pi_4^S(S^3 \wedge \mathrm{P}^2) \xrightarrow{(\gamma_3 \wedge 1_2)_*} \pi_4^S(\mathrm{P}^3 \wedge \mathrm{P}^2) \xrightarrow{(i' \wedge 1_2)_*} \pi_4^S(\mathrm{P}^4 \wedge \mathrm{P}^2) \longrightarrow 0.$$

We have $\pi_4^S(\mathrm{P}^3 \wedge \mathrm{P}^2) \cong \pi_4^S(S^3 \wedge \mathrm{P}^2) \oplus \pi_4^S(\mathrm{P}^2 \wedge \mathrm{P}^2) \cong (\mathbf{Z}_2)^3$ and $\pi_4^S(S^3 \wedge \mathrm{P}^2) = \mathbf{Z}_2\{\iota \wedge i_{1,2}\}$. By (2.4), we obtain

$$(\gamma_3 \wedge 1_2)(\iota \wedge i_{1,2}) = \gamma_3 \wedge i_{1,2} = (i_{2,3}\tilde{\eta}) \wedge i_{1,2}$$

and $(i_{2,4}\tilde{\eta}) \wedge i_{1,2} = 0$. So, by (4.1),

(4.2) $$((\Sigma i_{2,4})\tilde{\eta}_2) \wedge i_{1,4} = 0.$$

And we have $\pi_4^S(\mathrm{P}^4 \wedge \mathrm{P}^2) \cong (\mathbf{Z}_2)^2 \cong \pi_4^S(\mathrm{P}^4 \wedge \mathrm{P}^3)$, which contains $i_{1,4} \wedge ((i_{2,3})\tilde{\eta})$ as one generator.

Next we consider the exact sequence

$$\pi_4^S(\mathrm{P}^4 \wedge S^3) \xrightarrow{(1_4 \wedge \gamma_3)_*} \pi_4^S(\mathrm{P}^4 \wedge \mathrm{P}^3) \xrightarrow{(1 \wedge i')_*} \pi_4^S(\mathrm{P}^4 \wedge \mathrm{P}^4) \longrightarrow 0.$$

We have $\pi_4^S(\mathrm{P}^4 \wedge S^3) = \mathbf{Z}_2\{i_{1,4} \wedge \iota\}$ and $\pi_4^S(\mathrm{P}^4 \wedge \mathrm{P}^3) \cong \pi_4^S(\mathrm{P}^4 \wedge S^3) \oplus \pi_4^S(\mathrm{P}^4 \wedge \mathrm{P}^2) \cong (\mathbf{Z}_2)^3$. By (2.4),

$$(1_4 \wedge \gamma_3)(i_{1,4} \wedge \iota) = i_{1,4} \wedge \gamma_3 = i_{1,4} \wedge (i_{2,3}\tilde{\eta})$$

and $i_{1,4} \wedge (i_{2,4}\tilde{\eta}) = 0$. So, by (4.1),

$$i_{1,4} \wedge ((\Sigma i_{2,4})\tilde{\eta}_2) = 0. \tag{4.3}$$

Thus $\pi_4^S(\mathrm{P}^4 \wedge \mathrm{P}^4) \cong (\mathbf{Z}_2)^2$.

Since $i = i_{3,\infty} : \mathrm{P}^3 \hookrightarrow \mathrm{P}^\infty$ is an H-map, there is a commutative diagram

$$\begin{array}{ccc}
\Sigma \mathrm{P}^3 \wedge \mathrm{P}^3 & \xrightarrow{\Sigma i \wedge i} & \Sigma \mathrm{P}^\infty \wedge \mathrm{P}^\infty \\
\Big\downarrow h & & \Big\downarrow h_\infty \\
\Sigma \mathrm{P}^3 & \xrightarrow{\Sigma i} & \Sigma \mathrm{P}^\infty .
\end{array}$$

By the diagram, by (4.2) and by the fact that $h''\tilde{\eta}_3 = h'(\iota_3' \wedge i_2)\tilde{\eta}_3 = h'(\tilde{\eta}_2 \wedge i_2)$, we obtain

$$\begin{aligned}
(\Sigma i_{3,\infty})h''\tilde{\eta}_3 &= (\Sigma i_{3,\infty})h'(\tilde{\eta}_2 \wedge i_2) \\
&= h_\infty(\Sigma i_{2,\infty} \wedge i_{2,\infty})(\tilde{\eta}_2 \wedge i_2) \\
&= 0.
\end{aligned}$$

By Lemma 2.3.(i), $(\Sigma i_{2,\infty})\tilde{\eta}_2\eta_4 = (\Sigma i_{3,\infty})h'(\tilde{\imath}_4\eta_4)$. Hence, by Lemma 4.3, by (4.2) and (4.3), we have $(\Sigma i_{2,\infty})\tilde{\eta}_2\eta_4 = h_\infty(\Sigma i_{2,\infty} \wedge i_{2,\infty})(\tilde{\imath}_4\eta_4) = 0$.

Obviously the generators of $\pi_5(\Sigma \mathrm{P}^\infty \wedge \mathrm{P}^\infty)$ come from those of $\pi_5(\Sigma \mathrm{P}^3 \wedge \mathrm{P}^3)$. So, by Lemma 3.3.(ii) and its proof, the generators of $\pi_5(\Sigma \mathrm{P}^\infty)$ are determined. This completes the proof. □

5. Proof of Theorem 1.3

Let $\langle \gamma_3, i_{1,3} \rangle \in \pi_4(\mathrm{P}^3)$ be the Samelson product. Let $ad : \pi_5(\Sigma \mathrm{P}^3) \to \pi_4(\Omega\Sigma \mathrm{P}^3)$ be the adjoint isomorphism. The proof of the following fact was pointed out by Morisugi.

LEMMA 5.1. $(r \circ ad)([\Sigma\gamma_3, \Sigma i_{1,3}]) = \gamma_3\eta_3$.

PROOF. Let $i : S^1 \to SO(2)$ be the canonical homeomorphism. By (16.2) and (16.8) of [**6**], we have $q_*\langle i, \gamma_3 \rangle = \langle i, \eta_2 \rangle = \eta_2^2$. So $\langle \gamma_3, i_{1,3} \rangle = \gamma_3\eta_3$. Since the adjoint of $[\Sigma\gamma_3, \Sigma i_{1,3}]$ is the Samelson product $\langle E\gamma_3, Ei_{1,3} \rangle$ and $r : \Omega\Sigma \mathrm{P}^3 \to \mathrm{P}^3$ is an H-map,

$$(r \circ ad)([\Sigma\gamma_3, \Sigma i_{1,3}]) = r\langle E\gamma_3, Ei_{1,3} \rangle = \langle \gamma_3, i_{1,3} \rangle.$$

This completes the proof. □

We set $\beta_1 = (\Sigma^2 i_{3,4})\circ s$ for the inclusion s in (2.4). Since $p_{4,2}\circ i_{3,4} = i_4 \circ p_{3,2}$, we have ([**10**])

$$(\Sigma^2 p_{4,2})\beta_1 = i_6$$

and

$$2\beta_1 = -(\Sigma^2 i_{2,4})\tilde{\eta}_3.$$

We show

LEMMA 5.2. $(\Sigma i_{3,4})h''\tilde{\eta}_3 = \Sigma\gamma_4$ and $(\Sigma i_{2,4})\tilde{\eta}_2\eta_4 = 0$.

PROOF. By use of the isomorphism $\pi_5(\Sigma \mathrm{P}^5) \cong \pi_5(\Sigma \mathrm{P}^\infty)$ and by Lemma 4.4, we obtain $(\Sigma i_{3,5})h''\tilde{\eta}_3 = 0$ and $(\Sigma i_{2,5})\tilde{\eta}_2\eta_4 = 0$. In the exact sequence ($i = \Sigma i_{4,5}$)

$$\pi_6(\Sigma \mathrm{P}^5, \Sigma \mathrm{P}^4) \xrightarrow{\partial} \pi_5(\Sigma \mathrm{P}^4) \xrightarrow{i_*} \pi_5(\Sigma \mathrm{P}^5),$$

we have $\pi_6(\Sigma \mathrm{P}^5, \Sigma \mathrm{P}^4) \cong \mathbf{Z}$ and $\partial \hat{\iota}_5 = \Sigma\gamma_4$. So, by Lemma 4.4, we obtain $(\Sigma i_{3,4})h''\tilde{\eta}_3 = x\Sigma\gamma_4$ and $(\Sigma i_{2,4})\tilde{\eta}_2\eta_4 = y\Sigma\gamma_4$ for $x, y = 0$ or 1. We note that $p_{4,2}\circ\gamma_4 = i_4\eta_3$. So, by composing $\Sigma p_{4,2}$ to the first relation, we get that $xi_5\eta_4 = i_5 \circ \Sigma p_{3,2}\tilde{\eta}_3'' = i_5\eta_4$. Hence $x = 1$.

Suspending the second equality gives $y\Sigma^2\gamma_4 = (\Sigma^2 i_{2,4})\tilde{\eta}_3\eta_5 = 2\beta_1 \circ \eta_5 = 0$. This implies $y = 0$. This completes the proof. □

Now we show

THEOREM 5.3. $[\Sigma\gamma_3, \Sigma i_{1,3}] = (\Sigma i_{2,3})[\iota_3', i_3]\tilde{\eta}_3$ and $\pi_5(\Sigma \mathrm{P}^4) = \mathbf{Z}_2\{\Sigma\gamma_4\} \oplus \mathbf{Z}_2\{(\Sigma i_{2,4})\tau\} \oplus \mathbf{Z}_2\{(\Sigma i_{3,4})\beta\}$.

PROOF. In the exact sequence ($i = \Sigma i_{3,4}$)

$$\pi_6(\Sigma \mathrm{P}^4, \Sigma \mathrm{P}^3) \xrightarrow{\partial} \pi_5(\Sigma \mathrm{P}^3) \xrightarrow{i_*} \pi_5(\Sigma \mathrm{P}^4) \xrightarrow{j_*} \pi_5(\Sigma \mathrm{P}^4, \Sigma \mathrm{P}^3),$$

we see that $\pi_5(\Sigma \mathrm{P}^4, \Sigma \mathrm{P}^3) \cong \mathbf{Z}$ and $\pi_6(\Sigma \mathrm{P}^4, \Sigma \mathrm{P}^3) = \mathbf{Z}_2\{\hat{\eta}_4\} \oplus \mathbf{Z}_2\{[\omega_5, \Sigma i_{1,3}]\}$ by [**5**]. Since j_* is trivial, $\pi_5(\Sigma \mathrm{P}^4) = \mathrm{Im}\ (\Sigma i_{3,4})^*$. We have $\partial\hat{\eta}_4 = \Sigma\gamma_3 \circ \eta_4$ and $\partial[\omega_5, \Sigma i_{1,3}] = [\Sigma\gamma_3, \Sigma i_{1,3}]$. By Lemmas 3.3 and 5.1, there exist integers $a, b, c \in \mathbf{Z}_2$ satisfying

$$[\Sigma\gamma_3, \Sigma i_{1,3}] = \Sigma\gamma_3 \circ \eta_4 + (\Sigma i_{2,3})(a\tilde{\eta}_2\eta_4 + b\tau) + c\beta.$$

By composing the inclusion map $j : (\Sigma \mathrm{P}^3, *) \to (\Sigma \mathrm{P}^3, \Sigma \mathrm{P}^2)$ to this equation, by using [**4**] and [**9**], we obtain $[2\omega_4, \Sigma i_{1,2}] = 2\omega_4 \circ \hat{\eta}_3' + c[\omega_4, \Sigma i_{1,2}]$, where $\hat{\eta}_3' \in \pi_5(CS^3, S^3) \cong \pi_4(S^3)$. Since $[2\omega_4, \Sigma i_{1,2}] = 2\omega_4 \circ \hat{\eta}_3' = 0$, we have $c = 0$.

By Theorem 1.1 of [**11**], $\Sigma\tau = 0$. So, by (2.4), we conclude that

$$0 = \Sigma^2\gamma_3 \circ \eta_5 + (\Sigma^2 i_{2,3})(a\tilde{\eta}_3\eta_5 + b\Sigma\tau) = (1 + a)(\Sigma^2 i_{2,3})\tilde{\eta}_3\eta_5.$$

Since $(\Sigma^2 i_{2,3})\tilde{\eta}_3\eta_5 \neq 0 \in \pi_6(\Sigma^2 \mathrm{P}^3)$, we have $a = 1$. Hence the relation

$$[\Sigma\gamma_3, \Sigma i_{1,3}] = (\Sigma i_{2,3})[\iota_3', i_3]\tilde{\eta}_3 + b(\Sigma i_{2,3})\tau$$

holds. By Lemma 4.4, $0 = (\Sigma i_{3,4})[\Sigma\gamma_3, \Sigma i_{1,3}] = b(\Sigma i_{2,4})\tau$, and so $b = 0$. This leads to the first assertion. The second one is a direct consequence of the first and Lemma 3.3. This completes the proof. □

6. The group $[\Sigma P^4, \Sigma P^n]$

First we remark that the commutativity of the group $[\Sigma P^4, \Sigma P^n]$ is an immediate consequence of Lemma 2.4.

We easily obtain that $[M^n, S^{n-1}] = \mathbf{Z}_2\{\eta_{n-1}p_n\}$ for $n \geq 3$. So, by (2.3), we have

$$p_3(\overline{\Sigma\gamma_2}) = \eta_3 p_4. \tag{6.1}$$

We also obtain the following: $[M^4, S^2] = \mathbf{Z}_2\{\eta_2^2 p_4\}$; $[M^5, S^2] = \mathbf{Z}_4\{\eta_2\bar{\eta}_3\}$, $2(\eta_2\bar{\eta}_3) = \eta_2^3 p_5$. Then the following result overlaps with that of Corollary 10.6.7 of [**3**].

LEMMA 6.1. $[\Sigma P^4, S^2] = \mathbf{Z}_2\{\eta_2\bar{\eta}_3\Sigma p_{4,2}\}$.

PROOF. By use of the cofibration starting with $\overline{\Sigma\gamma_2}$, we have an exact sequence ($\gamma = \overline{\Sigma\gamma_2}$, $i = \Sigma i_{2,4}$, $p = \Sigma p_{4,2}$)

$$[M^4, S^2] \xleftarrow{\gamma^*} [M^3, S^2] \xleftarrow{i^*} [\Sigma P^4, S^2] \xleftarrow{p^*} [M^5, S^2] \xleftarrow{(\tilde{\eta}_3 p_5)^*} [M^4, S^2].$$

By (6.1), we have $(\eta_2 p_3) \circ \overline{\Sigma\gamma_2} = \eta_2^2 p_4$ and $(\eta_2^2 p_4)(\tilde{\eta}_3 p_5) = \eta_2^3 p_5 = 2(\eta_2\bar{\eta}_3)$. This completes the proof. □

Since $\overline{\Sigma\gamma_2} \circ p_{4,2} = 0$, we have $\tilde{\eta}_3\Sigma p_{4,3} = \tilde{\eta}_3 p_5 \Sigma p_{4,2} = \Sigma(\overline{\Sigma\gamma_2} \circ p_{4,2}) = 0$ and $\eta_4\Sigma p_{4,3} = p_4 \circ \tilde{\eta}_3\Sigma p_{4,3} = 0$ by (2.3). So

$$\tilde{\eta}_3\Sigma p_{4,3} = 0 \text{ and } \eta_4\Sigma p_{4,3} = 0. \tag{6.2}$$

By (2.1) and (6.1),

$$2\iota_3' \circ \overline{\Sigma\gamma_2} = i_3\eta_2 p_3 \circ \overline{\Sigma\gamma_2} = i_3\eta_2^2 p_4 = 2\tilde{\eta}_2 \circ p_4 = 0.$$

So there exists an extension $\overline{2\iota_3'} \in [\Sigma P^4, \Sigma P^2]$ of $2\iota_3'$. Then we show

LEMMA 6.2. **(i):** $p_3(\overline{2\iota_3'}) = \bar{\eta}_3\Sigma p_{4,2}$.
(ii): $[\Sigma P^4, \Sigma P^2] = \mathbf{Z}_4\{\overline{2\iota_3'}\} \oplus \mathbf{Z}_2\{\tau\Sigma p_{4,3}\}$ and $2(\overline{2\iota_3'}) \equiv i_3\eta_2\bar{\eta}_3\Sigma p_{4,2} \bmod \tau\Sigma p_{4,3}$.

PROOF. First we note that

$$[M^5, M^3] = \mathbf{Z}_2\{i_3\eta_2\bar{\eta}_3\} \oplus \mathbf{Z}_2\{\tilde{\eta}_2\eta_4 p_5\} \oplus \mathbf{Z}_2\{\overline{\Sigma\gamma_2}\tilde{\eta}_3 p_5\} \oplus \mathbf{Z}_2\{\tau p_5\}$$

and

$$[M^6, M^4] = \mathbf{Z}_2\{i_4\eta_3\bar{\eta}_4\} \oplus \mathbf{Z}_2\{\tilde{\eta}_3\eta_5 p_6\} \oplus \mathbf{Z}_2\{\delta p_6\}.$$

Furthermore, by the definitions, we note that $p_3\tau = 0$ and $p_4\delta = 0$. Now, by (2.1), (6.1), (6.2) and by the properties of Toda brackets,

$$\begin{aligned} p_3(\overline{2\iota_3'}) &\in p_3 \circ \{i_3\eta_2 p_3, \overline{\Sigma\gamma_2}, p_{4,2}\} \\ &= -\{p_3, i_3\eta_2 p_3, \overline{\Sigma\gamma_2}\} \circ \Sigma p_{4,2} \\ &\supset -\{p_3, i_3, \eta_2^2 p_4\} \circ \Sigma p_{4,2} \\ &\bmod p_{3*}[M^5, M^3] \circ \Sigma p_{4,2} = \{\eta_3^2 p_5 \Sigma p_{4,2}\} = 0. \end{aligned}$$

Since $2\bar{\eta}_3 = 2\iota_3 \circ \bar{\eta}_3$ and $\{p_4, i_4, 2\iota_3\} \ni \iota_4 \bmod 2\iota_4$,

$$\begin{aligned} \{p_4, i_4, 2\bar{\eta}_3\} &\supset \{p_4, i_4, 2\iota_3\} \circ \bar{\eta}_4 \\ &\ni \bar{\eta}_4 \\ &\bmod p_{4*}[M^6, M^4] + \pi_4(M^4) \circ 2\bar{\eta}_4 = 2\{\bar{\eta}_4\}. \end{aligned}$$

So we have $\{p_4, i_4, 2\bar{\eta}_3\} = \pm\bar{\eta}_4$. Since $\Sigma : [M^5, S^3] \to [M^6, S^4]$ is an isomorphism, we have $\{p_3, i_3, \eta_2^2 p_4\} = \pm\bar{\eta}_3$. Hence $p_3(\overline{2\iota_3'}) = \bar{\eta}_3\Sigma p_{4,2}$. This leads to (i).

We consider the exact sequence ($\gamma = \overline{\Sigma\gamma_2}$, $i = \Sigma i_{2,4}$, $p = \Sigma p_{4,2}$)

$$[M^4, M^3] \xleftarrow{\gamma^*} [M^3, M^3] \xleftarrow{i^*} [\Sigma P^4, M^3] \xleftarrow{p^*} [M^5, M^3] \xleftarrow{(\tilde{\eta}_3 p_5)^*} [M^4, M^3].$$

We have $(\tilde{\eta}_3 p_5)^*[M^4, M^3] = \mathbf{Z}_2\{\tilde{\eta}_2\eta_4 p_5\} \oplus \mathbf{Z}_2\{\overline{\Sigma\gamma_2}\tilde{\eta}_3 p_5\}$. By (2.1), we see $2\Sigma(\overline{2\iota_3'}) = 2\iota_4' \circ \Sigma(\overline{2\iota_3'}) = \Sigma(i_3\eta_2 p_3(\overline{2\iota_3'}))$. So, by (i), $2\Sigma(\overline{2\iota_3'}) = i_4\eta_3\bar{\eta}_4\Sigma^2 p_{4,2}$. Since $p_{4,2} \circ i_{3,4} = i_4 \circ p_{3,2}$, we obtain, in $[\Sigma^2 P^3, M^4]$,

$$i_4\eta_3\bar{\eta}_4\Sigma^2 p_{4,2} \circ \Sigma^2 i_{3,4} = i_4\eta_3\bar{\eta}_4 \circ \Sigma^2(i_4 \circ p_{3,2}) = 2\tilde{\eta}_3(\Sigma^2 p_{3,2}) \neq 0.$$

This implies that $i_4\eta_3\bar{\eta}_4\Sigma^2 p_{4,2} \neq 0$. This leads to the relation in (ii), completing the proof. □

We set $\alpha = 2\iota_{\Sigma P^3} - \Sigma\gamma_3 \circ \Sigma p_{3,2} = \pm(\Sigma i_{2,3})\tilde{\eta}_2\Sigma p_{3,2} + (\Sigma i_{1,3})\eta_2\Sigma q$. We show

LEMMA 6.3. *α is extendible to ΣP^4 and $[\Sigma P^4, \Sigma P^3] = \mathbf{Z}_4\{\bar{\alpha}\} \oplus \{\beta\Sigma p_{4,3}\} \oplus \mathbf{Z}_2\{(\Sigma i_{2,3})\tau\Sigma p_{4,3}\}$, where*

$$2\bar{\alpha} \equiv (\Sigma i_{1,3})\eta_2\bar{\eta}_3\Sigma p_{4,2} \bmod \{\beta\Sigma p_{4,3},\ (\Sigma i_{2,3})\tau\Sigma p_{4,3}\}.$$

PROOF. We consider the exact sequence ($\gamma = \Sigma\gamma_3$, $i = \Sigma i_{3,4}$, $p = \Sigma p_{4,3}$):

$$\pi_4(\Sigma P^3) \xleftarrow{\gamma^*} [\Sigma P^3, \Sigma P^3] \xleftarrow{i^*} [\Sigma P^4, \Sigma P^3] \xleftarrow{p^*} \pi_5(\Sigma P^3) \xleftarrow{(\Sigma\gamma)^*} [\Sigma^2 P^3, \Sigma P^3].$$

Since $\alpha \circ \Sigma\gamma_3 = 0$, α is extendible to ΣP^4 and $\operatorname{Ker}\gamma^* = \{\bar{\alpha}\}$. We see that $[\Sigma^2 P^3, \Sigma P^3] \cong [M^4, \Sigma P^3] \oplus \pi_5(\Sigma P^3)$. We note that $\Sigma p_{4,2} \circ \Sigma i_{3,4} = i_5\Sigma p_{3,2}$ and

$$((\Sigma i_{1,3})\eta_2\bar{\eta}_3\Sigma p_{4,2}) \circ \Sigma i_{3,4} = (\Sigma i_{1,3})\eta_2^2\Sigma p_{3,2} = 2((\Sigma i_{2,3})\tilde{\eta}_2\Sigma p_{3,2}) = 2\alpha.$$

So we conclude that $(2\bar{\alpha} - (\Sigma i_{1,3})\eta_2\bar{\eta}_3\Sigma p_{4,2}) \circ \Sigma i_{3,4} = 0$, and hence

$$2\bar{\alpha} - (\Sigma i_{1,3})\eta_2\bar{\eta}_3\Sigma p_{4,2} \in \pi_5(\Sigma P^3) \circ \Sigma p_{4,3}.$$

By Lemmas 2.3.(iii) and 3.3, we see that $\operatorname{Im}(\Sigma\gamma)^* = \mathbf{Z}_2\{h''\tilde{\eta}_3\} \oplus \mathbf{Z}_2\{(\Sigma i_{2,3})\tilde{\eta}_2\eta_4\} \oplus \mathbf{Z}_2\{(\Sigma\gamma_3)\eta_4\}$ and $\pi_5(\Sigma P^3) \circ \Sigma p_{4,3} = \{(\Sigma i_{2,3})\tau\Sigma p_{4,3}, \beta\Sigma p_{4,3}\}$. This completes the proof. □

We set $\varepsilon = \pm\tilde{\eta}_2\Sigma p_{3,2} + i_3\eta_2\Sigma q$ satisfying $(\Sigma i_{2,3})\varepsilon = 2\iota_{\Sigma P^3} - \Sigma\gamma_3 \circ \Sigma p_{3,2}$. Since $\pm\tilde{\eta}_2\Sigma p_{3,2} \circ \Sigma\gamma_3 = \pm 2\tilde{\eta}_2 = i_3\eta_2^2$ and $(i_3\eta_2\Sigma q) \circ \Sigma\gamma_3 = i_3\eta_2^2$, $\varepsilon \circ \Sigma\gamma_3 = (\pm\tilde{\eta}_2\Sigma p_{3,2} + i_3\eta_2\Sigma q) \circ \Sigma\gamma_3 = 0$. So there exists an extension $\bar{\varepsilon} \in [\Sigma P^4, \Sigma P^2]$ of ε. Then we obtain the following.

LEMMA 6.4. *$\bar{\varepsilon} \equiv \overline{2\iota_3'} \bmod \tau\Sigma p_{4,3}$ and $(\Sigma i_{2,3})\bar{\varepsilon}$ is taken as an extension of α.*

PROOF. We see

$$\overline{2\iota_3'} \circ \Sigma i_{2,4} = \bar{\varepsilon} \circ \Sigma i_{2,4} = i_3\eta_2 p_3.$$

So, by the proof of Lemma 6.2.(ii), we obtain the first assertion.

We see that

$$(\Sigma i_{2,3})\bar{\varepsilon} \circ \Sigma i_{3,4} = (\Sigma i_{2,3})(\pm\tilde{\eta}_2\Sigma p_{3,2} + i_3\eta_2\Sigma q) = \alpha.$$

So $(\Sigma i_{2,3})\bar{\varepsilon}$ is taken as an extension of α. This leads to the second assertion, completing the proof. □

We show the following.

LEMMA 6.5. *$[P^4, P^3] = 0$.*

PROOF. We consider the cofiber sequence

$$S^1 \xrightarrow{i_{1,4}} \mathrm{P}^4 \xrightarrow{p_{4,1}} \mathrm{P}^4_2 \xrightarrow{\phi} S^2 \longrightarrow \cdots .$$

So we have the exact sequence

$$0 \longleftarrow [\mathrm{P}^4, S^2] \xleftarrow{p_{4,1}{}^*} [\mathrm{P}^4_2, S^2] \xleftarrow{\phi^*} \pi_2(S^2) \longleftarrow 0.$$

Obviously ϕ is an extension of $2\iota_2$ and $[\mathrm{P}^4_2, S^2] = \{\phi\}$. So we conclude that $[\mathrm{P}^4, S^2] = 0$.

The fibering $q : \mathrm{P}^3 \to S^2$ induces the exact sequence of sets

$$[\mathrm{P}^4, S^1] \xrightarrow{i_{1,3*}} [\mathrm{P}^4, \mathrm{P}^3] \xrightarrow{q_*} [\mathrm{P}^4, S^2].$$

Obviously we see $[\mathrm{P}^4, S^1] = 0$. This leads to the assertion, completing the proof. □

We set $k = \Sigma i_2 \wedge i_2$. By (6.1) and (2.2),

$$kp_3 \circ \overline{\Sigma\gamma_2} = (\Sigma i_2 \wedge i_2) \circ \eta_3 p_4 = 2\tilde{\imath}_4 \circ p_4 = 0.$$

So there exists an extension $\overline{kp_3} \in [\Sigma\mathrm{P}^4, \Sigma\mathrm{P}^2 \wedge \mathrm{P}^2]$ of kp_3. Then we show

LEMMA 6.6. $[\Sigma\mathrm{P}^4, \Sigma\mathrm{P}^2 \wedge \mathrm{P}^2] = \mathbf{Z}_4\{\overline{kp_3}\}$ and $2\overline{kp_3} = k\bar{\eta}_3\Sigma p_{4,2}$.

PROOF. We consider the exact sequence induced from the cofibration starting with $\overline{\Sigma\gamma_2}$ ($i = \Sigma i_{2,4}$ and $p = \Sigma p_{4,2}$):

$$[M^4, \Sigma\mathrm{P}^2 \wedge \mathrm{P}^2] \xleftarrow{\overline{\Sigma\gamma_2}^*} [M^3, \Sigma\mathrm{P}^2 \wedge \mathrm{P}^2] \xleftarrow{i^*} [\Sigma\mathrm{P}^4, \Sigma\mathrm{P}^2 \wedge \mathrm{P}^2]$$
$$\xleftarrow{p^*} [M^5, \Sigma\mathrm{P}^2 \wedge \mathrm{P}^2] \xleftarrow{(\tilde{\eta}_3 p_5)^*} [M^4, \Sigma\mathrm{P}^2 \wedge \mathrm{P}^2].$$

Obviously we obtain the following: $[M^3, \Sigma\mathrm{P}^2 \wedge \mathrm{P}^2] = \mathbf{Z}_2\{kp_3\}$, $[M^4, \Sigma\mathrm{P}^2 \wedge \mathrm{P}^2] = \mathbf{Z}_2\{\iota'_3 \wedge i_2\} \oplus \mathbf{Z}_2\{\tilde{\imath}_4 p_4\}$ and $[M^5, \Sigma\mathrm{P}^2 \wedge \mathrm{P}^2] = \mathbf{Z}_2\{k\bar{\eta}_3\} \oplus \mathbf{Z}_2\{(\iota'_3 \wedge i_2)\tilde{\eta}_3 p_5\} \oplus \mathbf{Z}_2\{\tilde{\imath}_4 \eta_4 p_5\}$. This implies that

$$\mathrm{Im}\ (\tilde{\eta}_3 p_5)^* = \{(\iota'_3 \wedge i_2)\tilde{\eta}_3 p_5,\ \tilde{\imath}_4 \eta_4 p_5\}.$$

For $i' = \iota'_3 \wedge i_2$ and $p' = \iota'_3 \wedge p_2$, we note $k = i' \circ i_4$. Since $\overline{kp_3}$ is a representative of a Toda bracket $\{kp_3, \overline{\Sigma\gamma_2}, p_{4,2}\}$,

$$-p'\overline{kp_3} \in \{p', kp_3, \overline{\Sigma\gamma_2}\} \circ \Sigma p_{4,2}.$$

We note that $\{p', i', 2\iota'_4\} \ni \iota'_5 \bmod 2\iota'_5$. So, by (6.1), (2.3) and (2.1),

$$\begin{aligned}
\{p', kp_3, \overline{\Sigma\gamma_2}\} &\supset \{p', i', i_4 p_3 \overline{\Sigma\gamma_2}\} \\
&= \{p', i', i_4 \eta_3 p_4\} \\
&= \{p', i', 2\iota'_4\} \\
&= \pm\iota'_5 \\
\mathrm{mod}\ \ & p'_*[M^5, \Sigma\mathrm{P}^2 \wedge \mathrm{P}^2] + [M^4, M^5] \circ \Sigma(\overline{\Sigma\gamma_2}) = \{i_5 \eta_4 p_5\} = \{2\iota'_5\}.
\end{aligned}$$

Hence obtain

$$p'\overline{kp_3} = \pm\Sigma p_{4,2}.$$

By use of the EHP sequence, $\Sigma^\infty : [\Sigma\mathrm{P}^4, \Sigma\mathrm{P}^2 \wedge \mathrm{P}^2] \to \{\mathrm{P}^4, \mathrm{P}^2 \wedge \mathrm{P}^2\}$ is an isomorphism. By the fact that $p' = \Sigma(\iota'_2 \wedge p_2)$, by Lemma 4.1 and (6.2), we conclude that

$$2\overline{kp_3} = (i'(i_4\bar{\eta}_3 + \tilde{\eta}_3 p_5)p') \circ \overline{kp_3} = (k\bar{\eta}_3 p') \circ \overline{kp_3} + i'(\tilde{\eta}_3 p_5)p'\overline{kp_3} = k\bar{\eta}_3\Sigma p_{4,2}.$$

This completes the proof. □

Now we show the following.

THEOREM 6.7. $2\bar{\alpha} = (\Sigma i_{1,3})\eta_2\bar{\eta}_3\Sigma p_{4,2}$ and $2(\overline{2\iota_3'}) = 2\bar{\varepsilon} = i_3\eta_2\bar{\eta}_3\Sigma p_{4,2}$.

PROOF. By Lemmas 2.2 and 6.5, $h_* : [\Sigma P^4, \Sigma P^3 \wedge P^3] \to [\Sigma P^4, \Sigma P^3]$ is an isomorphism. By Lemma 3.2, $[\Sigma P^4, \Sigma P^3 \wedge P^3] \cong [\Sigma P^4, \Sigma P^2 \wedge P^2] \oplus [\Sigma P^4, M^6 \vee M^6]$. We see that

$$[\Sigma P^4, M^6 \vee M^6] \cong [\Sigma P^4, M^6] \oplus [\Sigma P^4, M^6] \cong \mathbf{Z}_2\{i_6\Sigma p_{4,3}\} \oplus \mathbf{Z}_2\{i_6\Sigma p_{4,3}\}.$$

So, by the proof of Lemma 3.3.(ii),

$$[\Sigma P^4, \Sigma P^3] = h'_*[\Sigma P^4, \Sigma P^2 \wedge P^2] \oplus \mathbf{Z}_2\{(\Sigma i_{2,3})\tau\Sigma p_{4,3}\} \oplus \mathbf{Z}_2\{\beta\Sigma p_{4,3}\}.$$

Hence, by Lemma 6.3,

$$\bar{\alpha} \equiv h'\overline{kp_3} \bmod \{(\Sigma i_{2,3})\tau\Sigma p_{4,3},\ \beta\Sigma p_{4,3}\}.$$

By Lemmas 6.6 and 2.3.(i), we conclude that

$$2\bar{\alpha} = 2h'\overline{kp_3} = h'k\bar{\eta}_3\Sigma p_{4,2} = (\Sigma i_{1,3})\eta_2\bar{\eta}_3\Sigma p_{4,2}.$$

By Lemmas 6.2 and 6.4, we obtain the last two relations. This completes the proof. □

By using the fact that $\pi_3(\Sigma P^4) = \mathbf{Z}_2\{(\Sigma i_{3,4})h'''\}$ and $\pi_4(\Sigma P^4) = \mathbf{Z}_4\{(\Sigma i_{2,4})\tilde{\eta}_2\}$, we obtain

$$[M^4, \Sigma P^4] = \mathbf{Z}_2\{(\Sigma i_{3,4})h''\} \oplus \mathbf{Z}_2\{(\Sigma i_{2,4})\tilde{\eta}_2 p_4\}.$$

Now we show

THEOREM 6.8. $[\Sigma P^4, \Sigma P^4] = \mathbf{Z}_8\{\iota_{\Sigma P^4}\} \oplus \mathbf{Z}_2\{(\Sigma i_{3,4})\beta\Sigma p_{4,3}\} \oplus \mathbf{Z}_2\{(\Sigma i_{2,4})\tau\Sigma p_{4,3}\}$ and $2\iota_{\Sigma P^4} \equiv (\Sigma i_{3,4})\bar{\alpha} \bmod \{(\Sigma i_{3,4})\beta\Sigma p_{4,3},\ (\Sigma i_{2,4})\tau\Sigma p_{4,3}\}$.

PROOF. We consider the exact sequence ($\gamma = \Sigma\gamma_3,\ i = \Sigma i_{3,4},\ p = \Sigma p_{4,3}$)

$$\pi_4(\Sigma P^4) \xleftarrow{\gamma^*} [\Sigma P^3, \Sigma P^4] \xleftarrow{i^*} [\Sigma P^4, \Sigma P^4] \xleftarrow{p^*} \pi_5(\Sigma P^4) \xleftarrow{\Sigma\gamma^*} [\Sigma^2 P^3, \Sigma P^4].$$

By Lemma 2.6.(ii), $[\Sigma P^3, \Sigma P^4] = \mathbf{Z}_8\{\Sigma i_{3,4}\} \oplus \mathbf{Z}_2\{(\Sigma i_{1,4})\eta_2\Sigma q\}$. We have $(\Sigma i_{1,4})\eta_2\Sigma q \circ \Sigma\gamma_3 = 2(\Sigma i_{2,4})\tilde{\eta}_2 \neq 0$. We see that $(2\iota_{\Sigma P^4} - (\Sigma i_{3,4})\bar{\alpha}) \circ \Sigma i_{3,4} = 2\Sigma i_{3,4} - (\Sigma i_{3,4})\alpha = 0$. So $2\iota_{\Sigma P^4} - (\Sigma i_{3,4})\bar{\alpha} \in (\Sigma p_{4,3})^*\pi_5(\Sigma P^4)$. Therefore, by Theorems 6.7 and 5.3,

$$4\iota_{\Sigma P^4} = (\Sigma i_{1,4})\eta_2\bar{\eta}_3\Sigma p_{4,2} \text{ and } 8\iota_{\Sigma P^4} = 0. \tag{6.3}$$

We see that $[\Sigma^2 P^3, \Sigma P^4] \cong [M^4, \Sigma P^4] \oplus \pi_5(\Sigma P^4)$. Hence, by Lemma 5.2 and Theorem 5.3, Im $(\Sigma^2\gamma_3)^* = \{(\Sigma i_{3,4})h''\tilde{\eta}_3,\ (\Sigma i_{2,4})\tilde{\eta}_2\eta_4\} = \{\Sigma\gamma_4\}$. Thus

$$\Sigma\gamma_4 \circ \Sigma p_{4,3} = 0. \tag{6.4}$$

Hence Theorem 5.3 determines the group $[\Sigma P^4, \Sigma P^4]$. This completes the proof. □

By the equation (6.4), we obtain the following.

COROLLARY 6.9. $(\Sigma i_{4,5})_* : [\Sigma P^4, \Sigma P^4] \to [\Sigma P^4, \Sigma P^5]$ is an isomorphism.

About the group $\epsilon(\Sigma P^4)$ of self-equivalences of ΣP^4, the following result is an immediate consequence of Theorem 6.8.

REMARK 6.10. $\epsilon(\Sigma P^4) \cong (\mathbf{Z}_8)^* \times \mathbf{Z}_2 \times \mathbf{Z}_2$, where $(\mathbf{Z}_8)^*$ means the unit group of $\mathbf{Z}_8$.

Tsukiyama advised the author to determine $\epsilon(\Sigma P^4)$.

References

[1] S. Araki and H. Toda, Multiplicative structures in mod q cohomology theories II, Osaka J. Math. **3**(1966), 81-120.

[2] M. G. Barratt, I. M. James and N. Stein, Whitehead products and projective spaces, J. Math. Mech. **9**(1960), 813-819.

[3] H. J. Baues, Homotopy Type and Homology, Oxford Mathematical Monographs, Oxford Sci. Publ.

[4] A. L. Blakers and W. S. Massey, Products in homotopy theory, Ann. of Math., **58**(1953), 192-201.

[5] I. M. James, On the homotopy groups of certain pairs and triads, Quart. J. Math. Oxford, (2) **5**(1954), 260-270.

[6] I. M. James, The topology of Stiefel manifolds, London Math. Soc. Lecture Note **24**, 1976.

[7] K. Morisugi, Hopf constructions, Samelson Products and suspension maps, Contemporary Mathematics **239**(1999), 225-238.

[8] J. Mukai, On the attaching map in the Stiefel manifold of 2-frames, Math. J. Okayama Univ. **33**(1991), 177-188.

[9] J. Mukai, Note on existence of the unstable Adams map, Kyushu J. Math. **49**(1995), 271-279.

[10] J. Mukai, Some homotopy groups of the double suspension of the real projective space $\mathbf{R}P^6$, Matemática Contemporânea **13**, 235-249 (1997).

[11] J. Mukai, Generators of some homotopy groups of the mod 2 Moore space of dimension 3 or 5, to appear in Kyushu J. Math.

[12] C. M. Naylor, Multiplications on $SO(3)$, Michigan Math. J. **13** (1966), 27-31.

[13] E. Rees, Multiplications on projective spaces, Michigan Math. J. **16** (1969), 297-301.

[14] H. Toda, Generalized Whitehead products and homotopy groups of spheres, J. Inst. Poly. Osaka City Univ. **3**(1952), 43-82.

[15] H. Toda, Composition methods in homotopy groups of spheres, Ann. of Math. Studies, **49**, Princeton, 1962.

[16] H. Toda, Order of the identity class of a suspension space, Ann. of Math., **78**(1963), 300-325.

[17] G. W. Whitehead, Elements of homotopy theory, GTM **61** (1978), Springer.

Mathematical Sciences, Faculty of Science, Shinshu University, Matsumoto, Nagano Pref. 390, Japan

E-mail address: `jmukai0@gipac.shinshu-u.ac.jp`

Contemporary Mathematics
Volume **274**, 2001

PHANTOM ELEMENTS AND ITS APPLICATIONS

JIANZHONG PAN AND MOO HA WOO

ABSTRACT. In our previous work [13], a solution of the Tsukiyama problem about self homotopy equivalence was found by using a generalization of phantom map. In this note, the fundamental result is established for such a generalization. This is the first time one can deal with phantom maps to the target space not satisfying finite type condition. An application to forgetful map is also discussed briefly.

1. INTRODUCTION

The main aim of this paper is to study phantom map, its generalization and applications. After the discovery of the first example of phantom map by Adams and Walker [1], the phantom map theory receives a lot of attention. The main aim of these previous studies is however to understand it, that is, the computation and the properties of phantom maps. The first application of phantom map theory was given by Harper and Roitberg [5],[14] who applied it to compute $SNT(X)$ and $Aut(X)$. Recently applications are also found by Roitberg [15] and Pan [12] where several conjectures of McGibbon were settled. On the other hand, a remarkable connection was established by Pan and Woo [13] between the Tsukiyama problem about self homotopy equivalence and a generalization of phantom map. A byproduct of this connection is that a special case of Tsukiyama problem is almost equivalent to the famous Halperin conjecture in rational homotopy theory [4].

1991 *Mathematics Subject Classification.* 55P10,55P60,55P62,55R10.

Key words and phrases. Phantom map, Forgetful map, Halperin conjecture.

The first author is partially supported by the NSFC project 19701032 and ZD9603 of Chinese Academy of Science and the second author is partially supported by the Korea Research Foundation made in the program year of (2000).

The following characterizations for a map between nilpotent spaces of finite type to be a phantom map are known in [9].

Theorem 1.1. *Let X, Y be nilpotent CW-complexes of finite type and Y be 1-connected. Then for any map $f : X \to Y$, the following are equivalent*

(i) f is a phantom map.

(ii) $\hat{e} \circ f \simeq *$ where $\hat{e} : Y \to \hat{Y}$ is the profinite completion.

(iii) $f \circ \tau \simeq *$ where $\tau : X_\tau \to X$ is the homotopy fiber of the rationalization.

On the other hand, in our previous paper [13], we generalized the concept of phantom map to that of g-phantom element and announced a theorem characterizing an element to be a phantom element which is Theorem 2.7 in [13]. In this paper, we will generalize further so that we can deal with the target space which is not of finite type.

Theorem 1.2. *Let X be a nilpotent CW-complex of finite type, Y be 1-connected such that $\pi_n(Y)$ is a reduced group for $n \geq 2$ and $g : X \to Y$ be any map. Then the following are equivalent:*

(i) $\alpha \in \pi_j(map_*(X,Y); g)$ is a g-phantom element.

(ii) $(\hat{e}_*)_*(\alpha) = 0$.

(iii) $(\tau^*)_*(\alpha) = 0$,

where $(\hat{e}_*)_* : \pi_j(map_*(X,Y); g) \to \pi_j(map_*(X,\hat{Y}); \hat{g})$ and $(\tau^*)_* : \pi_j(map_*(X,Y); g) \to \pi_j(map_*(X_\tau, Y); g_\tau)$.

Note that the assumption that Y is 1-connected is not a real restriction, since a map $f : X \to Y$ is phantom if and only if the corresponding map $\tilde{f} : X \to \tilde{Y}$ is phantom by an observation of Zabrodsky [17, p.135], where $\tilde{Y}$ is the universal covering of Y.

In this paper, we will give a complete proof of this theorem. As an application we have the following as a corollary of Proposition 3.3.

Corollary 1.3. *Let P be a 1-connected CW-complex with type F_0 such that P is finite dimensional or satisfies $H^*(P, \mathbb{Z}_p)$ locally finite over A_p for each prime p. Assume further that $\pi_n(Baut(P))$ is a reduced group for $n \geq 2$ and P satisfies one of the following.*

(i) P is rationally equivalent to Kähler manifold.

(ii) $H^*(P; Q)$ as an algebra has at most 3 generators.

(iii) P is rationally equivalent to G/U where G is a compact Lie group and U is a closed subgroup of maximal rank.

Then for all $m \geq 1$, any finitely generated abelian group H and every principal $K(H, 2m)$-bundle with total space homotopy equivalent to P, the associated forgetful map is injective.

The organization of this paper is as follows. In section 2, Theorem 1.2 will be proved and an application to forgetful map will be discussed in section 3. In this paper, we will use the following notations:

(i) H is a finitely generated abelian group.

(ii) $map(X, Y)$ is the space of continuous mappings from X to Y.

(iii) $map_*(X, Y)$ is the subspace of pointed mappings from (X, x_0) to (Y, y_0).

(iv) $l : X \to X_{(0)}$ is the rationalization.

(v) Let $\tau : X_\tau \to X$ be the homotopy fiber of l. Then $X_\tau \xrightarrow{\tau} X \to X_{(0)}$ is a cofibration up to homotopy.

(vi) $\hat{e}_p : Y \to \hat{Y}_{\mathbb{Z}_{p^\infty}}$ is Bousfield-Kan's p-completion and $\hat{e} = (e_2, e_3, \cdots)$ $: Y \to \hat{Y}$ where $\hat{Y} = \prod_p \hat{Y}_{\mathbb{Z}_{p^\infty}}$. Let Y_ρ be the homotopy fiber of $\hat{e}$.

The readers should refer to [13] for all other notations which are not explained here.

In concluding the introduction, we would like to give the following

Conjecture 1.4. *The condition that $\pi_n Y$ is a reduced group in this paper can be deleted.*

2. Phantom elements

Let us begin with definition.

Definition 2.1. Let X be a CW-complex, Y be a space and $g : X \to Y$ be any map. Then an element $\alpha \in \pi_j(map_*(X, Y); g)$ is called a *g-phantom element* if $(i_n^*)_*(\alpha) = 0$ for all $n \geq 0$ where $(i_n^*)_* : \pi_j(map_*(X, Y)) \to \pi_j(map_*(X^n, Y))$ is the homomorphism induced by the inclusion $i_n : X^n \to X$. Denoted by

$$Ph_j^g(X, Y) = \{\alpha \in \pi_j(map_*(X, Y); g) | \alpha \text{ is a } g\text{-phantom element}\}.$$

Obviously if g is the constant map and $j = 0$, then α is a g-phantom element if and only if it represents the homotopy class of a map which is a phantom map.

An abelian group is said to be *reduced* [7] if it has no nontrivial divisible subgroups, where an abelian group G is said to be divisible if, given any $x \in G$ and any nonzero integer n, there exists $y \in G$ such that $x = ny$.

Since any nontrivial divisible group is infinitely generated, thus any finitely generated abelian group is reduced.

The main aim of this section is to prove Theorem 1.2. Before doing this, let us give some results necessary to the proof .

Lemma 2.2. *Let Y be 1-connected such that $\pi_n(Y)$ is a reduced group for $n \geq 2$. Then we have*

(i) $\pi_j(\hat{Y}) = \prod_p Ext(\mathbb{Z}_{p^\infty}, \pi_j(Y))$.

(ii) $(\hat{e}_*)_* : \pi_j map_*(W, Y)_f \to \pi_j map_*(W, \hat{Y})_{\hat{f}}$ *is injective for a finite CW-complex W.*

Proof. The first statement follows from the fact that $Hom(\mathbb{Z}_{p^\infty}, B) = 0$ for a reduced group B since otherwise there will be a nontrivial divisible subgroup in B.

To prove the second statement, note that the induced map $\pi_n(Y_\rho) \to \pi_n(Y)$ is trivial since $\pi_n(Y)$ is a reduced group and $\pi_n(Y_\rho)$ is rational thus divisible by the arithmetic square [3] in Theorem 1.2. It follows that $\hat{e}_* : \pi_n(Y) \to \pi_n(\hat{Y})$ is injective and an easy induction argument using the homotopy exact sequence associated with the inclusion $X_n \to X_{n+1}$ shows what we want for $j > 1$. For $j = 1$, it can be proved by a similar argument used in the proof of Theorem 5.3 in [6, Ch.2]. □

Proposition 2.3. *Let X be a nilpotent space and Y be 1-connected such that $\pi_n(Y)$ is a reduced group for $n \geq 2$. Then the following hold:*

(i) $[\Sigma^j X_{(0)}, \hat{Y}] = *$ *and* $\tilde{H}^j(X_{(0)}, \pi_i(\hat{Y})) = 0$ *for all* $j, i \geq 0$.

(ii) $[\Sigma^j X_\tau, Y_\rho] = *$ *and* $\tilde{H}^j(X_\tau, \pi_i(Y_\rho)) = 0$ *for all* $j, i \geq 0$.

Proof. That $\tilde{H}^j(X_{(0)}, \pi_i(\hat{Y})) = 0$ follows from the fact that

$$Hom(A, B) = 0, Ext(A, B) = 0$$

for a rational group A and $B = \prod_p Ext(\mathbb{Z}_{p^\infty}, B')$, where B' is any abelian group. Then the equation

$$[\Sigma^j X_{(0)}, \hat{Y}] = \lim_{\leftarrow j} [\Sigma^j X_{(0)}, \hat{Y}^{(j)}]$$

implies the first statement, where $\hat{Y}^{(j)}$ is the jth term in Posinikov tower of $\hat{Y}$.

The equation about cohomology in second statement is true since $\pi_i(Y_\rho)$ is rational while the proof of another equation is similar to that in the first statement. □

Proposition 2.4. *Let X be a nilpotent space and Y be 1-connected such that $\pi_n(Y)$ is a reduced group for $n \geq 2$. Then the following hold :*

(i) $\tau^* : map_*(X, \hat{Y}) \overset{w}{\simeq} map_*(X_\tau, \hat{Y})$.
(ii) $\rho_* : map_*(X_{(0)}, Y_\rho) \overset{w}{\simeq} map_*(X_{(0)}, Y)$.
(iii) $\hat{e}_* : map_*(X_\tau, Y) \overset{w}{\simeq} map_*(X_\tau, \hat{Y})$.
(iv) $l^* : map_*(X_{(0)}, Y_\rho) \overset{w}{\simeq} map_*(X, Y_\rho)$.

Proof. The first and the last statements follow from Proposition 2.3 and the well known Zabrodsky Lemma. The second and the third statements follow from a $\lim^1$ argument for $j \geq 0$

$$* \to \lim_{\leftarrow n}{}^1 \pi_{j+1} map_*(Z, E_n) \to \pi_j map_*(Z, E) \to \lim_{\leftarrow n} \pi_j map_*(Z, E_n) \to *$$

where in the second statement, $Z = X_{(0)}$ and E_n is the nth term in the Moore-Postnikov tower of the map $\rho : Y_\rho \to Y$ while in the third statement, $Z = X_\tau$ and E_n is the nth term in the Moore-Postnikov tower of the map $\hat{e} : Y \to \hat{Y}$. In both case, the sequence $\{\pi_j map_*(Z, E_n)\}$ is a sequence consisting of isomorphisms and thus the $\lim^1$ is trivial and the desired isomorphism follows immediately. □

Proposition 2.5. *Let X be a nilpotent space and Y 1-connected such that $\pi_n(Y)$ is a reduced group for $n \geq 2$. Let $g : X \to \hat{Y}$ be any map. Then*

$$Ph_j^g(X, \hat{Y}) = *.$$

Proof. $Ph_j^g(X, \hat{Y})$ is the $\lim^1$ of a sequence of compact groups and continuous homomorphisms which is well known to be trivial. □

Proposition 2.6. *Let X, Y be two nilpotent spaces with Y 1-connected such that $\pi_n(Y)$ is a reduced group for $n \geq 2$. Then the following hold:*

(i) $map_*(X_\tau, Y) \stackrel{w}{\simeq} map_*(X, \hat{Y})$.
(ii) $map_*(X_{(0)}, Y) \stackrel{w}{\simeq} map_*(X, Y_\rho)$.

Proof. This is an easy consequence of Proposition 2.4. □

Proof of Theorem 1.2. The equivalence between the last two statements follows directly from the following commutative diagram

$$\begin{array}{ccc} \pi_j map_*(X, Y) & \xrightarrow{(\tau^*)_*} & \pi_j map_*(X_\tau, Y) \\ {\scriptstyle (\hat{e}_*)_*}\downarrow & & {\scriptstyle (\hat{e}_*)_*}\downarrow \\ \pi_j map_*(X, \hat{Y}) & \xrightarrow{(\tau^*)_*} & \pi_j map_*(X_\tau, \hat{Y}) \end{array}$$

where the bottom horizontal homomorphism and the right side vertical homomorphism are isomorphisms by Proposition 2.4.

Now assume the first statement, then we have $(i_n^*)_*(\hat{e}_*(\alpha)) = 0$ for all $n \geq 0$. It follows from Proposition 2.5 that $\hat{e}_*(\alpha) = 0$. The proof of another direction is similar to that in [13] using Lemma 2.2 instead of Sullivan's original result which is stated only for space of finite type. □

Remark 2.7. It is easy to see that the above proof follows from the same pattern as that given by Oda and Shitanda [11]. We give a proof here because Oda informed us that there were gaps in their proof and they don't know if the result is true or not. The similar proof applies also to the equivalent case which will be discussed in future publication.

As noted in [13], the natural question related to the application of phantom element to the forgetful map is

Question 2.8. *For two maps $f, g : X \to Y$, what is the relation between $Ph_j^g(X, Y)$ and $Ph_j^f(X, Y)$?*

Proposition 2.9. *Let X, Y be nilpotent CW-complexes such that*

$$[\Sigma^j X_\tau, Y] = [\Sigma^{j+1} X_\tau, Y] = 0.$$

If $g : X \to Y$ is a phantom map, then we have

$$Ph_j^g(X, Y) = \pi_j(map_*(X, Y); g).$$

Proof. The proof is the same as that in [13] . □

In our application we have to be able to compute $Ph^g_j(X,Y)$. Before giving this kind of result, recall that a CW-complex is called unstable if all the attaching maps vanish under suspension. It is Baues [2] who noted the following which is dual to Zabrodsky's integral approximation.

Theorem 2.10. *Let X be a 1-connected CW-complex. Then there is an unstable complex $\bar{X}$ and a rational equivalence $h : \bar{X} \to X$.*

Remark 2.11. Let X be an unstable CW-complex. Then it is easy to prove that $Ph^g_j(X,Y) = *$ for any map $g : X \to Y$.

Proposition 2.12. *Let X be a 1-connected CW-complex and Y 1-connected such that $\pi_n(Y)$ is a reduced group for $n \geq 2$. Suppose further that the component of $map_*(X,\hat{Y})$ consisting the constant map is weakly contractible and $g : X \to Y$ is a phantom map. Then*

$$Ph^g_j(X,Y) = \pi_j(map_*(X,Y); g) = \prod_{k>0} H^k(X, \pi_{k+j+1}(Y_\rho))$$

Proof. As first noted by Oda and Shitanda, the similar proof as in that of Theorem B of [17] leads to the following homotopy fibration

$$\bigcup_g map_*(X,Y)_g \to map_*(\bar{X},Y)_* \to map_*(\bar{X},\hat{Y})_*$$

where the union is over phantom maps g. On the other hand, different components of $\bigcup_g map_*(X,Y)_g$ are homotopy equivalent since $\bigcup_g map_*(X,Y)_g$ is the homotopy fiber of a map between two connected spaces. It follows that

$$Ph^g_j(X,Y) = \pi_j(map_*(X,Y); *) = \pi_j(map_*(X,Y_\rho); *) =$$

$$= [\Sigma^{j-1}X, \Omega Y_\rho] = \prod_{k>0} H^k(X, \pi_{j+k+1}(Y_\rho))$$

where $*$ is the constant map. □

We are ready to state results related to the applications. Before that we have another definition

Definition 2.13. Let A_p be the mod p Steenrod algebra. An unstable module M over A_p is called *locally finite* if and only if for any $x \in M$, only finite elements of A_p can act nontrivially on M.

Example 2.14. *Let P be a space such that $H^*(P, \mathbb{Z}_p)$ is locally finite over A_p. Then so is ΩP. In particular, if P is a finite CW-complex, then $H^*(\Omega P, \mathbb{Z}_p)$ is locally finite over A_p.*

Theorem 2.15. *Let $X = K(H, m+2)$ $(m \geq 1)$, $Y = Baut(P)$ such that $\pi_n(Y)$ is a reduced group for $n \geq 2$ and $g : X \to Y$ be any map where P is a 1-connected finite dimensional CW-complex or P satisfies that $H^*(P, \mathbb{Z}_p)$ is locally finite over A_p for each prime p. Then for $j \geq 1$ we have*

$$Ph_j^g(X, Y) = \pi_j(map_*(X, Y); g) = [\Sigma^j X, Y_\rho].$$

Proof. The proof is the same as that of the corresponding result in [13] using results of Zabrodsky [17] and Miller [10] or Theorem 8.8 in [16]. □

Similarly we have

Theorem 2.16. *Let $X = BG$, $Y = Baut(P)$ such that $\pi_n(Y)$ is a reduced group for $n \geq 2$ and $g : X \to Y$ be a phantom map. Then for $j \geq 1$ we have*

$$Ph_j^g(X, Y) = \pi_j(map_*(X, Y); g) = [\Sigma^j X, Y_\rho]$$

where G is a connected compact Lie group and P is a 1-connected finite dimensional CW-complex or P satisfies $H^*(P, \mathbb{Z}_p)$ to be locally finite over A_p for each prime p.

3. An Application to the forgetful map

Given a principal G-bundle $\pi : P \to B$, let

$$aut^G(P) = \{g | g : P \to P \text{ is a G-equivariant homotopy equivalence}\}.$$

and

$$aut(P) = \{g | g : P \to P \text{ is a homotopy equivalence}\}.$$

There is a natural map $f : aut^G(P) \to aut(P)$ which forgets the G-action. Let

$$Aut^G(P) = \pi_0(aut^G(P))$$

and

$$Aut(P) = \pi_0(aut(P)).$$

Then the map f induces a map

$$F : Aut^G(P) \to Aut(P)$$

which is called a *forgetful map* by Tsukiyama. The question posed by Tsukiyama in [8] is the following

Question 3.1. *When is the forgetful map F injective?*

One of the main results in [13] is the following

Theorem 3.2. *Let $\pi : P \to B$ be a principal G-bundle. Then there is an exact sequence*

$$\pi_1(aut(P)) \xrightarrow{\delta} \pi_1(map_*(BG, Baut(P)), c) \to Aut^G(P) \xrightarrow{F} Aut(P),$$

where $c : BG \to Baut(P)$ is determined by the principal bundle.

Combined with results in [13] , we have

Proposition 3.3. *Let P be as in Theorem2.16 . If*

$$\bigoplus_{i>1} \pi_{2i}(map(P_{(0)}, P_{(0)}); id) = 0,$$

then for all $m \geq 1$, any finitely generated abelian group H and every principal $K(H, 2m)$-bundle with total space homotopy equivalent to P, the associated forgetful map is injective .

We have also similar results for $K(H, 2m+1)$-bundle or G-bundle where G is a connected compact Lie group which will be omitted in this paper.

Unlike that in [13], there are no complete results if the group $\pi_1(map_*(BG, Baut(P)), c)$ is nontrivial although we know that it is still uncountable since the group $\pi_1(aut(P))$ itself may be uncountable too. Thus same results as in [13] can be obtained if the group $\pi_1(aut(P))$ is countable. This is so if $P = \Omega P'$ where P' is a finite complex. An interesting question is

Question 3.4. *If we consider the map δ in the exact sequence of Theorem 3.2, is it possible that the image of δ is always countable group?*

ACKNOWLEDGMENT. The authors are grateful to the referee for his careful reading and suggested improvements of the paper.

REFERENCES

[1] J.F. Adams and J. Walker, *An example in homotopy theory*, Proc. Camb. Phil. Soc. **60** (1960), pp. 699-700.
[2] H.J. Baues, *Rationale homotopietypen*, Manus.Math. **20**(1977), pp.119-131.
[3] E. Dror, W.G. Dwyer and D.M. Kan, *An arithmetic square for virtually nilpotent spaces*, Ill. J. Math. **21**(1977), pp.242-254.
[4] N. Dupont, *Problems and conjectures in rational homotopy theory*, Expos. Math. **12**(1994), pp. 323-352.
[5] J.R. Harper and J. Roitberg, *Phantom maps and spaces of the same n-type for all n*, J. Pure and Applied Algebra **80**(1992), pp.123–137.
[6] P. Hilton, G. Mislin and J. Roitberg, *Localization of nilpotent groups and spaces*, North Holland Math. Series 15, 1975.
[7] T.W. Hungerford, *Algebra*, Holt, Rinehart and Winston, INC, 1974.
[8] D.W. Kahn, *Some research problems on homotopy self-equivalences*, LN in Math. 1425(1990), pp.204–207.
[9] C.A. McGibbon, *Phantom maps*, Handbook of algebraic topology (I. M. James,ed.), North-Holland (1995), pp. 1209–1257.
[10] H. Miller, *The Sullivan conjecture on maps from classifying spaces*, Ann.Math. **120**(1984), pp.39-87.
[11] N. Oda and Y. Shitanda, *Localization, completion and detecting equivariant maps on skeletons*, Manuscript Math. **65**(1989), pp.1-18.
[12] J. Pan, *Having H-space structure is not a generic property*, submitted.
[13] J. Pan and M.H. Woo, Phantom maps and forgetful maps, to appear.
[14] J. Roitberg, *Note on phantom phenomena and group of self-homotopy equivalences*, Comment. Math. Helv. **66** (1991), pp. 448–457.
[15] J. Roitberg, *The Lusternik-Schnirelmann category of certain infinite CW-complexes*, Topology **39**(2000), pp.95–101.
[16] L. Schwartz, *Unstable modules over the Steenrod algebra and Sullivan fixed point set conjecture*, Chicago Lectures in Mathematics, 1994.
[17] A. Zabrodsky, *On phantom maps and a theorem of H.Miller*, Israel J. Math. **58**(1987), pp.129–143.

INSTITUTE OF MATH.,ACADEMIA SINICA, BEIJING 100080, CHINA
E-mail address: pjz@math03.math.ac.cn

KOREA UNIVERSITY, SEOUL 136-701, KOREA
E-mail address: woomh@kuccnx.korea.ac.kr

Contemporary Mathematics
Volume **274**, 2001

Homotopy equivalences of lens spaces of one-relator groups.

John W. Rutter

ABSTRACT For one-relator groups π with torsion we construct a lifting from automorphisms of π to homomorphisms between modified Lyndon resolutions of Z over $Z\pi$. This construction gives new results about the (2m+1)–dimensional lens spaces X^{2m+1} of the group π, and their 2m–skeleta X^{2m}: these finite (π,n)–complexes are defined by a sequence of residue classes $p_1,\dots,p_{m+1}$ mod$q$, respectively $p_1,\dots,p_m$ mod$q$. Possible (MacLane/Whitehead) k–invariants of these complexes coincide with the units in the ring $H^{n+1}(\pi;\pi_n)\cong Z_q$, where π_n is the appropriate homotopy group of a truncated Lyndon resolution, or, equivalently, is the n–th homotopy group of any (π,n)–complex: this extends the result obtained, in the case where π is finite cyclic, by Eilenberg and MacLane. We classify the homotopy types of these (π,n)–complexes X^n: all have the same homotopy type in case n is even, but in case n is odd we give examples where any two have the same homotopy type if and only if they have the same k–invariant up to sign. However the homotopy type of $X^{2m+1}(\mathcal{P}_\pi{:}k)\vee S^{2m+1}$ does not depend on k: this gives further examples of non-cancellation. We use these results to determine the homotopy trees of (π,n)–complexes where π is a one-relator group with torsion. Also we calculate $\mathcal{E}^*(X^{2m+1})$, the group of pointed homotopy self-equivalence classes, as a semi-direct product, and calculate $\mathcal{E}^*(X^{2m})$ as a semi-direct product "up to inner automorphisms". In particular $\mathcal{E}^*(X^{2m})$ is a semi-direct product in case π is finite cyclic. Some of these problems have been considered by Jajodia for X^2.

1 Introduction

Classic lens spaces (of finite cyclic groups) have been investigated extensively. We study here basic properties of odd-dimensional lens spaces of (finitely generated) one relator groups π with torsion and of their even dimensional skeleta. These lens spaces are (finite) (π,n)–complexes with the number of 1–cells being the number of generators of the presentation and having one cell in each other dimension up to n. In §2 we define the

2000 Mathematics subject classifiacation: Primary 55P10 (Homotopy equivalences). Secondary 55P15 (Classification of homotopy type), 22J06 (Cohomology of groups).

Key words and phrases: homotopy self-equivalence, self homotopy equivalence, one-relator group, Lyndon's resolution, lens space, 2-complex, cohomology of groups, (π,n)–complex.

(2m+1)–dimensional lens space $L(\mathcal{P}_\pi: 1, p_2, \ldots, p_{m+1})$ determined by a one-relator presentation $\mathcal{P}_\pi$ with q–torsion and a sequence of residue classes $p_2, \ldots, p_{m+1} \bmod q$. Here each p_r is prime to q. In Theorem 2.1, extending the result obtained, in the case where π is finite cyclic, by Eilenberg and MacLane, we show that the k–invariant (MacLane-Whitehead invariant) of this space is $k = p_2, \ldots, p_{m+1} \bmod q$, and that the k–invariant of its 2m–skeleton is $k = p_2, \ldots, p_m \bmod q$. In particular (Corollaries 2.2 and 2.3), for each unit k in $H^{2m+2}{}_\phi(\pi; \mathbf{Z}\pi.\langle R,q\rangle) \cong \mathbf{Z}_q$, there is a lens space $X^{2m+1}(\mathcal{P}_\pi : k)$ having k–invariant k where $p_2 = \ldots = p_m = 1$ and $p_{m+1} = k$. Similarly, for each unit k in $H^{2m+1}{}_\phi(\pi; \mathbf{Z}\pi.(R-1)) \cong \mathbf{Z}_q$, there is the 2m–skeleton $X^{2m}(\mathcal{P}_\pi : k)$, having k–invariant k, of an appropriate lens space, where $p_2 = \ldots = p_{m-1} = 1$ and $p_m = k$. In §3 we construct a lifting from automorphisms θ of π to θ–endomorphisms of the Lyndon resolution of π and more generally to θ–homomorphisms $f: X^{2m+1}(\mathcal{P}_\pi : k) \to X^{2m+1}(\mathcal{P}_\pi : h)$ at the level of chain complexes of the universal covers. Using this construction and a modification of it, we show in §4 that the spaces $X^{2m+1}(\mathcal{P}_\pi : k)$ and $X^{2m+1}(\mathcal{P}_\pi : h)$ have the same homotopy type if, and only if, there is a homotopy equivalence θ of π satisfying $\theta(R) = uR^t u^{-1}$ for some t prime to q, where $kh^{-1} = \pm t^{m+1} \bmod q$: we give examples where the homotopy type is completely determined by the k–invariant up to sign. We also show that $X^{2m}(\mathcal{P}_\pi : k)$ and $X^{2m}(\mathcal{P}_\pi : h)$ have the same homotopy type in general, and that

$$X^{2m+1}(\mathcal{P}_\pi : k) \vee S^{2m+1} \quad \text{and} \quad X^{2m+1}(\mathcal{P}_\pi : h) \vee S^{2m+1}$$

have the same homotopy type in general. We deduce the trees of homotopy types of (π,n)–complexes in the case where π is a one-relator group. In §5 we calculate the group $\mathcal{E}^*(X^{2m+1}(\mathcal{P}_\pi : k))$ of pointed homotopy self-equivalence classes as a semi-direct product, and we calculate the group $\mathcal{E}^*(X^{2m}(\mathcal{P}_\pi : k))$ as a semi-direct product "up to inner automorphisms". For non-simply-connected spaces, the group of homotopy self-equivalence classes has previously been calculated only for aspherical spaces where the group $\operatorname{aut}\pi$ is known, for certain spaces with only two non-vanishing homotopy groups and for certain (π,n)–complexes including quotients of spheres, and only in a few other general or isolated cases. In §6 we consider the special cases of these calculations in the case where π is finite cyclic and include a new result. The Appendix contains calculations of the cohomology of π, the specific isomorphisms of which we use in our previous calculations.

We use the following notation: $[\operatorname{aut}\pi_n]^\phi = \operatorname{aut}_{Z\pi_1} \pi_n$ is the group of $Z\pi_1$–automorphisms, and $H^n{}_\sigma(X;G)$ is the n-dimensional cohomology of $X \to M$ with the local coefficients determined by $\sigma: \pi_1 \to \operatorname{aut} G$: we identify

$H^n{}_\sigma(\pi;G)$ with $H^n{}_\sigma(K_1(\pi);G)$, where $M = K_1(\pi)$ is the Eilenberg-MacLane space.

2 The lens space of a one-relator group and 2-dimensional complexes with a single 2-cell.

Each finite connected 2–dimensional complex has the homotopy type of the cellular model of some finite presentation of the fundamental group π_1: for this statement we allow the possibility of trivial relations. In the case where X has only one 2–cell, which is attached non-trivially, we have that X has the homotopy type of the cellular model of a finite one-relator presentation. Any two finite 2–dimensional complexes, each with a single 2–cell, have the same homotopy type if, and only if, their fundamental groups are isomorphic [7, Theorem 1]. In the one-relator case, where the relator ρ is not a proper power, then either i) the relator is a primitive element, π is free and the cohomological dimension $\mathrm{cd}\pi = 1$, or ii) $\mathrm{cd}\pi = 2$ [20]/[21], and $\mathcal{P}_\pi$ is aspherical (see for example [1, Lemma page 382]).

A *(π,n)–complex* ($n \geq 2$) is an n–dimensional CW complex which has fundamental group π and whose universal covering complex $\tilde{X}$ is (n–1)–connected. Given that X is a (π,n)–complex, then the skeleton X^r is a (π,r)–complex for $2 \leq r \leq n$, and, if G operates discretely on X, then X/G is a (μ,n)–complex for some group μ. Of special interest is the class of 2–dimensional complexes, including all compact surfaces: all of these are $(\pi,2)$–complexes. We recall some of the properties of (π,n)–complexes: for further details see [15], [16] and [17]. For a (π_1,n)–complex X the Postnikov decomposition collapses to

$$X \to \dots \to X_{n+1} \to X_n \to X_1 = M = K_1(\pi_1(X)),$$

where $X_n = \bar{P}_h \to X_1$ is classified by the MacLane-Whitehead/Postnikov invariant (the first non-trivial Postnikov invariant) $h: M \to L^\phi{}_{n+1}(\pi_n)$, which is a section for $L^\phi{}_{n+1}(\pi_n) \to M$ (see [16, §2]) and which can canonically be regarded as an element of $H^{n+1}{}_\phi(M;\pi_n) \cong H^{n+1}{}_\phi(\pi_1;\pi_n)$. Here $L = L^\phi{}_{n+1} = L^\phi{}_{n+1}(\pi_n) = \tilde{M} \times_\phi K_{n+1}(\pi_n)$ is the twisted Eilenberg-McLane space determined by the (left) action $\phi: \pi_1 \to \mathrm{aut}\,\pi_n$ of the fundamental group, and $M = K_1(\pi_1)$.

The lens spaces of a one-relator group π with torsion and their skeleta are finite (π,n)-complexes: in case π is finite cyclic, these lens spaces are the classic odd–dimensional lens spaces. Let π be a one-relator group with torsion presented by $\mathcal{P}_\pi = \langle x_1,\dots,x_d \mid R^q \rangle$ with relator $\rho = R^q$ where $q \geq 2$, R is not a proper power, and $R^q \neq e$ in the free group $F^d = \langle x_1,\dots,x_d \rangle$; and let $\mathcal{P}_\pi = (S^1 \vee \dots \vee S^1) \cup_\rho e^2$ be the associated cellular model. The generalised lens space $L(\mathcal{P}_\pi : 1, p_2, \dots, p_{m+1})$, of dimension $(2n+1)$, of this presentation $\mathcal{P}_\pi = \langle x_1, \dots, x_d \mid R^q \rangle$, is represented as a CW complex with one cell in each dimension $2 \le s \le 2m+1$, such that $X^2 = \mathcal{P}_\pi$, X^{2r+1} is obtained from X^{2r} $(r \ge 1)$ by adding a $(2r+1)$–cell with attaching map the generator

$$1.(R^{s_{r+1}} - 1) = <R, s_{r+1}>.(R-1) = (R-1).<R, s_{r+1}>$$

of the left ideal $\pi_{2r}(X^{2r}) \cong Z\pi.(R^{s_{r+1}} - 1) = Z\pi.(R-1)$, and X^{2r} is obtained from X^{2r-1} $(r \ge 2)$ by adding a $2r$–cell with attaching map the generator $1.<R,q>$ of the left ideal $\pi_{2r-1}(X^{2r-1}) \cong Z\pi.<R,q>$. Here

$$<R,s> = (1 + R + \dots + R^{s-1}) \text{ and } <R,-s> = -R^{-s}<R,s> \quad (s \ge 1).$$

Each p_r is prime to q, which need not itself be prime, and therefore p_r is a unit in the ring Z_q; and s_r is the (multiplicative) inverse of the unit p_r. The cellular chain complex for the universal cover of $L(\mathcal{P}_\pi : 1, p_2, \dots, p_{m+1})$ is given by modifying the $(2m+1)$–dimensional truncation of Lyndon's resolution using the automorphisms $.<R, s_r>$ of $Z\pi.(R-1)$ as follows:

$$Z\pi \xrightarrow{.(R^{s_{m+1}}-1)} Z\pi \xrightarrow{.<R,q>} Z\pi \xrightarrow{.(R^{s_m}-1)} \dots\dots\dots \xrightarrow{.<R,q>} Z\pi \xrightarrow{.(R^{s_2}-1)} Z\pi \xrightarrow{\partial_2}$$

$$\{Z\pi\}^d \xrightarrow{\partial_1} Z\pi.$$

Thus $\partial_{2r+1} = .(R^{s_{r+1}} - 1) : C_{2r+1} \to C_{2r}$ $(r \ge 1)$, $\partial_{2r} = .<R,q> : C_{2r} \to C_{2r-1}$ $(r \ge 2)$, $\partial_1 = .(x_1 - 1, \dots, x_d - 1)$,

$$\partial_2 = .<R,q>.(\partial R/\partial x_1, \dots, \partial R/\partial x_d),$$

where $(\partial R/\partial x_1, \dots, \partial R/\partial x_d)$ is the Jacobian matrix in terms of Fox partial derivatives [6], where partial differentiation takes place in the free group F^d, ε is the augmentation $\sum n_i z_i \rightsquigarrow \sum n_i$, and $.T$ and $T.$ denote multiplication on the right, respectively left, by T.

In the general case considered above, for $m \ge 1$, $L(\mathcal{P}_\pi : 1, p_2, \dots, p_{m+1})$ has fundamental group π, and is a $(\pi, 2m+1)$–complex for which

$$\pi_{2m+1}(L(\mathcal{P}_\pi : 1, p_2, \dots, p_{m+1})) \cong Z\pi.<R,q>.$$

The $(2r+1)$–skeleton X^{2r+1} $(2m+1 \ge 2r+1 \ge 3)$ of $L(\mathcal{P}_\pi : 1, p_2, \dots, p_{m+1})$ is the generalised lens space $L(\mathcal{P}_\pi : 1, p_2, \dots, p_{r+1})$, and the $2r$–skeleton X^{2r} $(2m \ge 2r \ge 2)$ is a $(\pi, 2r)$–complex for which $\pi_{2r}(X^{2r}) \cong Z\pi.(R-1)$. These isomorphisms, which give the higher homotopy groups, follow from the extension of the above partial resolution taking $s_k = 1$ for $k > m+1$ and using the isomorphisms $\pi_n(X) \cong H_n(\tilde{X}) \cong \operatorname{im} C_{n+1}$, the first of which is valid for any (π, n)–complex, and the second of which follows from the exactness of the resulting modification of Lyndon's resolution.

The following theorem extends the result obtained by Eilenberg and MacLane [5, Theorem 27.1] in the case where $\pi = \mathbf{Z}_q$: their convention for the lens space differs from the one used here.

Theorem 2.1 The MacLane-Whitehead/Postnikov invariant k of

$$L(\mathcal{P}_\pi : 1, p_2, \dots, p_{m+1}) \quad (m \geq 1)$$

with respect to the base resolution is

$$k = p_{m+1}\, p_m \cdots p_3\, p_2 \in \mathbf{Z}_q \cong H^{2m+2}{}_{\phi}(\pi_1; \pi_{2m+1})\,.$$

Similarly the k–invariant of the $2m$–skeleton X^{2m} $(m \geq 2)$ of the lens space is

$$k = p_m p_{m-1} \cdots p_3 p_2 \in \mathbf{Z}_q \cong H^{2m+1}{}_{\phi}(\pi_1; \pi_{2m})\,.$$

Proof Consider the chain map from the truncated base resolution to the defining chain complex for the lens space given by the following commutative diagram, where f_0, f_1, and f_2 are the appropriate identity maps,

$$f_3 = f_4 = \boldsymbol{.}\langle R^{s_2}, p_2\rangle,\quad f_5 = f_6 = \boldsymbol{.}\langle R^{s_3}, p_3\rangle\langle R^{s_2}, p_2\rangle, \dots,$$
$$f_{2m-1} = f_{2m} = \boldsymbol{.}\langle R^{s_m}, p_m\rangle \dots \langle R^{s_3}, p_3\rangle\langle R^{s_2}, p_2\rangle, \text{ and}$$
$$f_{2m+1} = \boldsymbol{.}\langle R^{s_{m+1}}, p_{m+1}\rangle\langle R^{s_m}, p_m\rangle \dots \langle R^{s_3}, p_3\rangle\langle R^{s_2}, p_2\rangle.$$

$$\begin{array}{ccccccccccccc}
Z\pi & \xrightarrow{\ .\langle R,q\rangle\ } & Z\pi & \xrightarrow{\ .(R-1)\ } & Z\pi & \xrightarrow{\ .\langle R,q\rangle\ } & \cdots\cdots & \xrightarrow{\ .\langle R,q\rangle\ } & Z\pi & \xrightarrow{\ .(R-1)\ } & Z\pi & \xrightarrow{\ \partial_2\ } \\
\bar{k}\downarrow & & f_{2m+1}\downarrow & & f_{2m}\downarrow & & & & f_3\downarrow & & f_2\downarrow & \\
Z\pi.\langle R,q\rangle & \longrightarrow & Z\pi & \xrightarrow[\ .(R^{s_{m+1}}-1)\ ]{} & Z\pi & \xrightarrow[\ .\langle R,q\rangle\ ]{} & \cdots\cdots & \xrightarrow[\ .\langle R,q\rangle\ ]{} & Z\pi & \xrightarrow[\ .(R^{s_2}-1)\ ]{} & Z\pi & \xrightarrow[\ \partial_2\ ]{}
\end{array}$$

$$\begin{array}{ccc}
\{Z\pi\}^d & \xrightarrow{\ \partial_1\ } & Z\pi\,. \\
f_1\downarrow & & f_0\downarrow \\
\{Z\pi\}^d & \xrightarrow[\ \partial_1\ ]{} & Z\pi\,.
\end{array}$$

The k–invariant (see [4, §6]) is the composite map

$$\bar{k} = f_{2m+1} \circ (\boldsymbol{.}\langle R,q\rangle) = p_{m+1} p_m p_{m-1} \cdots p_3 p_2 \boldsymbol{.}\langle R,q\rangle,$$

which corresponds to the element $k \equiv p_{m+1}p_m p_{m-1} \cdots p_3 p_2$ of the cohomology group by Remark A.2 of the Appendix. The proof is similar in the case of the 2m–dimensional complex X^{2m}.

In each case the k–invariant is the image of the identity under the group epimorphism $[\hom(\pi_n, \pi_n)]^\phi \to H^{n+1}{}_\phi(\pi_1; \pi_n) \cong \mathbf{Z}_q$, and is an additive group generator of $\mathbf{Z}_q$, and therefore a unit in the ring $\mathbf{Z}_q$.

In view of this result,

$$L(\mathcal{P}_\pi : 1, p_2, \ldots, p_{m+1}) \text{ and } L(\mathcal{P}_\pi : 1, q_2, \ldots, q_{m+1}) \ (m \geq 1),$$

where $q_2 = q_3 = \ldots = q_m = 1$ and $q_{m+1} = k = p_{m+1} p_m \cdots p_3 p_2$, have the same k–invariant: we denote this latter space by $X^{2m+1}(\mathcal{P}_\pi : k)$. Similarly we denote by $X^{2m}(\mathcal{P}_\pi : k)$ the even skeleton where $q_2 = q_3 = \ldots = q_{m-1} = 1$ and

$$q_m = k = p_m p_{m-1} \cdots p_3 p_2 :$$

this has the same k–invariant as the 2m–skeleton of $L(\mathcal{P}_\pi : 1, p_2, \ldots, p_{m+1})$. Thus each of these (π,n)–complexes $L(\mathcal{P}_\pi : 1, p_2, \ldots, p_{m+1})$ and their even-dimensional skeleta have the same k–invariants as ones obtained by adding two, respectively one, cells to the universal "base" (2m–1)–dimensional lens space for which each $p_k = s_k = 1$.

Corollary 2.2 Let k be a unit in the ring $H^{2m+2}{}_\phi(\pi; \mathbf{Z}\pi.\langle R, q\rangle) \cong \mathbf{Z}_q$ $(m \geq 1)$. Then k is the k–invariant of the $(\pi, 2m+1)$–complex $X^{2m+1}(\mathcal{P}_\pi : k)$.

Corollary 2.3 Let k be a unit in the ring $H^{2m+1}{}_\phi(\pi; \mathbf{Z}\pi.(R-1)) \cong \mathbf{Z}_q$ $(m \geq 2)$. Then k is the k–invariant of the $(\pi, 2m)$–complex $X^{2m}(\mathcal{P}_\pi : k)$.

3. A lifting from automorphisms of π.

We now give a construction for lifting an automorphism θ of π to a θ–endomorphism of the standard Lyndon resolution and to θ–homomorphisms $X^{2m}(\mathcal{P}_\pi : k) \to X^{2m}(\mathcal{P}_\pi : h)$ and $X^{2m+1}(\mathcal{P}_\pi : k) \to X^{2m+1}(\mathcal{P}_\pi : h)$ at the chain complexes level.

We first note some further properties of one-relator groups. Let π be presented by $\mathcal{P}_\pi = \langle x_1, \ldots, x_d \mid R^q \rangle$ with only one relator $\rho = R^q$ where R is not a proper power, $q \geq 1$ and $R^q \neq e$ in the free group. We have [10, Theorem 4.12] that π has elements of finite order if, and only if, $q > 1$: that is if, and only if, the relator is a proper power. Moreover [10, Theorem 4.13] if the relator is a proper power then the elements of finite order are precisely those which are conjugates of a power of R. Also [9, Theorem 4] the conjugate subgroups $\mathbf{Z}_q$ and ${}^g(\mathbf{Z}_q)$ have trivial intersection unless g is a power of R in

π, where $\mathbf{Z}_q$ is the cyclic subgroup of π generated by R. Therefore ${}^g(R)=R^t$ if, and only if, $t\equiv 1 \bmod q$ and $g\in \mathbf{Z}_q$. We define

$$A_{m+1}=\{\theta\in \operatorname{aut}(\pi,\mathbf{Z}_q):\theta(R)=R^t, t^{m+1}\equiv\pm 1 \bmod q\} \text{ and}$$

$$B_{m+1}=\{\theta\in \operatorname{aut}\pi:\theta(R)=uR^tu^{-1}, t^{m+1}\equiv\pm 1 \bmod q, u\in \pi\}.$$

The following result is immediate.

Lemma 3.1 Let the group π be presented with two or more generators and one relator R^q, where R is not a proper power and $q>1$. Then π has trivial centre, $\operatorname{aut}\pi=\operatorname{aut}(\pi,\mathbf{Z}_q)\cdot\operatorname{innaut}\pi$ is the product of the two subgroups, and the isomorphism (left action) $\pi\to\operatorname{innaut}\pi$ induces the isomorphism

$$\operatorname{aut}(\pi,\mathbf{Z}_q)\cap\operatorname{innaut}\pi\cong\mathbf{Z}_q.$$

Furthermore, for each element θ of $\operatorname{aut}\pi$, we have $\theta(R)=uR^tu^{-1}$ where $t \bmod q$ and the left coset $u\mathbf{Z}_q$ are uniquely determined by θ. Also we have $B_{m+1}=A_{m+1}\cdot\operatorname{innaut}\pi$, and similarly $A_{m+1}\cap\operatorname{innaut}\pi\cong\mathbf{Z}_q$.

We denote by Γ the image of the homomorphism $\operatorname{aut}\pi\to\mathbf{Z}_q{}^*\cong\operatorname{aut}\mathbf{Z}_q$ given by $\theta\to t \bmod q$.

Lemma 3.2 Let R be a primitive element (one of a set of generators) in the free group $\langle x_1,\dots,x_d\rangle$, then $\Gamma=\operatorname{aut}\mathbf{Z}_q$.

For a (π_1,n)–complex we have the following exact sequence of groups, which is essentially a modification of part of the cohomology sequence of the mapping cone sequence given in [16, §2]: in the case where n is 2, this is a local-coefficient cohomology version of the Hopf Theorem (cf. [3, Theorem 2.2])

$$0\to H^n{}_\phi(\pi_1;\pi_n)\to H^n{}_\phi(X;\pi_n)\to[\hom(\pi_n,\pi_n)]^\phi\to H^{n+1}{}_\phi(\pi_1;\pi_n)\to 0.$$

Using the fact that $H^{n+1}{}_\phi(X;\pi_n)=0$, we obtain $H^{n+1}{}_\phi(X_n;\pi_n)=0$ by repeated application of [15, Theorem 3.1], and the sequence follows as noted. The group epimomorphism

$$h^*:[\hom(\pi_n,\pi_n)]^\phi\to H^{n+1}{}_\phi(\pi_1;\pi_n)$$

is given by $\alpha \rightsquigarrow h^*(\alpha)=\alpha_*(h)$ [15, §3], and is split by the homomorphism induced by $L^\phi{}_{n+1}(\pi_n)\to K_1(\pi_1)$. The kernel of h^* is a left ideal of the ring $[\hom(\pi_n,\pi_n)]^\phi$; it is also a right ideal if, for example, $H^{n+1}{}_\phi(\pi_1;\pi_n)$ is cyclic. In the case where $\ker h^*$ is a two sided ideal, h^* induces a (generally non-commutative) ring structure on $H^{n+1}{}_\phi(\pi_1;\pi_n)$, for which h is the multiplicative identity; and the group epimorphism then becomes a ring epimorphism (see also [4, Theorem 1] in cases where X is finite).

Given $\theta \in \operatorname{aut}\pi$, we now define a θ–homomorphism $f:\mathrm{Lres}_\pi\mathbf{Z}\to \mathrm{Lres}_\pi\mathbf{Z}$ of the standard Lyndon resolution $\mathrm{Lres}_\pi\mathbf{Z}$. By Lemma 6.1 we have $\theta(R)= uR^tu^{-1}$, where $t \bmod q$ and the left coset $u.Z_q$ in π are uniquely determined by θ. Let $\Theta:F^d\to F^d$ be an endomorphism of the free group F^d which lifts the automorphism θ. Then we have $\Theta(R)= KUR^tU^{-1}$, where K lies in the kernel of $p:F^d\to \pi$ and $p(U)= u$. We use below the free calculus in the free group F^d: in particular we use the fundamental formula of the free calculus and the chain rule [6, 2.3 and 2.6]. In order not to complicate notation, we use the same symbols for elements such as R, x_i and $\partial R/\partial x_i$ in the free group F^d and for their images in the group π. Using free partial differentiation in the free group, and then applying the homomorphism $p:F^d\to \pi$, we have $\partial/\partial x_i(ZR^qZ^{-1})= z.\langle R,q\rangle.\partial R/\partial x_i$ and

$$\partial/\partial x_i(ZR^{-q}Z^{-1})= -z.R^{-q}.\langle R,q\rangle.\partial R/\partial x_i,$$

where $p(Z)= z$. Since K is a product of elements of the form $ZR^{\pm q}Z^{-1}$, we have easily $\partial K/\partial x_i= w.\langle R,q\rangle.\partial R/\partial x_i$, where $w= p(W)$ and W is independent of i. We define $f:\mathrm{Lres}_\pi\mathbf{Z}\to \mathrm{Lres}_\pi\mathbf{Z}$ by

$$\begin{aligned}
&f_0(y) = \mathbf{Z}\theta(y)\\
&f_1(y_1, \dots , y_d) = (\mathbf{Z}\theta(y_1), \dots , \mathbf{Z}\theta(y_d)).(a_{ij}) \quad \text{where } a_{ij} = \partial\Theta(x_i)/\partial x_j\\
&f_2(y) = \mathbf{Z}\theta(y).\{\mathbf{Z}\theta(\langle R,q\rangle).w + u.\langle R,t\rangle\}\\
&f_{2r+1}(y) = \mathbf{Z}\theta(y).u.\langle R,t\rangle^{r+1} \qquad (r \geq 1)\\
&f_{2r}(y) = \mathbf{Z}\theta(y).u.\langle R,t\rangle^{r} \qquad (r \geq 2)
\end{aligned}$$

where $(z_1, \dots , z_d).(a_{ij}) = (\Sigma_i\, z_i.a_{i1}, \dots , \Sigma_i\, z_i.a_{id})$ as usual.

Lemma 3.3 $f:\mathrm{Lres}_\pi\mathbf{Z}\to \mathrm{Lres}_\pi\mathbf{Z}$ is a θ–endomorphism of the Lyndon resolution.

Proof. We have

$$\partial\Theta(R)/\partial x_i = \partial K/\partial x_i + \theta(1 - R).\partial U/\partial x_i + u.\langle R,t\rangle.\partial R/\partial x_i ,$$

and therefore

$$\begin{aligned}
f_1\partial_2(y) &= \mathbf{Z}\theta(y).\mathbf{Z}\theta(\langle R,q\rangle).\Theta(\partial R/\partial x_1,\dots,\partial R/\partial x_d).[\partial\Theta(x_i)/\partial x_j]\\
&= \mathbf{Z}\theta(y).\mathbf{Z}\theta(\langle R,q\rangle).(\partial\Theta(R)/\partial x_1,\dots,\partial\Theta(R)/\partial x_d)\\
&= \mathbf{Z}\theta(y).\mathbf{Z}\theta(\langle R,q\rangle).\{\mathbf{Z}\theta(1 - R).(\partial U/\partial x_1,\dots,\partial U/\partial x_d) +\\
&\qquad (w.\langle R,q\rangle + u.\langle R,t\rangle).(\partial R/\partial x_1,\dots,\partial R/\partial x_d)\}\\
&= \partial_2 f_2(y) ,
\end{aligned}$$

since $\mathbf{Z}\theta(\langle R,q\rangle).u = u.\langle R,q\rangle$. That $f_0\partial_1 = \partial_1 f_1$ is immediate from the fundamental formula, and the lemma follows easily.

We note that the chain map f is not uniquely determined by θ but depends on the choices of u within a left coset of $\mathbf{Z}_q$ and of the integer t within its mod q class, and further on the subsequent choices of U and of Θ. We also note, in the case where θ and Θ are the inner automorphisms given by $\theta(z)=uzu^{-1}$ and $\Theta(Z)=UZU^{-1}$, that

$$a_{ij}=(1-ux_iu^{-1}).\partial U/\partial x_j+u.\partial x_i/\partial x_j,$$

and that $f_r(y)=u.y$ for $r\geq 2$. In the case where θ is the identity, we can take $U=1$ and $f_r(y)=y$ for $r\geq 0$. A special case of the chain map f corrects [8, Lemma 1].

Next we modify this chain map to obtain a chain map from the chain complex of the universal cover of $X^{2m+1}(\mathcal{P}_\pi:k)$ to the chain complex of the universal cover of $X^{2m+1}(\mathcal{P}_\pi:h)$. We define f_i, for $0\leq i\leq 2m$, as in Lemma 3.3 and redefine

$$f_{2m+1}(y)=\mathbf{Z}\theta(y).u.<R,t>^m.<R^K,t>.<R^H,\sigma>$$

where K, H, and σ are chosen to satisfy $kK\equiv 1 \bmod q$, $hH\equiv 1 \bmod q$, and $\sigma k\equiv h \bmod q$. We have the lemma:

Lemma 3.4 f is a θ-homomorphism from the chain complex of the universal cover of $X^{2m+1}(\mathcal{P}_\pi:k)$ to the chain complex of the universal cover of

$$X^{2m+1}(\mathcal{P}_\pi:h).$$

We similarly redefine

$$f_{2m+2}(y)=\mathbf{Z}\theta(y).u.<R,t>^m.<R^K,t>.<R^H,\sigma>,$$

and we have the lemma:

Lemma 3.5 f is a θ–homomorphism from the chain complex of the universal cover of $X^{2m+2}(\mathcal{P}_\pi:k)$ to the chain complex of the universal cover of

$$X^{2m+2}(\mathcal{P}_\pi:h).$$

The chain homomorphisms of Lemmas 3.4 and 3.5 will depend on the choices of K, H, and σ within their mod q classes. In the case where $h=k=1$, the homomorphisms f_{2m+1} and f_{2m+2} can be taken to be the ones defined for the Lyndon resolution.

4 Homotopy type

We now classify the homotopy types of the spaces $X^n(\mathcal{P}_\pi:k)$ and determine the trees of homotopy types of (π,n)–complexes where π is a one-relator group with torsion.

Theorem 4.1 The lens spaces $X^{2m+1}(\mathcal{P}_\pi{:}k)$ and $X^{2m+1}(\mathcal{P}_\pi{:}h)$ have the same homotopy type $(m\geq 1)$ if, and only if, there is an integer t and an automorphism θ of π satisfying $\theta(R)=uR^tu^{-1}$ where $t^{m+1}h=\pm k \bmod q$.

Proof Suppose that there is an automorphism θ satisfying $\theta(R)=uR^tu^{-1}$ where $t^{m+1}h=\eta k \bmod q$ and $\eta=\pm 1$. We consider the chain map f of Lemma 3.4 from the chain complex of $X^{2m+1}(\mathcal{P}_\pi{:}k)$ to the chain complex of $X^{2m+1}(\mathcal{P}_\pi{:}h)$. For $m\geq 1$, we have

$$\begin{aligned} f_{2m+1}(y.\langle R,q\rangle) &= \mathbf{Z}\theta(y).u.\langle R,q\rangle.\langle R,t\rangle^m.\langle R^K,t\rangle.\langle R^H,\sigma\rangle \\ &= t^{m+1}\sigma\mathbf{Z}\theta(y).u.\langle R,q\rangle \quad (m\geq 1), \end{aligned}$$

where $\sigma k\equiv h \bmod q$. Since $t^{m+1}h=\eta k \bmod q$, where $\eta=\pm 1$, there is a unique $b\in \mathbf{Z}$ such that $t^{m+1}\sigma+bq=\eta$. We modify the chain map f to a chain map g by replacing f_{2m+1} by g_{2m+1} where

$$g_{2m+1}(y)= f_{2m+1}(y)+ b\mathbf{Z}\theta(y).u.\langle R,q\rangle.$$

Then $g_{2m+1}(y.\langle R,q\rangle)= \eta\mathbf{Z}\theta(y).u.\langle R,q\rangle$, and, since $\eta u.\langle R,q\rangle$ is uniquely determined by θ, the modified construction determines an isomorphism $\pi_{2m+1}(X^{2m+1}(\mathcal{P}_\pi{:}k))\cong \pi_{2m+1}(X^{2m+1}(\mathcal{P}_\pi{:}h)))$. Therefore g is a θ chain equivalence. Conversely let g be a chain map representing a homotopy equivalence from $X^{2m+1}(\mathcal{P}_\pi{:}k)$ to $X^{2m+1}(\mathcal{P}_\pi{:}h)$. Then the restriction of g_{2m+1} to

$$\mathbf{Z}\pi.\langle R,q\rangle\cong \pi_{2m+1}(X^{2m+1}(\mathcal{P}_\pi\,{:}k))$$

is, say, a θ–isomorphism γ to $\mathbf{Z}\pi.\langle R,q\rangle\cong \pi_{2m+1}(X^{2m+1}(\mathcal{P}_\pi{:}h))$, where $\theta(R)=uR^tu^{-1}$. The $\mathbf{Z}\pi$–homomorphism $\mathbf{Z}\pi \to \mathbf{Z}\pi.\langle R,q\rangle$ representing the Postnikov invariant k in

$$H^{2m+2}_{\ \ \phi}(\pi;\mathbf{Z}\pi.\langle R,q\rangle)= N/\{\langle R,q\rangle.\mathbf{Z}\pi.\langle R,q\rangle\}\cong \mathbf{Z}_q \quad (m \geq 1)$$

(see Theorem A.1 of the Appendix) is given by $1 \rightsquigarrow \tilde{k}.\langle R,q\rangle$ say, where $\tilde{k}.\langle R,q\rangle\in N$ and $\tilde{k}\equiv k \bmod q$. Using the construction f, of Lemma 6.2, but in this case for the automorphism θ^{-1}, we have that $(\theta,\gamma).k$ is determined by

$$\begin{aligned} 1 \rightsquigarrow g_{2m+1}(v.\langle R,s\rangle^{m+1}.\tilde{k}.\langle R,q\rangle) &= \tilde{k}s^{m+1}\langle R,q\rangle.u^{-1}.\beta= \\ &\qquad \tilde{k}s^{m+1}\theta(v).\kappa.\langle R,q\rangle \end{aligned}$$

and $\langle R,q\rangle.u^{-1}.\beta= \kappa.\langle R,q\rangle$, where $st=1 \bmod q$, $\theta(v)=u^{-1}$ and $\beta= g_{2m+1}(1)$. Since γ, the restriction of g_{2m+1}, gives an automorphism of $\mathbf{Z}\pi.\langle R,q\rangle$, we have easily that $\varepsilon(\beta)= \varepsilon(\kappa)= \pm 1$. Therefore, on applying the augmentation $\bmod q^2$ to $\tilde{k}s^{m+1}\theta(v).\kappa.\langle R,q\rangle \equiv \tilde{h}.\langle R,q\rangle$ in

$$H^{2m+2}_{\phi}(\pi;\mathbf{Z}\pi.\langle R,q\rangle) = N/\langle R,q\rangle.\mathbf{Z}\pi.\langle R,q\rangle,$$

since $(\theta,\gamma).$ is an isomorphism, we have $ks^{m+1} \equiv \pm h \bmod q$. Theorem 4.1 now follows by 2.1 and Theorem A.1.

Corollary 4.2 Let R be a primitive element of the free group $\langle x_1,\dots,x_d\rangle$, then the lens spaces $X^{2m+1}(\mathcal{P}_\pi{:}k)$ and $X^{2m+1}(\mathcal{P}_\pi{:}h)$ have the same homotopy type if and only if

$$kh^{-1} \equiv \pm\, t^{m+1}$$

where t is some integer prime to q.

Let n_q be the number of units in $\mathbf{Z}_q$. Then $t^{n_q} \equiv 1$ for each such unit t.

Corollary 4.3 Let $m = n_q - 1$, then $X^{2m+1}(\mathcal{P}_\pi{:}k)$ and $X^{2m+1}(\mathcal{P}_\pi{:}h)$ have the same homotopy type if and only if $h \equiv \pm\, k$.

Theorem 4.4 The spaces $X^{2m}(\mathcal{P}_\pi{:}k)$ and $X^{2m}(\mathcal{P}_\pi{:}h)$ have the same homotopy type $(m \geq 2)$.

Proof Let θ be an automorphism of π. We consider the chain map f of Lemma 3.4 from the chain complex of the universal cover of $X^{2m}(\mathcal{P}_\pi{:}k)$ to the chain complex of the universal cover of $X^{2m}(\mathcal{P}_\pi{:}h)$. We have

$$\begin{aligned} f_{2m}(y.(R-1)) &= \mathbf{Z}\theta(y).u.(R^t-1).\langle R,t\rangle^m.\langle R^K,t\rangle.\langle R^H,\sigma\rangle \\ &= \mathbf{Z}\theta(y).u.(R-1).\langle R,t\rangle^{m+1}.\langle R^K,t\rangle.\langle R^H,\sigma\rangle. \end{aligned}$$

Now $.\langle R,t\rangle$, $.\langle R^K,t\rangle$ and $.\langle R^H,\sigma\rangle$ determine automorphisms of

$$\mathbf{Z}\pi.(R-1) = \mathbf{Z}\pi.(R^K-1) = \mathbf{Z}\pi.(R^H-1)$$

with inverses $.\langle R^t,\tau\rangle$, $.\langle R^{Kt},\tau\rangle$ and $.\langle R^{H\sigma},\kappa\rangle$, where $\tau t \equiv 1 \bmod q$ and $\sigma\kappa \equiv 1 \bmod q$. Therefore f_{2m} determines a θ–isomorphism from

$$\pi_{2m}(X^{2m}(\mathcal{P}_\pi{:}k)) \cong H_{2m}(\tilde{X}^{2m}(\mathcal{P}_\pi{:}k)) \cong \mathbf{Z}\pi.(R-1)$$

to $\pi_{2m}(X^{2m}(\mathcal{P}_\pi{:}h)) \cong H_{2m}(\tilde{X}^{2m}(\mathcal{P}_\pi{:}h)) \cong \mathbf{Z}\pi.(R-1)$. Also

$$f_{2m}(y.(R-1)) = \mathbf{Z}\theta(y).u.(R^t-1).\langle R,t\rangle^m.\langle R^K,t\rangle.\langle R^H,\sigma\rangle = \dots = \dots$$

does not depend on the choices of t, K, H, and s within their $\bmod q$ classes, though it does depend on the choice of u within its left coset $u\mathbf{Z}_q$. We have proved that, for each automorphism θ of π, there is a θ chain equivalence from the chain complex of $X^{2m}(\mathcal{P}_\pi{:}k)$ to the chain complex of $X^{2m}(\mathcal{P}_\pi{:}h)$. The theorem is proved by taking θ to be the identity automorphism. We use the general construction given here in a later proof.

The group $\operatorname{aut}\pi_1 \times_\phi \operatorname{aut}\pi_n$ operates on the left of $H^{n+1}{}_\phi(\pi_1;\pi_n)$ as a group of automorphisms, essentially by

$$(\theta,\lambda).(\beta) = \lambda \circ \beta \circ \theta^{-1}:$$

in the case where $H^{n+1}{}_\phi(\pi_1;\pi_n)$ is cyclic this operation permutes the units (equivalently the additive group generators) of the ring. The two theorems above give the condition that the appropriate k–invariants lie in the same orbit under this action. We also have the following theorem.

Theorem 4.5 The spaces $X^{2m+1}(\mathcal{P}_\pi{:}k)\vee S^{2m+1}$ and $X^{2m+1}(\mathcal{P}_\pi{:}h)\vee S^{2m+1}$ have the same homotopy type $(m\geq 1)$.

Proof The chain complex of the universal cover of $X^{2m+1}(\mathcal{P}_\pi{:}k)\vee S^{2m+1}$ is obtained from the chain complex of the universal cover of $X^{2m+1}(\mathcal{P}_\pi{:}k)$ by replacing $\partial_{2m+1} = .(R^k-1)\colon \mathbf{Z}\pi \to \mathbf{Z}\pi$ by $(.(R^k-1),0)\colon \mathbf{Z}\pi+\mathbf{Z}\pi \to \mathbf{Z}\pi$. At the chain complex level we define a chain map from $X^{2m+1}(\mathcal{P}_\pi{:}k)\vee S^{2m+1}$ to $X^{2m+1}(\mathcal{P}_\pi{:}1)\vee S^{2m+1}$ by the identity in dimensions less than $2m+1$ and by

$$A = \begin{pmatrix} .\langle R,k\rangle & .\langle R,q\rangle \\ \varepsilon & .\langle R^k,K\rangle \end{pmatrix} : \mathbf{Z}\pi+\mathbf{Z}\pi \to \mathbf{Z}\pi+\mathbf{Z}\pi$$

in dimension $2m+1$, where $kK = q\varepsilon+1$. The resulting chain map is a chain isomorphism. The inverse of A is

$$\begin{pmatrix} .\langle R^k,K\rangle & -.\langle R,q\rangle \\ -\varepsilon & .\langle R,k\rangle \end{pmatrix}.$$

Therefore $X^{2m+1}(\mathcal{P}_\pi{:}k)\vee S^{2m+1}$ and $X^{2m+1}(\mathcal{P}_\pi{:}1)\vee S^{2m+1}$ have the same homotopy type.

We now determine, in the case where π is a one relator group, the trees $HT(\pi,n)$ of homotopy types of (π,n)–complexes. The Euler characteristic of $X^{2m}(\mathcal{P}_\pi{:}1)$ is $\chi(X^{2m}(\mathcal{P}_\pi{:}1)) = 2-d$ $(m\geq 1)$, and the tree $HT(\pi,2m)$ $(m\geq 2)$ is the single stalk

$$[X^{2m}] \to [X^{2m}\vee S^{2m}] \to [X^{2m}\vee 2S^{2m}] \to [X^{2m}\vee 3S^{2m}] \to \dots,$$

with root $X^{2m}(\mathcal{P}_\pi{:}1)$ at level $2-d$ and where level $t>2-d$ is precisely the homotopy type of $X^{2m}(\mathcal{P}_\pi{:}1)\vee(t-2+d)S^{2m}$. The Euler characteristic of $X^{2m+1}(\mathcal{P}_\pi{:}k)$ is $\chi(X^{2m+1}(\mathcal{P}_\pi{:}k)) = 1-d$ $(m\geq 1)$, and the tree $HT(\pi,2m+1)$ $(m\geq 1)$ has roots at level $1-d$ as determined by Theorem 4.1

$$\{[X^{2m+1}(\mathcal{P}_\pi{:}k)]\} \leftarrow [X^{2m+1}(\mathcal{P}_\pi{:}1)\vee S^{2m+1}] \leftarrow [X^{2m+1}(\mathcal{P}_\pi{:}1)\vee 2S^{2m+1}] \leftarrow \dots .$$

and where level $t<1-d$ is precisely the homotopy type of

$$X^{2m+1}(\mathcal{P}_\pi{:}1)\vee(1-d-t)S^{2m+1}.$$

5 Homotopy self-equivalences

We now calculate the group of pointed homotopy self-equivalence classes of $X^{2m+1}(\mathcal{P}_\pi:k)$ as a semi-direct product, and calculate the group of pointed homotopy self-equivalence classes of $X^{2m}(\mathcal{P}_\pi:k)$ as a semi-direct product "up to inner automorphisms" (see Lemma 3.1).

By obstruction theory, for an n–dimensional CW complex X the representation $\mathcal{E}^*(X) \to \mathcal{E}^*(X_n)$, between groups of pointed homotopy self-equivalence classes, is an isomorphism. Therefore, for these (π,n)–complexes, we have (see for example [17, Corollary 2.5]) the abelian-kernel extension

$$0 \to H^n{}_\phi(\pi_1;\pi_n) \to \mathcal{E}^*(X) \to \mathrm{stab}_k(\mathrm{aut}\,\pi_1 \times_\phi \mathrm{aut}\,\pi_n) \to 1 \tag{5.1}$$

induced by the representation of $\mathcal{E}^*(X)$ into the automorphisms of the homotopy groups of X. For a (π,n)–complex X, the group $\mathcal{E}^*(X)$ is thus represented onto the automorphism group of the algebraic n–type (π_1,π_n,k) of X, and the representation is an isomorphism if, and only if, $H^n{}_\phi(\pi_1;\pi_n)=0$. In the cases we consider here $H^n{}_\phi(\pi_1;\pi_n)=0$ (see Theorem A1 of the Appendix), so that $\mathcal{E}^*(X) \cong \mathrm{aut}(\pi_1,\pi_n,k) \cong \mathrm{stab}_k(\mathrm{aut}\,\pi_1 \times_\phi \mathrm{aut}\,\pi_n)$. Similarly, in these cases, $\mathcal{E}^*(X|Y)$, the set of pointed homotopy equivalence classes of maps from X to Y is represented bijectively onto the set

$$\mathrm{iso}((\pi_1(X),\pi_n(X),k_X),\quad (\pi_1(Y),\pi_n(Y),k_Y)).$$

Theorem 5.2 The representation into the automorphism groups of the homotopy groups of the generalised (2m+1)-dimensional lens space

$$L(\mathcal{P}_\pi : 1,p_2,\dots,p_{m+1}) \quad (m \geq 1)$$

induces an isomorphism

$$\mathcal{E}^*(L(\mathcal{P}_\pi : 1,p_2,\dots,p_{m+1})) \cong \mathrm{stab}_k(\mathrm{aut}_{Z\pi} \quad (Z\pi.\langle R,q\rangle)) \rtimes B_{m+1},$$

where $B_{m+1} = \{\theta \in \mathrm{aut}\,\pi : \theta(R) = uR^tu^{-1}, t^{m+1} \equiv \eta \bmod q, \eta = \pm 1, u \in \pi\}$, is the image of the projection $\mathrm{stab}_k(\mathrm{aut}\,\pi \times_\phi \mathrm{aut}(Z\pi.\langle R,q\rangle)) \to \mathrm{aut}\,\pi$, and where the semi-direct product is determined by the homomorphic section

$$B_{m+1} \to \mathrm{stab}_k(\mathrm{aut}\,\pi \times_\phi \ \mathrm{aut}(Z\pi.\langle R,q\rangle))$$

given by $\theta \rightsquigarrow (\theta,F)$ where $F(y.\langle R,q\rangle) = \eta Z\theta(y).u.\langle R,q\rangle$.

Proof The construction $\theta \rightsquigarrow g$ defined in the proof of Theorem 4.1 determines a homomorphism

$$B_{m+1} = \{\theta \in \mathrm{aut}\,\pi : \theta(R) = uR^tu^{-1}, t^{m+1} \equiv \pm 1 \bmod q\} \to \mathrm{aut}(Z\pi.\langle R,q\rangle)$$

into the automorphism group of $\mathbf{Z}\pi.<R,q> \cong \pi_{2m+1}(X^{2m+1}(\mathcal{P}_\pi : k))$, and hence a homomorphic partial section $B_{m+1} \to \mathrm{stab}_k(\mathrm{aut}\,\pi \times_\phi \mathrm{aut}\,\mathbf{Z}\pi.<R,q>)$ of the projection. That the image of the projection

$$\mathrm{stab}_k(\mathrm{aut}\,\pi \times_\phi \mathrm{aut}(\mathbf{Z}\pi.<R,q>)) \to \mathrm{aut}\,\pi$$

is $B_{m+1} = A_{m+1}.\mathrm{innaut}\,\pi$, follows also from the proof of Theorem 4.1.

We denote by $\mathbf{Z}_q$ the subgroup of π generated by R, and we denote by $\mathrm{aut}(\pi, \mathbf{Z}_q)$ the subgroup of $\mathrm{aut}\,\pi$ consisting of those automorphisms which extend automorphisms of $\mathbf{Z}_q$ (see Lemma 3.1).

Theorem 5.3 Let X^{2m} be the 2m-skeleton of the generalised (2m+1)-dimensional lens space $L(\mathcal{P}_\pi : 1, p_2, \ldots, p_{m+1})$ $(m \geq 2)$. Then the representation

$$\mathcal{E}^*(X^{2m}) \cong \mathrm{stab}_k(\mathrm{aut}\,\pi \times_\phi \mathrm{aut}(\mathbf{Z}\pi.(R-1))) \to \mathrm{aut}\,\pi$$

is onto, and has a homomorphic partial section

$$\mathrm{aut}(\pi, \mathbf{Z}_q) \to \mathrm{stab}_k(\mathrm{aut}\,\pi \times_\phi \mathrm{aut}(\mathbf{Z}\pi.(R-1)))$$

given by $\theta \rightsquigarrow (\theta, F)$ where

$$F(y.(R-1)) = \mathbf{Z}\theta(y).<R,t>.<R^{s_2},t>. \;\ldots\; .<R^{s_m},t>.(R-1),$$

and $\theta(R) = R^t$.

Theorem 5.4 Let X^2 be the cellular model of the group π which is presented by $\mathcal{P}_\pi = \langle x_1, \ldots, x_d \mid R^q \rangle$ with only one relator R^q where R is not a proper power, $q > 1$ and $R^q \neq e$ in the free group. Then the representation into the automorphism groups of the homotopy groups of X^2 induces the abelian-kernel extension [1]

$$0 \to H^2_\phi(\pi; \mathbf{Z}\pi.(R-1)) \to \mathcal{E}^*(X^2) \to \mathrm{stab}_k(\mathrm{aut}\,\pi \times_\phi \mathrm{aut}\,\mathbf{Z}\pi.(R-1)) \to 1.$$

Furthermore the projection $\mathrm{stab}_k(\mathrm{aut}\,\pi \times_\phi \mathrm{aut}(\mathbf{Z}\pi.(R-1))) \to \mathrm{aut}\,\pi$ is onto, and has a homomorphic partial section

$$\mathrm{aut}(\pi, \mathbf{Z}_q) \to \mathrm{stab}_k(\mathrm{aut}\,\pi \times_\phi \mathrm{aut}(\mathbf{Z}\pi.(R-1))).$$

The group $\mathrm{aut}\,\pi \times_\phi \mathrm{aut}(\mathbf{Z}\pi.(R-1))$ operates on the left of $H^2_\phi(\pi; \mathbf{Z}\pi.(R-1))$ as a group of automorphisms, by $(\theta, \lambda).(\beta) = \lambda \circ \beta \circ \theta^{-1}$. Alternatively we have the exact sequence

$$1 \to \mathcal{K}^*_{\pi_1}(X^2) \to \mathcal{E}^*(X^2) \to \mathrm{aut}\,\pi \to 1,$$

where $\mathcal{K}^*_{\pi_1}(X^2)$ is isomorphic to the group of units

[1]This corrects Corollary 2 of [7].

$$U\{H^2_\phi(X^2;Z\pi.(R-1)),\oplus\} = U\{H^2_\phi(X^2;Z\pi.(R-1)),\times\},$$

and we have the group extension

$$0 \to H^2_\phi(\pi;Z\pi.(R-1)) \to \mathcal{K}^*_{\pi_1}(X^2) \to \mathrm{stab}_k([\mathrm{aut}\,Z\pi.(R-1)]^\phi) \to 1\,.$$

Proof of Theorems 5.3 and 5.4. From the proof of Theorem 4.4, it follows that the construction $\theta \rightsquigarrow f$ determines a θ–automorphism of $\pi_{2m}(X^{2m}) \cong H_{2m}(\tilde{X}^{2m}) \cong Z\pi.(R-1)$. Further, on choosing $U=1$, this gives a (uniquely defined) homomorphism $T : \mathrm{aut}(\pi, Z_q) \to \mathrm{aut}(Z\pi.(R-1))$ into the automorphism group of $Z\pi.(R-1) \cong \pi_{2m}(X^{2m})$, and hence a homomorphic partial section $\mathrm{aut}(\pi,Z_q) \to \mathrm{stab}_k(\mathrm{aut}\,\pi \times_\phi \mathrm{aut}\,\pi_{2m}(X^{2m}))$, given by $\theta \rightsquigarrow (\theta, T(\theta))$, of the projection. Theorem 5.3 and Theorem 5.4 now follow by Exact sequence 5.1, [16, Theorem 2.4] and Theorem A.1.

Since the relevant cohomology groups are given (see the Appendix) by

$$H^2_\phi(\pi;Z\pi.(R-1)) = L\,/\,\langle R,q\rangle.(\partial R/\partial x_1,\dots,\partial R/\partial x_d).\{Z\pi\}^d.(R-1)$$

and

$$H^2_\phi(X^2;Z\pi.(R-1)) = Z\pi.(R-1)\,/\,\langle R,q\rangle.(\partial R/\partial x_1,\dots,\partial R/\partial x_d).\{Z\pi\}^d.(R-1),$$

the calculation of the group of homotopy self-equivalences of X^2 is, apart from the extension, now reduced to a purely algebraic one. Using the construction and notation given in the Appendix, we have that the action $(\theta,\lambda).$ is defined $\mathrm{mod}\langle R,q\rangle.(\partial R/\partial x_1,\dots,\partial R/\partial x_d).\{Z\pi\}^d.(R-1)$ by the function induced on $Z\pi.(R-1)$ by f_2: thus $\beta \rightsquigarrow (\langle R,q\rangle.Z\theta(z) + Z\theta(v.\langle R,s\rangle)).Z\theta(\beta).c$ where z, v, and s are determined for θ^{-1} as w, u, and t are for θ, and $\lambda(1.(R-1)) = c.(R-1))$.

We note that $\mathrm{stab}_k(\mathrm{aut}\,\pi_1 \times_\phi \mathrm{aut}\,\pi_2) \to \mathrm{aut}\,\pi_1(X)$ is not, in general, an isomorphism even with the restrictions of Theorem 5.3. See, for example, the case of the pseudo-projective planes given below.

In the case where the (non-trivial) relator R of the presentation $\mathcal{P}_\pi = \langle x_1,\dots,x_d \mid R\rangle$ is not a proper power, the cellular model X is aspherical as noted above, and therefore $\mathcal{E}^*(X) \cong \mathrm{aut}\,\pi$. Also, in the case where the relator is a primitive element, the group π is free and again X is aspherical. For completeness we consider next the remaining case of a finite 2-complex with only one 2-cell, which is attached trivially.

Theorem 5.5 Let $X = (S^1 \vee \dots \vee S^1) \vee S^2$ be the one-point union of d circles and a 2–sphere. Then $\mathcal{E}^*(X) \cong \mathrm{aut}\,F^d \times_\phi \mathrm{aut}\,ZF^d \cong U(ZF^d) \rtimes \mathrm{aut}\,F^d$ where the semi-direct product is determined by the homomorphic section

$$\mathrm{aut}\,F^d \rightsquigarrow \mathrm{aut}\,F^d \times_\phi \mathrm{aut}\,\mathbf{Z}F^d$$

given by $\theta \rightsquigarrow (\theta, \mathbf{Z}\theta)$.

Proof Since the cohomological dimension of a free group is one, and

$$\mathrm{aut}_{\mathbf{Z}F^d}\mathbf{Z}F^d \cong U(\mathbf{Z}F^d),$$

we have $\mathcal{E}^*(X) \cong \mathrm{aut}\,F^d \times_\phi \mathrm{aut}\,\mathbf{Z}F^d \cong U(\mathbf{Z}F^d) \rtimes \mathrm{aut}\,F^d$ as stated.

6 Special cases: quotients of spheres under a finite group action and pseudo-projective planes.

We consider spaces of the form $\Sigma^n/\mathbf{Z}_q$ where Σ^n is a finite complex which is a homotopy n–sphere and the finite group $\mathbf{Z}_q$ acts freely and cellularly. Given that $q \geq 3$ this can only occur in the case where n is odd. In the case where $q=2$, we have the real projective spaces $\mathbf{R}P^n = \Sigma^n/\mathbf{Z}_2$ $(n \geq 1)$. In the case where $n=2m+1$, $m \geq 1$ and $q \geq 3$, we have that $\Sigma^{2m+1}/\mathbf{Z}_q$ is a $(\pi, 2m+1)$–complex and is a simple space (ie. $\mathbf{Z}_q$ acts trivially on the higher homotopy groups). Also, using the covering property, we have

$$\pi_{2m+1}(\Sigma^{2m+1}/\mathbf{Z}_q) = \mathbf{Z}.$$

Notable among these quotient spaces are the (2m+1)–dimensional classic lens spaces $L(q{:}p_1,\dots,p_{m+1})$ $(q \geq 2, m \geq 1)$, where each p_r is prime to q, which need not itself be prime, and therefore p_r is a unit in the ring $\mathbf{Z}_q$ with inverse s_r say. The free action of $\mathbf{Z}_q$ on S^{2m+1} has generator $g{:}S^{2m+1} \to S^{2m+1}$ given by

$$g(z_1,\dots,z_{m+1}) = (R_1(z_1),\dots,R_{m+1}(z_{m+1})),$$

where $R_r(z) = ze^{2\pi i p_r/q}$ and $S^{2m+1} \subset \mathbf{C}^{m+1}$. This determines a free action of $\mathbf{Z}_q$ on S^{2m+1} and we define $L(q{:}p_1,\dots,p_{m+1}) = S^{2m+1}/\mathbf{Z}_q$ under this action (see also M.M. Cohen [2, §26]). The lens space $L(q{:}p_1,\dots,p_{m+1})$ can be represented as a CW complex with one cell in each dimension $0 \leq i \leq 2m+1$, such that the skeleton X^{2r+1} is the lens space $L(q{:}p_1,\dots,p_{r+1})$. The cellular chain complex for the universal cover is given as before by modifying the truncation of Lyndon's resolution using the automorphisms $.\langle R, s_r\rangle$ as follows:

$$\mathbb{Z}\mathbf{Z}_q \xrightarrow{.(R^{s_{m+1}}-1)} \mathbb{Z}\mathbf{Z}_q \xrightarrow{.\langle R,q\rangle} \mathbb{Z}\mathbf{Z}_q \xrightarrow{.(R^{s_m}-1)} \dots \to \mathbb{Z}\mathbf{Z}_q \xrightarrow{.(R^{s_2}-1)} \mathbb{Z}\mathbf{Z}_q \xrightarrow{.\langle R,q\rangle}$$

$$\mathbb{Z}\mathbf{Z}_q \xrightarrow{.(R^{s_1}-1)} \mathbb{Z}\mathbf{Z}_q.$$

It is usual to take $s_1=p_1=1$: we then have

$$L(\mathcal{P}_{Z_q}:1,p_2,\dots,p_{m+1}) = L(q:1,p_2,\dots,p_{m+1})\ ;$$

and, in the case where also $m=1$, we have the classic 3–dimensional lens space $L(q:p)=L(q:1,p)$. We note the standard isomorphism $\mathrm{aut}\mathbf{Z}_q \cong \mathbf{Z}_q^*$, given by $\theta \rightsquigarrow \theta(1)$, where $\mathbf{Z}_q^*$ is the group of units in the ring $\mathbf{Z}_q$.

Corollary 6.1

$$\mathcal{E}^*(L(q:p_1,p_2,\dots,p_{m+1})) \cong \{t\in \mathbf{Z}_q^*: t^{m+1} \equiv \pm 1\}\ \ (m\geq 1,\ q\geq 3),$$

and

$$\mathcal{E}^*(\mathbf{R}P^{2m+1}) \cong \mathbf{Z}_2\,.$$

Proof Since $\mathbf{Z}\mathbf{Z}_q.\langle R,q\rangle \cong \mathbf{Z}$, by making the obvious slight modification to the proof of Theorem 5.2, we have easily $\mathcal{E}^*(L(\mathcal{P}_\pi:1,p_2,\dots,p_{m+1})) \cong \mathrm{stab}_k(\mathrm{aut}\mathbf{Z} \rtimes \{t\in \mathbf{Z}_q^*: t^{m+1} \equiv \pm 1\})$, and the result is immediate.

The space X^{2m} is the 2m–skeleton of $L(q:p_1,p_2,\dots,p_{m+1})$, or, equivalently, is obtained by deleting a (suitable) point. We have $\pi_{2m}(X^{2m}) \cong \mathbf{Z}\pi/\mathbf{Z}$ where $\mathbf{Z} \to \mathbf{Z}\pi$ is given by $x \rightsquigarrow x.\langle R,q\rangle$. Let $U(\mathbf{Z}\mathbf{Z}_q/\mathbf{Z})$ be the group of units in the quotient ring. We have the canonical isomorphism

$$U(\mathbf{Z}\mathbf{Z}_q/\mathbf{Z}) \cong \mathrm{aut}_{\mathbf{Z}\mathbf{Z}_q}(\mathbf{Z}\mathbf{Z}_q.(R-1))$$

(see Lemma 6.2). The modq augmen-tation $\varepsilon_q : \mathbf{Z}\mathbf{Z}_q/\mathbf{Z} \to \mathbf{Z}_q$ is given by $\Sigma n_r R^r \rightsquigarrow \Sigma n_r \bmod q$, $U^1(\mathbf{Z}\mathbf{Z}_q/\mathbf{Z})$ is the subgroup of $U(\mathbf{Z}\mathbf{Z}_q/\mathbf{Z})$ consisting of units of augmentation 1, $\gamma_t = \mathbf{Z}\theta/\mathbf{Z}$ is the automorphism of $\mathbf{Z}\mathbf{Z}_q/\mathbf{Z}$ determined by θ where $\theta(R)=R^t$, and $v.x$ is the usual group ring multiplication. Since γ_t is an automorphism for $t\in \mathbf{Z}_q^*$, it does of course preserve units. We also note that the projection $\mathbf{Z}\mathbf{Z}_q \to \mathbf{Z}\mathbf{Z}_q/\mathbf{Z}$ induces the canonical isomorphism $U^1(\mathbf{Z}\mathbf{Z}_q) \cong U^1(\mathbf{Z}\mathbf{Z}_q/\mathbf{Z})$, where $U^1(\mathbf{Z}\mathbf{Z}_q)$ is the group of units having $\mathbf{Z}$–augmentation 1.

Since $\mathbf{Z}_q$ is abelian, there is a ring epimorphism

$$\mathbf{Z}\mathbf{Z}_q \to \mathrm{hom}_{\mathbf{Z}\mathbf{Z}_q}(\mathbf{Z}\mathbf{Z}_q.(R-1),\ \mathbf{Z}\mathbf{Z}_q.(R-1)),$$

given by $\alpha \rightsquigarrow T_\alpha$ where $T_\alpha(x.(R-1)) = x.\alpha.(R-1) = x.(R-1).\alpha$, whose kernel is the two-sided ideal $\langle R,q\rangle.\mathbf{Z}\mathbf{Z}_q$. More generally we consider the composite function

$$\mathbf{Z}_q^* \times \mathbf{Z}\mathbf{Z}_q \to \mathrm{aut}\mathbf{Z}_q \times_\phi \mathrm{aut}\mathbf{Z}\mathbf{Z}_q \to \mathbf{Z}_q^* \times_\phi \mathrm{aut}(\mathbf{Z}\mathbf{Z}_q.(R-1))$$

given by

$$(t,u) \rightsquigarrow (t, 1 \rightsquigarrow u) \rightsquigarrow (t, 1.(R-1) \rightsquigarrow u.\langle R,t\rangle.(R-1)).$$

We have the following Lemma, which provides an alternative way of stating the results in this case where $\pi=Z_q$.

Lemma 6.2. There is the group isomorphism

$$U\{ZZ_q/\langle R,q\rangle.ZZ_q\}\cong \mathrm{aut}_{ZZ_q}(ZZ_q.(R-1));$$

and, in the case where q and s are relatively prime, $\alpha=.\langle R,s\rangle$ determines an element of $\mathrm{aut}_{ZZ_q}(ZZ_q.(R-1))$. More generally we have the isomorphism

$$\{Z_q{}^*\times U\{ZZ_q/\langle R,q\rangle.ZZ_q\},\circ\}\cong \mathrm{aut}Z_q\times_\phi \mathrm{aut}(ZZ_q.(R-1)).$$

Corollary 6.3. In the case where $X= L(q:p_1,\dots,p_{m+1})$ $(m\geq 1)$, we have

$$\mathcal{E}^*(X^{2m}) \cong$$
$$\{(t,1.(R-1)\rightsquigarrow a.(R-1))\in Z_q{}^*\times_\phi \mathrm{aut}(ZZ_q.(R-1)):\ \varepsilon(a)=t^{m+1}\mathrm{mod}q\}$$
$$\cong \mathrm{aut}_{ZZ_q}(ZZ_q.(R-1))\rtimes Z_q{}^*\cong U^1(ZZ_q/Z)\rtimes Z_q{}^*,$$

where the (left) action for the second semi-direct product is given by $t\rightsquigarrow\gamma_t$. In particular $\mathcal{E}^*(RP^{2m})\cong Z_2$. Alternatively we have, in the general case,

$$\mathcal{E}^*(X^{2m}) \cong \{\{(u,t)\in U(ZZ_q/Z)\times Z_q{}^*:\ \varepsilon_q(u)\equiv t^m \mathrm{mod}q\},\circ\},$$

where the (left) composition $\circ$ is given by $(v,s)\circ(u,t)= (v.\gamma_s(u),st)$.

We note that, as m varies, the image of the representation $\mathcal{E}^*(X^{2m})\to Z_q{}^*\times_\phi \mathrm{aut}(ZZ_q.(R-1))$ varies, even though the images have the same naturally-induced semi-direct product form as indicated.

Proof Since $\pi_{2m+1}(\Sigma^{2m+1}/Z_q)\cong Z$ and $\pi_{2m}(X^{2m})\cong ZZ_q/Z$ where $Z\to ZZ_q$ is given by $r\rightsquigarrow r\langle R,q\rangle$, we have by Theorem 4.2,

$$\mathcal{E}^*(X^{2m})\cong \mathrm{stab}_h(\mathrm{aut}Z_q\times_\phi \mathrm{aut}(ZZ_q.(R-1))),$$

where h is the multiplicative identity of the ring $H^{2m+1}{}_\phi(\pi;Z\pi.(R-1))\cong Z_q$. Furthermore the projection $\mathrm{stab}_k(\mathrm{aut}Z_q\times_\phi \mathrm{aut}(ZZ_q.(R-1)))\to \mathrm{aut}Z_q$ is onto, and has a homomorphic partial section

$$\mathrm{aut}Z_q\to \mathrm{stab}_k(\mathrm{aut}Z_q\times_\phi \mathrm{aut}(ZZ_q.(R-1)))$$

given by $\theta\rightsquigarrow(\theta,F)$ where

$$F(y.(R-1))= Z\theta(y).\langle R^{s_1},t\rangle.\langle R^{s_2},t\rangle.\ \dots\ .\langle R^{s_m},t\rangle.\langle R,t\rangle.(R-1).$$

The result now follows from Theorem A.4 of the Appendix and Lemma 6.2.

In the case where m=1, we have $X^2=M_q=S^1\cup_q e^2$, the pseudo-projective plane, which is obtained from a disc by identifying q oriented equal projective plane, which is obtained from a disc by identifying q oriented equal segments making up the boundary S^1 of the disc: thus M_q is the 2–skeleton of the classic 3-dimensional lens spaces $L(q{:}p)$. Define '(left) composition', a twisting of the standard multiplication, in the group ring $\mathbf{ZZ}_q/\mathbf{Z}$ by $\kappa \circ \tau = \kappa . \gamma_{\varepsilon_q(\kappa)}(\tau)$.

Corollary 6.4 The representation of $\mathcal{E}^*(M_q)$ onto $\operatorname{aut}\pi_1(M_q) \cong \operatorname{aut}H_1(M_q) \cong \operatorname{aut}\mathbf{Z}_q \cong \mathbf{Z}_q{}^*$ determines a semi-direct product $\mathcal{E}^*(M_q) \cong U^1(\mathbf{ZZ}_q/\mathbf{Z}) \rtimes \mathbf{Z}_q{}^*$, with action given by $t \rightsquigarrow \gamma_t$. Moreover $\mathcal{E}^*(M_q)$ is isomorphic to the group of units in $(\mathbf{ZZ}_q/\mathbf{Z}, \circ)$.

Remark 6.5. Corollary 6.1 was proved by Olum [11] and [12, 7.4] (see also [19, Theorem 2A]). The results in Corollaries 6.1 and 6.3 for $\mathbf{R}P^n$ were given by Olum [11]. The last isomorphism of Corollary 6.3 was given by Schellenberg [18, Theorem 6.1]) (see also [13, Theorem ii]). Corollary 6.4 was given by Olum [12, Theorems 3.4 and 3.5, and 7.2] using an algebraic method. Rutter [14] subsequently gave a geometrical interpretation of Olum's results using obstruction theory, and indeed determined the sets (M_q, M_u) and the compositions $(M_u, M_w) \times (M_q, M_u) \to (M_q, M_w)$ for $q, u, w \geq 2$: the semi-group structure of (M_q, M_q) is somewhat more complicated than that of $\mathcal{E}^*(M_q)$ since $\theta_{\alpha(\kappa)}$ is not an isomorphism unless multiplication by $\alpha(\kappa)$ is in $\mathbf{Z}_q{}^*$.

Appendix

We include here relevant calculations of the cohomology of one-relator groups and the action upon them of automorphism groups. For any group π and any $\mathbf{Z}\pi$–module A, we define the cohomology $H^n(\pi; A)$ to be $H^n(\operatorname{res}\pi; A) = H^n(\hom_{\mathbf{Z}\pi}(\operatorname{res}\pi, A))$ where $\operatorname{res}\pi$ is a free $\mathbf{Z}\pi$–resolution of the trivial $\mathbf{Z}\pi$–module $\mathbf{Z}$. Up to equivalence, this definition is independent of the choice of resolution, and, for any group, the bar resolution $B(\mathbf{Z}\pi)$ may be used. For finitely presented one-relator groups with torsion the cohomology may be more readily calculated by using Lyndon's resolution. Recall that a $\mathbf{Z}\pi$–chain complex is of finite type if, as a $\mathbf{Z}\pi$–module, it is finitely generated in each dimension.

Let π be a one-relator group with torsion, presented by

$$\mathcal{P}_\pi = \langle x_1, \ldots, x_d \mid R^q \rangle$$

with relator R^q where R is not a proper power, $q>1$ and $R^q \neq e$ in the free group $F^d = \langle x_1, \ldots, x_d \rangle$. In this case we have Lyndon's left $\mathbf{Z}\pi$–module free augmented finite-type resolution of the trivial $\mathbf{Z}\pi$–module $\mathbf{Z}$:

$$\to Z\pi \xrightarrow{.<R,q>} Z\pi \xrightarrow{.(R-1)} Z\pi \xrightarrow{.<R,q>} Z\pi \xrightarrow{.(R-1)} Z\pi \xrightarrow{\partial_2} \{Z\pi\}^d \xrightarrow{\partial_1} Z\pi \xrightarrow{\varepsilon} Z \to 0\,.$$

There is a similar right $Z\pi$–module resolution of Z, with a corresponding chain complex. The cohomology groups $H^r{}_\phi(\pi;Z\pi)$ are zero for $r\geq 3$. We consider the cohomology groups of π with coefficients in the left $Z\pi$–module $Z\pi.(R-1)$ generated by $1.(R-1)$. Applying the functor

$$\hom_{Z\pi}(\ \ ,Z\pi.(R-1))$$

to Lyndon's resolution, we obtain the cochain complex of groups

$$\ldots \leftarrow Z\pi.(R-1) \xleftarrow{<R,q>.} Z\pi.(R-1) \xleftarrow{(R-1).} Z\pi.(R-1) \xleftarrow{\partial_2{}^*} \{Z\pi\}^d.(R-1) \xleftarrow{\partial_1{}^*} Z\pi.(R-1)\,,$$

where $\partial_1{}^* = (x_1-1,\ldots,x_d-1).$ and $\partial_2{}^* = <R,q>.(\partial R/\partial x_1,\ldots,\partial R/\partial x_d).$. Let

$$\begin{aligned} M &= \ker\{<R,q>.:Z\pi.(R-1)\to Z\pi.(R-1)\} \\ &= \ker\{<R,q>.:Z\pi\to Z\pi\} \cap \ker\{.<R,q>:Z\pi\to Z\pi\}, \\ L &= \ker\{(R-1).:Z\pi.(R-1)\to Z\pi.(R-1)\} \\ &= \ker\{(R-1).:Z\pi\to Z\pi\} \cap \ker\{.<R,q>:Z\pi\to Z\pi\}, \text{ and} \\ N &= \ker\{(R-1).:Z\pi.<R,q>\to Z\pi.<R,q>\} \\ &= \ker\{(R-1).:Z\pi\to Z\pi\} \cap \ker\{.(R-1):Z\pi\to Z\pi\}. \end{aligned}$$

Then we have $H^{2k+1}{}_\phi(\pi;Z\pi.(R-1)) \cong M/\{(R-1).Z\pi.(R-1)\}$ $(k\geq 1)$, and, similarly, $H^{2k}{}_\phi(\pi;Z\pi.<R,q>) \cong N/\{<R,q>.Z\pi.<R,q>\}$ $(k\geq 2)$.

Theorem A.1 The cohomology of π with coefficients in the left $Z\pi$–modules $Z\pi.(R-1)$ and $Z\pi.<R,q>$ is given by
i) $H^{2r+1}{}_\phi(\pi;Z\pi.(R-1)) \cong Z_q$ $(r\geq 1)$ induced by the map defined on $\ker\{<R,q>.:Z\pi\to Z\pi\} = (R-1).Z\pi$ by $(R-1).w \rightsquigarrow \varepsilon(w) \bmod q$ (see Remark A2),
ii) $H^{2r}{}_\phi(\pi;Z\pi.(R-1)) = L\,/\,<R,q>.Z\pi.(R-1) = 0$ $(r\geq 2)$,
iii) $H^2{}_\phi(\pi;Z\pi.(R-1)) = L\,/\,<R,q>.(\partial R/\partial x_1,\ldots,\partial R/\partial x_d).\{Z\pi\}^d.(R-1)$,
iv) $H^{2r+1}{}_\phi(\pi;Z\pi.<R,q>) = 0$ $(r\geq 1)$,
v) $H^{2r}{}_\phi(\pi;Z\pi.<R,q>) \cong Z_q$ $(r\geq 2)$ induced by the map defined on $\ker\{<R,q>.:Z\pi\to Z\pi\} = (R-1).Z\pi$ by $(R-1).w \rightsquigarrow \varepsilon(w) \bmod q$ (see Remark A2),

vi) $H^2_\phi(\pi;\mathbf{Z}\pi.\langle R,q\rangle) = N / \langle R,q\rangle.(\partial R/\partial x_1,\dots,\partial R/\partial x_d).\{\mathbf{Z}\pi\}^d.\langle R,q\rangle$.

Proof. Consider the following commutative diagram where all rows and columns are exact except the final column:

$$\begin{array}{ccccccccccccc}
 & & \cdot\langle R,q\rangle & & \cdot(R-1) & & \partial_2 & & \partial_1 & & \varepsilon & & \\
\dots\to & \mathbf{Z}\pi & \to & \mathbf{Z}\pi & \to & \mathbf{Z}\pi & \to & \{\mathbf{Z}\pi\}^d & \to & \mathbf{Z}\pi & \to & \mathbf{Z} & \to 0\\
 & \langle R,q\rangle.\downarrow & & \downarrow & & \downarrow & & \downarrow & & \downarrow & & \downarrow q & \\
\dots\to & \mathbf{Z}\pi & \to & \mathbf{Z}\pi & \to & \mathbf{Z}\pi & \to & \{\mathbf{Z}\pi\}^d & \to & \mathbf{Z}\pi & \to & \mathbf{Z} & \to 0\\
 & (R-1).\downarrow & & \downarrow & & \downarrow & & \downarrow & & \downarrow & & \downarrow 0 & \\
\dots\to & \mathbf{Z}\pi & \to & \mathbf{Z}\pi & \to & \mathbf{Z}\pi & \to & \{\mathbf{Z}\pi\}^d & \to & \mathbf{Z}\pi & \to & \mathbf{Z} & \to 0\\
 & \langle R,q\rangle.\downarrow & & \downarrow & & \downarrow & & \downarrow & & \downarrow & & \downarrow q & \\
\dots\to & \mathbf{Z}\pi & \to & \mathbf{Z}\pi & \to & \mathbf{Z}\pi & \to & \{\mathbf{Z}\pi\}^d & \to & \mathbf{Z}\pi & \to & \mathbf{Z} & \to 0\\
 & (R-1).\downarrow & & \downarrow & & \downarrow & & \downarrow & & \downarrow & & \downarrow 0 &
\end{array}$$

Using the Zig-zag Lemma A.3 below we obtain the isomorphisms:

$$\begin{aligned}
&(\ker\{\langle R,q\rangle.\} \cap \ker\{.\langle R,q\rangle\}) \,/\, \{(R-1).\mathbf{Z}\pi.(R-1)\}\\
&\qquad \cong (\ker\{\langle R,q\rangle.\} \cap \ker\{\varepsilon\}) \,/\, \{(R-1).\mathbf{Z}\pi.(R-1)\} \;.(x_1-1,\dots,x_d-1)\\
&\qquad \cong \mathbf{Z}\pi \,/\, (\ker\{(R-1).\} \cup \ker\{\varepsilon\})\\
&\qquad \cong \mathbf{Z} \,/\, q(\varepsilon(\mathbf{Z}\pi)) \cong \mathbf{Z}_q
\end{aligned}$$

where $\mathbf{Z}\pi \,/\, (\ker\{(R-1).\} \cup \ker\{\varepsilon\}) \to \mathbf{Z}_q$ is a left $\mathbf{Z}\pi$–module homomorphism induced by the evaluation. This proves i). The proofs of ii), iv) and v) are similar.

Remark A.2. The isomorphisms involving ε in i) and v) of Theorem A.1 relate to the isomorphism from the penultimate column (corresponding to dimension zero) to the final column. Of course, if $A.\langle R,q\rangle = \langle R,q\rangle.B$, then $\varepsilon(A) = \varepsilon(B)$. In the case where $\pi = \mathbf{Z}_q$, we note that, under Zig-Zag Lemma A.3, and modulo the appropriate subgroups, each element $w \in \mathbf{ZZ}_q$ is mapped by the identity under the sequence of moves $(R-1).$, $(.(R-1))^{-1}$, $\langle R,q\rangle.$, and $(.\langle R,q\rangle)^{-1}$, for example: thus, in this case, the isomorphism $H^{2r+1}{}_\theta(\mathbf{Z}_q;\mathbf{ZZ}_q.(R-1)) \cong \mathbf{Z}_q$ is given by the map induced on $M = \mathbf{ZZ}_q.(R-1)$, in dimension $2r+1$, by $A.(R-1) \rightsquigarrow \varepsilon(A) \bmod q$, and the isomorphism $H^{2r}{}_\theta(\mathbf{Z}_q;\mathbf{ZZ}_q.\langle R,q\rangle) \cong \mathbf{Z}_q$ is given by the map induced on $N = \mathbf{ZZ}_q.\langle R,q\rangle$, in dimension $2r$, by $A.\langle R,q\rangle \rightsquigarrow \varepsilon(A) \bmod q$. In the general case the isomorphisms to $\mathbf{Z}_q$ induced on the subgroups $\mathbf{ZZ}_q.(R-1)$ and $\mathbf{ZZ}_q.\langle R,q\rangle$ of M and N respectively are given by these same formulae. Thus each cohomology class can be represented by elements of these subgroups. Also in the general case the alternate elements $1 \in \ker\{(\langle R,q\rangle.)\circ(.(R-1))\}$ and

$1 \in \ker\{(.\langle R,q\rangle)\circ((R-1).)\}$ correspond, under zig-zagging, along a rising diagonal in the diagram in the proof of Theorem A.1, except that the appropriate element in dimension 1 is $(\partial R/\partial x_1, \ldots, \partial R/\partial x_d)$, and each maps to the generator 1 of $\mathbf{Z}_q$ under the zig-zag procedure.

Also, for q and s relatively prime, $.\langle R,s\rangle$ is a $\mathbf{Z}\pi$–automorphism of the coefficient group $\mathbf{Z}\pi.(R-1)$ with inverse $.\langle R^s,s'\rangle$ where s' is inverse to s mod q. Notice also the canonical isomorphisms $\mathbf{Z}\pi/(\mathbf{Z}\pi.\langle R,q\rangle) \cong \mathbf{Z}\pi.(R-1)$ and $\mathbf{Z}\pi/(\mathbf{Z}\pi.(R-1)) \cong \mathbf{Z}\pi.\langle R,q\rangle$.

Zig–zag Lemma A.3. Given the commutative diagram of exact sequences of groups and homomorphisms

$$\begin{array}{ccccc} & & P & & \\ & & \downarrow & & \\ Q & \rightarrow & V & \rightarrow & W, \\ & & \downarrow & & \downarrow \\ & & T & \rightarrow & U \end{array}$$

we have that the homomorphisms of the diagram induce the isomorphisms

$$(\mathrm{im}\{V\to T\} \cap \ker\{T\to U\})/\mathrm{im}\{Q\to T\} \cong$$
$$\ker\{V\to U\}/(\mathrm{im}\{P\to V\} \cup \mathrm{im}\{Q\to V\}) \cong$$
$$(\mathrm{im}\{V\to W\} \cap \ker\{W\to U\})/\mathrm{im}\{P\to W\}.$$

In case the fundamental group is finite cyclic and presented in the standard way by $\mathcal{P}_\pi = \langle R \mid R^q \rangle$, Lyndon's resolution takes on a particularly simple form, since $\partial_2 = .\langle R,q\rangle$ and $\partial_1 = .(R-1)$. Also, since the group is finite, we can extend Lyndon's resolution to a complete resolution $\{C_q\}_{q\in\mathbf{Z}}$. The cohomology groups are

$H^{2r-1}{}_\phi(\mathbf{Z}_q;\mathbf{ZZ}_q.(R-1)) \cong \mathbf{Z}_q$, given by $\sum n_i R^i.(R-1) \rightsquigarrow \sum n_i \bmod q$,
$H^{2m}{}_\phi(\mathbf{Z}_q;\mathbf{ZZ}_q.(R-1)) = 0$,
$H^{2r-1}{}_\phi(\mathbf{Z}_q;\mathbf{ZZ}_q.\langle R,q\rangle) = 0$, and
$H^{2r}{}_\phi(\mathbf{Z}_q;\mathbf{ZZ}_q.\langle R,q\rangle) \cong \mathbf{Z}_q$ for $r \geq 1$ (see Remark A.2).

Notice that, in this case, $\mathbf{ZZ}_q/(\mathbf{ZZ}_q.(R-1)) \cong \mathbf{ZZ}_q.\langle R,q\rangle = \ker(.(R-1)) = \mathbf{Z}.\langle R,q\rangle \cong \mathbf{Z}$, and that $\mathbf{ZZ}_q.(R-1) = \ker(.\langle R,q\rangle) \cong \mathbf{ZZ}_q/(\mathbf{ZZ}_q.\langle R,q\rangle)$, regarded as a $\mathbf{Z}$–module, has a (free) basis of dimension $q-1$ consisting of $1.(R-1), R.(R-1), \ldots, R^{q-2}.(R-1)$.

The action of $\mathrm{aut}\,\pi \times_\phi \mathrm{aut}(\mathbf{Z}\pi.(R-1))$ on $H^{2m+1}{}_\phi(\pi;\mathbf{Z}\pi.(R-1)) \cong \mathbf{Z}_q$ can be calculated using the construction f. Using the construction f, defined as above for the standard Lyndon resolution but here for the automorphism $\psi = \theta^{-1}$ rather than θ, we have that the action of $\mathrm{aut}\,\pi \times_\phi \mathrm{aut}(\mathbf{Z}\pi.(R-1))$ on

$$H^{2m+1}_{\theta}(\pi;\mathbf{Z}\pi.(R-1))\cong\mathbf{Z}_q$$

is determined by the endomorphism of $\hom_{Z\pi}(\mathbf{Z}\pi,\mathbf{Z}\pi.(R-1))\cong \mathbf{Z}\pi.(R-1)$ given by

$$\begin{aligned}(\theta,\lambda).\{a.(R-1)\} &= \mathbf{Z}\theta(f_{2m+1}(1).a).b.(R-1)\\ &= \mathbf{Z}\theta(v.\langle R,s\rangle^{m+1}.a).b.(R-1),\end{aligned}$$

where $b.(R-1) = \lambda(1.(R-1))$, $\Psi(R)= LVR^sV^{-1}$ and $st\equiv 1 \bmod q$. In the case where $V=1$, it follows that the element $1.(R-1)$, corresponding to the generator of $\mathbf{Z}_q$, is mapped to $\langle R^t,s\rangle^{m+1}.b.(R-1)$. Theorem A.4 below now follows using Remark A.3. The proof of Theorem A.5 is similar.

Theorem A.4 The action of $\operatorname{aut}\mathbf{Z}_q\times_{\phi}\operatorname{aut}(\mathbf{Z}\mathbf{Z}_q.(R-1))$ on

$$H^{2m+1}_{\phi}(\mathbf{Z}_q;\mathbf{Z}\mathbf{Z}_q.(R-1))\cong \mathbf{Z}_q \quad (m\geq 1)$$

is given by

$$(\theta,\lambda).(1)\equiv t^{-(m+1)}\varepsilon(b) \bmod q,$$

where $\theta(R)= R^t$ and $\lambda.(1.(R-1))= b.(R-1)$.

Theorem A.5 The action of $\operatorname{aut}\mathbf{Z}_q\times_{\phi}\operatorname{aut}(\mathbf{Z}\mathbf{Z}_q.\langle R,q\rangle) = \operatorname{aut}\mathbf{Z}_q\times_{\phi}\operatorname{aut}\mathbf{Z}$ on

$$H^{2m+2}_{\phi}(\mathbf{Z}_q;\mathbf{Z}\mathbf{Z}_q.\langle R,q\rangle)\cong \mathbf{Z}_q \quad (m\geq 0)$$

is given by $(\theta,\lambda).(1)\equiv t^{-(m+1)}\varepsilon(b)\equiv \pm\, t^{-(m+1)} \bmod q$, where $\theta(R)= R^t$ and $\lambda.(1.\langle R,q\rangle)= b.\langle R,q\rangle$.

The corresponding actions on the cohomology in the case of a general one-relator group with torsion are also indicated above.

References

[1] Cockroft W. H., On two dimensional aspherical complexes, Proc. London Math. Soc. 4 (1954) 375-384.

[2] Cohen M.M., A course in simple homotopy theory, Springer-Verlag, 1973.

[3] Dyer M. N., On the second homotopy module of two-dimensional CW complexes, Proc. Amer. Math. Soc. 55 (1976) 400-404.

[4] Dyer M. N., Projective k-invariants, Comm. Math. Helv. 51 (1976) 259-277.

[5] Eilenberg S. and MacLane S., Homology of spaces with operators II, Trans. Amer. Math. Soc. 65 (1949) 49-99.

[6] Fox R.H., Free differential calculus, Ann. Math. 57 (1953) 547-560.

[7] Jajodia S., On 2-dimensional CW-complexes with a single 2-cell, Pacific J. Math. 80 (1979) 191-203.

[8] Jajodia S., Homotopy classification of lens spaces for one-relator groups with torsion, Pacific J. Math. 89 (1980) 301-311.

[9] Karass A., Magnus W., and Solitar D., Elements of finite order in groups with a single defining relation, Comm. Pure Appl. Math. 13 (1960) 57-66.

[10] Magnus W., Karass A., and Solitar D., Combinatorial group theory, 2nd revised edition, Dover New York (1976).

[11] Olum P., Mappings of manifolds and the notion of degree, Ann. of Math. 58 (1953) 458-480.

[12] Olum P., Self-equivalences of pseudo-projective planes, Topology 4 (1965) 109-127.

[13] Plotnick S., Homotopy equivalence and free modules, Topology 21 (1982) 91-99.

[14] Rutter J. W., Homotopy classification of maps between pseudo-projective planes, Quaestiones Mathematicae, 11 (1988) 409-422.

[15] Rutter J. W., Whitney-sums (fibre-joins) in over space theory and obstruction theory for cohomology with local coefficients, Proc. Royal Soc. Edinburgh 115A (1990) 359-365.

[16] Rutter J. W., The group of homotopy self-equivalences of non-simply-connected spaces using Postnikov decompositions, Proc. Royal Soc. Edinburgh 120A (1992) 47-60.

[17] Rutter J. W., The group of homotopy self-equivalences of non-simply-connected spaces using Postnikov decompositions II, Proc. Royal Soc. Edinburgh 122A (1992) 127-135.

[18] Schellenberg B., The group of homotopy self-equivalences of some compact CW-complexes, Math. Ann. 200 (1973) 253-266.

[19] Smallen D., The group of self-equivalences of certain complexes, Pacific J. Math. 54 (1974) 269-276.

[20] Stallings J. R., Groups of cohomological dimension one, Ann. Math. 88 (1968) 312-334.

[21] Swan R. G., Groups of cohomological dimension one, J. Algebra 12 (1969) 585-610.

[22] Whitehead J.H.C.W., Combinatorial homotopy II, Bull. Amer. Math. Soc. 55 (1949) 453-496.

Division of Pure Mathematics,
Liverpool University,
Liverpool L69 3BX, England.

Contemporary Mathematics
Volume **274**, 2001

Principal S^1-bundles and forgetful maps

H. Shiga, K. Tsukiyama and T. Yamaguchi

ABSTRACT. We consider the forgetful map for principal S^1-bundles. It is shown that there are examples that forgetful maps are not monomorphisms.

1. Introduction

Let (P, q, B, G) be a principal G-bundle over the base space B. Then the topological group G acts on the total space P freely and $P/G = B$.

Let $aut_G(P)$ be the space of unbased G-equivariant self homotopy equivalences of P and $aut(B)$ be the space of unbased self homotopy equivalences of B. Then every element of $aut_G(P)$ induces a homotopy equivalence on the base space B.

We have the following commutative diagram

$$\begin{array}{ccccc} P & \xrightarrow{f} & P & \longrightarrow & E_G \\ {\scriptstyle q}\downarrow & & \downarrow{\scriptstyle q} & & \downarrow{\scriptstyle q_G} \\ B & \xrightarrow{\bar{f}} & B & \xrightarrow{k} & B_G \end{array}$$

and we have a Serre fibration (see[**1**],[**2**])

$$\Phi : aut_G(P) \to aut_k(B); \Phi(f) = \bar{f}, \tag{1.1}$$

where $aut_k(B) = \{\bar{f} \in aut(B); k\bar{f} \simeq k\}$ and k is a classifying map of the given principal G-bundle and (E_G, q_G, B_G, G) is a universal principal G-bundle.

Let

$$\mathcal{F}_G(P) = \pi_0(aut_G(P)) \tag{1.2}$$

$$\mathcal{F}(P) = \pi_0(aut(P)). \tag{1.3}$$

Then we consider the natural map

$$F : \mathcal{F}_G(P) \to \mathcal{F}(P); F([f]_G) = [f], \tag{1.4}$$

which forgets the G-action on P. We call this map a forgetful map(see [**7**]).

The sufficient condition for the injectivity of F is studied in [**7**], when the structure group of the given bundle is an Eilenberg-MacLane space $K(\pi, n) = G$.

1991 *Mathematics Subject Classification.* Primary 55P10; Secondary 55R10.

Key words and phrases. forgetful map, equivariant self homotopy equivalence, self homotopy equivalence, principal G-bundle.

Especially for the case that $G = K(Z,1) = S^1$, it is shown that $\pi_1(B) = 0$ and $\mathcal{E}_\#(B) = 1$ are sufficient conditions for the injectivity of F, by using the homotopy exact sequence of the above fibration (1.1) (see [**7**, Corollary 3.5]).

The following problem was posed in [**7**, Problem 1]:

For any principal S^1-bundle (P, q, B, S^1) with $\pi_1(B) = 0$, is the forgetful map

$$F : \mathcal{F}_{S^1}(P) \to \mathcal{F}(P)$$

a monomorphism?

In this note, we shall consider this problem. We show that there are three principal S^1-bundles where $KerF$ are not zero but finite, countable and uncountable, respectively.

The authors would like to thank the referee for kind comments.

2. Examples

We shall consider the product bundle

$$(S^3 \times Y, q \times id, S^2 \times Y, S^1) \tag{2.1}$$

of the Hopf bundle (S^3, q, S^2, S^1) and the trivial bundle $(Y, id, Y, *)$, where Y is a simply connected finite CW complex. We consider the forgetful map

$$F : \mathcal{F}_{S^1}(S^3 \times Y) \to \mathcal{F}(S^3 \times Y). \tag{2.2}$$

THEOREM 2.3. *Let $(S^3 \times Y, q \times id, S^2 \times Y, S^1)$ be the principal S^1-bundle as above. Then there is a space Y such that $KerF$ is countable.*

PROOF. Since $\mathcal{F}_{S^1}(S^3) = \pi_0(aut_{S^1}(S^3)) = 1$ ([**6**, Example 3.4]) and the structure group is S^1, we have a fibration of (1.1) associated with the above Hopf bundle(see [**1**, p.47])

$$map(S^2, S^1) \to aut_{S^1}(S^3) \xrightarrow{\Phi} aut(S^2)_{id}. \tag{2.4}$$

Set $Y = S^5 \times S^7$. Then $\pi_2(aut(Y)_{id}) \otimes Q = Q$ and $\pi_3(aut(Y)_{id}) \otimes Q = 0$ (cf. [**4**, Theorem 2]) and we have

$$\pi_0(aut(S^3 \times Y)) \cong \pi_0(map(Y, aut(S^3))) \times \pi_0(map(S^3, aut(Y)))$$

and

$$\pi_0(aut_{S^1}(S^3 \times Y)) \cong \pi_0(map(Y, aut_{S^1}(S^3))) \times \pi_0(map_{S^1}(S^3, aut(Y))).$$

Let $p_1 : S^3 \times Y \to S^3$ and $p_2 : S^3 \times Y \to Y$ be projections and put

$$T = \{f \in aut_{S^1}(S^3 \times Y) \mid [p_1 \circ f] = [p_1] \in \pi_0(map(Y, aut_{S^1}(S^3)_{id}), [p_2 \circ f] \in \pi_0(map_{S^1}(S^3, aut(Y)_{id})\},$$

LEMMA 2.5. *$KerF \subset \pi_0(T)$ and if $[f] \in \pi_0(T)$, there is an integer n such that $[f^n] \in KerF$.*

PROOF. Since $F([f]_G) = id$ if and only if $[p_1 \circ F(f)] = [p_1]$, $[p_2 \circ F(f)] = [p_2]$, $KerF \subset \pi_0(T)$. For $f \in T$, $[p_2 F(f^n)] = n[p_2 F(f)] \in \pi_3(aut(Y)_{id})$. Since $\pi_3(aut(Y)_{id}) \otimes Q = 0$, there is an integer n such that $[f^n] \in KerF$. □

Put

$$\overline{T} = \{\overline{f} \in aut(S^2 \times Y) \mid [\overline{p}_1 \circ \overline{f}] = [\overline{p}_1] \in \pi_0(map(Y, aut(S^2)_{id}), [\overline{p}_2 \circ \overline{f}] \in \pi_0(map(S^2, aut(Y)_{id})\},$$

where $\overline{p}_1 : S^2 \times Y \to S^2$ and $\overline{p}_2 : S^2 \times Y \to Y$ are projections.

For the fibration of (1.1) associated with (2.1)(see [**1**, p.47])

$$map(S^2 \times Y, S^1) \to aut_{S^1}(S^3 \times Y) \xrightarrow{\Phi} aut_k(S^2 \times Y), \tag{2.6}$$

we have

LEMMA 2.7. $\Phi_*(\pi_0(T)) = \pi_0(\overline{T})$.

PROOF. First we show $\Phi_*(\pi_0(T)) \subset \pi_0(\overline{T})$. For $f \in T$, there are commutative diagrams

$$\begin{array}{ccc} S^3 \times Y & \xrightarrow{p_1 \circ f} & S^3 \\ {\scriptstyle q\times id}\downarrow & & \downarrow{\scriptstyle q} \\ S^2 \times Y & \xrightarrow{\overline{p}_1 \circ \Phi(f)} & S^2 \end{array}$$

$$\begin{array}{ccc} S^3 \times Y & \xrightarrow{p_2 \circ f} & Y \\ {\scriptstyle q\times id}\downarrow & & \downarrow{\scriptstyle id} \\ S^2 \times Y & \xrightarrow{\overline{p}_2 \circ \Phi(f)} & Y, \end{array}$$

where q is the Hopf map. Then $[\overline{p}_1 \circ \Phi(f)] = [\overline{p}_1]$ and $[\overline{p}_2 \circ \Phi(f)] \in \pi_0(S^2, aut(Y)_{id})$ by these commutative diagrams. Hence $\Phi_*[f] = [\Phi(f)] = \Phi([p_1 \circ f], [p_2 \circ f]) = ([\overline{p}_1 \circ \Phi(f)], [\overline{p}_2 \circ \Phi(f)]) \in \pi_0(\overline{T})$.

Conversely, let $\overline{f} \in \overline{T}$. Since $[\overline{p}_1 \circ \overline{f}] = [\overline{p}_1] \in \pi_0(map(Y, aut(S^2)_{id}))$ and since there is the lifting $p_1 \in map(Y, aut_{S^1}(S^3))$ of $\overline{p}_1$ for the fibration (2.4), there is a lifting $g \in map(Y, aut_{S^1}(S^3))$ of $\overline{p}_1\overline{f}$ such that $[g] = [p_1]$. Set $[f] = ([g], [\overline{p}_2 \circ \overline{f} \circ (q \times id)])$, then $\Phi_*([f]) = ([\overline{p}_1 \circ \overline{f}], [\overline{p}_2 \circ \overline{f}]) = [\overline{f}]$. □

By the definition,

$$\pi_0(\overline{T}) \cong \pi_0(map(S^2, aut(Y)_{id}) \cong \pi_2(aut(Y)_{id}).$$

Then we can choose the element $\overline{f}$ of $\overline{T}$ such that $[\overline{p}_2 \circ \overline{f}] \otimes 1 \neq 0$ in $\pi_2(aut(Y)_{id}) \otimes Q$. From Lemma 2.7 there is an element $[f]$ of $\pi_0(T)$ such that $\Phi([f]) = [\overline{f}]$. From Lemma 2.5, there is an integer n such that $[f^n] \in KerF$. Remark that $[\Phi(f^{kn})] = kn[\overline{p}_2 \circ \overline{f}]$ in $\pi_2(aut(Y)_{id})$. Therefore $\{[f^{kn}]; k = 1, 2, \cdots\}$ is a countable infinite subset of $KerF$.

On the other hand, $KerF$ is countable, since

$$KerF \subset \pi_0(T) \cong \pi_0(\overline{T}) \cong \pi_2(aut(Y)_{id}) \subset [S^2 \times Y, Y],$$

where the right hand is at most countable ([**5**, Theorem 10.2]). This completes the proof of Theorem 2.3.

□

REMARK 2.8. If $Y = S^7$ in the theorem of above, $KerF$ is finite.

Next we shall consider the product bundle

$$(ES^1 \times X, q_{S^1} \times id, BS^1 \times X, S^1) \tag{2.9}$$

of the universal S^1-bundle $(ES^1, q_{S^1}, BS^1, S^1)$ and the trivial bundle $(X, id, X, *)$.

We consider the forgetful map

$$F : \mathcal{F}_{S^1}(ES^1 \times X) \to \mathcal{F}(ES^1 \times X). \tag{2.10}$$

THEOREM 2.11. *Let* $(ES^1 \times X, q_{S^1} \times id, BS^1 \times X, S^1)$ *be the principal* S^1*-bundle as above. Then there are three spaces for* X *such that* $KerF$ *are finite, countable and uncountable, respectively.*

PROOF. Let $X = K(H,m)(m > 1)$. $BS^1 = K(Z,2)$. By [**3**, Example 1],

$$\mathcal{F}(K(Z,2) \times K(H,m)) \cong \begin{bmatrix} Z_2 & H^m(K(Z,2),H) \\ 0 & autH \end{bmatrix}.$$

By ([**2**], [**6**]), $\mathcal{F}_{S^1}(ES^1 \times X) = \pi_0(aut_k(BS^1 \times X)) = \mathcal{F}_k(K(Z,2) \times K(H,m))$, where the classifying map k is the projection $BS^1 \times X \to BS^1$. It is easy to see that

$$\mathcal{F}_k(K(Z,2) \times K(H,m)) \cong \begin{bmatrix} 1 & H^m(K(Z,2),H) \\ 0 & autH \end{bmatrix}.$$

Since $\mathcal{F}(ES^1 \times X) = \mathcal{F}(K(H,m)) = autH$, we have

$$KerF \cong H^m(K(Z,2),H). \tag{2.12}$$

So, for even m, $KerF$ is finite if H is a finite abelian group and $KerF$ is countable if H is the group of integers.

Next let $X = S^3$ and consider the following product bundle

$$(ES^1 \times S^3, q_{S^1} \times id, BS^1 \times S^3, S^1). \tag{2.13}$$

By [**8**, Th. D], $[K(Z,2),S^3] = \mathbf{R}$. Define $\bar{f} : K(Z,2) \times S^3 \to K(Z,2) \times S^3$ by $\bar{f}(x,y) = (x, f(x)y)$ for $f \in [K(Z,2),S^3]$. Then we see that $[\bar{f}] \neq [\bar{g}]$ if $[f] \neq [g]$. Thus $\mathcal{F}_k(K(Z,2) \times S^3)$ contains $\mathbf{R}$. Since $\mathcal{F}_{S^1}(ES^1 \times S^3)$ is isomorphic to $\mathcal{F}_k(BS^1 \times S^3) = \mathcal{F}_k(K(Z,2) \times S^3)$, it is uncountable. On the other hand $\mathcal{F}(ES^1 \times S^3) \cong \mathcal{F}(S^3) = Z_2$ since ES^1 is contractible. Thus $KerF$ is uncountable. So we have an example that $KerF$ is uncountable, where the total space of a given principal S^1-bundle has the homotopy type of a finite CW complex.

□

References

[1] I. M. James, The space of bundle maps, Topology 2(1963), 45-59.

[2] H. Ōshima and K. Tsukiyama, On the group of equivariant self equivalences of free actions, Publ. Res. Inst. Math. Sci. Kyoto Univ. 22(1986), 905-923.

[3] A. J. Sieradski, Twisted self-homotopy equivalences, Pacific J. of Math. 34(1970), 789-802.

[4] S. B. Smith, Rational homotopy of the space of self-maps of complexes with finitely many homotopy groups, Trans. Amer. Math. Soc. 342(1994), 895-915.

[5] D. Sullivan, Infinitesimal computations in topology, Publ. I. H. E. S. 47(1978), 269-331.

[6] K. Tsukiyama, Equivariant self equivalences of principal fibre bundles, Math. Proc. Camb. Phil. Soc. 98(1985), 87-92.

[7] K. Tsukiyama, Equivariant homotopy equivalences and a forgetful map, Bull. Korean. Math. Soc. 36(1999), 649-654.

[8] A. Zabrodsky, On phantom maps and a theorem of H. Miller, Israel J. Math. 58(1987), 129-143.

DEPARTMENT OF MATHEMATICS, UNIVERSITY OF RYUKYUS, NISHIHARA, OKINAWA 903-01, JAPAN

E-mail address: shiga@sci.u-ryukyu.ac.jp

Department of Mathematics, Shimane University, Matsue, Shimane 690-8504, Japan
E-mail address: tukiyama@edu.shimane-u.ac.jp

University Education Center, University of Ryukyus, Nishihara, Okinawa 903-01, Japan
E-mail address: g993002@sci.u-ryukyu.ac.jp

Contemporary Mathematics
Volume **274**, 2001

Rational Type of Classifying Spaces for Fibrations

Samuel Bruce Smith

October 23, 2000

Abstract

We compute the rational cohomology, the rational homotopy Lie algebra and determine the rational homotopy type of the classifying space $Baut_1(X)$ for certain formal spaces X.

1. Introduction.

Let X be a simply connected CW complex of finite type and $aut_1(X)$ the identity component of the space of self-homotopy equivalences of X. The Dold-Lashof classifying space, $Baut_1(X)$, is the classifying space for orientable fibrations with fibre the homotopy type of X [2, 17, 1]. The cohomology of $Baut_1(X)$ thus gives characteristic classes for X-fibrations.

In Appendix 1 of [9], Milnor computed $H^*(Baut_1(S^n), \mathbb{Q})$ to be a polynomial algebra with a single positive degree generator. Thus $Baut_1(S^n)$ is a rational H-space. Milnor's result generalizes to products of even dimensional spheres. In fact, by a result due to Meier [8] and Thomas [19], $Baut_1(X)$ is a rational H-space whenever X is an F_0-space (see §2) whose rational cohomology admits only trivial negative degree derivations (Theorem 1, below).

When X is a product of odd spheres, X is itself a rational H-space. In this case, the rational homotopy type of $Baut_1(X)$ can be deduced from Sullivan's differential graded Lie algebra model for the classifying space [18] (see Theorem 2). To calculate $H^*(Baut_1(X), \mathbb{Q})$ for X a finite mixed product of spheres with no restrictions appears quite difficult. (See Example 5.2, below). Our purpose in this paper is to take a step in this direction. We determine the full rational homotopy type of the space $Baut_1(F)$ when F splits as a product of an F_0-space with no negative rational cohomology derivations and a rational H-space, under certain connectivity assumptions. In particular, we compute $H^*(Baut_1(F), \mathbb{Q})$ under these hypotheses and determine precisely when $Baut_1(F)$ is a formal space.

2. Main Results.

Let $\mathcal{A}$ be a simply connected graded algebra over $\mathbb{Q}$. Define a degree n derivation θ of $\mathcal{A}$ to be a linear self-map satisfying $\theta(\mathcal{A}^k) \subseteq \mathcal{A}^{k-n}$ and $\theta(xy) = \theta(x)y + (-1)^{n|x|}x\theta(y)$. The Lie bracket is the graded commutator: $[\theta_1, \theta_2] = \theta_1 \circ \theta_2 - (-1)^{|\theta_1||\theta_2|}\theta_2 \circ \theta_1$. Let $Der_+(\mathcal{A})$ denote the graded Lie algebra of all positive degree derivations of $\mathcal{A}$.

2000 *Mathematics Subject Classification.* Primary 55P62, 55P15.
Key words and phrases. Derivations; Coformality; Halperin's conjecture

If $\mathcal{A}$ factors, as graded algebra, in the form $\mathcal{A} = \mathcal{B} \otimes \mathcal{C}$ for simply connected graded algebras $\mathcal{B}$ and $\mathcal{C}$, we let $Der_+(\mathcal{C}, \mathcal{A})$ denote the subalgebra of $Der_+(\mathcal{A})$ consisting of all derivations of $\mathcal{A}$ which vanish on $\mathcal{B} \otimes 1$. We use this notation below in the case $\mathcal{A} = H^*(X \times Y, \mathbb{Q})$, $\mathcal{B} = H^*(X, \mathbb{Q})$, $\mathcal{C} = H^*(Y, \mathbb{Q})$. The factorization $\mathcal{A} = \mathcal{B} \otimes \mathcal{C}$ is just the rational Künneth Theorem.

Let L be a connected graded rational Lie algebra. The Lie cohomology of L, $H^*_{\text{Lie}}(L)$, is defined to be cohomology of the differential graded algebra $(\Lambda(sL), d_{[\ ,\ ]})$, where $\Lambda(sL)$ is the free graded algebra generated by the suspension of L and $d_{[\ ,\ ]}$ is dual to the Lie bracket [12, 10, 13]. In [12], Quillen proved that there exists a simply connected space $\|L\|$ satisfying $\pi_*(\Omega\|L\|) \otimes \mathbb{Q} \cong L$ and $H^*(\|L\|, \mathbb{Q}) \cong H^*_{\text{Lie}}(L)$. Specifically, let $(\mathcal{L}, \partial)$ be the minimal Lie algebra model for $(L, 0)$ and take $\|L\|$ to be the spatial realization of $(\mathcal{L}, \partial)$. By construction, $\|L\|$ is a *coformal* space; that is, the rational homotopy type of $\|L\|$ is a formal consequence of its rational homotopy Lie algebra.

A simply connected space X is an F_0-*space* if X has finite-dimensional rational homotopy and cohomology with

$$dim\pi_{even}(X) \otimes \mathbb{Q} = dim\pi_{odd}(X) \otimes \mathbb{Q} \quad \text{and} \quad H^{odd}(X, \mathbb{Q}) = 0.$$

The class includes (products of) spheres, complex projective spaces and homogeneous spaces G/H with $\text{rank}G = \text{rank}H$. In [5], Halperin conjectured $Der_+(H^*(X, \mathbb{Q})) = 0$ for all F_0-spaces X. This condition is equivalent to the collapsing of the rational Serre spectral sequence for all orientable fibrations with fibre X [8]. Halperin's conjecture has been confirmed in many special cases [6, 7, 14]; in particular, for the examples mentioned above.

Given a graded vector space Z, let $min(Z) = min\{n | Z^n \neq 0\}$ and $max(Z) = max\{n | Z^n \neq 0\}$. We prove the following results on the rational homotopy type of classifying spaces in Section 3:

Theorem 1 *Let X be an F_0-space with $Der_+(H^*(X, \mathbb{Q})) = 0$. Define integers d_{2n} for $n = 1, 2, \ldots,$ by*

$$d_{2n} = \sum_{k \geq n} dim(H^{2k-2n}(X, \mathbb{Q}))\,[dim(\pi_{2k-1}(X) \otimes \mathbb{Q}) - dim(\pi_{2k}(X) \otimes \mathbb{Q})].$$

Then

$$Baut_1(X) \simeq_{\mathbb{Q}} \prod_n K(\mathbb{Q}^{d_{2n}}, 2n).$$

Corollary 1 *Let X be as in Theorem 1. Let $\mathcal{L}_1(X)$ denote the oddly graded rational vector space having dimension d_{2n} in degree $2n-1$. View $\mathcal{L}_1(X)$ as an abelian Lie algebra. Then $Baut_1(X)$ is formal and coformal with*

$$\pi_*(\Omega Baut_1(X)) \otimes \mathbb{Q} \cong \mathcal{L}_1(X) \quad \text{and} \quad H^*(Baut_1(X), \mathbb{Q}) \cong \Lambda(s\mathcal{L}_1(X)).$$

Theorem 2 *Let Y be a simply connected rational H-space of finite type. Then $Baut_1(Y) \simeq_{\mathbb{Q}} \|Der_+(H^*(Y, \mathbb{Q}))\|$.*

Corollary 2 *Let Y be as in Theorem 2. Then $Baut_1(Y)$ is coformal with*

$$\pi_*(\Omega Baut_1(Y)) \simeq_{\mathbb{Q}} Der_+(H^*(Y, \mathbb{Q})) \quad \text{and}$$

$$H^*(Baut_1(Y), \mathbb{Q}) \cong H^*_{\text{Lie}}(Der_+(H^*(Y, \mathbb{Q}))).$$

Theorem 3 *Let $F = X \times Y$ where X and Y are as in Theorems 1 and 2, respectively. Suppose that*

$$min(\pi_*(X) \otimes \mathbb{Q}) + min(\pi_*(Y) \otimes \mathbb{Q}) \geq max(\pi_*(F) \otimes \mathbb{Q}) \quad and$$

$$max(\pi_*(X) \otimes \mathbb{Q}) \leq min(\pi_*(Y) \otimes \mathbb{Q}).$$

Then

$$Baut_1(F) \simeq_{\mathbb{Q}} Baut_1(X) \times \|Der_+(H^*(Y,\mathbb{Q}), H^*(F,\mathbb{Q}))\|.$$

Corollary 3 *Let F be as in Theorem 3. Then $Baut_1(F)$ is a coformal space with*

$$\pi_*(\Omega Baut_1(F)) \otimes \mathbb{Q} \cong \mathcal{L}_1(X) \times Der_+(H^*(Y,\mathbb{Q}), H^*(F,\mathbb{Q})) \quad and$$

$$H^*(Baut_1(F),\mathbb{Q}) \cong \Lambda(s\mathcal{L}_1(X)) \otimes H^*_{\text{Lie}}(Der_+(H^*(Y,\mathbb{Q}), H^*(F,\mathbb{Q}))).$$

In Section 4, we study the formality of spatial realizations of finite-dimensional graded Lie algebras. We obtain:

Theorem 4 *Let Y be a simply connected rational H-space with finite-dimensional rational homotopy. Then $Baut_1(Y)$ is formal if and only if*

i) $\pi_*(Y) \otimes \mathbb{Q}$ is concentrated in a single degree or

ii) $\pi_*(Y) \otimes \mathbb{Q}$ is concentrated in two degrees p and q with $p < q \leq 2p$, p odd, q even and $dim(\pi_p(Y) \otimes \mathbb{Q}) \leq dim(\pi_q(Y) \otimes \mathbb{Q})$ or

iii) $\pi_*(Y) \otimes \mathbb{Q}$ is concentrated in two degrees p and q with $p < q$, p odd, q even and $dim(\pi_p(Y) \otimes \mathbb{Q}) = 1$.

Theorem 5 *Let $F = X \times Y$ be as in Theorem 3. Then $Baut_1(F)$ is formal if and only if Y satisfies condition i), ii or iii) above.*

3. Rational Homotopy of Classifying Spaces for Fibrations.

A connected differential graded Lie algebra (*dgla*) (L, ∂) is a *model* for a simply connected complex X if the spatial realization of the Quillen minimal model for (L, ∂) is rationally equivalent to X. In this case, in particular, $H(L, \partial)$ is the rational homotopy Lie algebra of X. The following result is just a restatement of the definition of coformality:

Lemma 3.1 *Let X be a simply connected complex. Then X is coformal if and only if X has a dgla model (L, ∂) with $\partial = 0$.* □

In [18], Sullivan describes a dgla model for $Baut_1(X)$ when X is a simply connected complex of finite type. Let $(\mathcal{M}_X, d_X)$ be the Sullivan minimal model for X. Define a degree -1 derivation ∂_X of $Der_+(\mathcal{M}_X)$ by $\partial_X(\theta) = [d_X, \theta]$. Define a subalgebra $\widetilde{Der}_+(\mathcal{M}_X)$ of $Der_+(\mathcal{M}_X)$ by restricting, in degree 1, to derivations of $\mathcal{M}_X$ which vanish under ∂_X. The pair $(\widetilde{Der}_+(\mathcal{M}_X), \partial_X)$ is a dgla model for $Baut_1(X)$. For an indirect proof, use [4, Theorem 2] and the proof for the Schlessinger-Stasheff model [13].

Proof of Theorem 2. Since Y is a rational H-space, its Sullivan minimal model is just $(H^*(Y,\mathbb{Q}), 0)$. In the Sullivan dgla model for $Baut_1(Y)$, $\partial_X = 0$. The result follows from Lemma 3.1. □

We next recall some notation and results from [16]. We say a simply connected space X is *two-stage* if its Sullivan minimal model $(\mathcal{M}_X, d_X)$ can be written $\mathcal{M}_X = \Lambda(V_0) \otimes \Lambda(V_1)$, for finite-dimensional graded vector spaces V_0 and V_1 with $d_X(V_0) = 0$ and $d_X(V_1) \subseteq \Lambda(V_0)$. We may assume $d_X : V_1 \to \Lambda(V_0)$ is an injection. Write $V_0 = \mathbb{Q}(x_1, \ldots, x_m)$ and $V_1 = \mathbb{Q}(y_1, \ldots, y_n)$ so that $d_X(x_i) = 0$ and $d_X(y_j) = R_j(x_1, \ldots, x_m)$, a nontrivial polynomial without linear term in the x_i. By [3], X is formal if and only if the sequence $R_1, \ldots, R_n$ is regular in $\Lambda(V_0)$.

Define graded spaces $L_0(X)$ and $L_1(X)$ by setting

$$L_i^n(X) = \bigoplus_{k \geq 0} H^k(X, \mathbb{Q}) \otimes V_i^{n+k},$$

where $n > 0$ for $i = 0$ and $n \geq 0$ for $i = 1$. Let $L(X) = L_0(X) \oplus L_1(X)$. Thus $L(X)$ is spanned by elements $\alpha \otimes z_k$ where $z_k \in V_0 \oplus V_1$ is one of our basis vectors and $\alpha \in H^*(X, \mathbb{Q})$ is homogeneous of degree no more than $|z_k|$. The degree of $\alpha \otimes z_k$ is the difference $|z_k| - |\alpha|$.

Define a linear map $D : L_0(X) \to L_1(X)$ of degree -1 by

$$D(\alpha \otimes x_i) = \sum_{j=1}^{n} \alpha \cdot \left\{ \frac{\partial R_j}{\partial x_i} \right\} \otimes y_j,$$

where $\{P\}$ denotes the cohomology class in $H^*(X, \mathbb{Q})$ represented by an element $P \in \mathcal{M}_X$. Let

$$\mathcal{L}_0(X) = ker\{D : L_0(X) \to L_1(X)\} \text{ and}$$
$$\mathcal{L}_1(X) = cok\{D : L_0(X) \to L_1(X)\},$$

where $\mathcal{L}_1(X)$ is made connected by eliminating its elements of degree zero. By [16, Theorem 3.2], there is an isomorphism of graded vector spaces

$$\pi_*(\Omega Baut_1(X)) \otimes \mathbb{Q} \cong \mathcal{L}_0(X) \oplus \mathcal{L}_1(X)$$

for any two-stage space X.

Proof of Theorem 1. Let X be an F_0-space. Then, in the notation above, V_0 is evenly graded, V_1 is oddly graded and $\dim V_0 = \dim V_1$. Since $H^*(X, \mathbb{Q})$ is evenly graded, $L_0(X)$ is evenly graded and $L_1(X)$ is oddly graded. By hypothesis, X satisfies Halperin's conjecture. Meier [8] and Thomas [19] independently proved that Halperin's conjecture is true for an F_0-space X if and only if the space $Baut_1(X)$ has evenly graded rational homotopy groups. Thus, $\mathcal{L}_0(X) = 0$. The result now follows from the observation that any space with evenly graded rational homotopy is a rational H-space. □

Proof of Theorem 3. Let W be a finite-dimensional graded rational vector space such that $H^*(Y, \mathbb{Q}) \cong \Lambda(W)$. Write $W = \mathbb{Q}(w_1, \ldots w_k)$.

Repeat the above constuction for F. The Sullivan minimal model $(\mathcal{M}_F, d_F)$ is $\mathcal{M}_F = \Lambda(Z_0) \otimes \Lambda(Z_1)$ where $Z_0 = W \oplus V_0$, $Z_1 = V_1$, $d_F(Z_0) = 0$ and $d_F|Z_1 = d_X|V_1$. Computing the image of $D : L_0(F) \to L_1(F)$ directly, gives

$$\mathcal{L}_1(F) \cong \mathcal{L}_1(X) \oplus \bigoplus_{n>0} \bigoplus_{k>0} H^k(Y, \mathbb{Q}) \otimes V_1^{n+k}.$$

However, by our second degree hypothesis, $V_1^m = 0$ for $m \geq min\{W\}$. Thus $\mathcal{L}_1(F) = \mathcal{L}_1(X)$.

For the kernel of D, [16, Theorem 3.4] implies

$$\mathcal{L}_0(F) = \bigoplus_{n>0} \bigoplus_{k\geq 0} H^k(F, \mathbb{Q}) \otimes W^{n+k} \cong Der_+(H^*(Y, \mathbb{Q}), H^*(F, \mathbb{Q})),$$

as graded vector space. Using [16, Theorem 3.2] again, we conclude

$$\pi_*(\Omega Baut_1(F)) \otimes \mathbb{Q} \cong \mathcal{L}_1(X) \oplus Der_+(H^*(Y, \mathbb{Q}), H^*(F, \mathbb{Q})), \tag{1}$$

as graded vector space.

In §4 of [16], we identify cycle representatives in Sullivan's dgla model for the vector subspaces $\mathcal{L}_i(F)$, whenever F is formal and two-stage. Define an *elementary derivation* $P\partial z_k$ in $\widetilde{Der}_+(\mathcal{M}_F)$ to be the derivation which carries the basis element z_k of $Z_0 \oplus Z_1$ to $P \in \mathcal{M}_F$ and vanishes on the other basis elements. Here $|P| < |z_k|$ and, if the difference is 1, we need to restrict to the kernel of ∂_F. Define a linear surjection $p : \widetilde{Der}_+(\mathcal{M}_F) \to L(F)$ by $p(P\partial z_k) = \rho(P) \otimes z_k$ where $\rho : (\mathcal{M}_F, d_F) \to (H^*(F, \mathbb{Q}), 0)$ is a formalization for F. Define subspaces $D_0(F), D_1(F)$ and $B(F)$ of $\widetilde{Der}_+(\mathcal{M}_F)$ by

$$\begin{aligned} D_0(F) &= \text{Span}\{P\partial z_i \mid P \in \Lambda(Z_0), |P| < |z_i|, \ \ z_i \in Z_0\} \\ D_1(F) &= \text{Span}\{P\partial z_j \mid P \in \Lambda(V_0), |P| < |z_j|, \ \ z_j \in Z_1\} \quad \text{and} \\ B(F) &= \text{Span}\{P z_k \partial z_j | \ P \in \Lambda(Z_0), |P| + |z_k| < |z_j|, \ \ z_j, z_k \in Z_1\}. \end{aligned}$$

Note $\partial_F(D_1(F)) = 0$ and $\partial_F(B(F) \oplus D_0(F)) \subseteq D_1(F)$. Set

$$\begin{aligned} \mathcal{D}_0(F) &= ker\{\partial_F : B(F) \oplus D_0(F) \to D_1(F)\} \quad \text{and} \\ \mathcal{D}_1(F) &= cok\{\partial_F : B(F) \oplus D_0(F) \to D_1(F)\}. \end{aligned}$$

We view $\mathcal{D}_1(F)$ as a subspace of $\text{Der}_+(\mathcal{M}_F)$ by taking vector space complements in each degree.

By [16, Theorem 4.2], p induces an isomorphism $p : \mathcal{D}_1(F) \to \mathcal{L}_1(F)$ and a surjection $p : \mathcal{D}_0(F) \to \mathcal{L}_0(F)$ both of which correspond to the homology projection in $(\widetilde{Der}_+(\mathcal{M}_F), \partial_F)$. We use this result to complete the proof.

Our second degree hypothesis implies $\mathcal{D}_1(F) \cong \mathcal{L}_1(X)$, as above. Choose an additive space Z of cocycle representatives in $\mathcal{M}_X$ for $\widetilde{H}^*(X, \mathbb{Q})$. Define a subspace $\mathcal{D}_0'(F)$ of $\mathcal{D}_0(F)$ by

$$\mathcal{D}_0'(F) = \text{Span}\{P\partial w_k \mid P \in Z \oplus W, |P| < |w_k|, \ \ w_k \in W\}.$$

Our first degree hypothesis implies decomposables in $H^*(F, \mathbb{Q})$ involving elements from both $\widetilde{H}^*(X, \mathbb{Q})$ and $\widetilde{H}^*(Y, \mathbb{Q})$ occur in degrees $\geq max(W)$. It follows that

$$\mathcal{D}_0'(F) \cong Der_+(H^*(Y, \mathbb{Q}), H^*(F, \mathbb{Q})),$$

as graded Lie algebras.

Finally, note the subspace $\mathcal{D}_1(F) \oplus \mathcal{D}'_0(F)$ of cycles of $\widetilde{Der}_+(\mathcal{M}_F)$, is actually a subalgebra. Moreover, $\mathcal{D}_1(F)$ consists of indecomposable central elements of $\mathcal{D}_1(F) \oplus \mathcal{D}'_0(F)$. Thus we may write this subalgebra as $\mathcal{D}_1(F) \times \mathcal{D}'_0(F)$. By [16, Theorem 4.2] and Equation 1 above, the inclusion $(\mathcal{D}_1(F) \times \mathcal{D}'_0(F), 0) \hookrightarrow (\widetilde{Der}_+(\mathcal{M}_F), \partial_F)$, induces an equivalence on homology. Thus $(\mathcal{D}_1(F) \times \mathcal{D}'_0(F), 0)$ is a dgla model for $Baut_1(F)$. The result follows from Lemma 3.1. □

4. Formality of Spatial Realizations of Graded Lie Algebras.

Given a connected graded Lie algebra L over $\mathbb{Q}$, let $nil(L)$ denote the nilpotency of L. That is, $nil(L)$ is the length of the longest iterated bracket in L. We prove

Lemma 4.1 *Let L be a finite-dimensional connected graded Lie algebra over $\mathbb{Q}$. If $nil(L) > 2$ then $\|L\|$ is not formal.*

Proof. The Sullivan minimal model for $\|L\|$ is $(\Lambda(sL), d_{[\ ,\ ]})$ where $d_{[\ ,\ ]}$ is dual to the bracket in L [11, Proposition 3.3]. Suppose $\|L\|$ is formal. Then $\|L\|$ is formal with finite-dimensional rational homotopy and so hyperformal [3, Theorem 1]. In particular, $(\Lambda(sL), d_{[\ ,\ ]})$ is a two-stage dga which we may assume is in "normal form". That is, by [15, Lemma 3.3], we may assume $sL = V_0 \oplus V_1$ where $d_{[\ ,\ ]}(V_0) = 0$, $d_{[\ ,\ ]}(V_1) \subseteq \Lambda(V_0)$, $d_{[\ ,\ ]}|V_1$ is an injection and $d_{[\ ,\ ]}(sw) \neq d_{[\ ,\ ]}(\alpha)$ for any $sw \in V_1$ and decomposable $\alpha \in \Lambda(sL)$.

Choose $sw \in V_1$ homogeneous of minimal positive degree with $w = [[x,y],z]$ for $x, y, z \in L$. Since $d_{[\ ,\ ]}(V_0) = 0$ we may assume $v = [x,y] \in s^{-1}V_1$ and $z \in s^{-1}V_0$. Let $n = |w|$ and consider the nth Postnikov section, X_n, of $\|L\|$. The minimal model for X_n is $(\Lambda((sL)^{\leq n}), d_{[\ ,\ ]})$. Using the normal form above, we see that the element $d_{[\ ,\ ]}(sw)$ is a cocycle of $(\Lambda((sL)^{\leq n}), d_{[\ ,\ ]})$ which does not bound. Moreover, the element $sv \cdot sz$ is a nontrivial summand of this cocycle which does not bound. Thus $d_{[\ ,\ ]}(sw)$ represents an indecomposable element in $H^{n+1}(X_n, \mathbb{Q})$. By [15, Theorem 4.6], this contradicts the hyperformality of $\|L\|$. □

Lemma 4.2 *Let $F = \Lambda(a_i, b_{ij})$ be the free graded algebra generated by elements a_i of degree k and b_{ij} of degree l for $l \neq k$, $i = 1, \ldots n, j = 1, \ldots, m$. Set $R_j = \sum_{i=1}^{n} a_i b_{ij}$ for $j = 1, \ldots, m$. Then $R_1, \ldots, R_m$ is a regular sequence of F if and only if k and l are both even and $m \leq n$.*

Proof. If either k or l is odd then each R_j is a zero-divisor in F. If k and l are both even, consider the $m \times n$ matrix $B = (b_{ij})$, whose jth row of B represents R_j in the basis $\{a_1, \ldots, a_n\}$. If $m > n$ we can row reduce B to find a nontrivial linear combination $\beta_1 R_1 + \cdots + \beta_m R_m = 0$, where $\beta_j \in \Lambda(b_{ij})$. Thus the sequence is not regular when $m > n$.

Conversely, suppose k and l are even and $m \leq n$. If R_m is a zero divisor in $F/(R_1, \ldots, R_{m-1})$ then $\alpha R_m = \alpha_1 R_1 + \cdots + \alpha_{m-1} R_{m-1}$ for some $\alpha \notin (R_1, \ldots, R_{m-1})$. But this expression implies a nontrivial solution to the homogeneous system $BX = 0$ which is impossible. Since permutations of regular sequences are regular in a polynomial algebra, the regularity of the sequence follows by induction. □

Proof of Theorem 4. By Theorem 2, $Baut_1(Y) \simeq_{\mathbb{Q}} \|Der_+(H^*(Y, \mathbb{Q}))\|$. By hypothesis, $H^*(Y, \mathbb{Q})$ is a finitely generated free graded algebra. Thus $Der_+(H^*(Y, \mathbb{Q}))$

is a finite-dimensional graded connected Lie algebra. By [16, Theorem 5.3] (or direct calculation), $nil(Der_+(H^*(Y,\mathbb{Q})) \geq 2$ unless $\pi_*(Y) \otimes \mathbb{Q}$ is concentrated in either a single degree or in two degrees $p < q$ with either $q \leq 2p$ or p odd and $dim(\pi_p(Y) \otimes \mathbb{Q}) = 1$. In the former case, $Der_+(H^*(Y,\mathbb{Q}))$ is abelian and so $Baut_1(Y)$ is a rational H-space.

In the latter cases, write $H^*(Y,\mathbb{Q}) = \Lambda(w_1, \ldots, w_n, v_1, \ldots, v_m)$ where $|w_i| = p$ and $|v_j| = q$. Let $\alpha_i = 1\partial w_i, \beta_{ij} = w_i \partial v_j$ and $\gamma_j = 1\partial v_j$ Then $Der_+(H^*(Y,\mathbb{Q})) = \mathbb{Q}(\alpha_i, \beta_{ij}, \gamma_j)$ with $\gamma_j = [\alpha_i, \beta_{ij}]$, the only nontrivial brackets. Thus letting $a_i = s\alpha_i, b_{ij} = s\beta_{ij}$ and $c_j = s\gamma_j$, the Sullivan model for $Baut_1(Y)$ is just $(\Lambda(a_i, b_{ij}, c_j), d_{[\ ,\ ]})$ where $d_{[\ ,\ ]}(a_i) = d_{[\ ,\ ]}(b_{ij}) = 0$ and $d_{[\ ,\ ]}(c_j) = 1/nR_j$. Since $Baut_1(Y)$ has finite-dimensional rational homotopy, $Baut_1(Y)$ is formal if and only if the sequence $R_1, \ldots, R_m$ is regular in F ([3, Theorem 1] again). The result thus follows from Lemma 4.2. □

Proof of Theorem 5. By [16, Theorem 5.3],

$$nil(\pi_*(\Omega Baut_1(F)) \otimes \mathbb{Q}) = nil(\pi_*(\Omega Baut_1(Y)) \otimes \mathbb{Q})$$

when $F = X \times Y$, as hypothesized. Thus we are again reduced to the cases where $\pi_*(Y) \otimes \mathbb{Q}$ is concentrated in one or two degrees, as above. In the latter case, we choose an additive homogenous basis $\{z_1, \ldots, z_d\}$ for $H^*(X,\mathbb{Q})$ and set $\alpha_{ik} = z_k \partial w_i, \beta_{ij} = w_i \partial v_j$ and $\gamma_k = z_k \partial v_j$. Then $Der_+(H^*(Y,\mathbb{Q}), H^*(F,\mathbb{Q})) = \mathbb{Q}(\alpha_{ik}, \beta_{ij}, \gamma_{jk})$ with $\gamma_{jk} = [\alpha_{ik}, \beta_{ij}]$, the only nontrivial brackets. The Sullivan model for $Baut_1(F)$ is thus of the form $(\Lambda(s\mathcal{L}_1(X)), 0) \otimes (\Lambda(a_{ik}, b_{ij}, c_{jk}), d_{[\ ,\ ]})$ with $d_{[\ ,\ ]}(a_{ik}) = d_{[\ ,\ ]}(b_{ij}) = 0$ and $d_{[\ ,\ ]}(c_j) = 1/nR_{jk}$, where $R_{jk} = \sum_{i=1}^{n} a_{ik} b_{jk}$. The result follows, as above, from Lemma 4.2. □

5. Applications and Examples.

The following are consequences of our results and the proof of special cases of the Halperin conjecture.

5.1 Applications.

(1) Let $F = S^{2m_1} \times \cdots \times S^{2m_k} \times S^{2n_1+1} \times \cdots \times S^{2n_l+1}$, where $m_1 \leq \cdots \leq m_k$, $n_1 \leq \cdots \leq n_l$. If $l = 0$ then $Baut_1(F)$ is a rational H-space. If $k = 0$ then $Baut_1(F)$ is a coformal space which is formal if and only if all the n_j are equal.

When both $k, l > 0$, Theorem 3 applies provided $2m_k \leq n_1 + 1$ and $m_1 + n_1 \geq 2n_l + 1$. In this case, $Baut_1(F)$ is a coformal space having the rational H-space $Baut_1(S^{2m_1} \times \cdots \times S^{2m_k})$ as a rational factor. Moreover, $Baut_1(F)$ is formal if and only all the n_j are equal. This example generalizes directly to spaces F whose rational cohomology is a tensor product of free and truncated polynomial algebras. □

(2) Let X be an F_0-space with $Der_+(H^*(X,\mathbb{Q})) = 0$. Let $q \geq max(\pi_*(X) \otimes \mathbb{Q})$. Then $Baut_1(X \times K(V,q))$ is a rational H-space for any finite-dimensional rational vector space V. In particular, let $F = U(n)/U(n_1) \times \cdots \times U(n_k)$ or $F = Sp(n)/Sp(n_1) \times \cdots Sp(n_k)$ where $n - 1 \leq n_1 + \cdots + n_k \leq n$. Then $Baut_1(F)$ is a rational H-space. □

(3) If q is odd and W is another rational vector the space $Baut_1(X \times K(V,q) \times K(W, q+2))$ is coformal but not formal. In particular, if $F = U(n)/U(n_1) \times \cdots \times U(n_k)$ and $n - 2 \leq n_1 + \cdots n_k \leq n$ then $Baut_1(F)$ is coformal but not formal. □

Finally, we show the necessity of the hypotheses in Theorem 3.

5.2 Examples

(1) Let $F = S^2 \times S^3 \times S^7$, so that F violates the first but not the second degree hypothesis in Theorem 3. We show $Baut_1(S^2)$ is not a rational factor of $Baut_1(F)$. This implies that $Baut_1(F)$ is not coformal since $\pi_*(\Omega Baut_1(S^2)) \otimes \mathbb{Q}$ is a factor of $\pi_*(\Omega Baut_1(F)) \otimes \mathbb{Q}$ by [16, Theorem 5.3].

Note $Baut_1(S^2) \simeq_{\mathbb{Q}} K(\mathbb{Q}, 4)$. In the notation of §3, $V_0 = \mathbb{Q}(x_1), W = \mathbb{Q}(x_2, x_3)$ and $V_1 = \mathbb{Q}(y)$ where $|x_1| = 2, |x_2| = 3, |x_3| = 7, |y| = 3$ and $d_F(y) = x_1^2$. Let $\mathcal{D}$ denote the subalgebra of $\widetilde{Der}_+(\mathcal{M}_F)$ spanned by:

$$a = x_1\partial x_2, \quad b = x_1x_2\partial x_3, \quad c_1 = 1\partial y, \quad c_2 = 1\partial x_2, \quad c_3 = x_1^2\partial x_3$$

$$d_1 = y\partial x_3, \quad d_2 = x_2\partial x_3, \quad e = x_1\partial x_3, \quad f = 1\partial x_3.$$

The only boundary is $\partial_X(d_1) = -c_3 = -[a, b]$. Thus $\mathcal{D}$ is, in fact, a sub-dgla of $(\widetilde{Der}_+(\mathcal{M}_F), \partial_F)$. By [16, Theorems 3.1 and 4.2], $H(\mathcal{D}, \partial_X) \cong \pi_*(\Omega Baut_1(F)) \otimes \mathbb{Q}$. Thus the inclusion $(\mathcal{D}, \partial_X) \hookrightarrow (\widetilde{Der}_+(\mathcal{M}_F), \partial_F)$ induces a homology isomorphism: $(\mathcal{D}, \partial_X)$ is a dgla model for $Baut_1(F)$. Note that $[c_1, d_1] = [c_2, d_2] = f$. Let $\xi : (\mathcal{L}, \partial) \to (\mathcal{D}, \partial_X)$ denote the Quillen model for $(\mathcal{D}, \partial_X)$. For $i = 1, 2$, let $\bar{c_i}, \bar{d_i}$ in $\mathcal{L}$ denote the preimages of the indecomposables $c_i, d_i \in \mathcal{D}$. Then there exists an indecomposable $\bar{g} \in \mathcal{L}$ with $\partial(\bar{g}) = [\bar{c_1}, \bar{d_1}] - [\bar{c_2}, \bar{d_2}]$. By [11, Proposition 2.4], $H^4(Baut_1(F), \mathbb{Q}) = \mathbb{Q}(s\bar{c_1}, s\bar{c_2})$ and $s\bar{c_1} \cdot s\bar{d_1} = s\bar{c_2} \cdot s\bar{d_2} \in H^9(Baut_1(F), \mathbb{Q})$. Thus, since $H^*(Baut_1(F), \mathbb{Q})$ has no free factor generated in degree four, $Baut_1(S^2)$ is not a rational factor of $Baut_1(F)$. □

(2) Let $F = S^3 \times S^4 \times S^7$, so that F violates the second but not the first degree hypothesis in Theorem 3. In this case, it is easy to see that $Baut_1(S^4)$ is not a rational factor of $Baut_1(F)$. For $\pi_*(\Omega Baut_1(S^4)) \otimes \mathbb{Q} = \mathbb{Q}(1 \otimes y)$ where $y \in \pi_7(S^4) \otimes \mathbb{Q}$ is nontrivial. Using cycle representatives in Sullivan's model ([16, Theorem 4.2]), we have $1 \otimes y = q[1 \otimes x, \alpha \otimes y]$ in $\pi_*(\Omega Baut_1(F)) \otimes \mathbb{Q}$, where $x \in \pi_3(S^3) \otimes \mathbb{Q}$ and $\alpha \in H^3(S^3, \mathbb{Q})$ are nontrivial. Thus $\pi_*(\Omega Baut_1(S^4)) \otimes \mathbb{Q}$ is not a factor of $\pi_*(\Omega Baut_1(F)) \otimes \mathbb{Q}$. □

References

[1] G. Allaud, *On the classification of fibre spaces,* Math. Z. **92** (1966), 110-125.

[2] A. Dold and R. Lashof, *Principal quasi-fibrations and fibre homotopy equivalence of bundles,* Illinois J. Math. **3** (1959), 285- 305.

[3] Y. Felix and S. Halperin, *Formal spaces with finite dimensional rational homotopy,* Trans. Amer. Math. Soc. **270** (1982), 575-588.

[4] J.-B. Gatzinzi, *The homotopy Lie algebra of classifying spaces,* J. Pure and Appl. Alg. **120** (1997), 281-289.

[5] S. Halperin, *Finiteness in the minimal models of Sullivan,* Trans. Amer. Math. Soc. **230** (1977), 173-199.

[6] G. Lupton, *A note on the conjecture of S. Halperin,* Lecture Notes in Math., vol. 1440, Springer-Verlag (1990), 148-163.

[7] M. Markl, *Towards one conjecture on collapsing of the Serre spectral sequence,* Supplemento di Rendiconti del Circolo Matematico di Palermo, vol. 22, (1989), 151-159.

[8] W. Meier, *Rational universal fibrations and flag manifolds,* Math. Ann. **258** (1982), 329-340.

[9] J. Milnor, *On the characteristic classes for spherical fibrations,* Comm. Math. Helv. **43** (1968), 51-73.

[10] J. Neisendorfer, *Lie algebras, coalgebras and rational homotopy theory for nilpotent spaces,* Pacific J. of Math. (2) **74** (1978), 429-460.

[11] J. Neisendorfer and T. Miller, *Formal and coformal spaces,* Illinois J. Math. **22** (1978), 565-580.

[12] D. Quillen, *Rational homotopy theory,* Ann. Math. (2) **90** (1969), 205-295.

[13] M. Schlessinger and J. Stasheff, *Deformation theory and rational homotopy type,* Publ. Sci. I.H.E.S., to appear.

[14] H. Shiga and M. Tezuka, *Rational fibrations, homogeneous spaces with positive Euler characteristic and Jacobians,* Ann. Inst. Fourier Grenoble **37** (1987), 81-106.

[15] S. Smith, *Postnikov sections of formal and hyperformal spaces,* Proc. Amer. Math. Soc. **122** (1994) 893-903.

[16] ______, *Rational homotopy Lie algebra of classifying spaces for formal two-stage spaces,* J. Pure and Appl. Alg., to appear

[17] J. Stasheff, *A classification theorem for fibre spaces,* Topology **2** (1963), 239-246.

[18] D. Sullivan, *Infinitesimal computations in topology,* Publ. I.H.E.S. **47** (1977), 269-331.

[19] J. C. Thomas, *Rational homotopy of Serre fibrations,* Ann. Inst. Fourier **31** (1981), 71-90.

Department of Mathematics
Saint Joseph's University
Philadelphia, PA 19131 USA
email smith@sju.edu

Contemporary Mathematics
Volume **274**, 2001

Problems on Self-Homotopy Equivalences

Martin Arkowitz

1. Introduction

We give a list of open questions and problems on the subject of self-homotopy equivalences and related topics which have been submitted by the workshop participants and others. In many cases we briefly discuss the background of the problem and give some references to the literature. An earlier list of problems in this area was given by D. Kahn in 1989 [**Kah90**]. We list these problems in an appendix.

We begin by fixing our notation. For a based space X, $E(X)$ denotes the space of based homotopy equivalences of X into itself with the compact-open topology and $\mathcal{E}(X)$ denotes the group of based homotopy classes of based homotopy equivalences of X into itself. Then $\mathcal{E}(X)$ can be regarded as $\pi_0(E(X))$ or as the group of units of the monoid $[X, X]$. For any X, we define $\mathcal{E}_*(X)$ to be the subgroup of $\mathcal{E}(X)$ consisting of all α such that the induced homomorphism $\alpha_* : H_i(X) \to H_i(X)$ is the identity homomorphism id, for all i. If X is a finite dimensional CW-complex, then $\mathcal{E}_\#(X)$ is the subgroup of $\mathcal{E}(X)$ consisting of all α such that the induced homomorphism $\alpha_\# = \mathrm{id} : \pi_i(X) \to \pi_i(X)$, for all $i \leq \dim X$. Unless otherwise stated we will assume that all spaces are based, connected and have the based homotopy type of a CW-complex.

2. General Problems

(1) Realization: Given a group G, does there exist a space X such that $\mathcal{E}(X) \approx G$?

This problem has been around a long time and has been considered by many people (see [**Ark90**, p. 194] and [**Rut97**, p. 108]). As positive evidence in the case when G is finite, it has been observed in [**Ark90**, p. 175] that there always is a space X such that $G \subseteq \mathcal{E}(X)$, and Kojima has proved that $G \approx \pi_0(\mathrm{Diff} M)$ for some closed manifold M [**Koj88**, p. 297]. There are several variations of the realization problem. For example, if X is a 1-connected finite complex, then $\mathcal{E}(X)$ is finitely-presented by the Sullivan-Wilkerson Theorem, so one can ask if any finitely-presented group can be realized as $\mathcal{E}(X)$ for such an X. Similarly, Dror and Zabrodsky showed that $\mathcal{E}_\#(X)$ or $\mathcal{E}_*(X)$ is nilpotent for a finite dimensional CW-complex X [**DZ79**], and so one can ask if any nilpotent group can be realized as $\mathcal{E}_\#(X)$ or as $\mathcal{E}_*(X)$ for such an X. Arkowitz and Lupton have given examples of spaces X which are the rationalization of finite dimensional complexes such that $\mathcal{E}(X)$ is a finite group, and have inquired as to which finite groups can be so realized [**AL**, §5]. Pavesic has considered the stable group of equivalences $\mathcal{E}_S(X)$, i.e., the units of the ring of stable homotopy classes $\{X, X\}$, and has asked which finite cyclic groups G can be realized as $\mathcal{E}_S(X)$. He has shown that there must be strong restrictions on the order of G.

(2) (Rutter) Find an algorithm for calculating $\mathcal{E}(X)$ for a connected CW-complex X.

There are two general methods for calculating $\mathcal{E}(X)$. The first applies to 1-connected CW-complexes which are given a cellular or homology decomposition, and proceeds inductively over the stages of the decomposition [**Rut97**, §11]. The second applies to a Postnikov decomposition of a space and also proceeds inductively over the stages of the decomposition [**Rut97**, §13]. It would be useful to have a cellular or homology decomposition method in the case when X is not simply-connected.

3. Calculations

(3) (Rutter) Group Extensions: Determine the group extensions in the short exact sequences of groups which are used to calculate $\mathcal{E}(X)$.

Most calculations use one of the two methods mentioned in (2) and give rise to three term exact sequences or short exact sequences of not-necessarily abelian groups. These already appeared in the first paper on the group of equivalences by Barcus and Barratt [**BB58**]. The desired calculation of the middle group then depends on knowing the group extension. For a recent discussion of this, see Rutter [**Rut97**, §11, 13]. Note that special cases are known where the sequence splits.

4. Subgroups

Let $\mathcal{E}_\Omega(X)$ be the subgroup of all $\alpha \in \mathcal{E}(X)$ such that $\Omega\alpha = \mathrm{id}$ and $\mathcal{E}_{\#\infty}(X)$ be the subgroup of all $\alpha \in \mathcal{E}(X)$ such that $\alpha_\# = \mathrm{id} : \pi_i(X) \to \pi_i(X)$, for *all* i.

(4) (Maruyama, Roitberg) Genus: Is $\mathcal{E}_*$ or $\mathcal{E}_\#$ a genus invariant on the collection of 1-connected finite complexes?

Recall that two nilpotent complexes of finite type X and Y have the same (Mislin) genus if the localizations $X_{(p)} \equiv Y_{(p)}$ for all primes p. Similarly two nilpotent finitely-generated groups M and N have the same genus if $M_{(p)} \approx N_{(p)}$ for all primes p. We denote the genus set of X by $G(X)$ and the genus set of M by $G(M)$. Then $\mathcal{E}_*$ is a genus invariant means that $\mathcal{E}_*(X) \approx \mathcal{E}_*(Y)$ if $Y \in G(X)$. McGibbon and Moller showed that $\mathcal{E}$ is not a genus invariant [**MM92**] and Roitberg showed that $\mathcal{E}_*$ is not a genus invariant [**Roi96**]. In addition, Roitberg showed that $\mathcal{E}$ restricted to 1-connected finite complexes is not a genus invariant [**Roi96**]. The following is presented as weak evidence for an affirmative answer to (4). From work of Pickel we have that $G(M)$ is a finite set. By Maruyama's localization theorem [**Mar89**], if X and Y are 1-connected finite complexes and $Y \in G(X)$, then $\mathcal{E}_\#(Y) \in G(\mathcal{E}_\#(X))$. Thus there are finitely many possibilities for $\mathcal{E}_\#(Y)$ one of which is $\mathcal{E}_\#(X)$.

Let $\tau_X \subseteq \mathcal{E}_\#(X)$ be the torsion subgroup. Then Maruyama has suggested weaker versions of (4): (a) Does the set $G(\mathcal{E}_\#(X)/\tau_X)$ consist of one element? (b) If $Y \in G(X)$, is $\mathcal{E}_\#(X)/\tau_X \approx \mathcal{E}_\#(Y)/\tau_Y$? Maruyama has proved (b) for homotopy-associative H-spaces. Similar questions can be asked for $\mathcal{E}_*$.

(5) (Tsukiyama) Acyclic Spaces: If X is an acyclic finite complex, what is $\mathcal{E}_{\#}(X)$?

This is part of a general problem due to Kan of finding homotopy information about acyclic spaces [**Mil71**, Prob. 86]. It has been proved by Tsukiyama that if X is a simple acyclic space, then $\mathcal{E}_{\#\infty}(X) = 1$ [**Tsu75**].

(6) (Arkowitz, Pavesic) Is there a finite complex X such that $\mathcal{E}_{\Omega}(X) \neq \mathcal{E}_{\#\infty}(X)$?

Of course, it is always true that $\mathcal{E}_{\Omega}(X) \subseteq \mathcal{E}_{\#\infty}(X)$. Félix and Murillo have presented an example (due to Fred Cohen) of an infinite complex for which the inequality holds [**FM98**]. Another such example has been suggested by Roitberg by taking a non-trivial phantom map $g : K(\mathbb{Z}, 3) \to \Omega S^5$. If $X = K(\mathbb{Z}, 3) \times \Omega S^5$, then g determines $f : X \to X$, and $\mathrm{id} + f$ is in $\mathcal{E}_{\#\infty}(X)$ but not in $\mathcal{E}_{\Omega}(X)$. Other examples are similarly constructed using maps g of Eilenberg-MacLane spaces. In addition, for any finite complex X, $\mathcal{E}_{\#\infty}(X) \subseteq \mathcal{E}_{\#}(X)$, and Arkowitz and Maruyama have shown that there exist Moore spaces X with $\mathcal{E}_{\#\infty}(X) \neq \mathcal{E}_{\#}(X)$ [**AM98**]. Furthermore, Félix and Murillo have shown that for every n, there is a finite dimensional complex X and a map $f : X \to X$ such that $f_{\#} = \mathrm{id} : \pi_i(X) \to \pi_i(X)$, for $i \leq n$, but not for $i > n$.

(7) (Arkowitz-Maruyama [**AM98**, p. 153]) Stability: Let $\mathcal{E}_{\#n}(X)$ be the set of all α in $\mathcal{E}(X)$ such that $\alpha_{\#} = \mathrm{id} : \pi_i(X) \to \pi_i(X)$ for all $i \leq n$. If X is a finite complex, then is there an integer N such that

$$\mathcal{E}_{\#n}(X) = \mathcal{E}_{\#m}(X), \qquad \text{for all } m, n \geq N?$$

The answer is clearly yes if $\mathcal{E}_{\#i}(X)$ is finite for some i. Maruyama has proved the existence of N when X is a finite H_0-space (i.e., a space whose rationalization is an H-space) [**Mar**]. Félix and Thomas have proved it for the rationalization of any X [**FT99**]. (They claimed to have proved it in general, but, as they later noted, there is a mistake in the proof.)

(8) (Arkowitz) Localization: Is the natural homomorphism $\mathcal{E}_{\#\infty}(X) \to \mathcal{E}_{\#\infty}(X_{(p)})$ a p-localization homomorphism if X is a 1-connected finite complex?

For any $n \geq \dim X$, Maruyama has proved that $\mathcal{E}_{\#n}(X) \to \mathcal{E}_{\#n}(X_{(p)})$ is the p-localization homomorphism [**Mar89**]. Thus the result is true for any space X which satisfies the stability property of (7). Futhermore, it is not difficult to show that $\mathcal{E}_{\#\infty}(X_{(p)})$ is a p-local group. Thus it suffices to prove that the induced homomorphism $(\mathcal{E}_{\#\infty}(X))_{(p)} \to \mathcal{E}_{\#\infty}(X_{(p)})$ is an isomorphism. We note that Pavesic has given an example in his thesis of an infinite complex for which the natural homomorphism of (8) is not a localization homomorphism.

(9) (Scheerer-Tanré, Arkowitz-Lupton-Murillo) Upper Bound for Nilpotency: If X is a 1-connected finite complex, is the nilpotency of the groups $\mathcal{E}_{\#}(X)$ and $\mathcal{E}_{\#\infty}(X)$ bounded above by known numerical invariants of the homotopy type of X? In particular, is $\operatorname{nil} \mathcal{E}_{\#}(X) \leq \operatorname{cat} X - 1$?

Here cat is the Lusternik-Schnirelmann category of X (normalized so that suspensions have cat 1). Félix and Murillo have proved that $\operatorname{nil} \mathcal{E}_{\Omega}(X) \leq \operatorname{cat} X - 1$ [**FM98**] and that if Y is the rationalization of X, then $\operatorname{nil} \mathcal{E}_{\#\infty}(Y) \leq \operatorname{cat} Y - 1$ [**FM97**]. More recently, they have established the latter inequality with $\mathcal{E}_{\#}(Y)$

replacing $\mathcal{E}_{\#\infty}(Y)$. Furthermore, Scheerer and Tanré have shown that the solvability of $\mathcal{E}_{\#\infty}(X)$ is bounded above by s-cat$X - 1$, where s-cat is a spherical analogue of the Lusternik-Schnirelmann category [**ST99**]. They have conjectured that $\mathrm{nil}\,\mathcal{E}_{\#\infty}(X) \leq \text{s-cat}X - 1$. In [**ALM**] it is conjectured that $\mathrm{nil}\,\mathcal{E}_{\#}(X) \leq \mathrm{cl}_s X - 1$, where cl_s is a spherical analogue of cone length. Also, Murillo has suggested that an upper bound for $\mathrm{nil}\,\mathcal{E}_*(X)$ or $\mathrm{solv}\,\mathcal{E}_*(X)$ can be given in terms of an appropriate notion of cocat X.

(10) (Lupton) Let X be a 1-connected, elliptic space with positive Euler characteristic. Then is $\mathcal{E}_{\#}(X)$ finite?

Recall that X is elliptic if $\dim H_*(X;\mathbb{Q}) < \infty$ and $\dim \pi_*(X) \otimes \mathbb{Q} < \infty$, where dim denotes the total dimension of a graded vector space. Since Maruyama has shown that $\mathcal{E}_{\#}(X_{\mathbb{Q}}) \approx (\mathcal{E}_{\#}(X))_{\mathbb{Q}}$, the problem is equivalent to asking if $\mathcal{E}_{\#}(X_{\mathbb{Q}}) = 1$ for such spaces X. This problem is motivated by a conjecture of Halperin (cf. [**Lup98**] and references therein). Denote by $\mathrm{Der}_i H^*(X;\mathbb{Q})$ the vector space of derivations of degree i of $H^*(X;\mathbb{Q})$. Halperin's conjecture can then be phrased as follows: If X is a 1-connected, elliptic space with positive Euler characteristic, then $\mathrm{Der}_{<0} H^*(X;\mathbb{Q}) = 0$. The relation of this to problem (10) is as follows. By restricting a known correspondence between unipotent automorphisms and nilpotent derivations, it is possible to establish a similar correspondence between $\mathcal{E}_{\#}(X_{\mathbb{Q}})$ and the subspace of $\mathrm{Der}_0 H^*(X;\mathbb{Q})$ consisting of decomposable derivations. Then to answer (10) in the affirmative, it suffices to show that this subspace of $\mathrm{Der}_0 H^*(X;\mathbb{Q})$ is zero.

5. Products

(11) (Heath) Products of H-spaces or co-H-spaces: Extend the calculation of $\mathcal{E}(X \times Y)$, where X and Y are both H-spaces or both co-H-spaces [**Rut97**, pp. 104-107], to a product of three or more factors.

More specifically, consider a product of spheres $S^n \times S^m$ with $n > m \geq 1$. Let $H_{n,m}$ be the quotient of $\pi_{n+m}(S^n \times S^m)$ under Whitehead products and let $P_n : \pi_n(S^m) \to \pi_{n+m-1}(S^m)$ be defined by $P_n(\alpha) = [\alpha, \iota_m]$. Then it is known [**Saw75, Hea96**] that there is a split short exact sequence

$$1 \to \ker \Phi_2 \to \mathcal{E}(S^n \times S^m) \xrightarrow{\Phi_2} \mathcal{E}(S^n) \times \mathcal{E}(S^m) = \mathbb{Z}_2 \times \mathbb{Z}_2 \to 1,$$

where Φ_2 is obtained by composing with appropriate inclusions and projections. Using the fact that $S^n \times S^m$ is the mapping cone of a Whitehead product map, Heath has shown that $\ker \Phi_2$ is an extension of $H_{n,m}$ by $\ker P_n$ [**Hea96**].

For a product of three spheres $S^n \times S^m \times S^k$ with $n > m > k \geq 1$, there is still a split short exact sequence

$$1 \to \ker \Phi_3 \to \mathcal{E}(S^n \times S^m \times S^k) \xrightarrow{\Phi_3} \mathcal{E}(S^n) \times \mathcal{E}(S^m) \times \mathcal{E}(S^k) = \mathbb{Z}_2 \times \mathbb{Z}_2 \times \mathbb{Z}_2 \to 1,$$

and so $\mathcal{E}(S^n \times S^m \times S^k)$ is a semi-direct product of $\ker \Phi_3$ and $\mathbb{Z}_2 \times \mathbb{Z}_2 \times \mathbb{Z}_2$. However it is not clear how to describe $\ker \Phi_3$.

(12) (Pavesic) General Products: Find general methods to compute $\mathcal{E}(X \times Y)$; in particular, find general methods to calculate $\mathcal{E}(X \times X)$. More generally, if $X \times_A Y$ is the pull-back of $X \leftarrow A \to Y$, what is $\mathcal{E}(X \times_A Y)$?

As noted, the calculations to determine $\mathcal{E}(X \times Y)$ require special assumptions such as one or both of the spaces are H-spaces or co-H-spaces. At present, there seems to be no general method without strong restrictions on X and Y (see [**BH90, Pav99**]).

(13) (Pavesic) Let $\mathcal{E}_X(X \times Y) \subseteq \mathcal{E}(X \times Y)$ be the set of all homotopy equivalences of the form (p_X, f), where $p_X : X \times Y \to X$ is the projection and $f : X \times Y \to Y$, and let $\mathcal{E}_Y(X \times Y)$ be similarly defined as all homotopy equivalences of the form (g, p_Y). Is $\mathcal{E}(X \times Y)$ generated by $\mathcal{E}_X(X \times Y)$ and $\mathcal{E}_Y(X \times Y)$?

6. Miscellaneous

(14) (Booth) If X is a compactly-generated space, let $E(X)$ be the space of self homotopy equivalences of X with the compactly-generated version of the compact-open topology. Then $E(X)$ is a topological monoid under composition of homotopy classes. Are there any spaces X for which $E(X)$ does not have an inverse up to homotopy?

If X is a finite CW-complex, it follows from [**Sib69**] that $E(X)$ has a homotopy inverse. The answer to (14) may depend on the following considerations: Let $X \to E_\infty \to B_\infty$ be the universal fibration which classifies Hurewicz fibrations with fibre X, and let $E(X) \to \mathrm{Prin}_X E_\infty \to B_\infty$ be the associated principal fibration. Then $\mathrm{Prin}_X E_\infty$ is either contractible or weakly contractible. It is believed that no homotopy inverse for $E(X)$ exists in the latter case.

(15) (Rutter) If M is a manifold, let $\mathcal{H}(M)$ be the based isotopy classes of self homeomorphisms of M and let $\nu : \mathcal{H}(M) \to \mathcal{E}(M)$ be the natural homomorphism. If $\dim M \geq 4$, give conditions for ν to be an isomorphism.

There are examples to show that ν need not be a monomorphism or an epimorphism [**Rut97**, p. 28]. However, there is a large class of 3-manifolds for which ν is an isomorphism [**Rut97**, 6.1].

7. Appendix: 1988 Problem List

For general interest and purposes of comparison, we state Kahn's 1988 list of problems [**Kah90**] without comment or attribution.

1. Characterize those finite, connected complexes X for which $\mathcal{E}(X)$ is finitely-presented.

2. Is there a finite, connected complex X, with $\mathcal{E}(X)$ finitely-generated, but not finitely-presented?

3. Determine the spaces X for which $\mathcal{E}(X)$ is trivial.

4. Complete the calculation of $\mathcal{E}(X)$ for 1-connected rank two H-spaces X. Specifically, determine $\mathcal{E}(G_{2,-2})$.

5. Determine $\mathcal{E}(X)$ for non-simply-connected rank two H-spaces X. In particular, calculate the group for $X = RP^i \times S^j$, $i = 3, 7$ and $j = 1, 3, 7$ and for $X = RP^i \times RP^k$, $i = 3, 7$ and $k = 3, 7$.

6. Determine $\mathcal{E}(X)$ when X is the quaternionic projective space HP^n, $n > 2$ or the Cayley projective plane.

7. Calculate $\mathcal{E}(X)$ when X is a compact Lie group, a Kahler manifold, etc.

8. Determine the extension

$$1 \to \mathrm{Ext}(\pi, \pi_{n+1}(X)) \to \mathcal{E}(X) \to \mathrm{Aut}(\pi) \to 1$$

where X is a Moore space $M(\pi, n)$.

9. Determine the extensions for the short exact sequences for $\mathcal{E}(X)$, where X is a sphere bundle over a sphere.

10. Relate the nilpotency of the groups $\mathcal{E}_{\#}(X)$ to known numerical invariants of the space X.

11. What is the relation of $\mathcal{E}_{\#}(X)$ and $\mathcal{E}_{*}(X)$ for reasonable spaces X?

12. For a co-H-space X, study the subgroup $\mathcal{E}_{\text{co-H}}(X)$ of $\mathcal{E}(X)$ of co-H-maps which are homotopy equivalences. What is the suspension homomorphism $\mathcal{E}(X) \to \mathcal{E}_{\text{co-H}}(\Sigma X)$?

13. Let $E \to B$ be a principal G-bundle and let $\mathcal{E}_G(E)$ be the group of G-equivariant self fibre homotopy equivalences. Let $\mathcal{E}(E)_f$ be the group of free (i.e., unbased) self homotopy equivalences. When is the natural homomorphism

$$\mathcal{E}_G(E) \to \mathcal{E}(E)_f$$

a monomorphism?

14. Given a fibration $p : X \to B$ with a fixed section $s : B \to X$. Let $\mathcal{E}(p, s)$ be the group of homotopy classes of fibre homotopy equivalences $X \to X$ which are compatible with p and s. Study $\mathcal{E}(p, s)$ and relate it to the group of G-equivariant self equivalences of a free G-space or the self fibre homotopy equivalences of a fibration.

15. What is the relation between the groups $\mathcal{E}(X)_f$ and $\mathcal{E}(X)$? In particular, what is the homomorphism $\mathcal{E}(X) \to \mathcal{E}(X)_f$?

16. Determine the homotopy type of the identity component in X^X, especially when X has more than two non-vanishing homotopy groups.

17. In the case $\mathcal{E}(X)$ can be realized as homeomorphisms, what additional information can be obtained, for example, about the space X^X?

References

[AL] M. Arkowitz and G. Lupton, *Rational Obstruction Theory and Rational Homotopy Sets*, Math. Zeit., (to appear).

[ALM] M. Arkowitz, G. Lupton, and A. Murillo, *Subgroups of the Group of Self Homotopy Equivalences*, (preprint).

[AM98] M. Arkowitz and K.-I. Maruyama, *Self-Homotopy Equivalencs Which Induce the Identity on Homology, Cohomology or Homotopy Groups*, Top. Appl. **87** (1998), 133–154.

[Ark90] M. Arkowitz, *The Group of Self-Homotopy Equivalences – A Survey*, Groups of Self-Equivalences and Related Topics, Lecture Notes in Mathematics, vol. 1425, Springer-Verlag, 1990, pp. 170–203.

[BB58] W. Barcus and M. Barratt, *On the Homotopy Classification of the Extensions of a Fixed Map*, Trans. A. M. S. **88** (1958), 57–74.

[BH90] P. Booth and P. Heath, *On the Groups $\mathcal{E}(X \times_B Y)$ and $\mathcal{E}_B^B(X \times_B Y)$*, Groups of Self-Equivalences and Related Topics, Lecture Notes in Mathematics, vol. 1425, Springer-Verlag, 1990, pp. 17–31.

[DZ79] E. Dror and A. Zabrodsky, *Unipotency and Nilpotency in Homotopy Equivalences*, Topology **18** (1979), 187–197.

[FM97] Y. Félix and A. Murillo, *A Note on the Nilpotency of a Subgroup of Self Homotopy Equivalences*, Bull. L. M. S. **29** (1997), 486–488.

[FM98] Y. Félix and A. Murillo, *A Bound for the Nilpotency of a Group of Self Homotopy Equivalences*, Proc. A. M. S. **126** (1998), 625–627.

[FT99] Y. Félix and J. Thomas, *On Spaces of the Same Strong n-Type*, Homology, Homotopy and Appl. **1** (1999), 205–217.

[Hea96] P. Heath, *On the Group $\mathcal{E}(X \times Y)$ of Self Homotopy Equivalences of a Product*, Quaestiones Math. **19** (1996), 433–451.

[Kah90] D. Kahn, *Some Research Problems on Self-Homotopy Equivalences*, Groups of Self-Equivalences and Related Topics, Lecture Notes in Mathematics, vol. 1425, Springer-Verlag, 1990, pp. 204–207.

[Koj88] S. Kojima, *Isometry Transformations of Hyperbolic 3-Manifolds*, Top. Appl. **29** (1988), 297–307.

[Lup98] G. Lupton, *Variations on a Conjecture of Halperin*, Homotopy and Geometry, vol. 45, Banach Center Publications, 1998, pp. 115–135.

[Mar] K.-I. Maruyama, *Stability Properties of Maps between Hopf Spaces*, (preprint).

[Mar89] K.-I. Maruyama, *Localization of a Certain Subgroup of Self-Homotopy Equivalences*, Pacific J. Math. **136** (1989), 293–301.

[Mil71] R. Milgram, *Problems Presented to the 1970 A. M. S. Summer Colloquium on Algebraic Topology*, Algebraic Topology, Proceedings of Symposia in Pure Mathematics, vol. 22, American Mathematical Society, 1971, pp. 187–201.

[MM92] C. McGibbon and J. Moller, *How Can You Tell Two Spaces Apart When They Have the Same n-Type for All n?*, Adams Memorial Symposium on Algebraic Topology, London Math. Soc. Lecture Notes, vol. 175, Cambridge Univ. Press, 1992, pp. 131–143.

[Pav99] P. Pavesic, *Self Homotopy Equivalences of Product Spaces*, Proc. Roy. Soc. Edinburgh **129** (1999), 181–197.

[Roi96] J. Roitberg, *Genus and Symmetry in Homotopy Theory*, Math. Ann. **305** (1996), 381–386.

[Rut97] J. Rutter, *Spaces of Homotopy Self-Equivalences – A Survey*, Springer-Verlag, Berlin, 1997, Lecture Notes in Mathematics, vol. 1662.

[Saw75] N. Sawashita, *On the Group of Self-Equivalences of the Product of Spheres*, Hirosh. Math. J. **5** (1975), 69–86.

[Sib69] R. Sibson, *Existence Theorem for H-Space Inverses*, Proc. Camb. Phil. Soc. **65** (1969), 19–21.

[ST99] H. Scheerer and D. Tanré, *Variation zum Konzept der Lusternik-Schnirelmann-Kategorie*, Math. Nachr. **207** (1999), 183–194.

[Tsu75] K. Tsukiyama, *Note on Self Map Inducing the Identity Automorphism of Homotopy Groups*, Hirosh. Math. J. **5** (1975), 215–222.

Department of Mathematics, Dartmouth College, Hanover NH 03755 U. S. A.
E-mail address: `Martin.Arkowitz@Dartmouth.edu`

Selected Titles in This Series

(Continued from the front of this publication)

250 **Robert H. Gilman, Editor,** Groups, languages and geometry, 1999

249 **Myung-Hwan Kim, John S. Hsia, Yoshiyuki Kitaoka, and Rainer Schulze-Pillot, Editors,** Integral quadratic forms and lattices, 1999

248 **Naihuan Jing and Kailash C. Misra, Editors,** Recent developments in quantum affine algebras and related topics, 1999

247 **Lawrence Wasson Baggett and David Royal Larson, Editors,** The functional and harmonic analysis of wavelets and frames, 1999

246 **Marcy Barge and Krystyna Kuperberg, Editors,** Geometry and topology in dynamics, 1999

245 **Michael D. Fried, Editor,** Applications of curves over finite fields, 1999

244 **Leovigildo Alonso Tarrío, Ana Jeremías López, and Joseph Lipman,** Studies in duality on noetherian formal schemes and non-noetherian ordinary schemes, 1999

243 **Tsit Yuan Lam and Andy R. Magid, Editors,** Algebra, K-theory, groups, and education, 1999

242 **Bernhelm Booss-Bavnbek and Krzysztof Wojciechowski, Editors,** Geometric aspects of partial differential equations, 1999

241 **Piotr Pragacz, Michał Szurek, and Jarosław Wiśniewski, Editors,** Algebraic geometry: Hirzebruch 70, 1999

240 **Angel Carocca, Víctor González-Aguilera, and Rubí E. Rodríguez, Editors,** Complex geometry of groups, 1999

239 **Jean-Pierre Meyer, Jack Morava, and W. Stephen Wilson, Editors,** Homotopy invariant algebraic structures, 1999

238 **Gui-Qiang Chen and Emmanuele DiBenedetto, Editors,** Nonlinear partial differential equations, 1999

237 **Thomas Branson, Editor,** Spectral problems in geometry and arithmetic, 1999

236 **Bruce C. Berndt and Fritz Gesztesy, Editors,** Continued fractions: From analytic number theory to constructive approximation, 1999

235 **Walter A. Carnielli and Itala M. L. D'Ottaviano, Editors,** Advances in contemporary logic and computer science, 1999

234 **Theodore P. Hill and Christian Houdré, Editors,** Advances in stochastic inequalities, 1999

233 **Hanna Nencka, Editor,** Low dimensional topology, 1999

232 **Krzysztof Jarosz, Editor,** Function spaces, 1999

231 **Michael Farber, Wolfgang Lück, and Shmuel Weinberger, Editors,** Tel Aviv topology conference: Rothenberg Festschrift, 1999

230 **Ezra Getzler and Mikhail Kapranov, Editors,** Higher category theory, 1998

229 **Edward L. Green and Birge Huisgen-Zimmermann, Editors,** Trends in the representation theory of finite dimensional algebras, 1998

228 **Liming Ge, Huaxin Lin, Zhong-Jin Ruan, Dianzhou Zhang, and Shuang Zhang, Editors,** Operator algebras and operator theory, 1999

227 **John McCleary, Editor,** Higher homotopy structures in topology and mathematical physics, 1999

226 **Luis A. Caffarelli and Mario Milman, Editors,** Monge Ampère equation: Applications to geometry and optimization, 1999

225 **Ronald C. Mullin and Gary L. Mullen, Editors,** Finite fields: Theory, applications, and algorithms, 1999

For a complete list of titles in this series, visit the AMS Bookstore at **www.ams.org/bookstore/**.